Ambient Intelligence, Wireless Networking, and Ubiquitous Computing

For a listing of recent titles in the Artech House *Mobile Communications Series,* turn to the back of this book.

Ambient Intelligence, Wireless Networking, and Ubiquitous Computing

Athanasios Vasilakos
Witold Pedrycz

Editors

ARTECH HOUSE

BOSTON | LONDON
artechhouse.com

Library of Congress Cataloging-in-Publication Data
A catalog of this title is available from the Library of Congress.

British Library Cataloguing in Publication Data
A catalogue of this title is available from the British Library.

Cover design by Yekaterina Ratner

© 2006 ARTECH HOUSE, INC.
685 Canton Street
Norwood, MA 02062

All rights reserved. Printed and bound in the United States of America. No part of this book may be reproduced or utilized in any form or by any means, electronic or mechanical, including photocopying, recording, or by any information storage and retrieval system, without permission in writing from the publisher.

All terms mentioned in this book that are known to be trademarks or service marks have been appropriately capitalized. Artech House cannot attest to the accuracy of this information. Use of a term in this book should not be regarded as affecting the validity of any trademark or service mark.

International Standard Book Number: 1-58053-963-7
ISBN 13: 978-1-58053-963-0

10 9 8 7 6 5 4 3 2 1

List of Contributors

Chapter 1	Athanasios V. Vasilakos, Witold Pedrycz
Chapter 2	Giovanni Acampora and Vincenzo Loia, *Dipartimento di Matematica e Informatica, Università di Salerno, Italy*
Chapter 3	Norman M. Sadeh, Fabien L. Gandon, and Oh Byung Kwon, *School of Computer Science Carnegie Mellon University*
Chapter 4	Pasi Välkkynen, Lauri Pohjanheimo and Heikki Ailisto
Chapter 5	Tom Gross, *Faculty of Media, Bauhaus-University Weimar, Germany*
Chapter 6	Chandra Narayanaswami, *IBM TJ Watson Research Center, USA*
Chapter 7	Mario Cannataro, *University Magna Græcia of Catanzaro, Italy*; Domenico Talia, Paolo Trunfio, *DEIS, University of Calabria, Italy*
Chapter 8	Marius Portmann, *University of Queensland, Australia*; Sebastien Ardon, Patrick Senac, *ENSICA, Toulouse, France*
Chapter 9	D. Li, *Department of Computer Science, Central South University, Changsha*; H. Liu, *Department of Computer Science, Missouri State University, US*; Athanasios V. Vasilakos, University of Thessaly, Greece
Chapter 10	J. Caarls, P.P. Jonker, *Delft University of Technology*
Chapter 11	Nikolaos Georgantas, Valérie Issarny, *INRIA, UR Rocquencourt, Domaine de Voluceau, 78153 Le Chesnay Cedex, France*; Christophe Cerisara, *LORIA UMR 7503, 54506 Vandoeuvre-lès-Nancy, France*
Chapter 12	G. Andreoni, *Dipartimento di Bioingegneria, Politecnico di Milano, Italy* M. Anisetti, *Tecnologia dell'Informazione, Università degli Studi di Milano, Italy* B. Apolloni, *Dipartimento di Scienze dell'Informazione, Università degli Studi di Milano, Italy* V. Bellandi, *Tecnologia dell'Informazione, Università degli Studi di Milano, Italy* S. Balzarotti, *Laboratorio di Psicologia della Comunicazione, Università Cattolica Del Sacro Cuore, Italy* F. Beverina, *ST Microelectronics s.r.l., Agrate Brianza, Milano, Italy* M. R. Ciceri, *Laboratorio di Psicologia della Comunicazione, Università Cattolica Del Sacro Cuore, Italy* P. Colombo, *Laboratorio di Psicologia della Comunicazione, Università Cattolica Del Sacro Cuore, Italy* F. Fumagalli, *Dipartimento di Scienze dell'Informazione, Università degli Studi di Milano, Italy* G. Palmas, *ST Microelectronics s.r.l., Agrate Brianza, Milano, Italy* L. Piccini, *Dipartimento di Bioingegneria, Politecnico di Milano, Italy*

Chapter 13	M.J. O'Grady,
	Practice & Research in Intelligent Systems & Media (PRISM) Laboratory, Department of Computer Science, University College Dublin, Ireland.
	G.M.P. O'Hare,
	Adaptive Information Cluster (AIC), Department of Computer Science, University College Dublin, Ireland
	N. Hristova,
	Practice & Research in Intelligent Systems & Media (PRISM) Laboratory, Department of Computer Science, University College Dublin, Ireland.
	R. Tynan,
	Adaptive Information Cluster (AIC), Department of Computer Science, University College Dublin, Ireland
Chapter 14	Shivanajay Marwaha,
	Motorola Electronics Pte. Ltd., Singapore
	Dipti Srinivasan,
	Department of Electrical and Computer Engineering, National University of Singapore, Singapore,
	Chen Khong Tham,
	Department of Electrical and Computer Engineering, National University of Singapore,
	Athanasios Vasilakos,
	Professor, University of Thessaly, Greece
Chapter 15	Wenye Wang, *North Carolina State University*
	Janise Y. McNair, *University of Florida, Gainesville*
	Jiang (Linda) Xie, *University of North Carolina, Charlotte*
Chapter 16	Charalampos Karagiannidis,
	Department of Cultural Technology and Communication, University of the Aegean, Greece
	Athanasios Vasilakos,
	Department of Special Education, University of Thessaly, Greece
Chapter 17	Rutger Rienks, Anton Nijholt, and Dennis Reidsma,
	Human Machine Interaction (HMI), University of Twente, Enschede, Netherlands
Chapter 18	Mohamed Khedr,
	Arab Academy for Science and Technology. Communications and Electronic Department, Gamal Abdel Naser St. Alexandria, Egypt
Chapter 19	Menahem Friedman,
	Department of Physics, Nuclear Research Center—Negev, Israel; and Department of Information Systems Engineering, Ben-Gurion University of the Negev, Israel
	Moti Schneider,
	School of Computer Science, Netanya Academic College, Netanya, Israel
	Mark Last and Omer Zaafrany,
	Department of Information Systems Engineering, Ben-Gurion University of the Negev, Israel
	Abraham Kandel,
	Department of Computer Science and Engineering, University of South Florida
Chapter 20	Dimitri Plemenos,
	University of Limoges, France

Contents

Preface xvii

CHAPTER 1
Ambient Intelligence: Visions and Technologies 1

1.1 Introduction 1
1.2 Basic Functions and Devices of Ambient Intelligence 3
    1.2.1 IC Trends for Computing 5
1.3 The Processing Perspective 6
    1.3.1 The Fixed Base Network 6
    1.3.2 The Wireless Base Network 7
    1.3.3 The Sensor Network 7
1.4 The Communication Perspective 8
1.5 The Software Perspective 8
1.6 Computational Intelligence as a Conceptual and Computing Environment of AmI 10
1.7 Conclusions 11
    References 11

CHAPTER 2
Ambient Intelligence and Fuzzy Adaptive Embedded Agents 13

2.1 Introduction 13
2.2 Agents Meet AmI 14
2.3 Transparent Fuzzy Control 17
2.4 FML Environment Description 19
2.5 FML Agent Distribution 23
2.6 Adaptivity 26
2.7 Conclusions 32
    References 33

CHAPTER 3
Ambient Intelligence: The MyCampus Experience 35

3.1 Introduction 35
3.2 Prior Work 37
3.3 Overall System Architecture 38
3.4 A Semantic e-Wallet 40
3.5 Capturing User Preferences 46

3.6	Instantiating the MyCampus Infrastructure	49
	3.6.1 MyCampus Development Experience	53
3.7	Empirical Evaluation	54
3.8	Conclusions	57
3.9	Additional Sources of Information	58
3.10	Acknowledgments	58
	References	58

CHAPTER 4
Physical Browsing — 61

4.1	Introduction	61
4.2	Related Work	62
	4.2.1 Tangible User Interfaces	63
	4.2.2 Physical Browsing Research	63
4.3	Physical Browsing Terms and Definitions	66
	4.3.1 Physical Browsing	66
	4.3.2 Object	66
	4.3.3 Information Tag	66
	4.3.4 Link	67
	4.3.5 Physical Selection	67
	4.3.6 Action	67
4.4	Physical Selection Methods	67
	4.4.1 PointMe	68
	4.4.2 TouchMe	68
	4.4.3 ScanMe	69
	4.4.4 NotifyMe	70
4.5	Physical Browsing and Context-Awareness	70
4.6	Visualizing Physical Hyperlinks	72
4.7	Implementing Physical Browsing	73
	4.7.1 Visual codes	73
	4.7.2 Electromagnetic Methods	74
	4.7.3 Infrared Technologies	74
	4.7.4 Comparison of the Technologies	75
4.8	Demonstration Applications	75
	4.8.1 PointMe, TouchMe, ScanMe Demonstration	76
	4.8.2 TouchMe Demonstration Using a Mobile Phone	77
4.9	Conclusion	79
4.10	Acknowledgments	80
	References	80

CHAPTER 5
Ambient Interfaces for Distributed Workgroups: Design Challenges and Recommendations — 83

5.1	Introduction	83
5.2	Ambient Displays and Ambient Interfaces	84
5.3	Ambient Interfaces in TOWER	86
	5.3.1 TOWER	86

	5.3.2	Ambient Indicators	87
5.4	User Involvement		94
5.5	Recommendations for the Design of Ambient Interfaces		96
5.6	Conclusions		100
5.7	Acknowledgments		100
	References		100

CHAPTER 6
Expanding the Role of Wearable Computing in Business Transformation and Living — 103

6.1	Introduction		103
6.2	Wearable Computers—History and Present Status		104
6.3	Wearable Computing Applications		110
6.4	Factors Limiting the Impact of Wearable Computers		112
6.5	Factors Providing Positive Feedback Loop		115
6.6	Middleware Components for Accelerating Transformation		117
	6.6.1	Context Sensing	117
	6.6.2	Sensor Interfaces	118
	6.6.3	Data Logging and Analysis	119
	6.6.4	Energy Management and Awareness	120
	6.6.5	Suspend, Resume, and Session Migration Capabilities	120
	6.6.7	Device Symbiosis	121
	6.6.8	Privacy and Security	121
6.7	Conclusions		122
	References		122

CHAPTER 7
Grids for Ubiquitous Computing and Ambient Intelligence — 127

7.1	Introduction		127
7.2	Grid Computing		129
	7.2.1	Grid Environments	130
	7.2.2	The Open Grid Services Architecture	132
7.3	Towards Future Grids		133
	7.3.1	Requirements and Services for Future-Generation Grids	135
	7.3.2	Architecture of Future-Generation grids	136
7.4	Grids for Ubiquitous Computing and Ambient Intelligence		137
	7.4.1	Grids for Ambient Intelligence	139
7.5	Conclusion		140
7.6	Acknowledgments		140
	References		141

CHAPTER 8
Peer-to-Peer Networks—Promises and Challenges — 143

8.1	Introduction	143
8.2	Taxonomy of P2P Systems	145
8.3	P2P—The Promises	146
8.4	P2P—The Challenges	148

	8.4.1	Security	148
	8.4.2	Noncooperation—Freeriders	149
	8.4.3	Search and Resource Location	150
8.5	Unstructured P2P Systems		151
	8.5.1	Napster	151
	8.5.2	Gnutella	152
	8.5.3	Topology of Unstructured Peer-to-Peer Networks	154
8.6	Structured P2P Systems		157
	8.6.1	Background	157
	8.6.2	The Chord Distributed Lookup Protocol	158
	8.6.3	Pastry	161
8.7	Conclusions		162
	References		163

CHAPTER 9
Comparative Analysis of Routing Protocols in Wireless Ad Hoc Sensor Networks
167

9.1	Introduction		167
9.2	Communication Architecture		170
9.3	Design Factors		172
	9.3.1	Power Consumption	172
	9.3.2	Fault Tolerance	172
	9.3.3	Scalability	172
	9.3.4	Hardware Issues	173
9.4	Routing in Wireless Sensor Networks		173
	9.4.1	Flooding-Based Routing	173
	9.4.2	Gradient-Based Routing	176
	9.4.3	Hierarchical-Based Routing	179
	9.4.4	Location-Based Routing	181
9.5	IP Mobility Management in Wireless Sensor Networks		185
	9.5.1	IP Mobility Protocols	186
	9.5.2	Limitations of Mobile IP	188
9.6	Conclusions		188
	References		189

CHAPTER 10
Pose Awareness for Augmented Reality and Mobile Devices
193

10.1	Introduction		193
	10.1.1	Problem Description	195
	10.1.2	System Setup	196
10.2	Notation		197
10.3	Fusion Framework		197
	10.3.1	Quaternions	199
10.4	Kalman Filters		201
	10.4.1	Filter Variants	202
	10.4.2	General Setup	203
	10.4.3	Time Update	204

	10.4.4	Observation Update	206
	10.4.5	Coping with Lag	206
	10.4.6	Divergence Problems	207
	10.4.7	The Decentralized KF	208
	10.4.8	A Modular Kalman Filter	210
10.5	Camera Positioning		212
	10.5.1	Problem Specification	212
	10.5.2	Marker Layout	213
	10.5.3	Canny Edge Detection	213
	10.5.4	Contour Detection	215
	10.5.5	Rejecting Unwanted Contours	216
	10.5.6	Corner Detection	216
	10.5.7	Fitting Lines	217
	10.5.8	Determining the ID of the Marker	219
10.6	Determining the Pose from a Marker's Feature Points		220
	10.6.1 Coordinate Systems		220
	10.6.2	Camera Model	220
	10.6.3	Estimating the Pose	221
10.7	First Experiment		223
10.8	Calibration		224
	10.8.1	Camera Calibration	224
	10.8.2	Pattern Pose	224
	10.8.3	Camera Frame to Body Frame	225
10.9	Measurements		225
	10.9.1	Performance of the Subpixel Edge Detector	225
	10.9.2	The Dependence of the Pose Accuracy on the Viewing Angle	227
	10.9.3	The Dependence of the Pose Accuracy on the Location in the Image	229
10.10	Conclusions		233
	10.10.1	Pluggable Filter	233
	10.10.2	Usability for Mobile Devices	233
	10.10.3	Usability for Augmented Reality	233
10.11	Conclusions		235
	References		235

CHAPTER 11
Dynamic Synthesis of Natural Human-Machine Interfaces in Ambient Intelligence Environments — 237

11.1	Introduction		237
11.2	A Task Architectural Model for Ambient Intelligence		239
	11.2.1	Service/Resource Components	241
	11.2.2	Tasks	242
	11.2.3	The Task Synthesis Service	244
11.3	A Human-Machine Interface Functional Architecture for Ambient Intelligence		245
	11.3.1	Context-Awareness	247

	11.3.2	Natural Interaction	248
	11.3.3	Reusability	250
11.4	Synthesizing Natural Human-Machine Interfaces		251
	11.4.1	Dynamically Composable Human-Machine Interfaces	251
	11.4.2	Realization of the Scenario	253
11.5	Current Achievements and Future Perspectives		256
11.6	Conclusions		260
	References		260

CHAPTER 12
Emotional Interfaces with Ambient Intelligence — 263

12.1	Introduction		263
12.2	Background		264
	12.2.1	The General Framework	264
	12.2.2	The Key Role of Emotions	265
	12.2.3	The Emotion Physical Milieu	265
	12.2.4	The Emotion Psychological Milieu	266
	12.2.5	No Emotion without Cognition	267
12.3	Materials and Methods		269
12.4	The Data		269
12.5	The General Architecture		271
12.6	A Simulated Experiment		273
12.7	Current and Future Work		278
12.8	Conclusions		283
	References		284

CHAPTER 13
A Sense of Context in Ubiquitous Computing — 287

13.1	Introduction		287
13.2	Ubiquitous Computing: A Paradigm for the 21st Century		287
	13.2.1	Mobile Computing	288
	13.2.2	Wearable Computing	289
13.3	The Question of Context		289
	13.3.1	Some Definitions of Context	289
13.4	Reflections on Context in Mobile Computing		290
	13.4.1	Spatial Context	290
	13.4.2	User Profile	291
	13.4.3	Device Profile	291
	13.4.4	Environment	292
13.5	Wireless Advertising		292
	13.5.1	Emotions in Context	293
	13.5.2	Affective Computing Systems	293
	13.5.3	Introducing Ad-me	294
13.6	Ambient Sensors: Foundations of a Smart Environment		296
	13.6.1	Introducing the Intelligent Climitization Environment (ICE)	297
	13.6.2	Architecture	297
13.7	Conclusion		300

References 300

CHAPTER 14
Ad Hoc On-Demand Fuzzy Routing for Wireless Mobile Ad Hoc Networks — 303

- 14.1 Introduction — 303
- 14.2 Problem Statement — 304
  - 14.2.1 Limitations of Single Metric Single Objective Routing Schemes — 304
  - 14.2.2 Complexity in Multiobjective Routing in MANETs — 305
  - 14.2.3 Applicability of Fuzzy Logic for Multiobjective Routing in MANETs — 306
- 14.3 Brief Background of Fuzzy Logic — 307
- 14.4 Cost Function for MANET Multiobjective Routing — 308
  - 14.4.1 Objective 1 (O_1): Minimizing End-to-End Delay — 308
  - 14.4.2 Objective 2 (O_2): Maximizing Probability of Successful Packet Delivery — 308
  - 14.4.3 Objective 3 (O_3): Minimizing Total Battery Cost of the Route — 309
- 14.5 Ad Hoc On-Demand Fuzzy Routing (AOFR) — 312
  - 14.5.1 AOFR Route Discovery Phase — 313
  - 14.5.2 AOFR Route Reply Phase — 313
  - 14.5.3 Fuzzy Cost Calculation in AOFR — 313
- 14.6 Simulation Parameters — 317
  - 14.6.1 Mobility Model — 319
  - 14.6.2 Traffic Model — 319
  - 14.6.3 Energy Model — 320
- 14.7 Performance Evaluation — 320
  - 14.7.1 End-to-End Delay — 320
  - 14.7.2 Congestion Loss — 321
  - 14.7.3 Packet Delivery Fraction — 321
  - 14.7.4 Expiration Sequence — 321
  - 14.7.5 Normalized Routing Load — 322
  - 14.7.6 Route Stability — 323
  - 14.7.7 Packets-per-Joule — 323
- 14.8 Discussion — 324
- 14.9 Conclusions — 325
- References — 326

CHAPTER 15
Authentication and Security Protocols for Ubiquitous Wireless Networks — 329

- 15.1 Introduction — 329
- 15.2 System Architecture and Design Issues — 330
- 15.3 Authentication Architecture for Interworking 3G/WLAN — 332
  - 15.3.1 Mobile IP with AAA Extensions — 333
  - 15.3.2 Authentication Servers and Proxy — 334
  - 15.3.3 AAA and Inter-Domain Roaming — 335
- 15.4 Authentication in Wireless Security Protocols — 336

		15.4.1	Wired Equivalent Privacy 802.11 LANs	337

		15.4.1 Wired Equivalent Privacy 802.11 LANs	337
		15.4.2 Extensible Authentication Protocol and its Variants	337
		15.4.3 802.1x Authentication Protocol	337
		15.4.4 WiFi Protected Access and 802.11i	338
		15.4.5 Virtual Private Network	339
	15.5	Comparison Study of Wireless Security Protocols	339
		15.5.1 Security Policies	340
	15.6	Conclusions	341
		References	343

CHAPTER 16
Learning in the AmI: from Web-Based Education to Ubiquitous Learning Experiences — 345

16.1	Introduction	345
16.2	The Present Paradigm in Learning Technologies	346
16.3	The Emerging Paradigm: Mobile Learning	347
16.4	The Future Paradigm: Learning in the AmI	350
16.5	Enabling Technologies, Models, and Standards	351
	16.5.1 Personalization Technologies and Computational Intelligence	352
	16.5.2 Learning Technologies Standards	354
	16.5.3 Learning Theories and Models	355
16.6	Conclusions	356
	References	357

CHAPTER 17
Meetings and Meeting Support in Ambient Intelligence — 359

17.1	Introduction	359
17.2	What Are Meetings?	361
	17.2.1 Meeting Resources	362
	17.2.2 Meeting Process	362
	17.2.3 Meeting Roles	362
	17.2.4 Problems with Meetings	363
17.3	Technology: Mediation and Support	364
	17.3.1 The Virtuality Continuum	364
	17.3.2 Meetings in the Virtuality Continuum	365
	17.3.3 Technology and Meeting Resources	366
	17.3.4 Supporting Meeting Processes	368
	17.3.5 Supporting Meeting Roles	369
	17.3.6 Learning How to Respond	370
17.4	Projects on Meetings	371
	17.4.1 Recordings and Sensors	372
	17.4.2 Annotations and Layers of Analysis	372
	17.4.3 Applications and Tasks	373
17.5	Conclusions	374
	Acknowledgments	374
	References	374

CHAPTER 18

Handling Uncertain Context Information in Pervasive Computing Environments — 379

- 18.1 Extending Ontologies of Context with Fuzzy Logic — 380
- 18.2 The Fuzzy Ontology — 381
  - 18.2.1 The Membership Function Class — 381
  - 18.2.2 The Fuzzy Rule Class — 383
  - 18.2.3 The Similar Class — 383
  - 18.2.4 The Fuzzy Inference Class — 383
- 18.3 The Fuzzy Inference of Context — 384
  - 18.3.1 Defining Linguistic Terms and Generating Membership Functions — 387
  - 18.3.2 Building the Inductive Fuzzy Tree — 389
  - 18.3.3 Rule Generation and Inference Process — 392
- 18.4 Prototype Implementation: The Event-Notification Service — 396
  - 18.4.1 Fuzzy Inference in the Event-Notification Service — 396
  - 18.4.2 Semantic Inference in the Event-Notification Service — 396
- 18.5 Conclusions — 399
- References — 400

CHAPTER 19

Anomaly Detection in Web Documents Using Computationally Intelligent Methods of Fuzzy-Based Clustering — 401

- 19.1 Introduction — 401
- 19.2 The Problem — 403
- 19.3 Cosine-Based Algorithms — 404
  - 19.3.1 Crisp Cosine Clustering (CCC) — 405
  - 19.3.2 Fuzzy-Based Cosine Clustering (FCC) — 405
  - 19.3.3 Local Fuzzy-Based Cosine Clustering (LFCC) — 405
- 19.4 Fuzzy-based Global Clustering — 407
  - 19.4.1 A New Distance Measure — 407
  - 19.4.2 The General Scheme — 409
- 19.5 Application: Terrorist Detection System — 411
  - 19.5.1 The Experiment — 412
  - 19.5.2 The Results — 412
- 19.6 Conclusions — 413
- Acknowledgments — 414
- References — 414

CHAPTER 20

Intelligent Automatic Exploration of Virtual Worlds — 417

- 20.1 Introduction — 417
- 20.2 Why Explore Virtual Worlds? — 418
- 20.3 Simple Virtual World Understanding — 419
  - 20.3.1 Nondegenerated View — 419
  - 20.3.2 Direct Approximate Viewpoint Calculation — 420
  - 20.3.3 Iterative Viewpoint Calculation — 421

		20.3.4 Direct Exhaustive Viewpoint Calculation	422
20.4	What is Visual Complexity of a Scene?		422
20.5	How to Compute Visual Complexity		424
		20.5.1 Accurate Visual Complexity Estimation	424
		20.5.2 Fast Approximate Estimation of Visual Complexity	426
20.6	Virtual World Exploration		427
		20.6.1 Incremental Outside Exploration	428
		20.6.2 Viewpoint Entropy-Based Exploration	429
		20.6.3 Other Methods	431
20.7	Future Issues		431
		20.7.1 Online Exploration of Virtual Worlds	432
		20.7.2 Offline Exploration of Virtual Worlds	432
20.8	Conclusions		433
	References		435

Index   437

Preface

Ambient Intelligence (AmI) has recently emerged as a vibrant research endeavor shaping our view of future computing and its role in society. While the inception point of AmI could always be argued about, it becomes apparent today that AmI has emerged and established itself as a result of a vivid synergy between several essential components such as- ubiquitous or pervasive computing, intelligent systems, context -awareness and an ultimate appreciation of various social interactions formed between humans and systems.

The landscape of AmI is highly diversified coming with a plethora of concepts, algorithms, paradigms, and promising implementations as well as a variety of existing systems.

Hence building a detailed roadmap of AmI is perhaps a somewhat futile job. Yet what is presented in Figure P.1 is a viable attempt to offer a bird's eye view of the content of this volume.

Components	
AMBIENT	INTELLIGENCE
Smart materials	Media management & handling
MEMS tech. & sensor tech.	Natural interaction
Embedded Systems	Computational intelligence
Wearable Computing	Contextual awareness
Ubiquitous Communications	Emotional computing
Intelligent Interfaces	Ubiquitous learning experience
Adaptive software	Exploration of Virtual Worlds
Platform design	
Software & Service Architectures, Design, Engineering and Integration	
Experience prototyping	
INTEGRATION	

Figure P.1 The roadmap of the volume.

Given this overall framework, let us move on to show how the key concepts are reflected through the chapters.

The chapter entitled "Ambient Intelligence: Visions and Technologies" (Vasilakos and Pedrycz) offers a general overview of the area and identifies the key issues of synergy arising in the framework of ambient intelligence. The essential functional components are clearly defined along with their roles.

The first one deals with ubiquitous or pervasive computing. Its major contribution to the domain of AmI lies in the development of various ad hoc networking capabilities that exploit highly portable or else numerous, very-low-cost computing devices. The second pivotal area concerns intelligent systems research, which offers learning algorithms and pattern matching schemes, speech recognition and language translators, and gesture classification and situation assessment. The third component relates to the context awareness; here the ensuing research allows us track and position objects of all types and represent objects' interactions with their environments.

The chapter entitled "Ambient Intelligence and Fuzzy Adaptive Embedded Agents" (Acampora and Loia) focuses on the reverse interaction, (i.e., processes and mechanisms originated by devices) by showing how it realizes a collection of autonomous control services being able to minimize possible human effort and manage the environment in an automatic fashion. Through the AmI experience, the authors report how ubiquitous devices are able to find the suitable set of intelligent services and use them in a transparent way. The objective is accomplished by hybridizing the methodology of computational intelligence with the concepts of semantic Web .

In the sequel, "Ambient Intelligence: The MyCampus Experience" (Sadeh et al.) deliver an overview of the ambient intelligence conducted over the past five years by the MyCampus group at the Carnegie Mellon University. This project has drawn on multiple areas of expertise, looking at issues of usability while focusing on the development of an open Semantic Web infrastructure for context-aware and privacy-aware service provisioning.

P. Välkkynen et al. deliver a survey entitled "Physical Browsing," followed by description of the authors' own demonstration systems. The prevalent visions of ambient intelligence emphasize various forms of some natural interaction between user and functions and services embedded in the environment or available through mobile devices. In these scenarios, the physical and virtual worlds seamlessly gear into each other, making crossing the border between these worlds natural or even invisible to the user. The bottleneck in reaching these scenarios appears to be in the natural mapping between the physical objects and their virtual counterparts. The conjecture is that the physical browsing becomes an efficient vehicle for mapping this digital information and physical objects in our environment.

The study authored by Gross ("Ambient Interfaces for Distributed Workgroups: Design Challenges and Recommendations") offers an interesting survey of ambient interfaces, being followed by a description of the author's TOWER project and its application. Here an important awareness of information is provided by indicators that are in the physical environment of the user and can be noticed by the user, however they do not interrupt the user's everyday activities. On a whole, such ambient interfaces help the user bridging an acute gap between the electronic space in which they are working and the physical space they are really positioned in. In such a sense, ambient interfaces contribute to the design of intelligent workplaces.

Next, Narayanaswami in "Expanding the Role of Wearable Computing in Business Transformation and Livving" explores the role that can be played by wearable computing in transforming the way we do business and entertain ourselves. The chapter briefly covers the evolution of wearable computing and elaborates on the impact it has shown to date. Finally, the author examines some of the present limitations and discusses how they might be alleviated.

M. Cannataro et al. in their study entitled "Grids for Ubiquitous Computing and Ambient Intelligence" bring attractive and promising concepts of grids and show how they can be used in AmI. The grid is a new computing infrastructure that allows user to access remote computing resources and facilities that are not available at a single site. By linking users, computers, databases, sensors and other devices, the grid expands beyond the network paradigm offered by the Internet and the Web and brings a novel way of thinking about computing and usage of the existing resources.

An overview of P2P computing, its main challenges and perspectives are presented by Portman et al. in the chapter entitled "Peer-to-Peer Networks—Promises and Challenges." Here the authors state that the P2P is an important technology with a tremendous potential in a world that becomes increasingly connected and decentralised. It is shown that the P2P paradigm is especially well-suited for environments that are dynamic and cannot always rely on any centralized infrastructure. In this context, Ubiquitous Computing and Ambient Intelligence are the areas where the P2P concept will have a particularly visible role to play.

D. Lee et al. ("Comparative Analysis of Routing Protocols in Wireless Ad-Hoc Sensor Networks") deliver a survey on Routing Protocols in Wireless Ad -Hoc Sensor networks. A wireless ad hoc sensor network is a collection of low-cost, low-power, multifunctional sensor nodes that communicate unattended over wireless channel. The main advantage is that they can be deployed in almost any kind of terrain and any hostile environment which might be impossible to support when using traditional wired networks. In order to facilitate communication within the network, a routing protocol is used which to not only discover routes between sensor nodes, but also collect the relative information from the network. Due to limited resources of sensor nodes, it becomes a challenging issue to design a correct, scalable and efficient routing protocol which results in a robust, long live and low latency ad hoc sensor network. The authors examine routing protocols for ad hoc sensor networks and evaluate them in terms of a given set of parameters, as well as offer a thorough comparison and discuss their respective merits and drawbacks.

In the sequel, J. Caarls et al. ("Pose Awareness for Augmented Reality and Mobile Devices") deliver a description of their experiments on augmented reality devices, with emphasis on use of Kalman filters and quaternions. Here the context is not limited to the physical world around the user, but also incorporates the user's behavior, and terminal and network characteristics. The user carries a wearable terminal and a see-through display in which he can sense virtual visual information that augments reality. Augmented Reality differs from Virtual Reality in the sense that the virtual objects are rendered on a see-through headset. As with audio headphones, where one can hear sound in private, partly in overlay with the sounds from the environment, see-through headsets can do the same for visual information. The virtual objects are in overlay with the real visual world. The Augmented Reality can also be used to place visual information on otherwise empty places, such as white parts of walls of a museum. The 3D vector of position and orientation (or heading) is referred to as pose.

N. Georgantas et al. ("Dynamic Synthesis of Natural Human-Machine Interface in Ambient Intelligence Interfaces") based on the Web Services paradigm and the WSAMI middleware, elaborated a fine-grained task architectural model and an associated task of synthesis of middleware service, which provide for dynamic, situation-sensitive composition and reconfiguration of complex user tasks within AmI environments. The architectural model focuses especially on the synthesis of user interfaces, enabling multiple distributed interfaces in a task that may dynamically incorporate: (i) specialized, however reusable, UI back-end components; and (ii) generic UI front end-components that may be general-purpose I/O devices. The study shows how this approach effectively models user interfaces in AmI environments.

G. Andreoni et al. ("Emotional Interfaces with Ambient Intelligence") deliver a description of their experiments on the use of AmI hardware to detect and respond to human emotions and non-verbal signals. They set up an enabling system for eliciting and exploiting particular emotional states and analyzed, both theoretically and experimentally, the main features of the dynamical system realizing this cooperation between human and machine. They pointed out that attuning is a process and that automatically labeling with emotional states the non verbal signals produced by human is a too ambitious goal. Instead, they look for a cognitive system capable of managing the signals produced by humans within his emotional states.

M.J. O' Grady et al. ("A sense of Context in Ubiquitous Computing") deliver an overview of context in ubiquitous computing. The acknowledge that context itself is a rich and sometimes nebulous term. It is very much sensitive to a given situation and comes with the richness of sensory modalities. However the richer the sense of context, the greater the relevance and pertinence of the content or service delivered to the mobile user.

S. Marwaha et al. ("Ad Hoc on-Demand Fuzzy Routing for Wireless Mobile ad Hoc Networks") discuss an AOFR routing algorithm and explore the role of fuzzy logic in the development of adaptive multi-objective routing algorithm for Mobile Ad hoc NETworks (MANET). They discuss various limitations of single objective routing schemes, highlight the complexity involved in multi-objective routing and show how fuzzy logic can model the uncertainties involved in multi-objective routing in MANETs.

W. Wang et al. (in their contribution "Authentication and Security Protocols in Ubiquitous Wireless Networks") deliver an overview of issues of authentication and security in wireless environments. Wireless networks have evolved into a heterogeneous collection of network infrastructures providing a wide variety of options for user access such as Wi-Fi and cellular systems, which have facilitated ubiquitous wireless services, i.e., anytime, anywhere, and anything. However, the increasing concerns about security and universal access demand for efficient authentication and security protocols for mobile environments. The authors describe some detailed authentication procedures which are usually included in security protocols in order to demonstrate the interaction of security protocols at different layers with respect to data streams, delay, and throughput.

Next, C. Karagiannidis et al.("Learning in the AmI: from Web-based Education to Ubiquitous Learning Experience") explore the issues of learning in the AmI environment.They offer some interesting insights into several essential theoretical and technical aspects concerning the delivery of personalized, ubiquitous learning experiences, which are seamlessly integrated within our everyday activities.

R. Rienks et al. ("Meetings and Meeting Support in Ambient Intelligence"), deliver an overview of meetings and their everyday problems. They have shown that meetings all along the virtual continuum can benefit from the essential technologies of Ambient Intelligence. In particular, AmI systems will analyze information being conveyed and increase their proactive behavior in order to help realize meetings as comfortable, efficient and effective as possible.

M. Khedr in his study entitled ("Handling Uncertain Context Information in Pervasive Computing Environments") presents an approach to support levels of certainty in context-aware environment using inductive fuzzy reasoning over semantically modeled context. The author argues that representing the uncertainties found in context information with the use of ontologies will endow context-aware applications with the common understanding of vague contextual information and enhance the realization of the essential requirements of reusability and extensibility.

M. Friedman et al. ("Anomaly Detection in Web Documents Using Computationally Intelligent Methods of Fuzzy-Based Clustering") design several clustering mechanisms for detecting anomalies in web documents and offers techniques of content monitoring and its analysis carried out in the AmI environment. The existence of such tools could help identifying illegal activities on the Internet and possibly prevent future disasters.

D. Plemenos ("Intelligent Automatic Exploration of Virtual Worlds") explores a concept of the global Virtual World forming an integral part of the Am environment. The focus of this study is on a visual exploration of fixed unchanging virtual worlds.

We have been very fortunate having a number of eminent contributors who as being the leaders in the field contributed immensely to this volume. Our sincere thanks go the reviewers who did a superb job providing the authors with a critical yet highly constructive and helpful feedback. Undoubtedly, this feedback was essential in improving the quality of the volume in many tangible and important ways. Last but not least, our thanks go to the highly professional staff at the Artech House, especially Barbara Lovenvirth, whose diligence was instrumental in keeping the project on track.

Thanos Vasilakos
Witold Pedrycz
June 2006

CHAPTER 1
Ambient Intelligence: Visions and Technologies

Athanasios V. Vasilakos and Witold Pedrycz

1.1 Introduction

In an ambient intelligent environment [1, 2], people are surrounded by networks of embedded intelligent devices that provide ubiquitous information [3], communication, services, and entertainment. Furthermore, the devices seamlessly adapt themselves to users and even anticipate their needs. Ambient intelligent environments present themselves quite differently compared to contemporary handheld or stationary electronic devices. Electronics will be integrated into clothing, furniture, cars, houses, offices, and public places. This calls for solving an important problem of developing new concepts of user interfaces that allow for a natural interaction with these environments. As succinctly stated by Weiser [43], "the most profound technologies are those that disappear. They weave themselves into the fabric of everyday life until they are indistinguishable from it." A promising approach is the one in which users interact with their digital environments in the same way as they interact with each other. Reeves and Nass were the first to formulate this novel interaction equivalence, and they called it the media equation [5].

MIT's Oxygen project [16] and IBM's effort on pervasive computing [7] are similar approaches addressing the issue of integration of networked devices into peoples' backgrounds. Ambient intelligence (AmI) aims at taking the integration even one step further by realizing environments that are sensitive and responsive to the presence of people. The focus is on the users and their experiences from a consumer electronics perspective, which introduces several new basic problems related to natural user interaction and context-aware architectures [8] supporting human-centered information, communication, service, and entertainment. For a treatment of these novel distinguishing factors, we refer to the book *The New Everyday* [19].

AmI covers a whole world of underlying technologies used to process information: software, storage, displays, sensors, communication, and computing. To identify such vastly different devices that are needed to realize ambient intelligent environments we first introduce a scenario that facilitates the elicitation of a number of essential ambient intelligent functions from which device requirements can be determined.

AmI vision requires an intensive and carefully planned integration of many different and highly advanced technologies.These technologies may include energy-efficient, high-performance computing platforms; powerful media processing hardware and software; intelligent sensors and actuators, and advanced user-interface designs (vision, speech).

Ôhe highly dynamic AmI environment,t along with tightening cost and time-to-market constraints for various AmI products, requires that the enabling technologies for products be highly scalable in almost every aspect. For instance, an interactive, multiplayer gaming device must be able to seamlessly adapt to constantly changing networking conditions; a new generation of a high-end residential gateway must be introduced in the market without the necessity to redesign the processing infrastructure from scratch.

In a heterogeneous environment, refer to Figure 1.1, of wireless networks (2.5G-GPRS, 3G-WCDMA, hotspots and enterprise networks–Wireless LANs), a service platform solves interoperability issues and enables seamless roaming. Federation between various service platform operators enables users or terminals to authenticate on various access networks not used before. This ensures that consistent and coherent service access is provided, independent of the current network domain (federated service platform solutions).

In peer-to peer (P2P) distributed Internet applications [10], the resources of a large number of autonomous participants will be harnessed in order to carry out the system's function. In many cases, peers form self-organizing networks that are layered over the top of conventional Internet protocols and have no centralized structure. Grid systems [24] will arise from collaboration between smaller, better connected groups of users with a more diverse set of resources to share.

This chapter explores scalability of various AmI enabling technologies. The most interesting challenges lie in advanced techniques pushing the current state-of-the-art technologies to the limits. Figure 1.2 shows a conceptual picture of the AmI infrastructure. AmI will provide a very open environment with processing power present in almost any device. Thus, compute platforms are obviously a core AmI enabling technology. Many AmI devices, particularly server-like systems, will require extremely high processing power. Other, mainly mobile, AmI devices will need (high) processing power at a low energy cost. Section 1.3 explores scalable processing in an AmI environment in more detail.

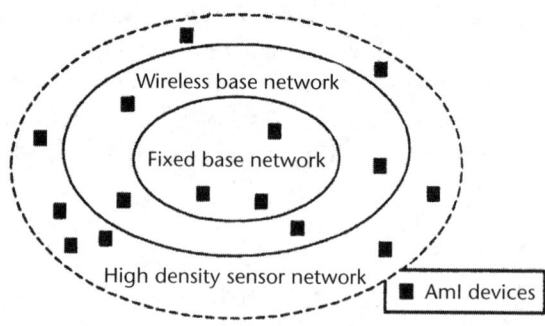

Figure 1.1 AmI environments.

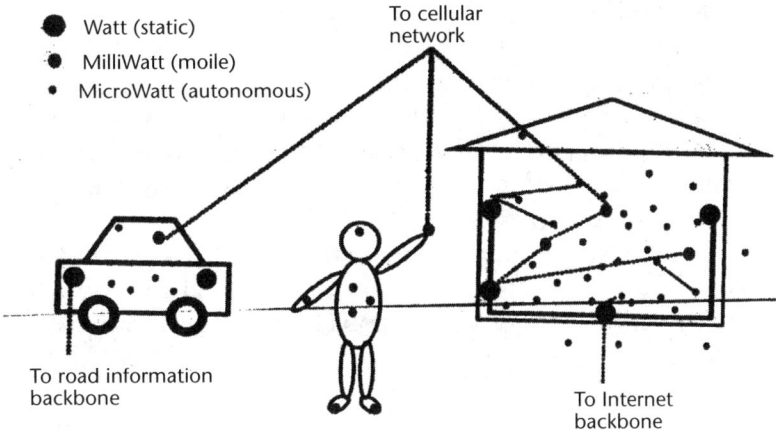

Figure 1.2 AmI processing and communication infrastructure.

A second crucial enabling technology for AmI is communication and network technology. AmI builds on ubiquitous computing, as envisioned by the late Mark Weiser [4, 12], who felt that computing technology would soon fade into and seamlessly blend with physical environments. To achieve pervasive computing, a communication infrastructure is needed; it must be omnipresent, flexible, and reliable. The base infrastructure will consist of a very high bandwidth fixed base network, augmented with future generation high-bandwidth wireless networks. To guarantee the "anywhere at any time" aspect of communication, the base infrastructure can be complemented with sensor networks consisting of many inexpensive, low-power, low-bandwidth, densely placed communication nodes. Section 1.4 discusses issues in high-density wireless networks in more detail. A third technology is the embedded software infrastructure required for AmI systems; it must be flexible and fully decentralized. Media and signal-processing algorithms must be highly scalable in order to cope with the very heterogeneous and dynamic environment. Novel programming models are required as well. Scalability from the software point of view is the topic of Section 1.5. In Section 1.6 we touch upon the technology of computational intelligence (CI) and demonstrate how it could support the major functionality of the AmI constructs.

1.2 Basic Functions and Devices of Ambient Intelligence

The previous scenario contains three basic ambient intelligent functions. First, the environment is context aware, which means there are sensors integrated into the environment that communicate events that can be combined to determine meaningful states and actions, such as person identification, position detection, and query interpretation. Second, audio, video, and data can be streamed wirelessly to any access device present in the environment, thus enabling ubiquitous wireless access to information, communication, services, and entertainment. Third, users in an ambient intelligent environment interact with their surroundings through natural modalities, such as speech, gesture, and tactile movements, thus enabling hands-free interaction with their environment.

These basic functions of ambient intelligence not only apply to home environments; they also apply to other environments such as mobile spaces (car, bus, plane, and train) public spaces (office, shop, and hospital) and private spaces (clothing). They support a variety of human activities, including work, security, healthcare, entertainment, and personal communications.

There are also miniature computers, which are super fast computers known as molecular computers. They are so small, they could be woven into clothing. We can potentially get the computational power of 100 workstations on the size of a grain of sand.

Smart Dust [13], tiny wireless microelectromechanical sensors (MEMS) will detect everything from light to vibrations. These "motes" would gather scads of data, run computations, and communicate that information using two-way band radio between motes in distances approaching 1,000 feet.

All devices that process information need energy to do so. Depending on the availability of energy resources in a device, the amount of information processing for a given technology is constrained. The availability of energy in a device is therefore the discriminating factor for the distribution of ambient intelligence functionality in a network of devices. The following three generic classes of devices are defined.

1. Autonomous devices that empower themselves autonomously over a full lifetime. They extract the required energy from the environment by scavenging light or electro-magnetic energy, and mechanical energy from vibrations or from temperature differences by means of thermo-electric generators. Examples are all kinds of tags and sensors. These autonomously empowered devices are called micro-Watt nodes.
2. Portable devices that use rechargeable batteries with typical autonomous operational times of a few hours and standby times of several days. Examples are personal digital assistants, mobile phones, wireless monitors, portable storage containers, and intelligent remote controls. These battery-powered devices are called milli-Watt nodes.
3. Static devices that have quasi-unlimited energy resource, (mains powered or combustion engine). Examples are large flat displays, recording devices, home servers, and large storage and computing devices. These net-empowered devices are called Watt nodes.

Examples of these device classes are already found today but are not yet ambient intelligent because they are not part of an integrated intelligent network.

The energy availability as well as the power dissipation of a device is not constant over time; different operating modes are defined depending on the device activity and energy resources. Large variations in peak to average power dissipation are encountered quite often. It is clear that for all types of devices the energy management is a key function, since it determines the potential for information processing.

Everyday devices such as alarm clocks, wristwatches, key chains, and even refrigerator magnets are made more intelligent through a new hardware and software platform (small personal object technology-SPOT [14]) that is small enough to scale down to the sizes required by such devices.

1.2.1 IC Trends for Computing

The IC technology road maps for computing are well described by the ITRS road maps [15]. Some of the trends are indicated in the graphs shown in Figures 1.3 and

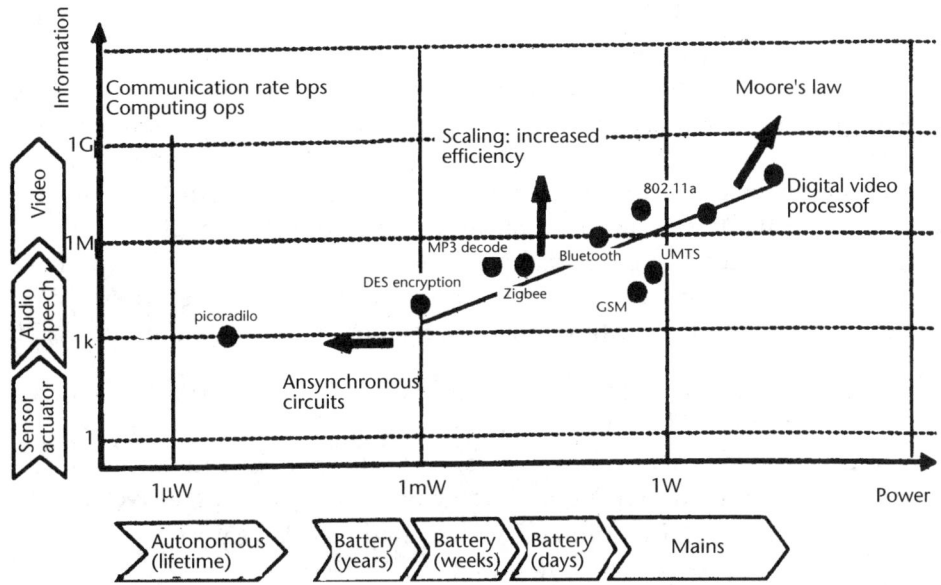

Figure 1.3 Major IC trends for computing (*From:* [16]).

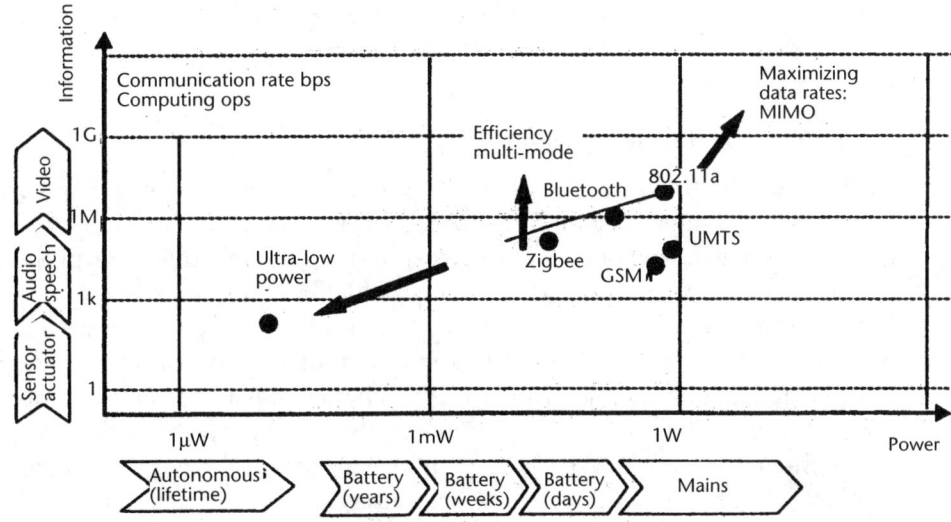

Figure 1.4 IC trends for communication (*From:* [16]).

1.4. The continued device scaling of CMOS technology results in a further increasing availability of computational power.

As described by Moore's law, this increased computational power is exploited in all kinds of parallel architectures. The increased power efficiency of new technologies is particularly interesting for the battery-powered and autonomous devices but is countered by the increased leakage currents in these technology generations. For ultralow power asynchronous circuits are very attractive.

1.3 The Processing Perspective

Ambient intelligence devices are expected to provide scalable processing power at every level of the networked infrastructure. Flexibility, both at deployment time (i.e., programmability) and in field (i.e., reconfigurability), is required to support scalability: Most AmI devices will be programmable and reconfigurable, in order to provide adequate performance under widely varying conditions. Furthermore, truly scalable computation must be cheap, reliable, and energy efficient. Even assuming no slowdown in technology scaling, designing highly flexible devices that meet cost, reliability, and energy efficiency requirements is going to be extremely challenging. In fact, as technology scales down, it becomes increasingly unwieldy [17]:

1. Logic gates get faster, but wires do not, and performance is increasingly interconnect dominated.
2. Power density (in active state) and leakage power (in idle state) increase significantly even if supply voltage is down-scaled aggressively.
3. Signal-to-noise ratio and circuit reliability decrease.
4. Design complexity scales up.

To address the scalability challenge in view of technology and design complexity limitations, hardware architectures for AmI will have widely varying characteristics depending on the network tier where they are deployed. We briefly overview trends in devices for the three network types as shown in Figure 1.2.

1.3.1 The Fixed Base Network

Devices for the fixed network infrastructure are rapidly evolving toward parallel architectures. Single-chip multiprocessors are a natural evolution of current single-processor solutions in an effort to support truly scalable data processing. The ever-increasing ratio between wire delay (which remains constant, even in the most optimistic assumptions) and devices switching speed (which scales with technology) has become a fundamental bottleneck in designing high-performance integrated systems [18]. The only viable answer to this challenge is to localize computation and emphasize parallelism. For this reason, even general-purpose single-chip processors for high-end servers are becoming increasingly distributed and highly parallel. Explicit parallelism is sought to avoid architectural bottlenecks (such as global register files and instruction fetch and decode logic) and to ensure long-term scalability across several technology generations. Most newly designed high-performance processors are highly parallel architectures, with multiple program counters [19]. Simultaneous multithreading is emerging as the new architectural leitmotif [20]. Moreover, hardware designers routinely resort to multiple execution units; distributed register files are representative instances of this trend [21]. One step further, next-generation architectures are focusing on parallelism not only at the microcity level multi-processing becomes multistreaming: streaming supercomputers are under development [22] where computation and storage are seen as successive stages of an information processing pipeline (a stream), and communication is made explicit as much as possible.

1.3.2 The Wireless Base Network

When considering the wireless base network, the push toward highly parallel multi-processing architectures is even stronger, coupled with a trend toward single-chip integration. This convergence toward multi-processor system-on-chip (MPSoC) platforms is also motivated by the quest for scalability. Energy efficiency and cost constraints are much tighter than for high-performance servers, and the computational requirements can be matched only by resorting to heterogeneous architectures (in contrast to homogeneous general-purpose processors), which provide computational power tailored to a specific class of applications [23–25]. Designing such heterogeneous application-specific (AS) MPSoCs is a challenging task because their complexity is comparable to that of most aggressive general-purpose platforms, yet their time-to-market and profitability windows are much shorter and focused.

To tackle the design challenges of ASMPSoCs, architectures and design flows are undergoing a profound revolution. Computational elements must be programmable (with application-specific programming paradigms, such as fine-grained bit level, control-dominated word level, data flow, etc.) to provide the much needed post fabrication programmability and in-field adaptation. Additionally, they should be standardized as much as possible to ensure software portability and to facilitate interfacing with other computational elements, on-chip and off-chip memories, and input-output devices. The communication infrastructure provides the connective fabric for the computational and storage element. Current shared-medium (bus-based) standardized fabrics are not scalable, and scalable network-on-a-chip (NoC) communication architectures are under active exploration and development [26].

Next-generation wireless base network devices will include tens to hundreds of application-specific processors (e.g., MAC accelerators, digital modems, cryptoprocessors), as well as several general-purpose processing cores (e.g., FPGA fabrics, DSP cores, VLIW multimedia processors, FP coprocessors, RISC controllers, etc.), a significant amount of on-chip storage, and various peripheral units. All these devices will be connected by an NoC.

1.3.3 The Sensor Network

The nodes of high-density sensor networks will be power and cost constrained. Scalability will be achieved by compromising to some degree flexibility at the node level and focusing on minimum energy consumption and minimum cost. A sensor network will contain a large number of simple nodes (with reduced flexibility) connected to the wireless base network. AmI will therefore be obtained as a result of a very large number of simple nodes communicating through distributed, adaptive protocols.

1.4 The Communication Perspective

The communication protocols in the AmI environment will support energy-quality scalability and application-aware design [25, 27]. The communication quality could be defined as a combination of several important metrics:range, reliability, latency, and throughput—and can be chosen by an application through a basic application programming interface(API). A power aware middleware layer will bridge the gap between the performance parameters and the hardware for energy scalability. The middleware manager must be empowered with hardware energy models for the digital processing circuits and radio transceiver, allowing this layer to select the minimum-energy hardware setting for the performance level commanded through the API [28].

Today wireless communication protocols suffer from several shortcomings: They are general purpose, and they are not designed for energy-efficient communication (i.e., a reliable transport and bandwidth fairness mechanisms of TCP are overhead for applications that do not require them).

In the AmI environment, protocols should be tuned to the target hardware and application. The protocols will be tuned for unidirectional data propagation in high-density microsensor networks. Microsensor nodes, which gather and relay observations about stimuli in the environment (people, vehicles, etc.) to a central base station, have the lowest performance and highest lifetime requirements of all AmI devices. Therefore the pertinent protocol is application aware. We move to a new era of routing protocols: from address-oriented to data-centric ones [29]. Data-centric technologies perform in-network aggregation of data to yield energy-efficient and application-specific dissemination.

1.5 The Software Perspective

The vision of AmI requires a fundamental paradigm shift on system architectures, programming models, and algorithm designs in order to integrate client/server model with the peer-to-peer model, address-based routing, routing with name-data centric routing, and location-transparent computing with location-aware computing, fixed, and centralized processing with adaptive and in-network processing.

AmI systems, as with embedded sensor networks, need software infrastructure support different from the current Internet. Node configurations and directories, global routing tables, and sequencing of events may be extremely hard to set up in AmI systems. AmI systems will be formed in ad hoc ways or reconfigured from existing systems.

1.5 The Software Perspective

A scalable AmI software infrastructure also needs effective ways to manage tasks and resources. Sensing and computing, deeply embedded into our natural surroundings, greatly improve the way in which raw data is collected,but also introduces a large amount of unpredictability in task requirements and resources availability.

Signal processing algorithms should be scalable, and individual AmI devices should rely on as little global coordination as possible. Algorithms should have some degree of autonomy and operate asynchronously. Local collaboration groups will enable scalability.

Another important issue about AmI systems is how to program them. Traditional programming technologies do not scale. Morever, time and space (location) have been removed from traditional programming technologies and focused on the level of abstraction. Given that, AmI programming technologies should introduce space and time without suffering in abstraction. The development of formal programming models and the use of software synthesis technologies to automatically generate the interactions of algorithm components is a key challenge. The automated software synthesis is a critical step toward the scalability of programming AmI systems (Figure 1.5).

Polymorphic active networks [30] ,unlike traditional IP networks, seek to reduce the rigidity of the service definition by making the infrastructure programmable and hope to speed up the deployment of new network services. The flexible programmability offered by active networks (AN) architectures can be broadly classified into two categories: packet programmability (programs and data) and node programmability(active extensions programs downloaded into active nodes for extending their functionality). The programming languages used for packet and node programs in various AN architectures have varying degrees of expressive powers (polymorphic-type languages).

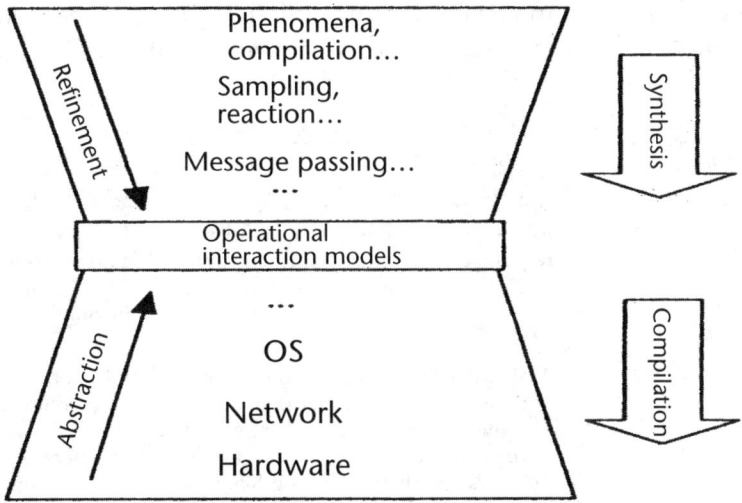

Figure 1.5 A programming methodology for embedded systems. (*From:* [1]).

1.6 Computational Intelligence as a Conceptual and Computing Environment of AmI

The discussion presented so far clearly demonstrates that AmI comes with a strong need for a comprehensive, computationally feasible and highly versatile environment. As far as the underlying technologies are concerned, this becomes even more profound as directly presented by the European Information Society Technology (IST) Advisory Groups where the contributing technologies of AmI have been clearly spelled out. We encounter a long list of fundamental components.

At the system end, there are smart materials, MEMS technology along with sensor technology, embedded systems, ubiquitous communications, and adaptive software.

At the user end, there are media management and handling, natural interaction, contextual awareness, and emotional computing

The fundamental AmI features such as embedding, context awareness and personalization, adaptive properties of system and learning call for the underlying advanced information technology facilitating their efficient realization. When looking at the existing alternatives, the paradigm of computational intelligence (CI) directly comes into play. CI, in its history going back to the beginning of the 1990s, embraces three fundamental domains of endeavor: granular computing, neural network, and evolutionary (or being more general) biologically inspired optimization. Granular computing and the technology of fuzzy sets, in particular are aimed at the realization and processing at a certain most suitable level of abstraction (which helps avoid or significantly reduce possible computing and communication burden). Neurocomputing offers a suite of flexible architectures and topologies of distributed computing architectures along with a variety of learning schemes that are of vital interest when building adaptive systems. The vast array of biologically inspired optimization vehicles (in particular, genetic algorithms, evolutionary programming, and so on) is of paramount interest when realizing a structural optimization of the systems. These three major technologies of CI are placed vis-à-vis the key features of AmI so that the relationships and the underlying support delivered by the CI environment become clearly delineated, as shown in Table 1.1.

Table 1.1 Major technologies of CI versus the Fundamental Features of AmI

Fundamental Requirement of AmI	Brief Description	Role and Contributions of the CI Technology
Context awareness	Ami system should possess abilities to recognize users, and situational context (both the environment and users).	Fuzzy sets offer an ability to capture context, carry out partial (flexible) matching, deal with incompletely specified and sensed environment.
Personalization	AmI environment adjusts to the needs of individual users (rather than adhering to some "generic" user). There is a need to cope with radically different needs of the users and their abilities to communicate with the system.	Formalizing relevance feedback and monitoring/evaluation of the continuously changing needs of the users is accomplished through the use of information granules and some learning of the original system.

Table 1.1 (continued)

Fundamental Requirement of AmI	Brief Description	Role and Contributions of the CI Technology
Adaptive properties and learning capabilities	It is required that in response to the user's request and evaluation of its current performance, the system can adjust itself.	Learning abilities are necessary, hence, the use of neurocomputing (endowed with various learning mechanisms) and evolutionary optimization is of high relevance.
Embedding	Devices are part of the network a way of embedding and assuring effective mechanisms of communication.	Fuzzy sets support strategy for their inner flexible control-based policy and mechanisms of interaction.

In essence, Table 1.1 clearly visualizes an important and appealing ways in which the requirements are conceptualized and algorithmically handled by the corresponding CI technologies.

1.7 Conclusions

The emergence of AmI will have a tremendous impact on everyday life. It becomes apparent that AmI systems will radically differ from today's systems. We have to redesign embedded software and hardware. AmI requires scalability of processing, communication, and software infrastructure in many aspects. The design community faces a lot of interesting challenges and opportunities for the years to come.

References

[1] Basten, T., M.C.W. Geilen, H.W.H.de Groot, *Ambient Intelligence: Impact on Embedded System Design*, Boston MA: Kluwer Academic Publishers, 2003.

[2] ISTAG, Scenarios for Ambient Intelligence in 2010. Feb. 2001. http://www.cordis.Iu/ist/istag.htm.

[3] Hightower, J., G. Borriello, "Location Systems for Ubiquitous Computing," *IEEE Computer*, 33, 8, 2001, 57–66.

[4] Weiser, M., "The computer of the Twenty-First Century," *Scientific American* 165(3), 94–104,1991.

[5] Reeves, B. and C. Nass, *The Media Equation*, Cambridge University Press, Cambridge, Massachusetts, 1996.

[6] Dertouzos, M., The Future of Computing, *Scientific American* Vol. 281, No. 2, pp. 52–55, 1999

[7] Satyannarayanan, M., Pervasive computing, vision and challenges, *IEEE Personal Communications*, August, 2001.

[8] Abowd, G. et al. Context-Awareness in Wearable and Ubiquitous Computing. *Proceedings of the 1st International Symposium on Wearable Computers*, 1997.

[9] Aarts, E., Marzano S. *The New Everyday:Visions of Ambient Intelligence*, 010 Publishing, Rotterdam,The Netherlands, 2003.

[10] Ripeanu, M. Peer-to-peer architecture case study: Gnutella network. In *2001 International conference on P2P computing* (August 2001).

[11] Foster, I., "The Anatomy of the Grid: Enabling Scalable Virtual Organizations," Lecture Notes in Computer Science, 2150, 2001.

[12] Weiser,M. Some computer science issues in ubiquitous computing, *Communications of the ACM*, 36, 7, 1993, 75–85.

[13] http://www.cmputerword.com/mobiletopics/mobile/story/o,10801,79572,00.html(smart dust).

[14] http://www.winsupersite.com/showcase/spot_preview.asp(Smart Personal Object Technology-SPOT).

[15] ITRS road map (2003), http://public.itrs.net/.

[16] Aarts, E., Roovers, R., "IC Design Challenges for Ambient Intelligence," DATE 2003, pp. 10002–10007.

[17] T. Karnik, S. Borkar, V. De, "Sub-90nmTechnologies-Challenges and Opportunities for CAD. *Proc. IEEE/ACM ICCAD'02*, Nov. 2002, pp. 203–206.

[18] R. Ho, K. Mai, M. Horowitz, "The Future of Wires" *Proc. IEEE*, Vol. 89, No. 4, Apr. 2001, 490–504.

[19] A. Moshovos, G. Sohi, "Microarchitectural Innovations:Boosting Microprocessor Performance Beyond Semiconductor Technology Scaling." *Proc. IEEE*, Vol. 89, No. 11, Nov. 2001, 1560–1575,.

[20] D. Deleganes, J Douglas, B. Kommandur, M.Patyra, "Designing a 3 GHz, 130 nm, Intel Pentium 4 processor" *IEEE symp. On VLSI Circuits*, pp. 130–133, Jun. 2002.

[21] V. Zynban, P. Kogge, "Inherently Lower-power Hgh-performance Superscalar Architectures." *IEEE Transactions on Computers* Vol. 50, No. 3, Mar. 2001, 268–285,.

[22] B. Khailany, et al., "Imagine: Media Processing with Stream." *IEEE Micro* Vol. 21, No. 2, Mar.-Apr. 2001.

[23] T.Koyama, K. Inoue, H Hanaki, M.Yasue, E. Iwata, "A250-MHz Single-chip Multiprocessor for Audio and Video Signal Processing." *IEEE J of Solid-State Circuits* Vol. 36, No. 11, Nov. 2001, 1768–1774.

[24] H. Zhang, et al., "A 1-V Heterogeneous Reconfigurable DSP IC for Wireless Besebend Digital Signal Processing." *IEEE J of Solid-State Circuits* Vol. 35, No. 22, Nov. 2000, 1697–1704,.

[25] Lin D., Liu H., Vasilakos A., "Comparative Analysis of Routing Protocols in Wireless Ad Hoc Sensor Networks," in *Ambient Intelligence, Wireless Networking, Ubiquitous Computing*, Norwood, MA: Artech House, 2006, Chapter 9.

[26] L.Benini, G.De Micheli, "Networks on Chip: A New SOC Paradigm" *IEEE Computer* Vol. 35, No. 1, Jan. 2002, 70–78.

[27] Marwaha S.,Srinivasan D.,Tham C.K.,Vasilakos A. "Ad Hoc On- demand Fuzzy Routing for Wireless Mobile Ad Hoc Networks", in *Ambient Intelligence,Wireless Networking, Ubiquitous Computing*, Norwood, MA: Artech House, 2006, Chapter 14.

[28] R.Min and A. Chandrakasan, "A Framework for Energy-Scalable Communication in High-Density Wireless Networks." *Proc. ISLPED 2002*, Aug. 2002.

[29] C.Intanagonwiwat, R.Giovindan, and D. Estrin, "Directed Diffusion: A Scalable and Robust Cmmunication Paradigm for Sensor Networks" *Proc. MobiCom'00, 6th International Conference*, Aug. 2000, Boston, MA, pp. 56–67.

[30] Cambell et al. A survey of programmable network, *ACM SIGCOMM Computer Communication Review*, Vol. 29, No. 2, pp. 7–23, April 1999.

CHAPTER 2
Ambient Intelligence and Fuzzy Adaptive Embedded Agents

Giovanni Acampora and Vincenzo Loia

2.1 Introduction

Intelligent environments are opening unprecedented scenarios, where people interact with electronic devices embedded in environments that are sensitive and responsive to the presence of people. The term *ambient intelligence* (or AmI) reflects this tendency, gathering best results from three key technologies: ubiquitous computing, ubiquitous communication, and intelligent user friendly interfaces. The emphasis of AmI is on greater user friendliness, more efficient services support, user empowerment, and support for human interactions. People are surrounded by smart pro-active devices that are embedded in all kinds of objects: an AmI environment is capable of recognizing and responding to the presence of different individuals, working in a seamless, unobtrusive, and often invisible way.

When designing AmI Environments, different methodologies and techniques have to be used, ranging from materials science, business models, network architectures, up to human interaction design. Independently from the wide range of approaches, some key technologies characterize AmI, such as:

- *Embedding*. Devices are (wired or unwired) plugged into the network. The resulting system consists of multiple devices, computer equipment, and software systems that must interact among them. Some of the devices are "simple" sensors others are "actuator" owning a crunch of control activity on the environment (central-heating, security systems, lightning systems, washing machines, refrigerators, etc.). The strong heterogeneity makes difficult a uniform policy-based management.
- *Context awareness*. This term appeared for the first time in [1], where the authors defined context as location, identities of nearby people and objects, and changes to those objects. Many research groups have been investigating context-aware applications, but there is no common understanding what context and context awareness exactly means. Roughly, the system should own a certain ability to recognize people and the situational context.
- *Personalization*. AmI environments are designed for people, not generic users. This means that the system should be so flexible and intelligent to tailor its

communication facilities so to increase the usability of the system by enhancing our senses. For this reason, the interaction is based on a rich integration of advanced results arising from computer vision, speech recognition, and other sensory and multimodal interfaces.

- *Adaption.* The system, being sensible to the user's feedback, is capable to modify the corresponding actions have been or will be performed. Generally speaking, adaptive systems are designed to deal with changing environmental conditions whilst maintaining performance objectives. Over the years, the theory of adaptive systems has evolved from relatively simple and intuitive concepts to a complex multifaceted theory. AmI contexts are so complex and heterogeneous to be considered as complex as a natural system, complex behaviors emerge as a result of nonlinear spatio-temporal interactions among a huge number of devices and services at different level of abstraction.

In all of the previous issues, the functional and spatial distribution of components and tasks is a natural thrust to employ the agent paradigm to design and implement AmI environments. AmI is a logically decentralized subsystem which can be viewed as an adaptive agent involved in parallel (semi)local interactions. These interactions give rise to:

1. User's actions detection by means of the network of sensor-agents;
2. Control activities in the environment by means of the pool of distributed actuator-agents;
3. Behavioral interactions with the people by means of user-interface agents.

Even though we report an increasing tendency in adopting the agent paradigm in AmI applications, a challenge remains still to solve: how the agent paradigm can support the efficient dynamic reconfiguration of control services balancing hardware constraints with an abstract and uniform approach. This chapter is devoted to discussing this problem and we demonstrate how is possible to solve it demanding agents to work together with other methodologies and techniques, in particular markup languages and fuzzy control strategies.

2.2 Agents Meet AmI

Artificial intelligence (AI) techniques have frequently been used to tackle the most difficult automation problems. This is particularly true in industrial applications where conventional software and teams of operators were unable to cope with the demands of rapidly changing, complex environments. This explain the richness of decision-support systems reported in scientific and industrial literature, designed to assist with different aspects of the control engineer's job. Apart the different approaches, a "first" phase, ranging from the late 1970s to early 1990s characterized by using AI in industrial application ends with a unanimous learned lesson: the difficulty in balancing effective realization of multicontrol modeling with design simplicity [2, 3]. This difficulty is due to the gap between the physical and logical distribution of components and operations of the "real" system to manage and the monolithic approach adopted to realize the intelligent (expert) system. Since 1985, a new tendency emerged in the field of AI: agent-based approaches become the alternative to mainstream knowledge-base systems. An agent is an entity capable of carrying out goals as a component of a wider community of agents that interact and cooperate on each other. Agents work exploiting concurrent and distributed processing performing some activities autonomously. Autonomy improves system modularization by organizing the modeling of the system in terms of delegation and distribution activities. Autonomy of agents evolve with the life of the system: when the agent shows an ability to improve its behavior over time, then the agent is said to be adaptive. These features influence the design choices of recent intelligent systems, also due to these unstoppable trends:

- Computers are smaller, more powerful, and cheaper (powerful machines are no longer prohibitively expensive);
- Improvements in network technology (real advantage of distributed computing);
- Emergence of object-oriented programming paradigms (active objects, delegation, reusing and asynchronous message passing naturally support agent-based design).

Since agents embody inner features that underlie adaptive, robust and effective behavior, it is more natural for designers to mix together heterogeneous techniques to better represent and handle, at different levels of granularity and complexity, dynamic environments. Furthermore, the possibility to integrate useful pieces of intelligence into embedded artifacts makes a realizable scenario of ubiquitous computing. This vision has strongly stimulated the research community in envisaging agents stemmed with physical devices. These agents are named *embedded* agents, that is—agents that run in an embedded system or device. Kaelbling and Rosenschein [4] refer to embedded agents (or situated automata) in a finite-state machine structure as shown in Figure 2.1.

Considering Figure 2.1, sensory input, and some of the agent's tasks, are processed by an update function which produces a state vector of binary-valued values, representing information coming from previous actions and sensing. In Figure 2.1, for example, there is a component whose value is 1 only when the perceptual system (a mobile robot) determines that the path ahead is free of obstacles, whereas the other component equal to 1 means the robot "believes" that the lights are on in room 17. Roughly, the components of the state vector can be viewed as the agent's beliefs about its world. The effectiveness of agent decision depends on the quality of the knowledge of its world and on the robustness of its inference mechanism. One solution is to enrich the environment with as more as possible sensors (in fact, large sensory sets are an issue for ubiquitous computing), but the drawback is the explosion of data and, consequently, the difficulty in finding relevant data [5]. The problem cannot be solved only in terms of size of information: even in case of a space full of sensors, we have to remember that people act in a nondeterministic way and that AmI is known to be designed for people and not for users (AmI prefers particularization over generalization). This leads to a huge amount of data, so

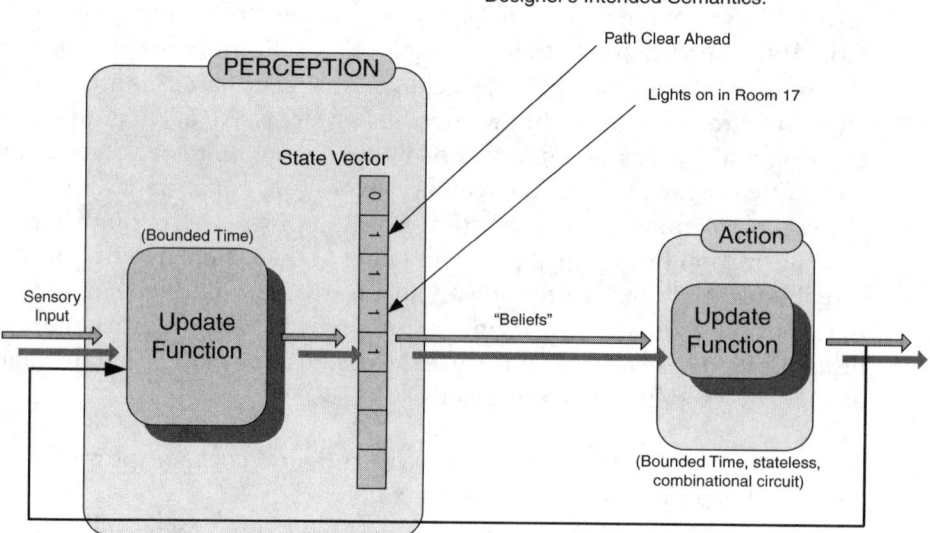

Figure 2.1 Embedded agent (From: [4]).

immense to become useless, unless of complex data processing that difficulty can be harmonized for near run-time responses.

Fuzzy logic can play a fundamental role to cope with this difficult problem, due to its ability in bringing together human expertise and dealing with the uncertainties and imprecisions typical of complex environments. Often, people make decisions based on rules: From an operating viewpoint this approach, is extremely useful, since instead of representing knowledge in a static way, rule-based systems, consisting of IF-THEN forms, represent knowledge in terms of rules that indicate what to do in different situations. In a similar way, fuzzy rules tend to mimic expert behaviors using a series of IF-THEN statements: the decision and the means of choosing that decision are replaced by fuzzy sets and the rules are replaced by fuzzy rules. This simple but powerful idea explains why fuzzy control is the most successful and dynamic branch of fuzzy logic system technology, in terms of both theoretical research and practical application [6]. Fuzzy logic controller (FLC) is used whenever conventional control methods are difficult to apply or show limitations in the performances—for example, due to complex system structure. The FLC approach is closer to the human approach to a control design, thanks to its ability to determine outputs for a given set of inputs without using conventional, mathematical models. FLC, using set of control rules and membership functions, converts linguistic variables into numeric values required in most applications, revealing robustness with regard to noise and variations of the system parameters. Nowadays, the availability of the high connectivity enables designers to propose new fuzzy control approaches that can utilize the additional networked computing power to execute tasks control more effectively [7]. In AmI context, the fuzzy sets and, particularly, the fuzzy logic controllers are useful to dynamically change the actuator setting by minimizing the user's impact usually necessary to improve the ideal habitability conditions. Within this aim, first works on applying agent technology enriched with fuzzy control facility to AmI applications [8] have clearly proved the remarkable benefits in terms of generalization, complexity reduction, and flexibility. Common issues shared by these works are to use multiple agents to control sub-parts of the environment using fuzzy rules that link sensors and effectors and to employ a (supervised or unsupervised) learning mechanism to dynamically create new rules. In such a framework, it appears clear that only those who succeed in making heterogeneous systems communicate through standard, controlled, and secure protocols will be able to furnish innovative solutions and satisfy users' needs. A strong obstacle to this goal is the overall complexity that increases as control and autonomy are decentralized and heterogeneous. Agent technology as well as fuzzy logic are of course very useful, but an additional effort must be made in proving further abstraction on the physical world to control, varying the perception level and control management depending on what kind of abstraction layer we need to face with.

2.3 Transparent Fuzzy Control

Since Zadeh's coining of the term fuzzy logic [9] and Mamdani's early demonstration of FLC [10], an enormous amount of progress has been made by the scientific community in the theoretical as well as application fields of FLC. Trivially, a fuzzy control allows the designer to specify the control in terms of sentences rather than equations by replacing a conventional controller, say, a proportional-integral-derivative (PID) controller with linguistic IF-THEN rules. Three basic parts characterize a typical FLC: (1) an input signal fuzzification (continuous input signals are transformed into linguistic variables), (2) a fuzzy engine (rules are processed with an inference paradigm), and (3) a defuzzification part (precise and physically interpretable values are computed for the control variables). Thus, the design of FLC include the following modules:

- Fuzzy knowledge base;
- Fuzzy rule base;
- Inference engine;
- Fuzzification subsystem;
- Defuzzification subsystem.

The fuzzy knowledge base contains the knowledge used by human experts. The fuzzy rule base represents the set of relations between fuzzy variables defined in the controller system. The inference engine is the fuzzy controller component able to extract new knowledge from the fuzzy knowledge base and fuzzy rule base.

From a technologic point of view, fuzzy control deals with the implementation of a controller on a specific hardware by using a (public or legacy) programming language useful to address in a (less or more) high-level fashion the hardware constraints. This situation may cause development time, very expensive and extremely complex if we envisage the possibility to collect different controllers in terms of an active and cooperative virtual organization, as happens in most of AmI applications. To achieve this objective, we choose XML-derived technologies. Extensible Markup Language (XML) [11] is a simple, very flexible text format derived from

SGML (ISO 8879). Originally designed to meet the challenges of large-scale electronic publishing, nowadays XML plays a fundamental role in the exchange of a wide variety of data on the Web, allowing designers to create their own customized tags, enabling the definition, transmission, validation, and interpretation of data between applications, devices, and organizations. If we use XML, we take control and responsibility for our information, instead of abdicating such control to product vendors. This is the motivation under our proposal: Fuzzy Markup Language, namely "FML" [12, 13]. Figure 2.2 gives a complete perspective of the central role of FML for obtaining transparency in fuzzy control.

The real motivation under FML is to free control strategy from the device. By using FML, we can markup our information according to our own vocabularies; extensible stylesheet language transformations (XSLT) [14] enable us to transform

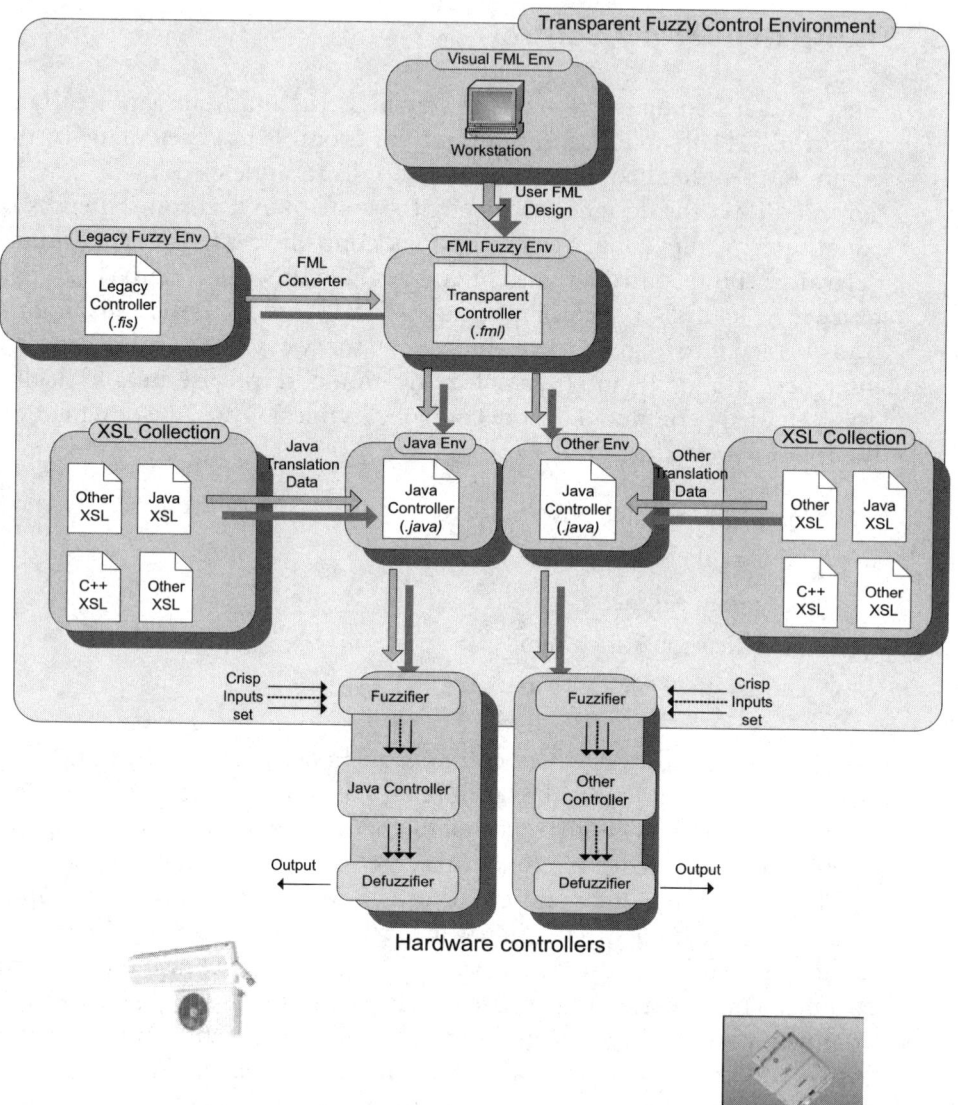

Figure 2.2 FML framework.

information marked up in FML from one vocabulary to another. As shown in Figure 2.2, the XSLT module FML, using an XSL [15] file containing the translation description, is able to convert control description into an executable form (in our case Java) for the hardware. With the XML framework, it is necessary to define a DTD [16] in order to assess the legal building blocks of an FML document. The production side FML scenario is depicted in Figure 2.3. The compilation of the FML program (left side of Figure 2.3) is performed in accordance to several layers; the DTD guarantees the validation process, whereas XSLT layers take the responsibility to produce the executable module in forms of high-level, portable programs. This is done thanks to the existence of the XSL collection library, designed for any supported languages (right side).

2.4 FML Environment Description

Essentially, an FML program consists of:

- Tags;
- Attributes;
- Values.

Tags are used (also in a nested way) to model the several parts of a fuzzy controller. Considering the above mentioned fuzzy concepts, we can give a tree-based structure to represent these conceptual components, as shown in Figure 2.4. Each node in a fuzzy controller tree represents a tag, while the relation father-child represents a nested relation between related tags. The role of the attributes and values is to generate a specific instance of a fuzzy controller. Currently, FML can model two

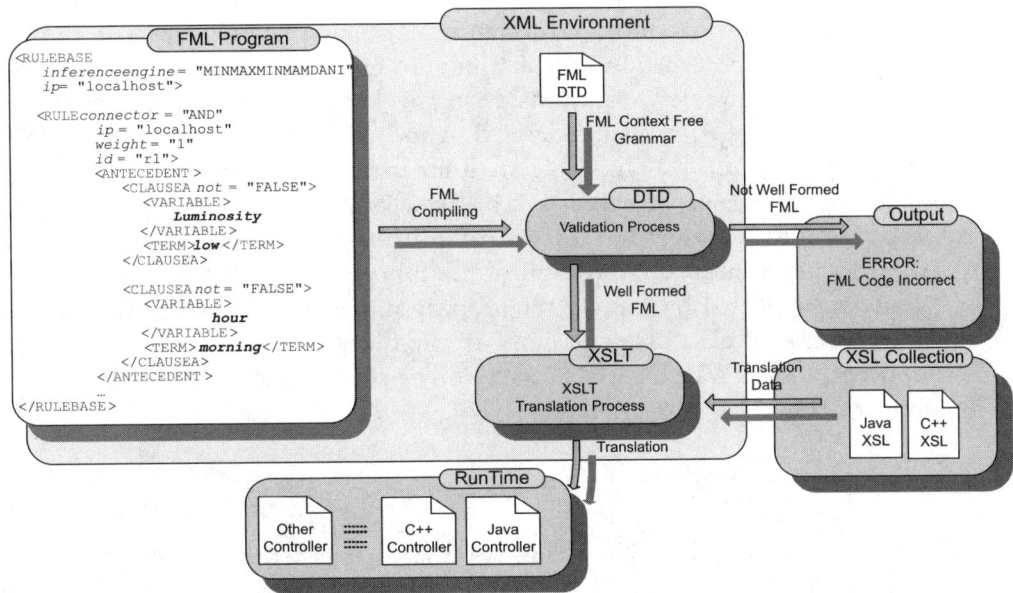

Figure 2.3 FML in XML environment (DTD + XSLT).

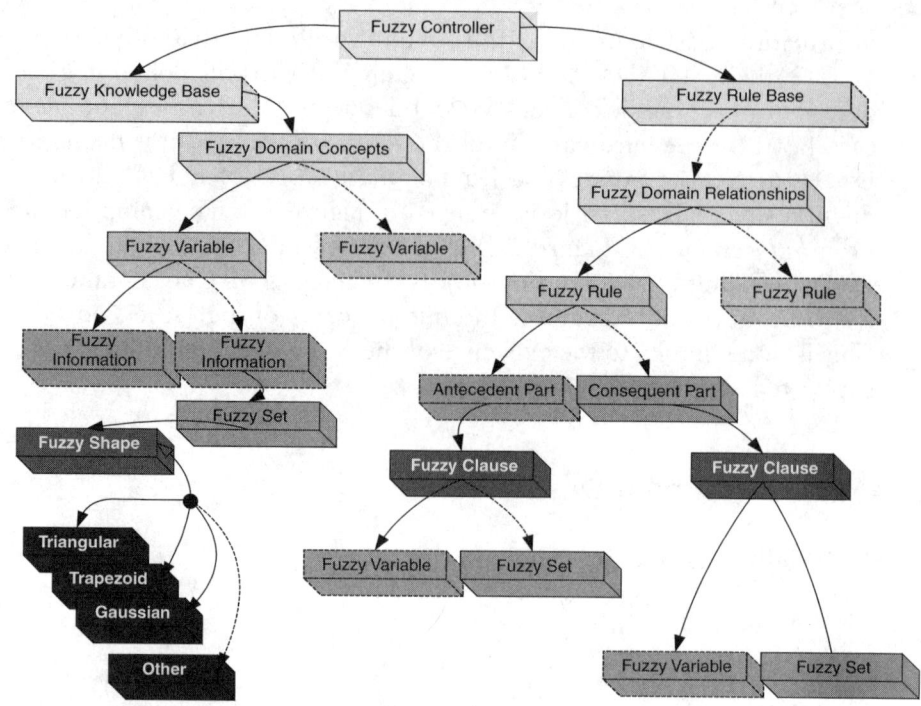

Figure 2.4 Fuzzy controller tree (Mamdani).

popular typologies of fuzzy controllers: Mamdani and TSK. For simplicity, we refer in this chapter to a Mamdani fuzzy controller designed to regulate a lighting system.

In order to model the controller node of a fuzzy tree, the FML tag <FUZZYCONTROL>is employed. Such a tag represents the root tag of the FML program—that is, the opening tag of each FML program. <FUZZYCONTROL> uses three tags: type, defuzzifyMethod and ip. The type attribute permits to specifies what kind of fuzzy controller designing. The attribute defuzzyfyMethod specifies the defuzzification method; the attribute ip indicates the location of the controller over the computer network. Considering the left side of Figure 2.4, the fuzzy knowledge base of the controller is defined by means of the tag <KNOWLEDGEBASE>. <KNOWLEDGEBASE> uses the attribute ip in order to situate the knowledge base in a precise place of the network. The fuzzy knowledge base consists of a set of concepts used to represent the real-world information in a fuzzy way. These concepts are then used in the composition of the fuzzy rule base. Of course, the fuzzy concepts are modeled by using a set of fuzzy informations such as the parameters of the universe of discourse, the name of the fuzzy concepts or the scale of the fuzzy concept. <KNOWLEDGEBASE> exploits a set of nested tags to model this kind of information:

- <FUZZYVARIABLE>;
- <FUZZYTERM>;
- A set of tags defining a shape of fuzzy sets (see the following text).

<FUZZYVARIABLE> defines the fuzzy concept—for example, "temperature"; <FUZZYTERM> defines a linguistic term describing the fuzzy concept—for example, "low temperature"; the set of tags defining the shapes of fuzzy sets is related to fuzzy terms. The attributes of <FUZZYVARIABLE> tags are: name, scale, domainLeft, domainRight, type, ip. The name attribute defines the name of the fuzzy concept—for instance, temperature; scale is used to define the scale used to measure the fuzzy concept—for instance, Celsius degree; domainLeft and domainRight are used to model the universe of discourse of fuzzy concept, that is, the set of real values related to the fuzzy concept, [e.g., 0°, 40°]; the position of fuzzy concept into rule (consequent part or antecedent part) is defined by type attribute; ip defines the position of fuzzy knowledge base in the computer network.

<FUZZYTERM> uses one attribute, name used to define the linguistic value associate with the fuzzy concept.

Fuzzy shape tags, used to complete the definition of fuzzy concept, are:

- <TRIANGULARSHAPE>;
- <LINEARSHAPE>;
- <TRAPEZOIDSHAPE>;
- <SSHAPE>;
- <ZSHAPE>;
- <PISHAPE>;

Every shaping tag uses a set of attributes that defines the real outline of the corresponding fuzzy set. The number of these attributes depends on the chosen fuzzy set shape.

Considering as example, the luminosity in defining our fuzzy control, we can model the knowledge base, and in particular the external luminosity input as follows:

```
<!DOCTYPE FUZZYCONTROL SYSTEM "fml.dtd">
<FUZZYCONTROL defuzzymethod = "CENTROID"  ip = "localhost"
    type = "MAMDANI">
<KNOWLEDGEBASE ip = "localhost">
  <FUZZYVARIABLE
    domainleft = "0" domainright = "1"
    ip = "localhost" name = "Luminosity" scale = "Lux"
    type = "INPUT">

    <FUZZYTERM name="low">
      <PISHAPE
        param1 = "0.0" param2 = "0.45">
      </PISHAPE>
    </FUZZYTERM>

    <FUZZYTERM name="MEDIUM">
      <PISHAPE
        param1 = "0.4999999999999994"
        param2 = "0.4499999999999996">
      </PISHAPE>
```

```
        </FUZZYTERM>

        <FUZZYTERM name="HIGH">
          <PISHAPE
            param1 = "0.5501" param2 = "1">
          </PISHAPE>
        </FUZZYTERM>
      </FUZZYVARIABLE>
      ...
    </KNOWLEDGEBASE>
```

This FML code is useful to understand how is possible to associate a fuzzy control activity (knowledge base and eventually the rule base) on a single host (in the previous example, localhost). A special tag that can also be used to define a fuzzy shape is <USERSHAPE>. This tag is used to customize fuzzy shape (custom shape). Custom shapes modelling is performed via a set of <POINT> tags, which lists the extreme points of geometric area defining the custom fuzzy shape. Obviously, the attributes used in <POINT> tag are x and y coordinates.

The right subtree is used to define another basic component of a fuzzy controller: the rule base. The fuzzy rule base defines the relationships among fuzzy concepts as IF antecedent part THEN consequent part format. A relationship is modeled by means of the tag <RULEBASE>. The <RULEBASE> tag uses two attributes: inferenceEngine and ip. The former is used to define inference operator type: MinMaxMinMamdani or LarsonProduct. The latter defines the network location of the set of rules used in the fuzzy controller. Another tag is fundamental to complete a fuzzy rule definition, the tag <RULE>. The attributes used by <RULE> are: id, connector, weight, ip. The id attribute identifies the rule; connector is used to define the logical operator used to tie the different clauses in the antecedent part; weight defines the importance of the rule during inference engine time; ip defines the location of the rule in the computer network. As shown previously, each logic rule is composed by two parts: antecedent and consequent. As inputs are received by the running fuzzy system, the rule base is evaluated.

The definition of antecedent and consequent rule parts is obtained using <ANTEDENT> and <CONSEQUENT> tags, respectively. The antecedent and consequent part of a rule can be defined as a conjunction or disjunction of fuzzy clause (defined by connector attribute of <RULE>).

<CLAUSEA> and <CLAUSEC> tags are used to model the fuzzy clauses in antecedent and consequent part, respectively. Fuzzy systems allows use of the complemented clause in the antecedent part. In order to treat the operator "not" in fuzzy clauses, <CLAUSEA> and <CLAUSEC> uses the attribute "not." To complete the definition of the fuzzy clause <VARIABLE>, <TERM>, and <TSKPARAM> have to be used. In particular the pair <VARIABLE>, <TERM> is used to define fuzzy clauses in antecedent and consequent parts of Mamdani controllers rules and in antecedent parts of TSK controllers rules. Alternatively, the pair <VARIABLE>, <TSKPARAM> is used to model the TSK functions in the consequent part of TSK controllers rules.

In order to support the designer in the writing of FML programs, a graphical environment allows as to design a fuzzy controller without knowing the details of FML. In this way, the designer is supported in the operative steps by interacting with the underlying system in a visual environment. In particular, the FML editor module

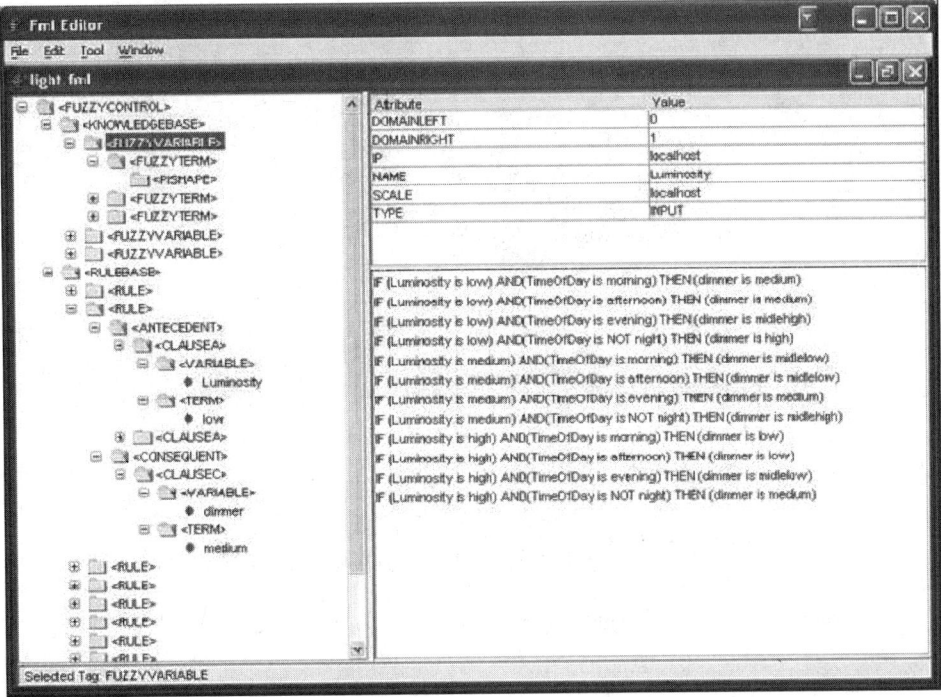

Figure 2.5 FML Visual Environment.

creates different pieces of the hierarchical representation of FML controller adding the appropriate attributes using dialog boxes and without writing FML code, as shown in Figure 2.5.

2.5 FML Agent Distribution

Our goal is to exploit the agent paradigm using mobile capability, in order to design, in a high-level fashion, the effective distribution of control activity and controller devices necessary to address ambient intelligence frameworks. The real advantage obtained from the agent distribution is the possibility to distribute the fuzzy rule base of a generic controller on many hosts and then apply a parallel inference fuzzy engine on the network. For real-size applications it is very easy to produce very large and complex rule bases that require a great amount of processor time. For fuzzy set theory, the inference is performed by manipulating membership functions, often demanding a large computational overhead. This prevents many applications from being implemented in real time without special hardware. This problem can be overcome exploiting the multiprocessor capability, available in any networked environment, managed by FML-oriented agents.

In order to integrate the mobile agent technology in the FML framework, four types of agents are necessary: creator agent, stationary fuzzy agents, inference agent, and registry agent.

The interactions between Creator Agent, Stationary Fuzzy Agents, and Registry Agent are shown in Figure 2.6. A basic definition of the agents follows.

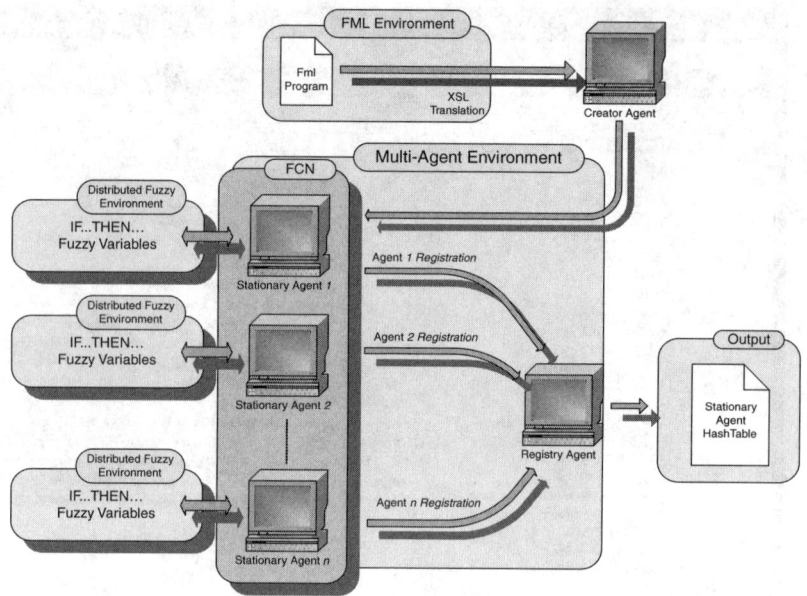

Figure 2.6 Multiagent system creation.

Creator agent is the agent produced as the result of the conversion from FML to Java via XSLT transformation. The code generated by such translation represents the core of mobile FML-based agents. This agent permits to instantiate the fuzzy entity defined in FML code and to distribute these objects. In particular, the creator agent serves to: embody the fuzzy contents defined in FML source code, and to apply the fuzzy rules on specific hosts available in the network.

Stationary fuzzy agents are defined to receive the fuzzy objects (fuzzy variables and fuzzy rules) generated by the creator agent. The distribution of the several stationary fuzzy Agents implies a collection of operative autonomous entities able to perform asynchronous and concurrent execution. The set of stationary fuzzy agents defines the fuzzy control network (FCN).

Registry agent is delegated to manage the FCN topology. It can be viewed as a fuzzy routing table that allows the other agents to migrate in the FCN in order to use

Registry Hash Table		
Temperature	Variable	192.168.0.35
Luminosity	Variable	192.168.0.24
Dimmer	Variable	192.168.0.17
If … Then	Rule	192.168.0.35
If … Then	Rule	192.168.0.56

Figure 2.7 Registry hash table.

the distributed information in an automatic way. The data structure used by the registry agent to model the FCN is represented by a hash table. The name of the fuzzy concepts defined in the FML programs represents the hash table *key*, while the hash *value* is defined as a record containing the following information: ip address and fuzzy concept type (variable or rule). An example of registry hash table is given in Figure 2.7.

Once realized by the FCN, each stationary agent is able to host other migrating agents and apply the fuzzy operator on distributed fuzzy information. In particular, inference agents is a mobile agent able to migrate on every host of FCN and, using local resources, execute a partial inference, obtain a partial inferred result, come back on source host, and apply a fuzzy union operator on partial results in order to compute the final control value. The inference agent migrates in the FCN using a parallel approach in order to minimize the inference time. In particular, the inference agent is cloned on starting host and migrated on FCN host in parallel. The inference operators used in this stage are the classical ones, such as Mamdani MinMaxMin and Larson product, while the fuzzy union operator represents a classical defuzzification method, such as centroid, average maximum, center average, mean of maximum, etc. Figure 2.8 shows the interactions between Inference Agent and other agents necessary to achieve the prefixed goal.

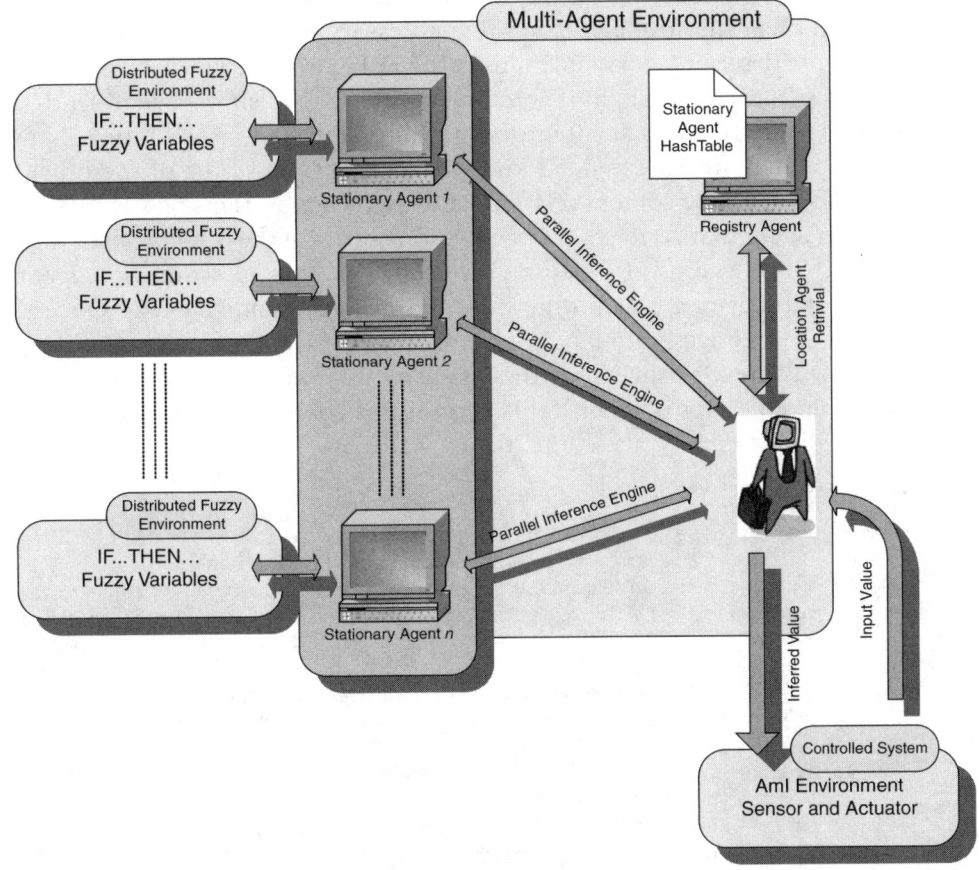

Figure 2.8 Multi-agent fuzzy system.

2.6 Adaptivity

The adaptivity provided by our framework consists in an automatic generation of fuzzy rules reflecting new user behaviors in a proactive environment. Once generated, the fuzzy rules are coded in FML language in order to exploit the distribution of transparent fuzzy control through the mobile agents. In particular, our home automation network is viewed as the set of computational nodes able to manipulate the sensors and devices actuators. Each node is identified with an agent entity that may provide two basic behaviors: learning mode and control mode. Control mode is essentially the control activity performed on FML, as explained previously. The learning mode provides the basic features for adaptive functionalities. This goal is achieved by the employment of two agents: acquisition and context agent.

The acquisition agents represent a smart access point to the corresponding domestic device (thermostat, light switch, light lamp, etc.). The context agent works as a supervisor: It receives information on users' habits from the acquisition agents and adapts the overall fuzzy rulebase taking into account the new user actions. The context agent uses the information received about the user's action message together with the context updating messages to: (1) generate a new fuzzy rule, (2) convert this rule into FML, and (3) update the overall fuzzy rulebase. In particular, the context agent uses the fuzzy concepts together with fuzzy operator as maximum of value fuzzification in order to translate the numerical information received from the acquisition agent into linguistic value necessary to compose the fuzzy rule.

In the descriptive example illustrated in Figure 2.9, the user switches on a lamp by using the light switch. The corresponding acquisition agent sends a message (label 1) to the context agent informing it about the new user's action. Now, it is necessary to acquire an extended knowledge about the status of the other devices in order to obtain the context related to the user's action. This task is performed by the context agent that contacts the acquisition agents of the other devices (message with label 2) and remains in a waiting status. The contacted acquisition agent reads the status of its device and sends this information to the context agent. Once all the reply messages (messages with label 3) are obtained, the context agent is able to compose the new context arising from the last user's action. The fuzzification of the user's action and the environment status permits us to obtain a single fuzzy rule coding this state. More formally, the context agents execute the following steps.

```
contextAgent
1. Wait messages from the acquisition agents (Message 1);
2. On the receipt of a message, namely actionMsg:
   a. Set actionVariable = actionMsg.getContent();
   b. Broadcasts a message to each sensor device acquisition
      agent;
   c. Waiting for all reply;
   d. On the receipt of i^th reply, namely replyMsgi:
        i. Set sensorVariable[i] = replyMsgi.getContent();
3. Set fuzzifiedActionVariable = fuzzify(actionVariable);
4. for each i: fuzzifiedSensorVariable =
   fuzzify(sensorVariable[i]);
5. createFMLRule(fuzzifiedActionVariable ,
   fuzzifiedSensorVariable);
6. go to 1;
```

2.6 Adaptivity

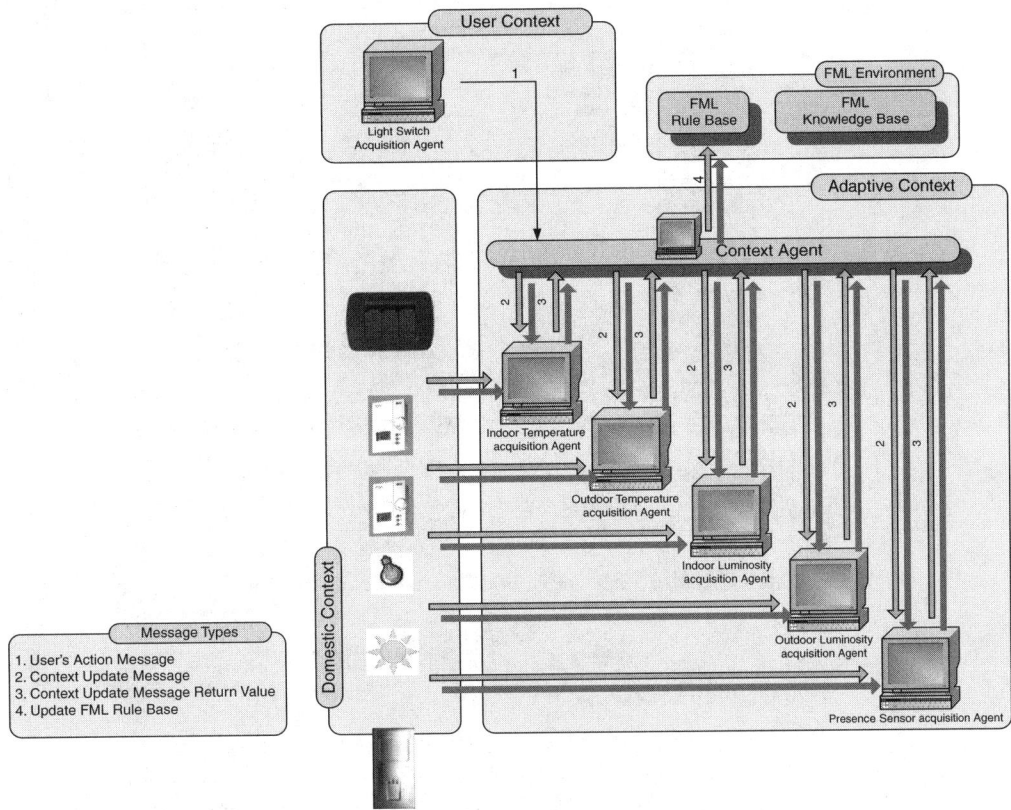

Figure 2.9 Adaptive framework message sending.

The actuator and sensor acquisition agent can be modelled, respectively, as follows.

```
actuatorAcquisitioAgent
1. Listen events on the related actuator device;
2. On the status change:
   a. Create a new message: actionMsg;
   b. Set the content of actionMsg:
      actionMsg.setContent(sensorStatus);
   c. Send actionMsg to ContextAgent;
3. go to 1.

sensorAcquisitioAgent
1. Wait messages from Context Agent;
2. On the receipt of Context Agent message:
   a. Read the actual status from sensor device;
   b. Create a new message: replyMsg;
   c. Set the content of replyMsg: replyMsg.setContent(status);
   d. Send replyMsg to ContextAgent;
3. go to 1;
```

An example of fuzzy rule generation showing the results from the interaction between context agent and acquisition agent is given in Figure 2.10. In this example, the user sets the indoor luminosity value at 52 Lux, and the lighting acquisition

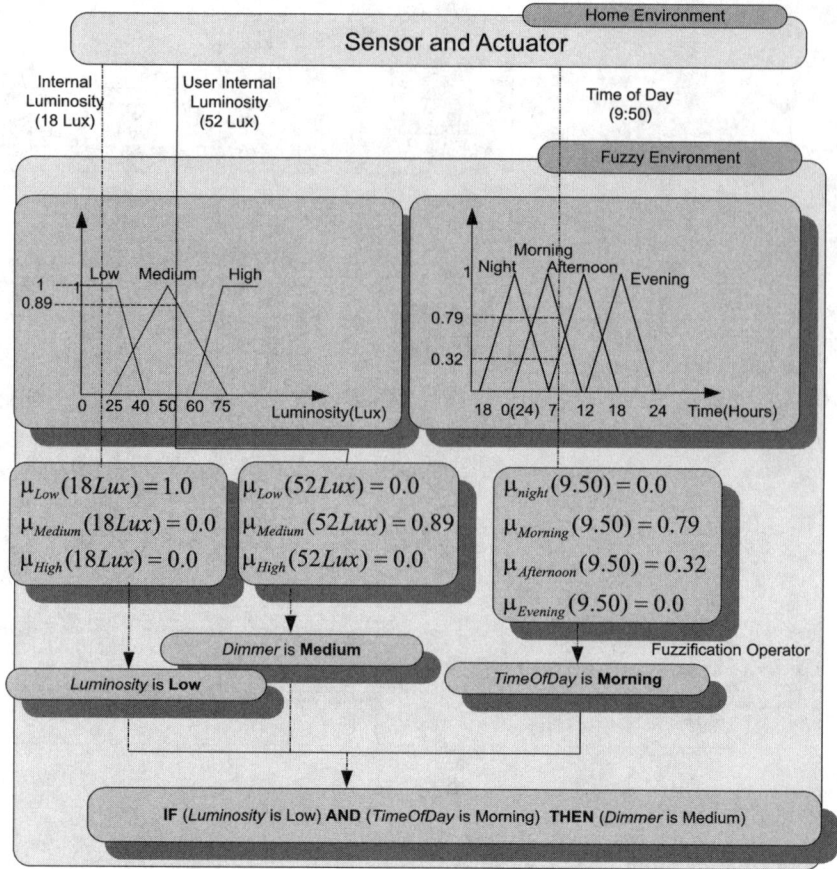

Figure 2.10 Rule generation example.

agents sends a message reporting the new temperature value to its context agent. The other acquisition agents send the actual status of the other devices (timer and luminosity sensors) to the context agent. At this point, the context agent synthesizes the numerical information received from acquisition agent in fuzzy concepts and generates the fuzzy rule. In particular, the information received from the thermostat agent is used to generate the consequent fuzzy rule and the information received from other sensor agents is used to generate the antecedent fuzzy rule. This allows composing the fuzzy rule:

IF (*luminosity* is Low) AND (TimeOfDay is Morning) THEN Dimmer is Medium

and the related FML code:

```
<RULEBASE
  inferenceengine = "MINMAXMINMAMDANI"
ip = "localhost">
  <RULE connector = "AND" ip = "localhost" weight = "1" id = "r1">
    <ANTECEDENT>
      <CLAUSEA not = "FALSE">
        <VARIABLE> Luminosity </VARIABLE>
        <TERM> low </TERM>
```

```xml
        </CLAUSEA>

        <CLAUSEA not = "FALSE">
          <VARIABLE> hour </VARIABLE>
          <TERM> morning </TERM>
        </CLAUSEA>
      </ANTECEDENT>

      <CONSEQUENT>
        <CLAUSEC not = "FALSE">
          <VARIABLE> dimmer </VARIABLE>
          <TERM> medium </TERM>
        </CLAUSEC>
      </CONSEQUENT>
    </RULE>
  ...
</RULEBASE>
```

The Java code related to the previous FML code (obtained via XSLT) is the following:

```java
public void fuzzyLuminosity () throws FuzzyException {
  NameInput.add("Luminosity");
  luminosity = new FuzzyVariable("Luminosity", 0, 1,"localhost");
  luminosity.addTerm("low", new PIFuzzySet(0.0, 0.45));
  luminosity.addTerm("medium",
    new PIFuzzySet(0.4999999999999994,0.4499999999999996));
  luminosity.addTerm("high",
    new PIFuzzySet(1.0000499999999999,0.4499499999999996));
}

public void fuzzyora () throws FuzzyException {
  NameInput.add("timeOfDay");
  ora = new FuzzyVariable("ora",0,24,"localhost");
  ora.addTerm("morning",
    new TriangleFuzzySet(0,7,12));
  ora.addTerm("afternoon",
    new TriangleFuzzySet(7,12,18));
  ora.addTerm("evening",
    new TriangleFuzzySet(12,18,24));
  ora.addTerm("night",
    new TrapezoidFuzzySet(0,7,18,24));
}
public void fuzzydimmer () throws FuzzyException {
  NameOutput.add("dimmer");
  dimmer= new FuzzyVariable("dimmer",0,1,"localhost");
  dimmer.addTerm("low",
    new PIFuzzySet(0.0,0.225));
  dimmer.addTerm("midlelow",
    new PIFuzzySet(0.2499999999999997,0.2249999999999998));
  dimmer.addTerm("medium",
    new PIFuzzySet(0.5,0.2249999999999998));
  dimmer.addTerm("midlehigh",
    new PIFuzzySet(0.75,0.2249999999999998));
  dimmer.addTerm("high",
    new PIFuzzySet(1.0,0.22500000000000003));
}

public void createRuleBase() throws FuzzyException {
```

```
rule_r1 = new FuzzyRule();
rule_r1.addAntecedent(new FuzzyValue(luminosity, "low"));
rule_r1.addAntecedent(new FuzzyValue(ora, "morning"));
rule_r1.addConclusion(new FuzzyValue(dimmer, "medium"));

rule_r2 = new FuzzyRule();
rule_r2.addAntecedent(new FuzzyValue(luminosity, "low"));
rule_r2.addAntecedent(new FuzzyValue(ora, "afternoon"));
rule_r2.addConclusion(new FuzzyValue(dimmer, "medium"));

rule_r3 = new FuzzyRule();
rule_r3.addAntecedent(new FuzzyValue(luminosity, "low"));
rule_r3.addAntecedent(new FuzzyValue(ora, "evening"));
rule_r3.addConclusion(new FuzzyValue(dimmer, "midlehigh"));

rule_r4 = new FuzzyRule();
rule_r4.addAntecedent(new FuzzyValue(luminosity, "low"));
rule_r4.addAntecedent(new FuzzyValue(ora,
 "night").fuzzyComplement());
rule_r4.addConclusion(new FuzzyValue(dimmer,"high"));
rule_r5 = new FuzzyRule();

rule_r5.addAntecedent(new FuzzyValue(luminosity, "medium"));
rule_r5.addAntecedent(new FuzzyValue(ora, "morning"));
rule_r5.addConclusion(new FuzzyValue(dimmer, "midlelow"));
rule_r6 = new FuzzyRule();
rule_r6.addAntecedent(new FuzzyValue(luminosity, "medium"));
rule_r6.addAntecedent(new FuzzyValue(ora, "afternoon"));
rule_r6.addConclusion(new FuzzyValue(dimmer, "midlelow"));

rule_r7 = new FuzzyRule();
rule_r7.addAntecedent(new FuzzyValue(luminosity, "medium"));
rule_r7.addAntecedent(new FuzzyValue(ora, "evening"));
rule_r7.addConclusion(new FuzzyValue(dimmer, "medium"));

rule_r8 = new FuzzyRule();
rule_r8.addAntecedent(new FuzzyValue(luminosity, "medium"));
rule_r8.addAntecedent(new FuzzyValue(ora,
"night").fuzzyComplement());
rule_r8.addConclusion(new FuzzyValue(dimmer, "midlehigh"));

rule_r9 = new FuzzyRule();
rule_r9.addAntecedent(new FuzzyValue(luminosity, "high"));
rule_r9.addAntecedent(new FuzzyValue(ora, "morning"));
rule_r9.addConclusion(new FuzzyValue(dimmer, "low"));

rule_r10 = new FuzzyRule();
rule_r10.addAntecedent(new FuzzyValue(luminosity, "high"));
rule_r10.addAntecedent(new FuzzyValue(ora, "afternoon"));
rule_r10.addConclusion(new FuzzyValue(dimmer, "low"));

rule_r11 = new FuzzyRule();
rule_r11.addAntecedent(new FuzzyValue(luminosity, "high"));
rule_r11.addAntecedent(new FuzzyValue(ora, "evening"));

rule_r11.addConclusion(new FuzzyValue(dimmer, "midlelow"));

rule_r12 = new FuzzyRule();
rule_r12.addAntecedent(new FuzzyValue(luminosity, "high"));
```

2.6 Adaptivity

```
     rule_r12.addAntecedent(new FuzzyValue(ora,
"night").fuzzyComplement());
     rule_r12.addConclusion(new FuzzyValue(dimmer, "medium"));
```

Figure 2.11 gives a view of the overall architecture considering the new layer "Learning mode" that embodies the adaptive functionality.

In this way, we can distribute fuzzy rule base in the network and exploit distributed processing by minimizing inference. The issues of delocalization and concurrency are delegated to the high-level interaction model managed by stationary fuzzy agent, registry agent, and inference agent. The set of stationary fuzzy agent is used to manipulate in a distributed way the concept coded in FML program

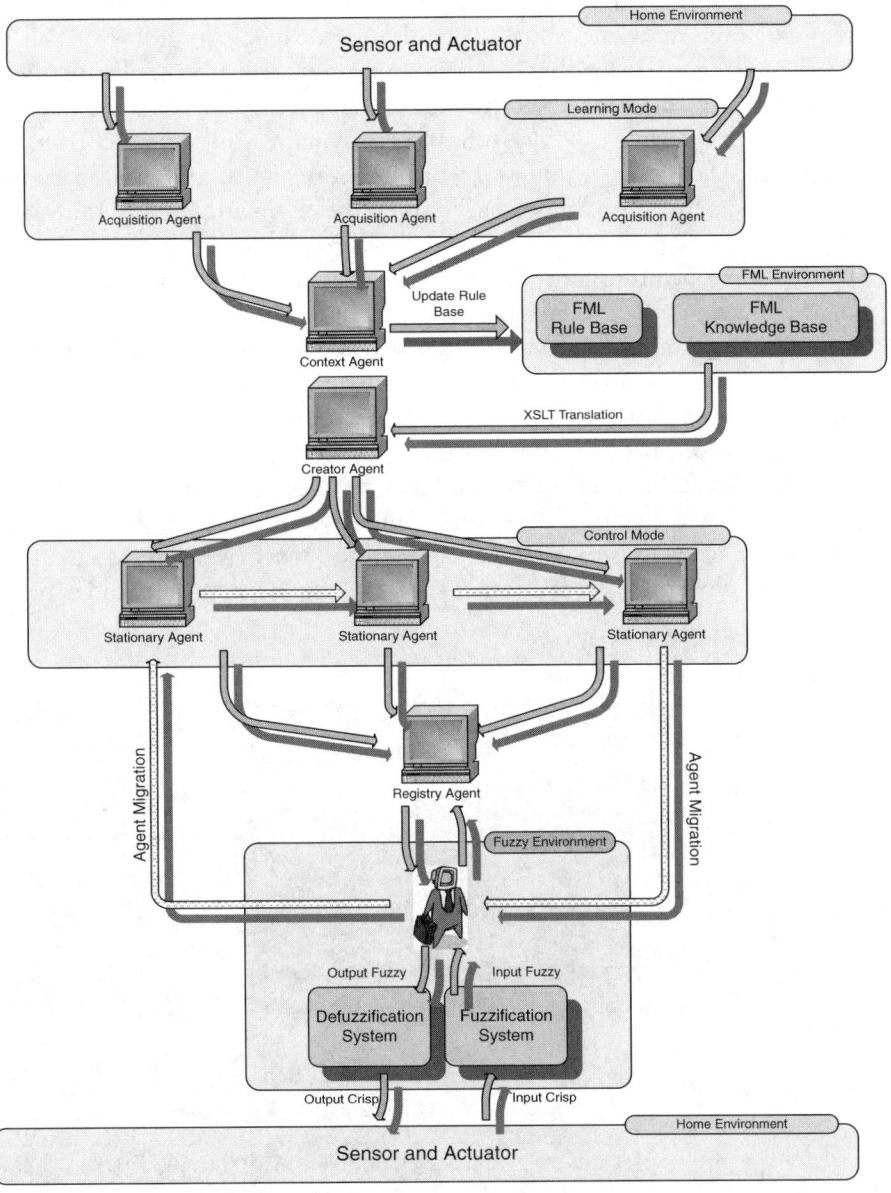

Figure 2.11 Adaptive framework based on FML.

and modeled in the distributed fuzzy controller. Due to the delocalization of rules, it is necessary to collect the partial inference results. This is done by the migration of inference agent that gathers the partial results available on each stationary agent. Just to give you an idea of the gain from shifting from a centralized to a decentralized mode, we give in Figure 2.12 a test evaluation done by spreading the control over three computational nodes. On the axis x we report the number of fuzzy rules evaluated, the required inference time, expressed in milliseconds, is on axis y.

2.7 Conclusions

When building AmI environments many challenges arise. The system must have a way to interact with its users to obtain feedback avoiding to be too intrusive. Explicit or implicit feedback is needed to take decisions that must be made in near real time. To reach this objective, agents have many advantages. The agents communicate with another one by exchanging asynchronous or synchronous, goal-driven messages. Decision making and learning strategies are facilitated by that each agent owns perspective only on a small part of the environment and takes decisions about this sub-view. Decentralization and concurrent execution allow us to better respect the real time commitment. Respecting this trend, so widely recognized in the design of advanced intelligent environments, the work discussed focuses on a proposal of a new markup language, FML, which allows designer to abstract the (fuzzy) control strategy from low-level features of sensors and actuators and to exploit the benefits of mobile agent computation as a flexible and efficient way to connect adaptive services and applications to hardware.

Recently, there has been a tendency to merge the virtual and the physical space, enabling new forms of interaction: Humans are immersed in the agent's world and vice versa according to a virtuality continuum where android agents, embedded in the user's physical world, interact with the user with human and synthetic agents. In

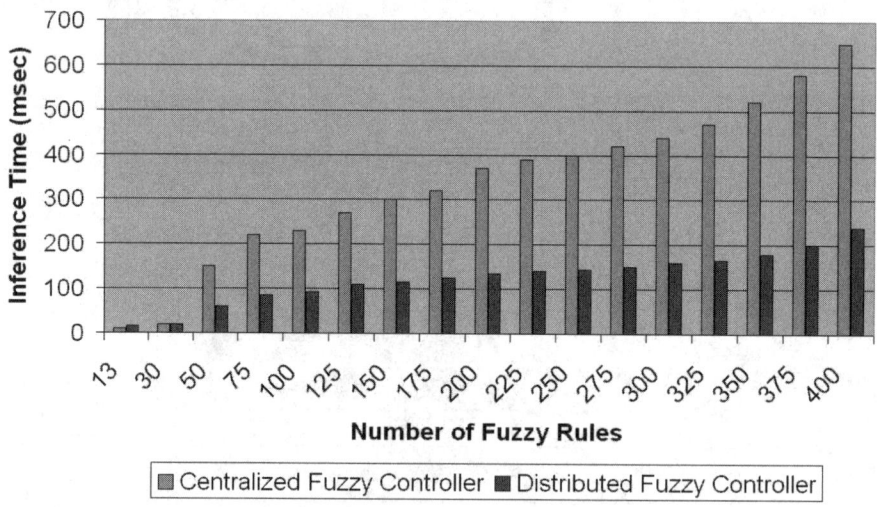

Figure 2.12 Decentralized vs centralized inference.

2.7 Conclusions

this direction, our intention is to extend FML toward other capabilities, such as human behaviour modeling, in order to be able to describe more complex AmI scenarios [17].

References

[1] Schilit, B., M. Theimer, "Disseminating Active Map Information to Mobile Hosts," *IEEE Network*, Vol. 8, No. 5, pp. 22–32, September 1994.

[2] Wright, M. L., M. W. Green, G. Fiegl, and P.F. Cross, "An Expert System for Real-Time control," *IEEE Software*, Vol. 3, No. 2, pp. 16–24, March 1986.

[3] Leitch, R., R. Kraft, and R. Luntz, "RESCU: a Real-time Knowledge Based System For Process Control," *Control Theory and Applications, IEE Proceedings D*, Vol. 138, No. 3, pp. 217–227, May 1991.

[4] Kaelbling, L. P., and Rosenschein, S. J., "Action and Planning in Embedded Agents," *Robotics and Autonomous Systems*, Vol. 6, Nos. 1-2, pp. 35–48, June 1990.

[5] Hagras, H., V. Callaghan, M. Colley, and G. Clarke, "A hierarchical fuzzy-genetic multi-agent architecture for intelligent buildings online learning, adaptation and control," *Information Sciences*, Vol. 150, No. 1-2, pp. 33–57, March 2003.

[6] Ying, H., "Fuzzy system technology: a brief overview," *IEEE Circuits and Systems Society, CAS Magazines*, pp. 28–37, 2000.

[7] Hagras, H., et al., "Creating an Ambient-Intelligence Environment Using Embedded Agents," *IEEE Intelligent Systems*, Vol. 19, No.6, pp. 12–20, December 2004

[8] Rutishauser, U., J. Joller, and R. Douglas, "Control and learning of ambience by an intelligent building", *IEEE Transactions on Systems, Man & Cybernetics*, Vol. 35, No. 1, pp.121–132, January 2004.

[9] Zadeh, L. A., "Fuzzy sets," *Information and Control*, Vol. 8, pp. 338–353, 1965.

[10] Mamdani, E. H., "Applications of fuzzy algorithms for simple dynamic plants," *Proceedings of IEE*, Vol. 121, pp.1585–1588, 1974.

[11] "Extensible Markup Language (XML) 1.0 III Edition", W3C Recommendation,

[12] Acampora, G., and V. Loia, "Fuzzy Control Interoperability for Adaptive Domotic Framework," *Proceedings of 2nd IEEE International Conference on Industrial Informatics*, June 24–26 2004, Berlin, Germany.

[13] Acampora, G., V. Loia, "Fuzzy Control Interoperability and Scalability for Adaptive Domotic Framework," *IEEE Transaction on Industrial Informatics*, Vol. 1, No. 2, pp. 97–111, 2005.

[14] "XSLT Requirements Version 2.0," W3C Working Draft 14 February 2001, http://www.w3.org/TR/2001/WD-xslt20req-20010214.

[15] "Extensible Stylesheet Language (XSL) Version 1.1," W3C Working Draft 17 December 2003.

[16] "DTD XML Specification", W3C Reference Report, http://www.w3.org/XML/1998/06/xmlspec-report-v21.htm.

[17] Acampora, G., et al., "Hybrid Computational Intelligence for Ambient Intelligent Environments," *Proceedings of the 3rd Atlantic Web Intelligence Conference*, June 6–9, 2005; Lodz, Poland, LNAI 3528, pp. 26–31, 2005.

CHAPTER 3
Ambient Intelligence: The MyCampus Experience

Norman M. Sadeh, Fabien L. Gandon, and Oh Byung Kwon

3.1 Introduction

Increasingly, application developers are looking for ways to provide users with added levels of convenience and ease of use through functionality that is capable of capturing the context within which they operate. This may involve knowing where the user is located, the task he or she is currently engaged in, eating preferences, colleagues, and a variety of other contextual attributes. While there are many sources of contextual information, they tend to vary from one user to another and also over time. Different users may rely on different location tracking functionality provided by different cell phone operators; they may use different calendar systems, and so on. Traditionally, context-aware applications and services have been hardwired to predefined sources of contextual information (e.g., relying on a particular set of sensors and protocols to track a user's location). As a result, they remain prohibitively expensive to build and maintain and are few and far between. We argue that what is needed is a more open environment, where context-aware applications can automatically discover and access a user's personal resources such as calendar or location tracking functionality. This can be done by viewing each source of contextual information (or personal resource) as a Web service. Unfortunately, current Web services standards such as UDDI [1] or WSDL [2] are not sufficient when it comes to describing a user's personal resources and to enabling automated access to them by context-aware applications. Another challenge, as we move towards more open platform for access to a user's personal information, revolves around privacy issues. Users should be able to retain control over who has access to their personal information under different conditions. For instance, I may be willing to let my colleagues see where I am or access my calendar activities between 8 a.m. and 5 p.m. on weekdays but not over the weekend. In addition, I may want to fine-tune the granularity of the answer provided to a given query, depending on the context of that query. For instance, I may be willing to disclose the room that I am in to some people but only the city where I am to others. In fact, I may even want to give different answers to different people, telling my secretary I am off to see my dentist, while telling my customers I am busy in a meeting.

Over the past 5 years, the MyCampus group at Carnegie Mellon University has been developing and experimenting with ambient intelligence technologies aimed at enhancing everyday life. The project has drawn on multiple areas of expertise, combining the development of an open semantic Web infrastructure for reusable, context-aware service provisioning with an emphasis on issues of privacy and usability. In this chapter, we provide an overview of our semantic Web infrastructure for ambient intelligence. Within this infrastructure, each source of contextual information (e.g., a calendar, location tracking functionality, collections of relevant user preferences, organizational databases) is represented as a semantic Web service. A central element of our infrastructure is its semantic e-Wallets. Each Semantic e-Wallet acts as a directory of contextual resources for a given user, while enforcing privacy preferences. Privacy preferences enable users to specify what information can be provided to whom in different contexts (access control). They also allow users to specify what we call obfuscation rules—namely rules that control the accuracy or inaccuracy of the information provided in response to different queries under different conditions.

We have validated our infrastructure in the context of several ambient intelligence environments: an environment aimed at enhancing everyday campus life at Carnegie Mellon University (CMU), a prototype context-aware museum tour guide developed for a museum in Taiwan, and several context-aware office applications. Thanks to our semantic Web infrastructure, each of these environments allows for the development of a growing collection of context-aware applications (often implemented as agents) capable of dynamically leveraging contextual information. This generally includes accessing location information, calendar activities, and a variety of other contextual attributes and preferences relevant to each environment. At CMU, students access the MyCampus environment from PDAs over the campus's 802.11 wireless LAN. Empirical results obtained with a group of students over a period of several days are summarized at the end of this chapter along with a brief discussion of case-based reasoning (CBR) functionality developed to learn context-sensitive user preferences.

The remainder of this chapter is organized as follows. Section 3.2 provides a brief overview of the state of the art in context awareness, privacy, Web services and the semantic Web, emphasizing limitations of the work reported so far in the literature. In Section 3.3, we provide an overview of our open semantic Web infrastructure for context-aware service provisioning and privacy. Section 3.4 focuses more specifically on the semantic e-Wallet and includes a high-level scenario outlining its operation in response to a query about the current location of a user. Section 3.5 discusses issues relating to capturing user preferences. Section 3.6 discusses our experience tailoring and deploying our infrastructure in support of different ambient intelligence environments. This includes examples of context-aware applications (or agents) our team has developed. Section 3.7 reports on an evaluation study conducted with a number of users on Carnegie Mellon's campus along with a discussion of CBR functionality to learn context-sensitive user preferences. Section 3.8 summarizes what we view as some of the main contributions of our work so far and briefly discusses our ongoing research.

3.2 Prior Work

Prior efforts to develop context-aware applications are many. Early work in context awareness includes the Active Badge System developed at Olivetti Research Lab to redirect phone calls based on people's locations [3]. The ParcTab system developed at the Xerox Palo Alto Research Center in the early 1990s relied on PDAs to support a variety of context-aware office applications (e.g., locating nearby resources such as printers, posting electronic notes in a room, etc.) [4, 5]. Other relevant applications that have emerged over the years range from location-aware tour guides to context-aware memory aids. More recent research efforts in context awareness include MIT's Oxygen [6], CMU's Aura [7], and several projects at Berkeley's GUIR [8, 9] to name just a few.

While early context-aware applications relied on ad hoc architectures and representations, it was quickly recognized that separating the process of acquiring contextual information from actual context-aware applications was key to facilitating application development and maintenance. Georgia Tech's Context Toolkit represents the most significant effort in this direction [10, 11]. In the Context Toolkit, widgets act as wrappers that provide access to different sets of contextual information (e.g., user location, identity, time, and activity), while insulating applications from context acquisition concerns. Each user (as well as other relevant entities such as physical objects or locations) has a context server that contains all the widgets relevant to it. This is similar to our notion of e-Wallet, which serves as a directory of all personal resources relevant to a given user (e.g., relevant location tracking functionality, relevant collections of preferences, access to one or more calendar systems, etc.). Our semantic e-Wallet, however, goes one step beyond Dey's Context Toolkit. It makes it possible to leverage much richer models of personal resources: what personal information they give access to, when to access one rather than the other, and how to go about accessing these resources. In addition, it includes access control and obfuscation functionality to enforce user privacy preferences. This richer model is key to supporting automated discovery and access of a user's personal resources by agents. In other words, while the Context Toolkit focuses mainly on facilitating the development of context-aware applications through offline, reuse and integration of context-aware components (i.e., widgets), our architecture emphasizes real-time, on-the-fly queries of personal resources by context-aware agents. These queries are processed through several layers of functionality that support automated discovery and access of relevant personal resources subject to user-specified privacy preferences.

The notion of e-Wallet as introduced in systems such as Microsoft's .NET Passport is not new. However, current implementations have been limited to storing a very small amount of information and offer very restricted control to the user when it comes to specifying what information can be made available to different services. For instance, in Passport, users can specify whether or not they are willing to share parts of their profiles with all participating sites but cannot distinguish between different participating sites. Our notion of semantic e-Wallet lifts these restrictions and allows users to control access to any of their personal resources. It also allows for multiple sources of similar information (e.g., multiple calendars or multiple location tracking functionality) and for functionality that can dynamically select which of these resources to tap based on the context and the nature of the query at hand (e.g., using your car's GPS system when you are driving and your cell phone operator's location tracking functionality when you are not).

Our notion of semantic e-Wallet extends recent efforts to develop rich languages for capturing user privacy preferences such as P3P's APPEL language [12]. It does so by making it possible to leverage any number of domain ontologies and by allowing for preferences that relate to any number of contextual attributes. In addition, it allows users to specify obfuscation rules through which they control the level of accuracy (or inaccuracy) at which their contextual information is disclosed to different parties under different conditions. This includes telling some people which room you are in, while simply telling others whether you are at work or not or whether you are in town or not. It also includes scenarios where you might want to pretend you are in one place, while you are really elsewhere.

Last but not least, while the security community has developed powerful languages to capture access control privileges, such as the Security Assertion Markup Language (SAML) [13], the XML Access Control Markup Language (XACML) [14], and the Enterprise Privacy Authorization Language (EPAL) [15], these languages, like P3P, do not take advantage of semantic Web concepts. Our work builds directly on recent efforts aimed at moving the Web from an environment where information is primarily made available for human consumption to one where it is annotated with semantic markup that makes it understandable to software applications. These efforts are part of a long-term vision generally referred to as the semantic Web [16, 17]. They have already resulted in a succession of semantic markup languages [18, 19], as well as early efforts to define Web service ontologies and markup in the context of languages such as DAML-S [20], OWL-S [21], and WSMO [22]. In our work, we have relied on the use of DAML+OIL [DAML01] and more recently OWL [19] to represent contextual information (e.g., location, calendar activities, social and organizational relationships, etc.) and privacy preferences and on semantic Web service concepts to support the automated discovery and access of personal and public resources. While a number of recent efforts concurrent to our work have looked at different aspects of privacy in ambient intelligence environments [9, 23–27], to the best of our knowledge, the MyCampus project was the first to introduce a semantic Web architecture for context-aware service provisioning and privacy. In addition, the project distinguishes itself by its unique blend of a strong technology-push approach to Ambient Intelligence (e.g., semantic e-Wallets, case-based reasoning, agent technologies), with a strong emphasis on usability.

3.3 Overall System Architecture

We consider an environment where users rely on an open set of task-specific applications (or agents) to assist them in the context of different activities (e.g., scheduling meetings with colleagues, reminding them of purchases they need to make, arranging trips or filtering incoming messages) [28, 29]. Some of the agents may be public or semipublic resources (e.g., an agent to help locate nearby printers in a building); others may be applications the user is subscribing to or applications downloaded on

3.3 Overall System Architecture

a PDA or an onboard computer. To function, each agent needs to access some information about its user as well as possibly other users. Access to a user's personal (or contextual) information is controlled by that user's e-Wallet subject to privacy (enforcing) rules. The e-Wallet manager (or simply e-Wallet) serves as a repository of static knowledge about the user—just like .NET Passport, except that here knowledge is represented using OWL. In addition, the e-Wallet contains knowledge about how to access more information about the user by invoking a variety of resources, each represented as a Web service. This knowledge is stored in the form of rules that map different contextual attributes onto one or more possible service invocations, enabling the e-Wallet to automatically identify and activate the most relevant resources in response to queries about the user's context (e.g., accessing the user's calendar to find out about his or her availability, or consulting one or more location tracking applications in an attempt to find out her or her current location). User-specified privacy rules, also stored in the e-Wallet, ensure that information about the user is only disclosed to authorized parties, taking into account the context of the query. They further adjust the accuracy or inaccuracy of the information provided in accordance with so-called "obfuscation" preferences.

Figure 3.1 provides an overview of our semantic Web environment. It illustrates a situation where access is from a PDA over a wireless network. However, our architecture extends to fixed Internet scenarios and more generally to environments where users can connect to the infrastructure through a number of access channels

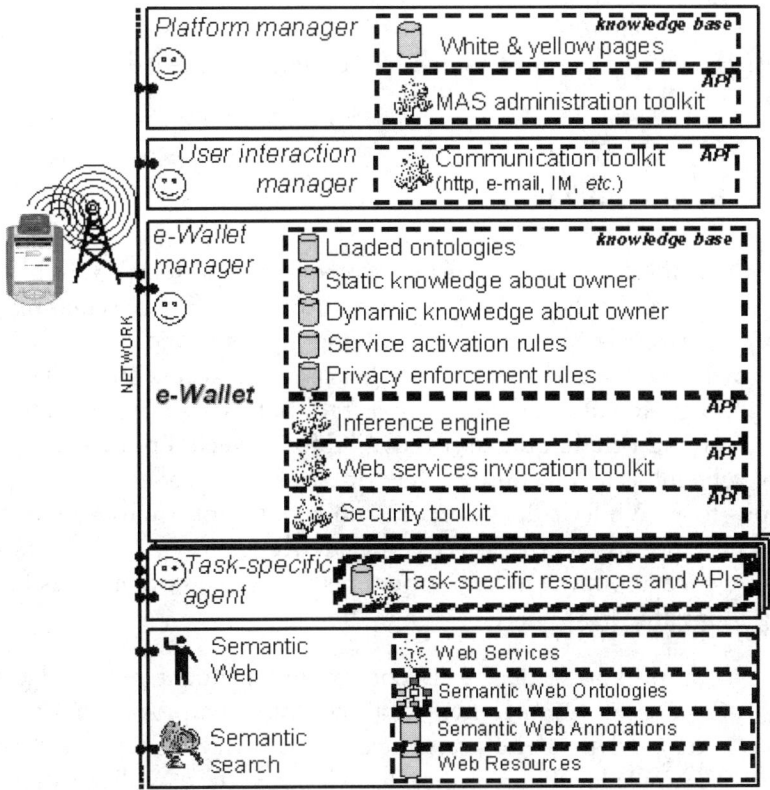

Figure 3.1 myCampus architecture: a user's perspective; the smiley faces represent agents.

and devices—information about the particular access device and channel can actually be treated as part of the user's context and be made available through his or her e-Wallet [30]. As can be seen in Figure 3.1, other key elements of our architecture include the following:

- One or more platform managers that build on top of directory facilitators and agent management systems, as defined in FIPA [31]. They manage the agents running at their sites, and maintain white and yellow page directories of these agents and the services they provide.
- User interaction managers that are responsible for interactions with the user. This includes managing login sessions as well as interactions with the user's agents and e-Wallet. Because different users interact with different sets of agents, this also includes the dynamic generation of interfaces for interacting with these agents and the customization of these interfaces to the current interaction context (e.g., particular access device). Communication with the User interaction manager typically takes place through a number of APIs, (e.g., an instant messaging API, an HTTP/HTML API, etc.).

Clearly, agents are not limited to accessing information about users in the environment. Instead, they also typically access public Web services, semantic Web annotations, public ontologies, and other public resources. On CMU's campus, where we have deployed myCampus, this includes access to a variety of services, such as 23 restaurant Web services or a pubic weather forecasting Web service.

In the following sections, we focus on the e-Wallet functionality. Additional details on myCampus and some of the agents we have deployed can be found in [28, 29, 32, 33].

3.4 A Semantic e-Wallet

The e-Wallet is a central element of our semantic Web architecture for context awareness and privacy. It provides a unified and secure semantic interface to all the user's personal resources [34], enabling agents in the system, whether working for the owner of the e-Wallet or for other users, to access and, when appropriate, modify information about the user subject to that user's privacy preferences (e.g. not just determining whether the user is available between 3 p.m. and 4 p.m. but also possibly scheduling a meeting at that time). The e-Wallet is not a static information repository. While it does contain some static information about the user, it is an agent acting as clearinghouse and gatekeeper for a user's personal resources. Its knowledge about the user, personal resources, and preferences falls into the following four categories:

1. Static knowledge. This context-independent knowledge typically includes the user's name, her email address, employer, home address and context-independent preferences (e.g., I like spicy vegetarian cuisine). This knowledge, like all other knowledge in the e-Wallet, can be edited by the user via the user interaction manager.

2. Dynamic knowledge. This is context-sensitive knowledge about the user, often involving a variety of preferences, such as; When driving, I don't want to receive instant messages.

3. Service invocation rules. These rules help leverage information resources external to the e-Wallet—both personal and public. They effectively turn the e-Wallet into a semantic directory of personal resources that can be automatically discovered and accessed to process incoming queries. Specifically, service invocation rules provide a mapping between contextual attributes and personal resources available to access these attributes, viewing each personal resource as a semantic Web service. An example of one such mapping is a rule indicating that a query about the user's current activity can be answered by accessing his or her Microsoft Outlook calendar. We have developed Web service wrappers for a variety of personal resources such as Microsoft Outlook Calendar or location tracking functionality. Service invocation rules are not limited to providing a one-to-one mapping between contextual attributes and personal resources. Instead, they can leverage rich ontologies of personal resources, enabling the e-Wallet to select among a number of possible personal resources based on availability, accuracy, and other relevant considerations. For instance, in response to a query about the user's location, the rules can specify that, when the user is driving, the best method available is the GPS in his or her car. If the user is at work and his or her wireless-enabled PDA is on, the user's location can be obtained using location tracking functionality running over the enterprise's wireless LAN. If everything else fails, the user's calendar might have some location information. Finally, it should be noted that to answer queries about the user, additional mapping rules that support automated discovery and access of public services may also be needed. For instance, a query such as "Tell me whether Fabien is in a sunny place right now" will typically require accessing Fabien's location as well as a public weather service.

4. Privacy preferences. These preferences encapsulate knowledge about what information the user is willing to disclose to others under different conditions. These preferences themselves fall into two categories:

   - Access control rules. These rules simply express who has the right to see what information under different conditions (e.g., My location should only be visible to members of my team during week days between 8 a.m. and 5 p.m.).

   - Obfuscation rules. Often, user privacy preferences are not black-and-white but rather involve different levels of accuracy or inaccuracy: Obfuscation by abstraction is about abstracting some details about the user's current context, such as telling people whether or not the user is in town without giving an exact location. Obfuscation by falsification is about scenarios where the user may not want to appear as if he or she is withholding information but would rather provide false information. For instance, a user may not want to reveal his or her true email address to a Web service for fear of getting spammed.

All of this knowledge (including rules [35]) is represented in OWL, referring to a number of relevant ontologies (e.g., ontologies about contextual attributes, personal resources, anad more specific knowledge such as cuisine types, and food preferences, message types, and message filtering preferences).

A scenario will help illustrate the key steps an e-Wallet goes through when processing an incoming query (Figure 3.2). For the sake of argument, we consider a query submitted by a user (Norman) to the e-Wallet of a second user (Fabien) to find out about that second user's current location. We assume that Fabien has specified that only his colleagues can access his location and only when he is on campus. In addition, when on campus, Fabien is only willing to disclose the building he is in but not the actual room.

The main steps involved in processing this query are as follows:

1. Asserting the query's context. As a first step, facts about the context of the query are asserted—namely, they are loaded into the e-Wallet's inference engine for possible use as part of inferences to be made in processing the query. In our example, one such assertion is that the sender of the query is Norman.

2. Asserting elementary information needs and the need to go through an authorization process. Here the query is translated into an aggregate goal that includes (1) a combination of elementary information needs—in our example the need to find Fabien's location, along with (2) a requirement to go through an authorization process. The authorization process, which is distributed across some of the following steps, results in the request being either denied or cleared, the latter possibly following the application of obfuscation rules. In our example, the authorization goal requires checking that Norman is entitled to having access to Fabien's location and that the level of resolution at which the query is answered is compatible with Fabien's privacy preferences.

3. Prechecking whether the query is allowable. A first check is performed to see whether the query is allowable based on access rights considerations. In our example, Fabien has specified that only his colleagues can access his location and only when he is on campus. In this example, for the sake of simplicity, we will assume that Norman is indeed a colleague of Fabien's, and this fact is

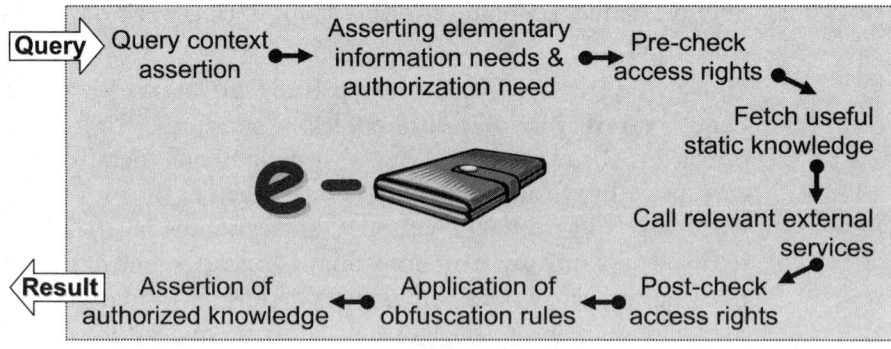

Figure 3.2 Main steps involved in processing a query submitted to an e-Wallet.

prestored in Fabien's e-Wallet. The other access right condition specified by Fabien namely, that he has to be on campus, still needs to be checked. This information however is not available locally. So the e-Wallet postpones the verification of this condition.

4. Checking the e-Wallet's local knowledge base. Some queries can be answered in whole or in part, using facts in the e-Wallet's local knowledge base, which, as we have seen in Section 3.3, contains both static (context-independent) and dynamic (context-sensitive) knowledge about the user. In our particular example, Fabien's location, which changes all the time, is not known locally by his e-Wallet. Accordingly, the e-Wallet needs to turn to outside sources of personal information to answer the query (see next step).

5. Invoking personal resources as Web services. When local knowledge is not sufficient to answer a query, the e-Wallet turns to its service invocation rules to identify external resources that might help answer the query. This may involve accessing one or more of the user's personal resources, such as the calendar and/or one or more trusted public services. In our example, the campus where Fabien works has a wireless LAN that supports location tracking. This functionality can be invoked by the e-Wallet to obtain Fabien's location. The actual invocation takes place through the Web service invocation toolkit introduced in Figure 3.1. In this particular case, the campus's location tracking service reports back that Fabien is in room Smith 234, which is on Carnegie Mellon's campus.

6. Postchecking whether the query is allowable. Now armed with knowledge of Fabien's location, the e-Wallet can check the second access right condition specified by Fabien namely, that he be on campus. In our example, this condition is met since Smith Hall is on Carnegie Mellon's campus. For the sake of simplicity, we assume that Fabien's e-Wallet possesses facts about buildings on Carnegie Mellon's campus. Otherwise, another external service would have to be identified and invoked to obtain this information. Fabien's e-Wallet has now established that Norman's request is allowable. This does not mean however that the authorization process required as part of the goals set in step 2 has been fully completed. Obfuscation rules may still need to be applied.

7. Application of obfuscation rules. Returning Fabien's room information obtained from the location tracking functionality would violate his obfuscation policy of just disclosing the building he is in. Accordingly, using its knowledge of rooms and buildings on Carnegie Mellon's campus, the e-Wallet looks for the building that room Smith 234 is in, namely Smith Hall.

8. Generating an answer. The query has now been fully processed and an acceptable answer generated. This answer (e.g. Fabien is in Smith Hall) can be returned to Norman.

As shown in Figure 3.3, we developed a three-layer implementation of our e-Wallet:

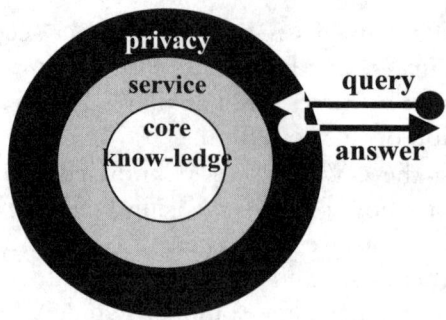

Figure 3.3 Three-layer architecture.

- *Core layer*. At the most basic level, the e-Wallet's knowledge includes an OWL metamodel, required to interpret OWL statements. In addition, it maintains both static (context-independent) and dynamic (context-dependent) knowledge about the user. This knowledge is obtained by loading available annotations about the user along with relevant ontologies and is currently completed using forward-chaining reasoning—to avoid having to infer the same facts over and over again.
- *Service layer*. This completes the e-Wallet's core knowledge with invocation rules that map information retrieval goals about contextual attributes onto external service invocations. These are modeled as backward-chaining rules. Given an information retrieval goal such as "Give me Fabien's location," they help identify and invoke one or more relevant information resources, each modeled as a Web service.
- *Privacy layer*. The outer layer is referred to as the privacy layer, as this is where privacy (enforcing) rules are applied. *Only authorized knowledge can be sent in response to queries*. Authorized knowledge is generated by applying privacy enforcing rules to service knowledge and core knowledge, thereby ensuring that information about the user is only disclosed to authorized parties and in accordance with relevant obfuscation rules.

Privacy enforcing rules are encoded as backward-chaining rules. These rules map needs for authorized knowledge onto needs for service knowledge from the service layer to be postprocessed subject to the privacy enforcing rules. Upon receiving an incoming query, the e-Wallet generates a need for authorized knowledge required to answer the query. This need in turn typically triggers needs for service knowledge and core knowledge, eventually resulting either in the generation of authorized knowledge that can be returned in response to the query or in an exception, if the query is found unallowable (e.g., an unauthorized party requesting the user's location or trying to schedule a meeting in his or her calendar). Security is directly enforced through the typing of knowledge representation structures.

The current implementation of our e-Wallet is based on JESS [36], a high-performance Java-based rule engine that supports both forward and backward chaining—the latter by verifying "needs for facts" as facts themselves, which in turn trigger forward-chaining rules. The e-Wallet's knowledge base is initialized with: (1) a model of RDF [37] triples as a template for unordered facts, (2) a model of special-

3.4 A Semantic e-Wallet

ized triples used in our three layers (core triples, service triples, and authorized triples) along with associated migration rules between the layers, and (3) an OWL metamodel.

Additional knowledge is loaded into the e-Wallet by translating OWL input files into JESS assertions and rules, using a set of XSLT stylesheets [38] (Figure 3.4). The OWL input files include ontologies and annotations that are transformed into (core) triple assertions, forward-chaining rules [35] (used to complete knowledge at the core layer) and service invocation rules and privacy enforcing rules (both represented as backward-chaining rules). The XSLT templates act as metarules which generate the body, the head, and typing used by the JESS rules.

As shown in Figure 3.5, *privacy enforcing rules* are defined using three tags: the content of the target tag describes the piece of knowledge to which this rule applies; the content of the check tag describes the conditions under which read access is granted; the content of the revision tag describes the obfuscation to be applied before migrating triples to the authorized layer. Note that, at the time of writing, our e-Wallet also supports limited write access rules.

As shown in Figure 3.6 the service rules have three child tags: the content of the output tag describes the piece of knowledge that this rule can produce; the content of the precondition tag describes the knowledge needed for calling the service; the content of the call tag describes the function to trigger and its parameters.

The body of each service rule requires a need for a particular piece of information (e.g., Fabien's location) along with the availability of a specific set of arguments (e.g., knowledge of the IP address of Fabien's PDA). When these conditions are matched, the rule fires and calls the service. (Figure 3.7 depicts the semantic Web service used to support location tracking over CMU's wireless LAN).

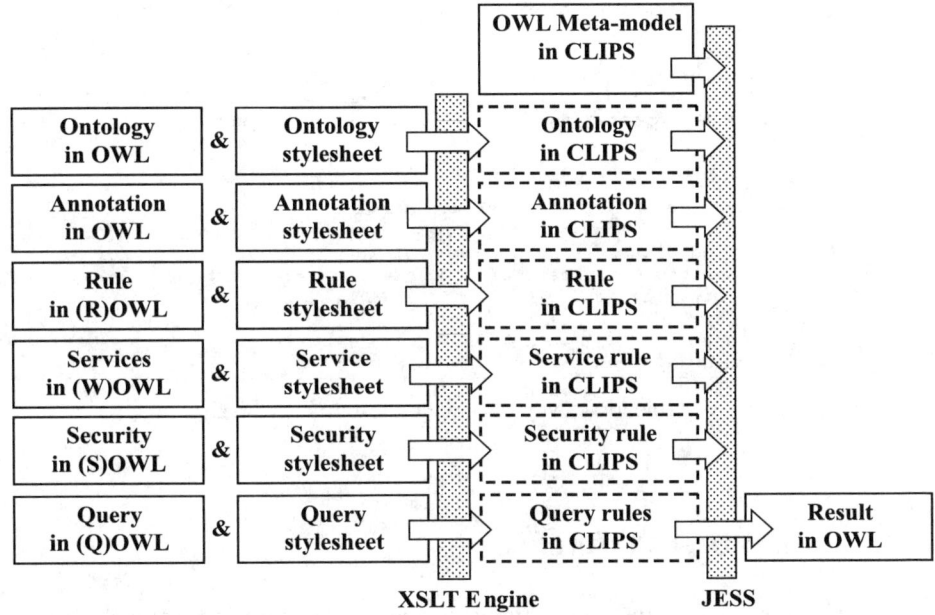

Figure 3.4 High-level flows and processes in the e-Wallet.

```
<sowl:ReadAccessRule>
  <rdfs:label>people can only know whether or not I am on campus</rdfs:label>
  <sowl:target>
    <mc:Person rdf:about="&variable;#owner">
      <mc:location rdf:resource="&variable;#location"/>
    </mc:Person>
  </sowl:target>
  <sowl:check>
    <rowl:And>
      <rowl:condition>
        <mc:E-Wallet rdf:about="&variable;#e-Wallet">
          <mc:owner>
            <mc:Person rdf:about="&variable;#owner"/>
          </mc:owner>
        </mc:E-Wallet>
      </rowl:condition>
      <rowl:condition>
        <mc:Place rdf:about="http://www.cmu.edu">
          <mc:include rdf:resource="&variable;#location" />
        </mc:Place>
      </rowl:condition>
      <rowl:not-condition>
        <qowl:Query rdf:about="&variable;#query">
          <qowl:sender rdf:resource="&variable;#owner" />
        </qowl:Query>
      </rowl:not-condition>
    </rowl:And>
  </sowl:check>
  <sowl:revision>
    <mc:Person rdf:about="&variable;#owner">
      <mc:location rdf:resource="http://www.cmu.edu"/>
    </mc:Person>
  </sowl:revision>
</sowl:ReadAccessRule>
```

Figure 3.5 Privacy rule obfuscating the location of the owner.

```
<wowl:ServiceRule wowl:salience="50">
  <rdfs:label>provide activity status for a person</rdfs:label>
  <wowl:output>
    <mc:Person rdf:ID="&variable;#person">
      <mc:has_activity rdf:resource="&variable;#activity" />
    </mc:Person>
  </wowl:output>
  <wowl:precondition>
    <mc:Person rdf:ID="&variable;#owner">
      <mc:PDA_endpoint>&variable;#endpoint</mc:PDA_endpoint>
    </mc:Person>
  </wowl:precondition>
  <wowl:call>
    <wowl:Service wowl:name="call-web-service">
      <wowl:qname>http://mycampus/PDAService#</wowl:qname>
      <wowl:endpoint>&variable;#endpoint</wowl:endpoint>
      <wowl:method>GetCurrentWeekAppointments</wowl:method>
      <wowl:user_id>&variable;#owner</wowl:user_id>
    </wowl:Service>
  </wowl:call>
</wowl:ServiceRule>
```

Figure 3.6 Service rule for activity-tracking invocation in WOWL.

3.5 Capturing User Preferences

As should be clear by now, our framework based on semantic Web technologies is capable of capturing a broad range of user preferences that may refer to any relevant

3.5 Capturing User Preferences 47

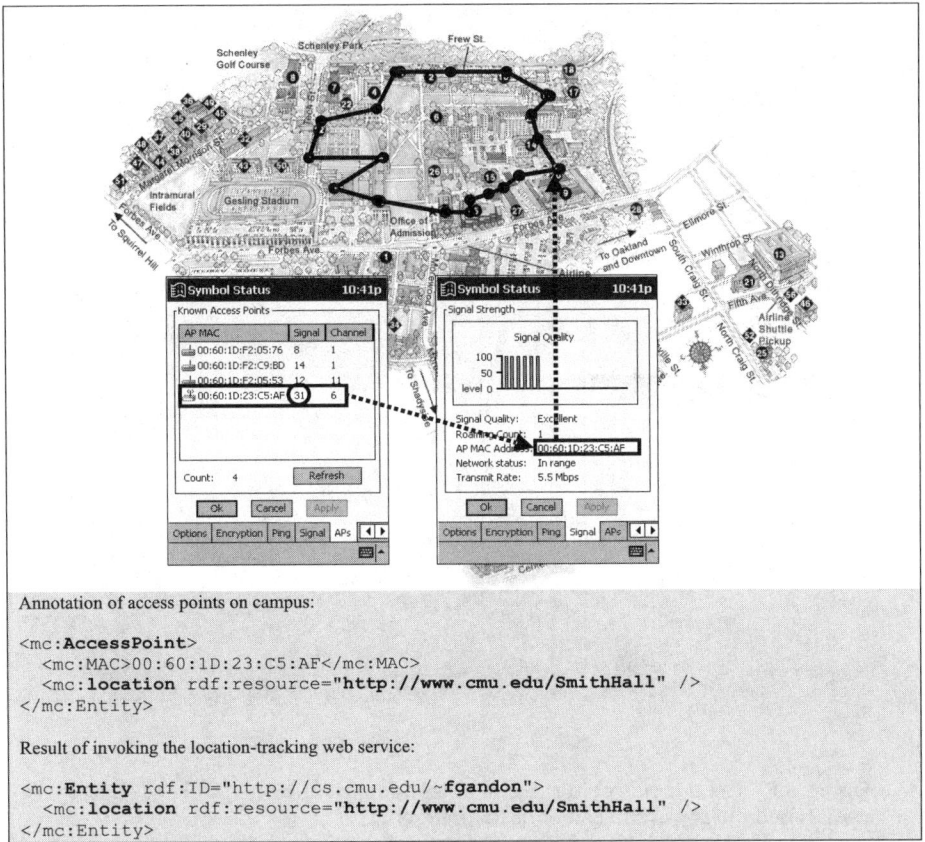

Figure 3.7 Semantic web service for location-tracking over CMU's wireless LAN.

set of OWL ontologies. This is true for message filtering preferences, food preferences, privacy preferences, scheduling preferences, and so on. One approach to capturing these preferences is to develop a variety of special-purpose editing tools that enable users to specify their preferences with regard to predefined sets of ontologies. For instance, each time a user subscribes to (or acquires) a new task-specific agent, he or she might be prompted by a special-purpose editor to select from a predefined set of preference options. The same could be done to capture predefined sets of privacy preferences. However, a key objective in our architecture has been to provide for an open environment, where new sources of contextual information, new contextual ontologies, and new agents can be introduced over time. Supporting the capture of user privacy preferences in this broader context ideally requires a general-purpose privacy preference editor that can refer to any relevant source of contextual information and any relevant contextual ontology. Figure 3.8 shows screenshots of such a general-purpose privacy preference editor, to capture both access control preferences and obfuscation preferences. While the editor is clearly too complex to be placed in the hands of lay users, it can be made available to system administrators, who can use it to capture a user's individual preferences (as well as relevant company policies in the case of employees). The editor uses XSLT stylesheets and allows users (typically system administrators) to browse (Figure

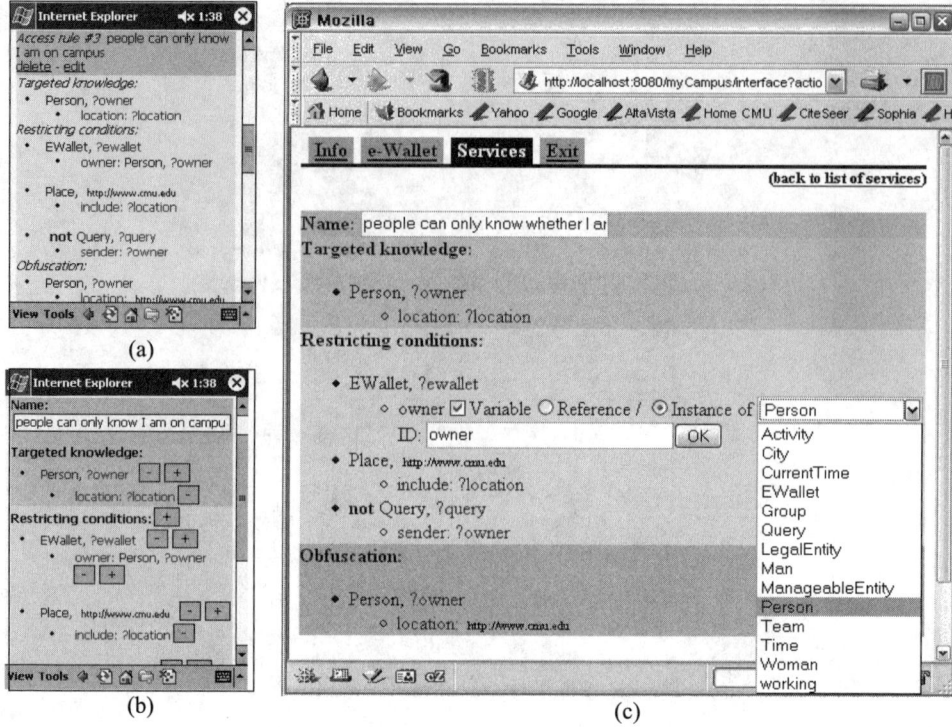

Figure 3.8 Generic rule editor that enables administrators to (a) browse and (b) (c) edit OWL-based privacy and confidentiality preferences.

3.8a) and edit privacy rules and policies (Figure 3.8b and 3.8c). It allows users to create new rules as well as edit and delete existing ones. The editor enables users to express privacy/confidentiality policies that relate to concepts and properties defined in currently available ontologies namely—ontologies loaded in the e-Wallet. The editor takes into account the OWL metamodel as the user edits rules. For instance, it will restrict the instantiation of a given concept to be within the range of a given property, as specified using the OWL "ObjectProperty" construct [19].

Our experience developing and evaluating context-aware applications, as discussed in Section 3.6 and 3.7, has shown that users often have complex and nuanced privacy preferences. Putting general-purpose interfaces such as the one shown in Figure 3.8 in their hands and expecting them to specify their preferences is simply unrealistic. General-purpose preference editors can at best be given to well-trained system administrators. In addition, users often do not even know their privacy preferences until actually confronted with a situation. Accordingly, we believe that a more pragmatic approach to capturing user preferences will often require the introduction of learning technology capable of progressively developing preference models based on user feedback. In Section 3.7, we report on our experience using case-based reasoning (CBR) to learn a user's context-sensitive message filtering preferences. While preliminary, empirical results obtained so far are rather encouraging, and we are starting work aimed at generalizing this functionality in the context of privacy preferences.

3.6 Instantiating the MyCampus Infrastructure

The MyCampus infrastructure has been instantiated in the context of several prototype Ambient Intelligence environments, including the following:

- An environment aimed at enhancing everyday campus life at Carnegie Mellon University;
- A museum tour guide environment developed for the National Museum of Natural Science in Taiwan;
- A smart office environment.

In the following text, we briefly review the instantiation of MyCampus developed for Carnegie Mellon's campus. Details of our museum tour guide instantiation can be found in [33] and details on our work on smart office applications are provided in [30].

The MyCampus infrastructure was originally conceived to enhance everyday campus life at Carnegie Mellon University through the incremental development of an open collection of context-aware applications. Campuses can be viewed as everyday life microcosms, where users (e.g., faculty, staff, and students) engage in a broad range of activities, from attending lectures and studying to socializing, having meals, making purchases, attending movies, and so on. Such a campus is representative of many of the challenges involved in successfully developing and deploying ambient intelligence applications. Over the years, the MyCampus team has worked with different groups of users to identify, design, and refine applications that could help them in the context of their daily activities. Most of these applications are implemented as context-aware agents in JADE [39]. They can automatically discover the e-Wallets of relevant users and access their contextual information subject to privacy preferences specified by these users in their e-Wallets. Sources of contextual information are wrapped as Web services. Shared ontologies ensure that the information provided by these services is understood by the agents. Some of these applications are briefly described in the following list.

- Context-aware recommender services. Several such applications have been developed and experimented with over the past few years. They include recommender services for places to eat, nearby movies, and public transportation. Many of these services extend beyond the geographical area covered by the campus.

  Screenshots of one such service, namely a context-aware Restaurant Concierge, are shown in Figure 3.9. The Restaurant Concierge makes suggestions on where to eat based on a number of preferences (e.g., types of cuisine, budget), as well as contextual considerations such as where the user is, how far he or she is from a given restaurant, how much time has until his or her next meeting, the weather, and so on. The concierge operates as a public service and obtains the user's preferences and contextual information by querying his or her e-Wallet. For instance, in response to a query about its user's location, the e-Wallet checks the user's privacy preferences and contacts location tracking functionality wrapped as a web service. A total of 23 Web services are used to

Figure 3.9 Screenshots of the Restaurant Concierge.

provide information (e.g., cuisine, location, menu, etc.) about different places where users can eat on or around campus. The "eat here!" button helps collect feedback from users to determine the extent to which current settings properly capture the users' preferences as well as to help assess the overall usefulness of the application.

- Context-aware message filtering services. The idea behind this application is that users do not necessarily want to see right away messages sent to them. In it simplest form, the service allows a user to specify preferences as to when he or she wants to see different types of messages based on the nature of the message (i.e., subject and sender), as well as based on activities specified in his or her calendar (see Figure 3.10). For instance, a user can specify that messages sent to her while she is in class should only be shown when her class is over. Here again the user's calendar is wrapped as a web service. Calendar information is obtained via the user's e-Wallet, which in turn invokes the user's calendar web service. The application also allows users to select among different delivery channels depending on their contexts. In addition, users can provide feedback to help the system refine the preferences they originally entered. This latter functionality, which relies on case-based reasoning, has proved particularly effective, as discussed in Section 2.7.

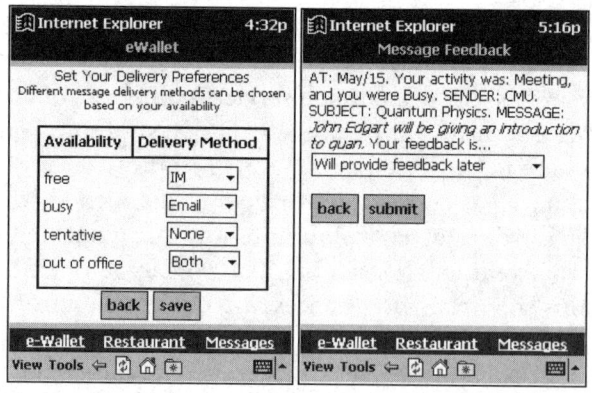

Figure 3.10 Screenshots of an early version of the context-aware message filtering service.

- Context-aware reminder applications. Several variations of this application have been developed over time, each helping to remind users about tasks they have to perform in relation to their location and possibly other contextual attributes (e.g., time of day, other calendar activities). Examples include a shopping list reminder application to tell users about items they have to purchase when they get within the vicinity of a store where the item can be found and an application that reminds students to pick up assignments when they have time or are near the place where the assignment is available. Some of these applications have proved more useful than others, in part due to the limited number of places where items can be purchased on campus.

- Context-sensitive crime alert application. Different areas of campus and its vicinity are more prone to incidents than others, depending on the time of day. This application was an experiment aimed at trying to warn users as they entered areas where an incident might have been recently reported. While such an application would a priori seem useful to people who venture in areas they are not familiar with, it did not prove very popular with campus users it was tested with, in part, we believe, due to the fairly low number of incidents on and around campus.

- Collaboration applications. One application enables people to selectively share PowerPoint presentations subject to access control policies specified in their e-Wallets. The slides can be viewed by users on nearby projectors, using their PDAs as a remote to flip from one slide to the next (see Figure 3.11).

- Community applications. Community applications have proved extremely popular among campus users and have been shown to have the potential to genuinely enhance everyday campus life. Applications we have developed range from mundane calendar scheduling applications and people locators to more sophisticated virtual, context-aware poster applications. Our calendar scheduling application and people locator application enable users to specify privacy policies, both in the form of access control preferences and obfuscation preferences. For instance, calendar scheduling users can customize their

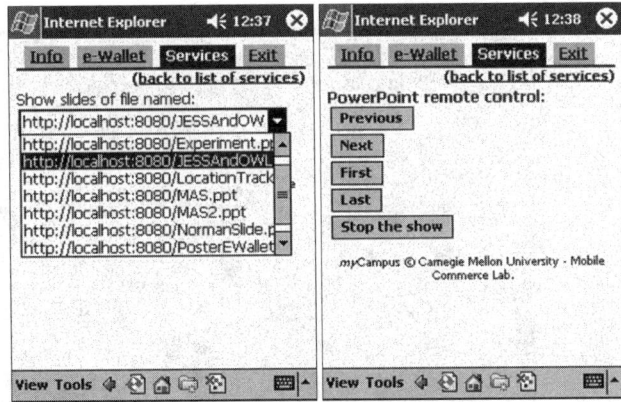

Figure 3.11 A remote slide controller enables users to selectively share presentations with others and allow them to use their PDAs to view the slides on nearby projectors.

preferences to disclose different schedule details to different people (e.g., indicating they are busy to some versus disclosing the actual activity they already have scheduled versus pretending they are unavailable). Figure 3.12 illustrates the messages exchanged when Jim invokes the calendar scheduling application (from the Interface client on his PDA) for a meeting with Mary. The application accesses both their calendar information subject to privacy preferences they each specified in their e-Wallets.

Similarly, access control and obfuscation policies specified in a user's e-Wallet enable him or her to selectively disclose a location to different people (e.g., classmates, roommates, friends, etc.) under different conditions and with different levels of accuracy (Figure 3.13).

An example of a more sophisticated community application is a context-aware poster service (called InfoBridge), enabling students to annotate, post, and retrieve posters based on both interests and contextual information. This application stemmed from two observations:

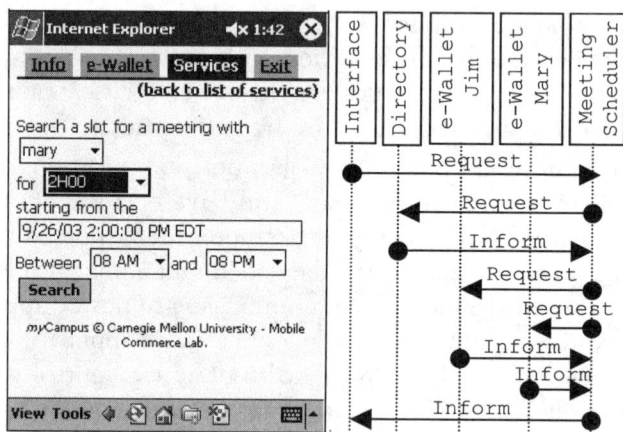

Figure 3.12 Scheduling meetings subject to e-Wallet-based privacy policies.

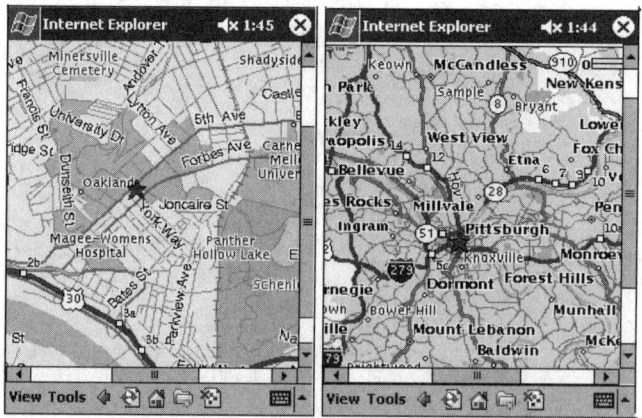

Figure 3.13 People locator with different obfuscation preferences: one user is willing to disclose the city block she is on while the other only discloses the city.

3.6 Instantiating the MyCampus Infrastructure

Placing posters all around campus is tedious, expensive, and not particularly pleasing to the eye.

People's centers of interest can often be correlated with different contextual attributes. For instance, students majoring in different areas will typically have different daily routes through campus (Figure 3.14)

In InfoBridge, users can publish virtual posters that are annotated with information about the type of activity advertised (i.e., type and topic) as well as relevant contextual information. Users with an interest in the topic or whose contextual attributes match the announcement, will have it added to their collection of posters for them to check at their convenience (Figure 3.15).

A typical InfoBridge usage scenario is summarized in Figure 3.16.

3.6.1 MyCampus Development Experience

When relevant contextual information has already been wrapped in the form of Web services, the time it takes to build a first prototype is rather minimal (e.g., from

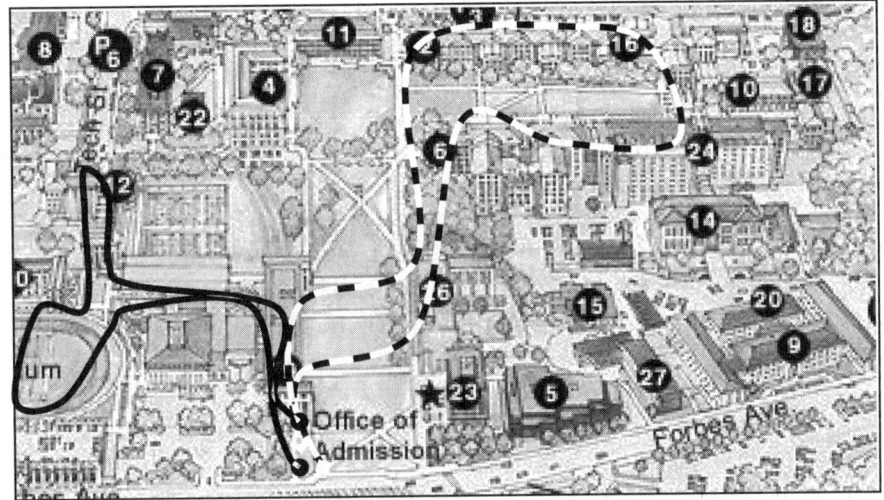

Figure 3.14 Typical routes of a design student and a computer science student at Carnegie Mellon.

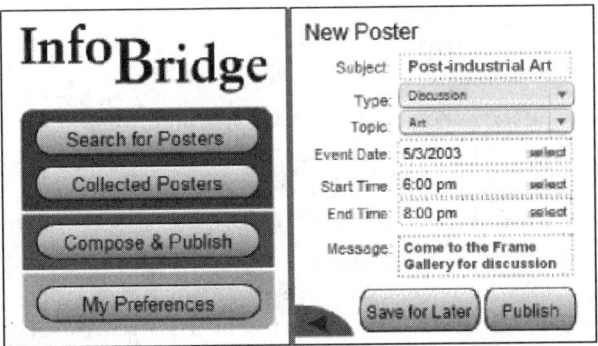

Figure 3.15 Screenshots of InfoBridge.

> *Jane knows that a group of students organizes a debate on free software at the cafeteria at 4PM. She is currently at the cafeteria and uses her PDA to place there a virtual poster containing information on the event. Thomas, who is an enthusiastic defender of Linux, receives the poster immediately even though he is not at the cafeteria. Later, John passes by the cafeteria. While John has not indicated interest or disinterest for the subject, he also receives the poster as he gets close to the cafeteria. His PDA rings/vibrates. He can consult the poster immediately or ignore it for the moment and look at it later along with other posters he will have collected as he moved around campus. When he later views the poster he can opt to add the event to his diary, to ignore it or specify that he does not want to receive any more posters on this topic. Once the date of the debate is passed, the poster is automatically purged.*

Figure 3.16 Typical InfoBridge usage scenario.

a few hours to a few days depending on the sophistication of the underlying logic). In situations where sources of contextual information had not been wrapped, our experience has often been that the resulting services end up being reused across other applications. As a result, once time and resources had been invested in developing our infrastructure and in wrapping an initial set of context services, we found that most of our time was spent in identifying, designing, evaluating, and refining services with groups of users. One lesson we have learned over and over again is that the time it takes to turn a seemingly promising idea into a working and usable prototype should never be underestimated. While several of our prototypes went through multiple significant refinement cycles and were eventually seen as being quite useful by users, few of them would really qualify as being truly ready for prime time. Another key lesson from our work with members of the Carnegie Mellon campus community has been their constant concerns about privacy. While many of them see benefits in our context-aware applications, they also all express their desire to control who has access to their contextual information, as we had anticipated. They all generally see value in the flexibility and expressiveness of our semantic e-Wallet technologies, though most require assistance with the e-Wallet settings. In Section 3.7, we report in more detail on a series of experiments conducted with MyCampus users at Carnegie Mellon University, looking at the benefits of context-aware ambient intelligence applications.

3.7 Empirical Evaluation

The MyCampus environment and its applications have been evaluated through a series of experiments in which users were observed and data collected over periods of several days at a time. This data was complemented by additional in-depth interviews with users to elicit additional information and develop a better understanding of how they interacted with the environment, what worked and what did not. In this section we briefly summarize results of a 3-day experiment conducted in early 2003 with a group of 11 users. The experiment involved observing how students interact with their e-Wallets to specify different sets of preferences and evaluate the benefits

3.7 Empirical Evaluation

of two specific applications: a version of our context-aware Restaurant Concierge and a context-aware message filtering agent. This was complemented by an additional study to evaluate the applicability of case-based reasoning (CBR) to learning individual users' context-aware message filtering preferences.

Over a period of 3 days, each user's message filtering agent was used to process a total of 44 messages, and the Restaurant Concierge was systematically used by participants in the experiment to decide where to eat, selecting from a total of 23 Web services created for restaurants on or near campus. Messages involved a mix of campus-specific news (e.g., announcements of talks, meetings, social events, movies, etc.) and general news (e.g., news headlines, weather forecasts, etc.). Users were allowed to specify both static and context-sensitive message filtering preferences (a priori preferences) for different categories of messages (e.g., messages about social events should be placed in my mailbox for later inspection, emergency messages should be shown to me right away, general news messages should be shown to me when I'm not busy, and I don't care for sport related news). As part of the experiments, participants were later asked to review each individual message they had received and indicate what the ideal filtering action for that message should have been—a posteriori preference (see Figure 3.17). By collecting a posteriori preferences for individual messages we were able to determine how well these preferences were captured by the more limited set of options available to users when they configured their message filtering agent ("a priori preferences"). As can be seen in Figure 3.17, actual messages sent as part of the experiment covered a wide range of topics. A posteriori preferences for seemingly related topics are often different (e.g., movies and symphony), illustrating the complex and nuanced nature of many user

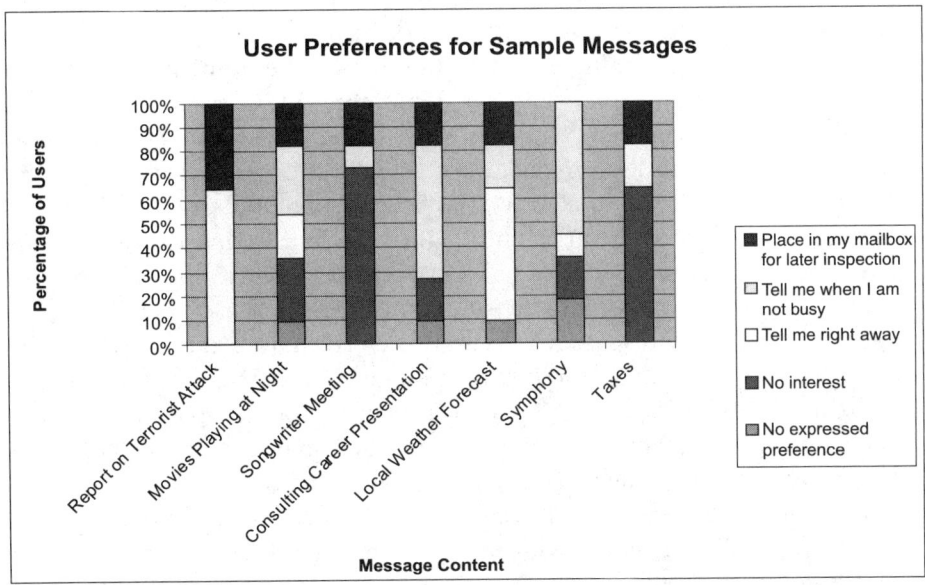

Figure 3.17 A posteriori message filtering preferences as collected from user feedback for a sample of 7 (out of 44) messages sent over a 3 day periods. Statistics in this figure are for a sample of 11 users.

preferences and the difficulty in capturing these preferences through a limited set of a priori preference options.

Analysis of results collected during the experiment confirmed the difficulty of properly capturing user preferences based on a limited set of a priori preference options. Participants indicated they were only satisfied with the way in which a little over 50% of the messages they had received had been processed. At the same time, results showed that performance would have been even worse had it not been for the context-sensitive options users were allowed to specify as part of their a priori preferences. On the whole, 70 % of the messages processed as desired had benefited from the presence of context-sensitive elements in the users' a priori preferences.

The complexity of users' actual (a posteriori) preferences, as illustrated in Figure 3.17, begs the question of whether these preferences could possibly be learned over time for individual users, thereby enabling the message filtering agent to progressively improve its decisions. To test this hypothesis, a simple case-based reasoning (CBR) [40, 41] module was implemented to attempt to learn more nuanced context-sensitive message filtering preferences for individual users based on their feedback. In CBR, past cases are used to guide decision making in new situations. In the message filtering agent, a case corresponds to a particular message and the a posteriori preference expressed by the user for processing that message. Cases are collected based on user feedback. As new messages come in, prior cases can be retrieved and matched against the new message according to various attributes (indices). Closely matching cases can be used to help decide what to do with the new message. Indices used to retrieve and match cases in the implementation of our CBR module included: the type of message (e.g., meeting, class announcement, food specials, etc.), the sender of the message, the user's current calendar activity, the user's current location, and the weather. Cases were matched, using Aha's Nearest Neighborhood Algorithm [42], as adapted by Cercone and colleagues [43]. Experiments were conducted, using the first 33 messages received by each individual user as training cases and the remaining 11 messages as test cases. Results could only be computed for a subset of participants since contextual attributes in some user logs had been corrupted. While more extensive experiments would need to be carried out to clearly establish the potential of CBR, our results showed a significant improvement in the quality of the filtering decisions, which jumped from an average accuracy of a little over 50% (with a priori preferences) to an accuracy of over 80% with the CBR module—accuracy here is measured as the percentage of filtering decisions that exactly match the a posteriori preference indicated by the user for each message. These results suggest that context-sensitive message filtering preferences may be too complex to be correctly specified upfront and that, instead, user feedback may need to be collected over time to develop finer models. These results are illustrated in Figure 3.18, where we show the improvement observed with the CBR module over the version of our message filtering agent relying on a priori user preferences. The results are for the two most extreme participants in our study, one who was nearly satisfied already with a priori preferences and one whose satisfaction significantly improves with CBR.

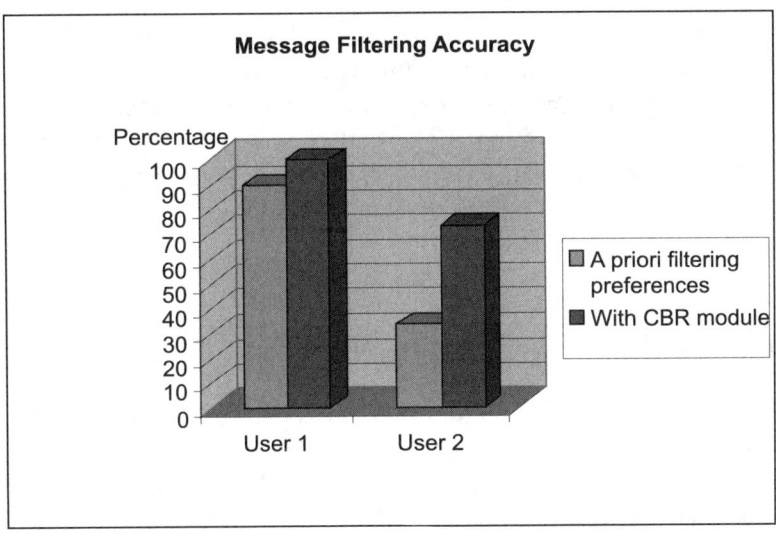

Figure 3.18 Comparing the accuracy of context-sensitive message filtering with a priori user preferences and with the CBR module for two extreme participants.

3.8 Conclusions

In this chapter we presented an overview of ambient intelligence work conducted over the past five years by the MyCampus group at Carnegie Mellon University. The project has drawn on multiple areas of expertise, looking at issues of usability while focusing on the development of an open semantic Web infrastructure for context-aware and privacy-aware service provisioning. We reviewed key architectural elements of the MyCampus semantic Web infrastructure, with a special emphasis on its introduction of semantic e-Wallets which act as both clearinghouses and gatekeepers to a user's personal information. Another important element of this infrastructure is the way in which sources of contextual information are modeled as semantic Web services. These services can automatically be identified and accessed to supplement an e-Wallet's local knowledge about its user. Our experience tailoring this architecture to different environments (everyday campus life applications, office applications, and a museum tour guide) has shown that the reusability of sources of contextual information modeled as semantic Web services can substantially reduce the time it takes to develop and refine new context-aware applications. This is particularly helpful in environments, where the objective is to develop a growing collection of context-aware applications over time—as has been the case with our work on Carnegie Mellon's campus. Despite these benefits and the positive feedback obtained from a number of users on the value of ambient intelligence applications we developed, our work with users has also shown that it typically takes numerous iterations before an application can be considered ready for prime time.

Our work with users at Carnegie Mellon University has also confirmed our initial intuition that privacy issues are central to user acceptance. It has shown that users' privacy preferences are often complex and nuanced. Capturing these preferences as well as other relevant context-sensitive preferences remains a major impediment to the broad acceptance of Ambient Intelligence technologies. Our initial work using case-based reasoning suggests that some of these preferences can be learned over time, based on user feedback, thereby alleviating the initial burden that would be placed on users if they had to specify all their preferences upfront. This is an area we are continuing to research.

3.9 Additional Sources of Information

Additional information, including a video on the MyCampus project, can be found at http://www-2.cs.cmu.edu/~sadeh/mycampus.htm. Some of the code developed as part of the project has also been released on SemWebCentral (see http://projects.semwebcentral.org/projects/rowl/). This includes:

- ROWL—Rule language in OWL and translation engine for JESS;
- A standalone version of our semantic e-Wallet, including a development environment.

3.10 Acknowledgments

The work reported herein has been supported in part under DARPA contracts F30602-02-2-0035 (DAML initiative) and F30602-98-2-0135. Additional support has been provided by IBM, HP, Symbol, Boeing, Amazon, Fujitsu, the EU IST Program (SWAP project), and the ROC's Institute for Information Industry. Over the years, the MyCampus project has benefited from the contributions of a number of participants. Besides the authors, past and present members of the project include: Jinghai Rao, Enoch Chan, Linh Van, Hirohiko Yamamoto, Kazuaki Takizawa, Yoshinori Shimazaki, Rahul Culas, Huntington Howe, Paul Ip, Ruth Lee, Mithun Sheshagiri, Shih-Chun Chou, Wen-Tai Hsieh, Matt Chang, Andrew Li, Polly Ng, Sumat Chopra, Srini Utpala, and Wilson Lau. The U.S. government is authorized to reproduce and distribute reprints for governmental purposes notwithstanding any copyright notation thereon.

References

[1] OASIS, "UDDI: Executive Overview: Enabling Service Oriented Architecture," http://uddi.org/pubs/uddi-exec-wp.pdf, 2004.

[2] W3C, "Web Services Description Language (WSDL) 1.1, Note 15" March 2001, http://www.w3.org/TR/wsdl.

[3] Want, R.,et al., "The Active Badge Location System," *ACM Transactions on Information Systems,* Vol. 10, No. 1, 1992, 91–102.

[4] Schilit, W., "A System Architecture for Context-Aware Mobile Computing," Ph.D. diss., Columbia University, 1995.

[5] Schilit, B., N. Adams, R. Want, "Context-Aware Computing Applications." *Proc. of the Workshop on Mobile Computing Systems and Applications*, IEEE Computer Society, Santa Cruz, CA, 1994, 85–90.

[6] Dertouzos, M., "The Future of Computing," *Scientific American*, August 1999.

[7] Garlan, D., et al., "Project Aura: Towards Distraction-Free Pervasive Computing," *IEEE Pervasive Computing, Special Issue on Integrated Pervasive Computing Environments*, Vol. 1, No. 2, April–June 2002, 22–3.

[8] Hong, J., and J. Landay, "A Context/Communication Information Agent," *Personal and Ubiquitous Computing*, Vol. 5, No. 1, 2001, 78–81.

[9] Hong, J., "An Architecture for Privacy-Sensitive Ubiquitous Computing," unpublished PhD diss., University of California at Berkeley, Berkeley, CA, 2004.

[10] Dey, A. K., and G. D. Abowd, "Toward a Better Understanding of Context and Context-Awareness," *GVU Technical Report* GIT-GVU-99-22, College of Computing, Georgia Institute of Technology, 1999.

[11] Dey, A., et al., "An Architecture to Support Context Aware Computing," *GVU Technical Report*, GIT-GVU-99-23, College Computing, Georgia Institute of Technology, Nov. 2000.

[12] W3C, "The Platform for Privacy Preferences 1.0 (P3P1.0) Specification, Recommendation 16," April 2002, http://www.w3.org/TR/P3P/.

[13] OASIS, "Security Assertion Markup Language (SAML), Technology Reports," April 14 2003, http://xml.coverpages.org/saml.html.

[14] OASIS, "Extensible Access Control Markup Language (XACML), Technology Reports," March 28 2003, http://xml.coverpages.org/xacml.html.

[15] Schunter, M., and C. Powers, "The Enterprise Privacy Authorization Language (EPAL 1.1), IBM Research Laboratory," http://www.zurich.ibm.com/security/enterprise-privacy/epal/.

[16] Berners-Lee, T., J. Hendler, and O. Lassila, "The Semantic Web," *Scientific American*, May 2001.

[17] Hendler, J., "Agents on the Web," *IEEE Intelligent Systems*, Vol. 16, No. 2, March–April 2001, pp. 30–37.

[18] DAML Joint Committee, "DAML+OIL language," March 27, 2001, http://www.daml.org/2001/03/daml+oil-index.html.

[19] W3C, "OWL Web Ontology Language Reference, Working Draft" March 31, 2003, http://www.w3.org/TR/owl-ref/.

[20] DAML Services Coalition, "DAML-S: Web Service Description for the Semantic Web," First International Semantic Web Conference, ISWC'02, Sardinia, Italy, LNCS 2342, 2002, 348-363.

[21] DAML Services Coalition, "OWL-S: Semantic Markup for Web Services", http://www.daml.org/services/owl-s/1.1/overview/, 2004.

[22] Feier, Cristina and Domingue, John. "WSMO Primer," WSMO, April 2005, available at http://www.wsmo.org/TR/d3/d3.1/v0.1/#S44.

[23] M. Ackerman. "Privacy in Pervasive Environments: Next Generation Labeling Protocols". *Pervasive and Ubiquitous Computing*, Vol 8, 2004, 430–439, http://dx.doi.org/10.1007/s00779-004-0305-8.

[24] Hengartner U., and P, Steenkiste, "Implementing access control to people location information," *9th ACM Symposium on Access Control Models and Technologies (SACMAT'04)*, Yorktown Heights, June 2004.

[25] Kagal L., T. Finin, and A. Joshi, "A policy language for a pervasive computing environment," In *Collection of IEEE 4th International Workshop on Policies for Distributed Systems and Networks*, June 2003.

[26] Langheinrich, M., "A privacy awareness system for ubiquitous computing environments," *Proceedings of Ubicomp*, 2002, pages 237–245.

[27] Lederer, S., et al., "Towards Everyday Privacy for Ubiquitous Computing," Technical Report UCB-CSD-03-1283, *Computer Science Division, University of California*, Berkeley, October 20, 2003. http://www.cs.berkeley.edu/projects/io/publications/privacy-techreport03a.pdf

[28] Sadeh, N., et al., "Creating an Open Agent Environment for Context-aware M-Commerce," in *Agentcities: Challenges in Open Agent Environments*, Burg, et al., Springer Verlag, 2003, pp. 152–158.

[29] Gandon, F., and N. Sadeh, "Semantic Web Technologies to Reconcile Privacy and Context Awareness," *Web Semantics Journal,* Vol. 1, No. 3, 2004.

[30] Miller, N., et al., "Context-Aware Computing Using a Shared Contextual Information Service," *Pervasive 2004*, Vienna, April 2004.

[31] FIPA, "Specifications (2002)," http://www.fipa.org/repository/fipa2000.html.

[32] Sheshagiri, M., N. Sadeh and F. Gandon, "Using Semantic Web Services for Context-Aware Mobile Applications" *MobiSys 2004 Workshop on Context Awareness*, Boston, June 2004.

[33] Chou, S-C., et al., "Semantic Web Technologies for Context-Aware Museum Tour Guide Applications", *International Workshop on Web and Mobile Information Systems (WAMIS'05)*, IEEE Computer Society, 2005.

[34] Gandon, F., and N. Sadeh, "A Semantic e-wallet to Reconcile Privacy and Context Awareness," *In Proceedings of the Second International Semantic Web Conference (ISWC03)*, Florida, October 2003.

[35] Gandon, F., and N. Sadeh, "ROWL: Rule language in OWL and translation engine for JESS," *Mobile Commerce Laboratory*, Carnegie Mellon University, 2004, http://mycampus.sadehlab.cs.cmu.edu/public_pages/ROWL/ROWL.html.

[36] Friedman-Hill, E., "JESS in Action: Java Rule-based Systems," *Manning Publications Company*, June 2003, http://herzberg.ca.sandia.gov/jess/.

[37] W3C, "RDF Vocabulary Description Language 1.0: RDF Schema, Working Draft 23," January 2003, http://www.w3.org/TR/rdf-schema/.

[38] W3C, "XSL Transformations (XSLT) Version 1.0, Recommendation 16" November 1999, http://www.w3.org/TR/xslt.

[39] Bellifemine, F., et al., "JADE : A While Paper," *EXP magazine*, Telecom Italia, Vol. 3, No. 3, September 2003. available at: http://exp.telecomitalialab.com/upload/articoli/V03N03Art01.pdf

[40] Kolodner, J., *Case-Based Reasoning*. Morgan Kaufmann Publishers, San Mateo, CA, 1993.

[41] Leake, D., *Case-Based Reasoning—Experiences, Lessons and Future Directions*. AAAI Press/The MIT Press, 1996.

[42] Aha, D. W., Kibler, D., and Albert, M. K., "_Machine Learning*, Vol. 6, 1991, 37–66.

[43] Cercone, N., An, A., and Chan C., "Rule-Induction and Case-based reasoning: hybrid architectures appear advantageous," *IEEE Transactions on Knowledge and Data Engineering*, Vol. 11, No. 1, 1999, pp. 166–174.

CHAPTER 4
Physical Browsing

Pasi Välkkynen, Lauri Pohjanheimo, and Heikki Ailisto

4.1 Introduction

Physical browsing is a means of mapping digital information and physical objects of our environment. It is analogous to the World Wide Web: The user can physically select, or click, links in the nearby environment. In physical browsing, the user can access information or services about an object by physically selecting the object itself. The enabling technology for this are tags that contain the information—for example, a Web addresses—related to the objects to which they are attached. This user interaction paradigm is best introduced with a simple scenario.

> Joe has just arrived on a bus stop on his way home. He touches the bus stop sign with his mobile phone and the phone loads and displays him a Web page, which tells him the expected waiting times for the next buses so he can best decide which one to use and how long he must wait for it. While he is waiting for the next bus, he notices a poster advertising a new interesting movie. Joe points his mobile phone at a link in the poster and his mobile phone displays the Web page of the movie. He decides to go see it in the premiere and clicks another link in the poster, leading him to the ticket reservation service of a local movie theater.

Mobile phones have become ubiquitous, and they have become versatile mobile computing platforms with access to diverse wireless services, especially with their World Wide Web and messaging capabilities. However, these small, portable devices have limited input capabilities, hindering the ease and convenience of use.

Passive radio frequency identification (RFID) tags and visual tags such as barcodes are simple, economic technologies for identifying objects. These tags can store some information—for example, an identification number, or more recently, a universal resource locator (URL). Until recently these tags have required specialized readers that have usually been connected to desktop or portable PCs and used mainly in logistics. The readers are getting smaller, making it feasible to integrate them into smaller mobile terminals such as PDAs or mobile phones, and thus more closely integrating the tags with the other functionalities of these mobile devices.

As can be seen in the next section physical browsing as a concept is not new. Various ways to associate physical and digital entities have been suggested and implemented [1–5]. By combining the aforementioned two technologies, ubiquitous mobile terminals and cheap tagging systems, it is finally possible to augment our physical environment in a grand scale. As the prices of tags are dropping and without a need for batteries or maintenance we will be able to tag practically anything we want, truly linking the physical and digital worlds. The momentum for physical browsing will thus be created from two directions: a need for easier and more convenient access to digital information, both WWW and local services, and the emergence of enabling technologies—affordable RFID (and visual) tagging technologies and their convergence with mobile phones and PDAs. The first RFID readers have already been released for mobile phones, and cameras able to read visual tags are becoming standard equipment in them.

As a user interaction paradigm, physical browsing is very much analogous to browsing the World Wide Web. A tagged environment includes links to information about the environment and objects in it. By selecting these links with a mobile terminal, the user clicks these physical links, and a terminal displays information related to the objects or activates other services related to them.

We first discuss the previous work done on the topic of combining physical and digital worlds, especially previous physical browsing research. Then we define central terms related to physical browsing and discuss physical selection, the equivalent of clicking a physical link in more detail. We then discuss how physical browsing relates to context awareness and the issues in visualizing the physical hyperlinks. We describe the most prominent technologies for implementing physical browsing, and we look at two demonstration systems we have built.

4.2 Related Work

Physical browsing is akin to Weiser's vision of ubiquitous computing [6] in the sense that in both concepts computational devices are brought to the real, physical world to augment it. Physical browsing can be seen as one user interaction concept in the wider picture of ubiquitous computing—one in which the user controls the interaction between him or her and the world with a mobile terminal. This view is slightly different from the visions of calm or disappearing computing that are connected to ubiquitous computing and included in Weiser's vision. Weiser strove for implicit interaction, whereas in physical browsing the interaction is very explicit. In practice, the most dramatic difference is in the implementation. The traditional view of ubiquitous computing emphasizes intelligence in the environment to make interaction implicit, which puts tremendous requirements on the infrastructure and the power of the appliances. Since physical browsing is based on more explicit interaction the environment can be augmented in extremely economic and simple ways—by augmenting objects with simple and cheap links to information. We see room and need for both views. After all, the important thing is to combine the strengths of the physical and digital worlds.

Coupling between digital and physical objects is thus a central concept of physical browsing. Augmented reality research has explored this area by using virtual reality techniques and devices to display digital information in the physical world, often in visual or aural form. Another approach, which is sometimes used in tangible user interfaces, is to augment the physical objects and make them act as containers and operators of information [7]. In physical browsing, we use the latter approach. We take first a look at some research on the tangible user interfaces and how they relate to physical browsing, and then on some research projects that are close to our view of physical browsing.

4.2.1 Tangible User Interfaces

Wellner has developed DigitalDesk [8], which is an augmented desk supporting various interactions between electronic and paper documents and is among the first systems creating associations between digital information and physical objects. Ishii and Ullmer have proposed the concept of tangible bits [9], which similarly bridge the gaps between the digital world of information and the physical world of our everyday environments. One of the key concepts of tangible user interfaces is thus association of physical objects to their digital counterparts. A good example of this is mediaBlocks [10], which are small, tagged wooden blocks that can be used as containers, transports, and controls for digital media. The approach of Ishii/Ullmer could be seen as making digital information tangible whereas our approach in physical browsing is slightly different, to create a simple, intuitive mapping between the existing physical objects and the (often existing) information about them.

In their later work, Ullmer and Ishii have redefined tangible user interface to mean *no distinction between input and output* [11], which creates a clear distinction between tangible user interfaces and physical browsing. Physical browsing is more about telling the mobile terminal which physical object we are interested in, rather than directly manipulating the object in question, as is the case in truly tangible or graspable user interfaces. It is not necessarily interacting with smart objects other than the terminal (although the terminal may interact with a smart object); instead, the objects in the environment are simply augmented to support the association between physical and digital.

Fishkin has proposed taxonomy for tangible user interfaces [12] and relaxed Ullmer and Ishii's later definition. The taxonomy involves two axes, embodiment and metaphor. The first axis answers: "To what extent does the user think of the state of computation as being embodied within a particular physical housing?" The alternatives are full (output device is input device), nearby (output takes place near input object), environmental (output is around the user) and distant (output is "over there"). The second axis of the taxonomy is metaphor, which Fishkin quantifies as none, noun (analogy between physical shape and information contained), verb (analogy in the act being performed) or noun and verb. In physical browsing the embodiment can be seen in one sense as "full"; the link and the information it points to seem to be inside the physical object. The metaphor axis is noun: The object physically resembles the information it contains.

4.2.2 Physical Browsing Research

Want and colleagues [13] state that people are at their most effective when they are using familiar everyday objects and that the desktop metaphor fails to capture the ease of use and flexibility of those objects. They propose bridging the gap between digital and physical by activating the everyday objects instead of using the metaphor, connecting physical objects with their virtual counterparts via various types of tags. Want and coworkers call these kinds of user interfaces physically-based user interfaces and state that the purpose of these interfaces is to seamlessly blend the affordances and strengths of physically manipulatable objects with virtual environments and artifacts, thereby leveraging the particular strengths of each. They have combined everyday physical objects, RFID tags, and portable computing to create several example applications. Their user interaction method is simple; the user waves the tagged object near his orher tablet PC, which has a tag reader, and then some service is launched in the PC. Some sample applications they have built include sending email messages via augmented business cards, linking a dictionary to a translator program, and opening a digital document via a physical document.

Kindberg and colleagues [2, 3] have created Cooltown, in which people, places, and physical objects are connected to corresponding Web sites. Their vision is that both worlds would be richer if they were connected to each other and that standard Web technologies are the best way to connect them. The user interaction theme of Cooltown is based on physical hyperlinks, which the users can collect to easily access services related to people, places, and things. Cooltown utilizes infrared (IR) and RFID technologies to transmit these links to the users' mobile terminals. In Cooltown museum the visitor can gather links related to the display pieces and view WWW pages related to them. In Cooltown conference room the users can collect a link to the room at its door and access, for example, the printers and projectors of the room via the link. Kindbergand coworkers have also created eSquirt, a drag-and-drop equivalent for physical environments. With eSquirt the users can collect links and "squirt" them to other devices—for example, squirting a previously collected link to a projector to display the corresponding Web page.

Ljungstrand and colleagues have built WebStickers [7, 14], a sample system to demonstrate their token-based interaction [1]. WebStickers is not a mobile physical browsing system, but a desktop-based system to help users better manage their desktop computer bookmarks. However, WebStickers illustrates well the association between digital and physical worlds by using tags. The idea of WebStickers is to make physical objects (tokens) act as bookmarks by coupling digital information (URLs) to them. The physical objects can be organized, stored, and handled outside the desktop in many ways that are not possible within the desktop environment. WebStickers can also provide the user with cognitive cues of the content of the page in more natural ways than textual lists of bookmarks.

WebStickers uses barcode stickers to store links to WWW pages in the form of an ID number, which, after reading, is coupled to a URL in a database. The users can associate the barcode stickers with Web addresses by themselves, print their own barcodes and even associate existing barcodes on products to URLs. When the barcode is read with a reader connected to the desktop computer, the browser can open the page the URL points to or display a list of addresses if several addresses are associated with one WebSticker.

Rekimoto and Ayatsuka have introduced CyberCode [4], a two-dimensional visual tagging system as part of their augmented reality research. They see visual codes as a potential tagging technology because of the increasing availability of cheap digital cameras in mobile devices and mobile phones. They have implemented several interesting applications and interaction techniques using CyberCode, some of which are possible only with a visual tagging technology. They have for example studied the possibility of transmitting links in a regular TV program, so that the user can point the camera of the mobile terminal towards the TV screen and access the visual tag that way. In addition to using CyberCodes as simple links to digital information, they have also implemented drag-and-drop operations similar to Cooltown's eSquirt and used visual codes for identifying the location and orientation of objects in their augmented reality environments.

Toye and colleagues [15] have developed an interaction technique for controlling and accessing site-specific services with a camera-equipped mobile phone, visual tags, and public information displays. The users can interact with local mobile services by aiming and clicking on tags using their camera phones. Toye and coworkers [15] have developed a mobile phone application, which, when activated, displays the view from the camera on the phone screen, highlighting tags it recognizes. The user can then click the link by clicking a button on the mobile phone when a tag is highlighted. This system can also utilize nearby public displays as a kind of "touch screen" in addition to the mobile phone screen—something that is only possible using visual tags.

GesturePen by Swindells and colleagues [16] is a pointing device for connecting devices. GesturePen works as a stylus for a PDA device, but it also includes an IR pointer, which can read IR tags attached to electronic devices and transmit the tag information to the PDA. The tag contains the identification of the device it is attached to so that the user can select the device for communication by pointing at it, instead of the traditional list, that mobile phones display when they scan the environment for Bluetooth devices. The motivation of Swindellsand coworkers is making the selection process easier, since it is more natural for people to think the device they want to communicate with as "that one there" instead of a menu item in a list.

In addition to the aforementioned research projects, there are also some commercial ventures utilizing physical browsing. Hypertag[1] is a commercial physical browsing system based on mobile phones and IR tags, which can send, for example, WWW addresses to the phones. Integrated RFID readers are appearing in mobile phones—for example, Nokia has released an RFID kit[2] and a Near Field Communication (NFC) shell[3] for their mobile phones. NFC [17, 18] is a technology for short-range (centimeters) wireless communication between devices. The devices are typically mobile terminals and consumer electronics devices, but also smart cards and low-frequency RF tags can be read with NFC readers. There are four applications, as defined by Philips: (1) "Touch and Go" allows simple data gathering from the environment or using the mobile terminal as a ticket or an access code; (2) "Touch and Confirm" lets the user confirm an interaction, (3) "Touch and Connect" helps in linking two NFC-enabled devices, making the device discovery and exchanging communication parameters easier, and finally (4) "Touch and Explore" allows the user or device to find out what services or functions are available in the target device. Interaction in Near Field Communication is based on virtually making the objects touch each other.

1. http://www.hypertag.com.
2. http://www.nokia.com/nokia/0,,55738,00.html.
3. http://www.nokia.com/nokia/0,,66260,00.html.

All these projects or products implement some specific parts and aspects of physical browsing. In the following sections we introduce a generic definition for physical browsing, what steps physical browsing consists of, methods for selecting the link, and technologies that can be used to implement physical selection.

4.3 Physical Browsing Terms and Definitions

4.3.1 Physical Browsing

The term physical browsing has first been introduced by Kindberg [3]. He describes it as: Users obtain information pages about items that they find and scan. Holmquist and colleagues have introduced a taxonomy for physical objects that can be linked with digital information and call it token-based access to digital information [1]. The precise definition of token-based access to digital information is a system where a physical object (token) is used to access some digital information that is stored outside the object, and where the physical representation in some way reflects the nature of the digital information it is associated with. This is a basis for what we call physical browsing—accessing digital information about physical objects and from the objects themselves. We can thus define physical browsing as a system where a mobile terminal and a physical object are used to access some digital information that is stored outside the object, and where the physical representation typically[4] in some way reflects the nature of the digital information it is associated with.

4.3.2 Object

Holmquist and colleagues [1] classify generic objects that can be associated with any type of digital information as containers. A container does not have to have any physical properties representing the information it contains. If there is a physical resemblance between the object and the information it contains, it is called a token. Token is thus more closely tied to the information it represents. In Cooltown [2], the object can also be a place or a person. We take that broader view of objects, including, for example, environments and people in addition to artifacts.

4.3.3 Information Tag

We have defined information tag [19] as a small and inexpensive unique identifier, which is attached to a physical object but has limited or no interaction with the object itself; contains some information, which is typically related to the object; can be read from near vicinity.

Holmquist's and coworkers' [2] token or container is thus the physical object itself, while information tag is the technical device that augments the token or container.

---

4. The physical object does not necessarily have to reflect the information, as Holmquist et al., [1] note in their definition of a container.

4.3.4 Link

The information tag provides the user a *link* to digital information about the object. This link may be an ID, which is mapped into some other link type (URL in the WebStickers example) a direct Web address, or a phone number as a link to a person, just to mention a few possibilities.

4.3.5 Physical Selection

The basic sequence of physical browsing can be divided into the following phases.

1. The user discovers a link in his or her environment and wants to access the information it leads to;
2. The user selects the link with his or her mobile terminal;
   The link activates an action in the mobile terminal.

Physical selection covers the second phase in physical browsing. In physical selection, the user tells the mobile terminal which link he or she wants to access—that is which tag he or she wants the terminal to read.

Our user interaction paradigm for physical browsing is based on three simple actions for selecting objects: pointing, touching, and scanning, which we also call PointMe, TouchMe, and ScanMe and which we describe in more detail in the following section. All these are initiated by the user, which corresponds to our view of ambient intelligence in which the user controls the interaction through a mobile terminal. In addition to these user-initiated selection methods, there is also NotifyMe, a selection method in which the selection is initiated by the tag or the mobile terminal.

4.3.6 Action

After the mobile terminal has read the link, some kind of action happens. Some example actions are opening a Web page, placing a phone call, or turning on the lights of the room. An important distinction is that this action is separate from the selection method. Actions can also be grouped into categories, as in the NFC user interaction paradigm [18]: for example Touch & Explore, and Touch & Connect.

We seek to create an association between physical browsing and more traditional Web browsing, so we will be using vocabulary that is more closely related to the existing Web browsing vocabulary, hence the term physical browsing itself instead of token-based access to digital information.

4.4 Physical Selection Methods

In this section, we describe the user-initiated physical selection methods, pointing, touching, and scanning. In addition to them, we discuss NotifyMe, a tag or terminal initiated selection method.

4.4.1 PointMe

Pointing is a natural way to access visible links that are farther away than in touching range. In PointMe, the user selects the tag by aligning his or her mobile device toward the intended tag. For pointing, a direct line of sight to the tag is needed, meaning that it does not work with tags embedded under the surface of the object. The tag can be accessed if it is within the range of the reader, typically at most a couple of meters for passive RFID tags, but enough to access tags within a room, for example. If the implementation technology is not directional as itself (for example, visual tags or infrared), the tag must have a sensor to detect when it is pointed. One possibility to implement this is adding a pointing beam to the mobile terminal as seen in Figure 4.1. When the sensor of the tag detects the light (visible or invisible), it knows it is being pointed and should respond.

In environments where tags are close to each other, pointing may be an ambiguous selection method. It is possible that the user hits the wrong tag or more than one tag with the pointing beam. The optimal width for the pointing beam is thus a compromise between easier aiming (wider beam is easier to point with) and probability for multiple hits (wider beam hits several tags more probably). Multiple hits may be presented in the same way as in the ScanMe method (see Section 4.4.3).

4.4.2 TouchMe

Touching is another natural way to access a visible link. In TouchMe, the tag is selected by bringing the mobile terminal close to it, virtually touching the tag. Touching is the most unambiguous physical selection method, and while it lacks in range, it is a powerful method when there are many tags close to each other making accurate pointing difficult. While pointing can be seen as selecting "that one there," touching is about "this one here."

Figure 4.1 PointMe. The user points a mobile terminal at a link in a movie poster. The mobile terminal reads the tag in the poster and displays the Web page of the movie.

Most previous physical browsing systems have used touching as their selection method. This is partly due to the short range of current RFID tags and readers and visual tagging systems. It is also easiest with regards to the user interface

4.4.3 ScanMe

In ScanMe selection method the user uses a mobile terminal to read all the tags in the environment. ScanMe is at its most useful when the user either knows with which object he or she wants to interact but does not know the exact location of the link or does not know which links are available in the environment. After scanning, a result of the scan is presented in the mobile terminal as seen in Figure 4.2.

ScanMe is similar to establishing a Bluetooth connection in current mobile phones. The phone is set to search for Bluetooth devices in the environment and it displays the result of the search as a list, which allows the user to select the target device from the GUI. In an ambient intelligence setting in which dozens or hundreds of objects are augmented the list quickly becomes too long to effectively navigate, which demonstrates the need for single-object selection methods such as TouchMe and PointMe.

ScanMe presents challenges for the design of the tags and the overall infrastructure. One question is how to map the information from tags to the visual presentation in the GUI, for example, the link texts in the Figure 4.2. Should the tag dedicate a part of its memory to the description information or should the description reside in a remote server?

Another challenge in physical selection is what to do after the tag is read. Should it be displayed in the mobile terminal for confirmation before activating the link? Different actions call for different confirmation policies; for example, simple physical Web browsing becomes quickly very tedious if every selection has to be followed by a confirmation to really open the page. On the other hand, a phone call to an expensive service number by accident would not be wanted. The selection methods

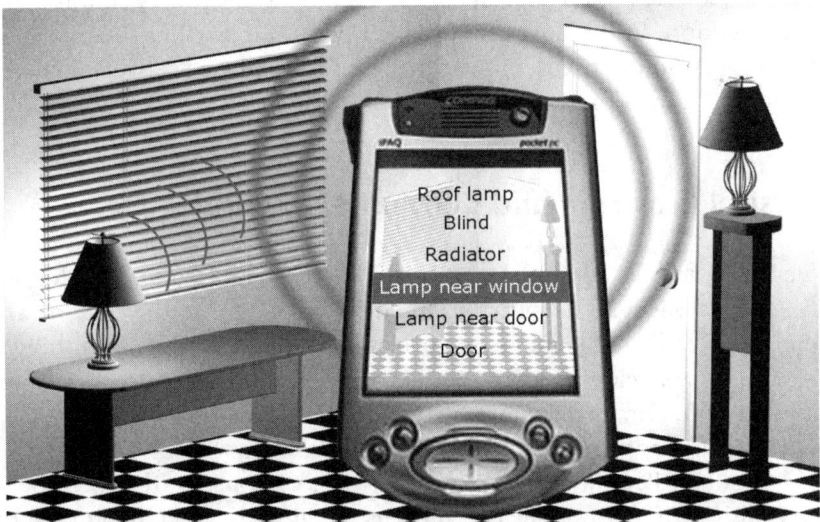

Figure 4.2 ScanMe. The user scans the whole room and all the links are displayed in the GUI of the mobile terminal.

also have an effect on the need for confirmation. If the user is pointing or touching a link, the user knows he or she wants to interact with that specific link, so confirmation may not be needed. However, when he is scanning an environment, he or she may not know which services are available and only wants to list them. In that case, opening Web pages or activating other actions is not what the user had in mind.

4.4.4 NotifyMe

In addition to the user-initiated selection methods, active tags can also start the communication. We call this tag-initiated selection NotifyMe. In NotifyMe, the tag sends its contents to the terminal, possibly based in the context it actively monitors, and the terminal displays the information to the user without the user having to select the tag. Another similar case is when the reader constantly, or based on context, reads tags from its environment. This is similar to the ScanMe selection method but the selection is not initiated by the user but by the terminal device. The terminal can then, based on the preferences of the user, display services and information it thinks are interesting to the user.

It is clear that the user should have the ultimate control of which services get pushed by NotifyMe to his or her terminal. A simple scenario involves the user walking through a shopping mall, past several small stores. If every advertisement these stores send to his or her mobile phone gets through, the simple stroll through the mall becomes a rather daunting task.

Siegemund and Flörkemeier [20] divide interaction between users and smart objects to interaction initiated by user (explicit, user in control) and interaction initiated by smart objects (implicit, user not so much in control). The latter is based on context information, and they call it invisible association. They state that explicit and implicit association methods may not be adequate in their pure form in a pervasive computing setting with a massive amount of smart objects, so a hybrid form is needed. They call this hybrid form invisible preselection, in which only probable candidates for user interaction are presented to the user for explicit selection. Physical selection can be seen as a complement to their pre-selection method, as a visible pre-selection. In the next section, we investigate further the relationship of physical browsing and context from the point of how physical browsing can help in setting the context information.

4.5 Physical Browsing and Context-Awareness

Dey [21] defines context as information that can be used to characterize the situation of an entity. The entity in his definition can be the user or another person, a place, a physical object or even an application. The important thing is that the entity is considered relevant to the interaction between the user and the context aware application. Dey considers a system context-aware, if it can use the context to provide the user information or services relevant to the task the user is involved in.

Pradhan [22] has explored one aspect of context, the location of the user, as a customization parameter for Web services in the CoolTown project []. Location can be represented as a point in a coordinate system (for example, latitude and longitude), or as a hierarchical location (in Pradhan's example Palo Alto is part of California, which in turn is part of the United States). Another representation for a location could be proximity to an object or place, which is often a more useful concept in determining context than the absolute location. Pradhan has defined semantic location as an orthogonal form of location, which can be represented as a universal resource identifier (URI). In CoolTown, different technologies for determining the semantic location have been explored. Of those technologies, encoding URLs corresponding to places in barcodes or CoolTown beacons and reading them can be seen as physical browsing. Pradhan illustrates this usage with an example in which the user is traveling in a bus and receives the URL of the bus to a PDA. With the URL and the coordinate location of the bus itself the user has access to customized Web services relevant to his situation in the bus.

Schmidt and Van Laerhoven [23] state that the traditional concept of explicit human-computer interaction is in contrast to the vision of invisible or disappearing computing that emphasizes implicit interaction. They propose context as a means of moving the interaction toward implicit interaction, so that the user does not need to explicitly input context information. However, perfect and always reliable context-awareness in dynamic real-world settings seems to be a very difficult, if not impossible, goal to obtain. Instead, there may exist a balance between the totally implicit interaction and traditional human-computer interaction. We propose that a simple, natural interaction such as physically pointing or touching an object—physical selection—could be an acceptable compromise between the reliability of explicit interaction and the naturalness of implicit interaction. The entities of Dey's context definition [21] can be detected by physical browsing, for example, by reading a tag attached to an artefact, the user tells the mobile terminal or application that he or she is interested of interacting with the object. The terminal or application can then be reasonably sure that the user is located near the object in question and that the object has some relevance to the user at that situation.

Physical browsing can thus be harnessed to help determine at least a part of context for context-aware applications. In the traditional context-aware computing approach, a large amount of sensor data is collected and then the context reasoned from the measurement data. In the MIMOSA project [24] we have defined several scenarios that utilize physical browsing in explicitly setting the context for a mobile terminal. For example, in one scenario, the user sets his or her mobile phone in silent mode by simply touching with the phone a sign that tells people entering the lecture room to silence their phones. In other MIMOSA scenarios, the users use many different mobile applications to interact with services in their environments. In our scenarios the user reads a "context tag" that explicitly tells the mobile terminal to start a service that is relevant to the situation, in the golf scenario example the application is a golf swing analyser application. This explicit interaction model also keeps the user in control all the time. In MIMOSA we are also using methods similar to physical browsing to detect when the context changes and either to start an application or change the state of a running application. For example when a golf player picks up a new golf club, the tag reader detects the change and starts measuring the club movements in order to analyse the swing.

It should be noted, however, that physical browsing is only one aspect to determining context information. While it has many benefits such as explicitness, reliability and ease of implementation, there are also disadvantages. The users must act to give the context information to the system, which may become too laborious if it has to be done too frequently. In addition, physical browsing can by no means give all the information needed for context awareness in all situations, but it could be a useful method to aid in determining context in some situations.

4.6 Visualizing Physical Hyperlinks

One challenge in physical browsing, especially in the physical selection phase is the visualization of the hyperlinks in a physical environment. Some tagging systems are inherently visual, for example, barcodes and matrix codes must be visible for the reader device to be of any use, and their visual appearance and data content are inseparable. RFID tags on the other hand are not so visible and they can even be inserted inside objects. This is sometimes an advantage since the tag will not interfere with the aesthetics of the object it augments, but from the user's point of view, it can present a problem.

The main challenges of visualizing physical hyperlinks are as follows.

1. How can a user know there is a link in a physical object or in an environment?
2. How can a user know in which part of the object the link is if he or she knows of its existence?
3. How can a user know which action or functionality will follow from activating the link?
4. How can a user know how a link can be selected if the link does not support all selection methods?

Currently, links are widely available in WWW pages. Optimally, the links are clearly marked as links and the link anchor is a word, a short phrase or image describing the link destination. WWW links are thus usually visible and include information about the action. In desktop WWW, browsing users have also learned to guess from the context what happens when they select a link; that is, from the text or images surrounding the link, or the broader context of the link, such as the whole page.

Physical hyperlinks should follow the same kinds of conventions. The link should be visible, its visual appearance should preferably follow some similar convention as the widely used underlining and link color serve in desktop WWW. The context of the link—the physical object it augments and the environment of the object—is a powerful cue for the user, but does not necessarily tell anything about the action that follows from the selection.

The link should thus optimally include information about the following.

1. The presence of a link;
2. The selection methods supported;
3. The action that follows selection.

This is a lot of information to be included in a small visual presentation, and the orthogonality of the second and third requirements makes the number of combinations large.

Another issue to consider is the aesthetics of the link. Not many of us would be happy to have barcodes all over our homes and even more beautifully designed tags may not gain any more popularity in certain environments. One strength of RFID tags is that they may be embedded under the surface of the object so they do not disturb the visual appearance of the object, making it impossible to detect the tag by human senses.

4.7 Implementing Physical Browsing

The three main alternatives for implementing physical browsing are visual codes, infrared communication and electro-magnetic methods. Wired communication methods are left out, since they require clearly more actions from the user than the physical selection paradigm implies.

4.7.1 Visual codes

The common barcode is by far the most widely used and best-known machine-readable visual code. It is a one-dimensional code consisting of vertical stripes and gaps, which can be read by optical laser scanners or digital cameras. Another type of visual code is a two-dimensional matrix code, typically square shaped and containing a matrix of pixels [25]. Optical character recognition (OCR) code consists of characters, which can be read by humans and machines. Special visual code, called SpotCode, is a means to implement novel applications for mobile phones [15]. For example, information or entertainment content can be downloaded by pointing the camera phone at the special circular SpotCode, which initiates a downloading application in the phone.

Visual tags are naturally suitable for unidirectional communication only, since they are usually printed on a paper or other surface and the data in them cannot be changed afterward [7]. The tags, usually printed on paper or plastic, are very thin and can thus be attached almost anywhere. The most significant differences between barcode, matrix code, and OCR are in the information density of the tag and the processing power needed to perform the image recognition. Barcodes have typically less than 20 digits or characters, while matrix tags can contain a few hundred characters. The data content of an OCR is limited by the resolution of the reading device (camera) and the available processing power needed for analyzing the code. Visual codes do not have any processing capability, and they do not contain active components, thus their lifetime is very long and they are inexpensive. The reading distance ranges from contact to around 20 centimeters with hand held readers and it could be up to several meters in the case of a digital camera, depending on the size of the code and resolution of the camera. By nature, visual codes are closer to the pointing class than touching type of selection.

Barcodes are widely used for labeling physical objects, especially in retail commerce and logistics. There are already a myriad of barcode readers, even toys, on the market. Software for interpreting barcodes in camera pictures is also available. The presence of barcodes in virtually all commercial products makes it possible to use them as links to various information sources ranging from manufacturer's product data to Greenpeace's boycott list.

4.7.2 Electromagnetic Methods

Barcodes notwithstanding, radio frequency identifiers are the most widely employed machine-readable tags used for identifying real-world objects. RFID-based tag technology has been so far used mainly for logistics and industrial purposes. The solutions have been typically vendor specific. Recently, there has been renewed interest in the use of RFID tags, partly due to strong academic and industrial alliances, most notably Electronic Product Code[5] and NFC Forum.

RFID systems incorporate small electronic tags that communicate with a compatible module called reader [26]. The communication may be based on a magnetic field generated by the reader (inductive coupling), or capacitive coupling which operates for very short distances. Longer operating ranges, even several meters, can be achieved by long-range RFID tags based on UHF (ultra high frequency) technologies [27]. The tags are typically passive, which means that they receive the energy needed for the operation from the electromagnetic field generated by the reader module, eliminating the need for a separate power supply. Additionally, there are active RFID tags which incorporate a separate power supply for increasing the operating range or data processing capability. RFID technology can be applied for physical selection by integrating a tag in the ambient device and a reader in the mobile device or vice versa.

An RFID tag typically contains an antenna and a chip for storing and possibly processing data. Tags can be unidirectional or bidirectional. Unidirectional tags are typically used in public environments, where the contents of the tags can only be read. Bidirectional tags are used when the user can freely change the contents of the tag. Most widely used communication frequencies are 125 kHz and 13.56 MHz, which is favored by the NFC Forum. Reading distances range from few millimeters to several meters. Applications include logistics (palette identification), antitheft devices, access cards and tokens, smart cards, and vehicle and railway car identification.

Certain technologies based on magnetic induction [28] and radio frequency wireless communication [29] particularly aimed for short ranges have been presented. These technologies can be applied for identification tags, especially when long reading ranges in order of several meters are needed.

4.7.3 Infrared Technologies

Infrared is widely used in local data transfer applications such as remote control of home appliances and communication between more sophisticated devices, such as laptops and mobile phones. In the latter case, the IrDA standard is widely accepted, and it has a high penetration in PC, mobile phone, and PDA environ-

---

5. http://www.epcglobalinc.org.

ments. Due to the spatial resolution inherent to the IR technology, IR is a potential technology for implementing physical selection applications based on the pointing concept.

An IR tag capable of communicating with a compatible reader module in the mobile device would consist of a power source, an IR transceiver and a microcontroller. The size of the tag depends on the implementation and intended use, but the smallest tags could easily be attached practically anywhere. The data transfer can be unidirectional or bidirectional. The operation range can be several meters, but a free line of sight (LOS) is required between the mobile device and the ambient device. In the IrDA standard, the specified maximum data rate is 16 Mbit/s and the guaranteed operating range varies from 0.2 to 5 meters, depending on the version used. One possible problem of IrDA, concerning especially the ambient device, is its high power consumption. For reducing the mean power consumption and thus extending the lifetime of the battery, if used, the IR tags can be woken up by the signal from the reader module [30, 31]. It is also possible that the tag wakes up periodically for sending its identification signal to the mobile device in its operating range.

In general, IR technologies are very commonplace. Many home appliances can be controlled by their IR remote controller. Several mobile phones and laptops incorporate an IrDA port, and with suitable software, they can act as tag readers. Components and modules are also available from several manufacturers.

4.7.4 Comparison of the Technologies

The most potential commercial technologies for implementing physical selection are compared in Table 4.1. Bluetooth is included for reference since it is the best-known local wireless communication technology. Obviously, exact and unambiguous values are impossible to give for many characteristics and this is why qualitative descriptions are used instead of numbers. When a cell in the table has two entries, the more typical, standard or existing one is without parenthesis, and the less typical, non-standard or emerging one is in parenthesis.

When considering the existing mobile terminals, it can be concluded that Bluetooth and IrDA are widely supported both in smart phones and PDAs; camera phones support visual codes to certain extent, and RFID readers are just emerging to mobile phones. Both barcode and RFID readers are available as Bluetooth connected accessories. Barcodes are cheap and extensively used; other visual tags are also cheap but not standardized; IrDA tags are virtually nonexistent and RFID tags are still rare and somewhat costly but gaining popularity due to EPC and NFC launches.

4.8 Demonstration Applications

We have created demonstrations as proofs of concept for physical browsing, and to study the user interaction in physical browsing. In this section, we describe our RFID emulation demonstration, which implements all three selection methods, and a genuine RFID and mobile phone-based system.

Table 4.1 Comparison of Potential Commercial Technologies for Physical Selection (Bluetooth included as a reference)

	Visual code	IrDA	RFID, inductive	RFID, UHF	Bluetooth
Selection concept	PointMe (TouchMe)	PointMe	TouchMe	ScanMe (TouchMe) (PointMe)	ScanMe
Data transfer type	Unidirectional	Bidirectional	Unidirectional (bidirectional)	Unidirectional	Bidirectional
Data rate	Medium	High	Medium	Llow-medium	High
Latency	Very short	Medium	Short	Sshort	Long
Operating range	Short-long	Medium (long)	Short (medium)	Medium-long	Long
Data storage type	Fixed	Dynamic	Fixed (dynamic)	Fixed (dynamic)	Dynamic
Data storage capacity	Limited	Not limited	Limited (not limited)	Llimited (not limited)	Not limited
Data processing	None	Yes	Limited	Limited	Yes
Unit costs	Very low	Medium	Low	Low	Medium-high
Power consumption	No	Medium	No (low)	No (low)	Medium-high
Interference hazard	No	Medium	Low-medium	Medium-high	Medium-high
Support in PDAs or mobile phones	Some (camera phones)	Yes	Emerging	No (emerging)	Some

4.8.1 PointMe, TouchMe, ScanMe Demonstration

To evaluate the physical selection methods and other aspects of physical browsing, we have built a prototype system emulating UHF RFID tags that can be read from a distance. The system supports the three selection methods, PointMe, TouchMe, and ScanMe.

The physical browsing system consists of a reader and tags. Because pointable UHF RFID tags with photo sensors are still under development, we have used Soap-Box units [] to emulate them. The remote SoapBoxes acting as tags communicate with the tag reader via RF communication. They are equipped with proximity sensors based on IR reflection to detect touching, but the IR receivers within the sensors detect pointing by IR as well. The SoapBoxes also contain illumination sensors, which may optionally be used for pointing by visible light, for example, laser.

As the mobile terminal, we use an iPAQ PDA with a central SoapBox as tag reader (Figure 4.3.). The black box on the right side is the central SoapBox, which communicates with the tags and it includes an IR LED for pointing. Between the PDA and the SoapBox is the battery case with a laser pointer that aids in pointing. When the user presses the red button of the SoapBox half way in, the laser pointer is activated to show where the invisible IR beam will hit when the button is fully pressed. The width of the IR beam is in the range of about 15–30 centimeters in a typical pointing distance of one to two meters, so the user has to get the laser indicator only near the tag he or she wants to point to.

Figure 4.3 The mobile terminal for physical browsing.

Pointing and touching display the page or other service immediately after the tag contents are read, assuming only one tag was hit. Scanning displays all found tags as hyperlinks in the graphical user interface (GUI) of the PDA. The user can then continue by selecting the desired tag from the GUI with the PDA stylus.

We have augmented physical objects with tags to demonstrate and explore the possibilities of physical browsing. Some actions we have associated with the tags are browsing Web pages, downloading a movie via a tag in a poster, sending email messages and adding an entry about an event into the calendar application of the PDA. The inner workings of our system are not exactly equivalent to passive UHF RFID tags, but the user experience is the same and this system can be used to explore the user interface and applications enabled by real RFID tags.

4.8.2 TouchMe Demonstration Using a Mobile Phone

For testing the TouchMe paradigm using a mobile phone, we have built another prototype in collaboration with Nokia. The prototype is comprised of an RFID reader device attached to a Nokia 6600 mobile phone and a set of tags. The reader module replaces the back cover of the phone as seen in Figure 4.4. Tags for this prototype can accommodate 63 characters, which is enough for most of the URLs and other identifiers. Nokia 6600 uses a Symbian operating system with open programming interfaces, making it possible to explore various scenarios using the RFID reader.

Although the design of the reader is proprietary, it shares many common characteristics with NFC devices. Reading range of the reader is 1 to 2 cm, which makes it suitable for TouchMe scenarios. Latency in communication and data transfer with the tag are negligible, making the touch action easy and convenient to use. In the prototype, the reader is connected to the mobile phone via Bluetooth connection, which is activated by pressing the button, located in the RFID cover. In this phone model, Bluetooth is the only feasible way for third-party hardware to transfer data from the reader to the phone. This causes high power consumption while the

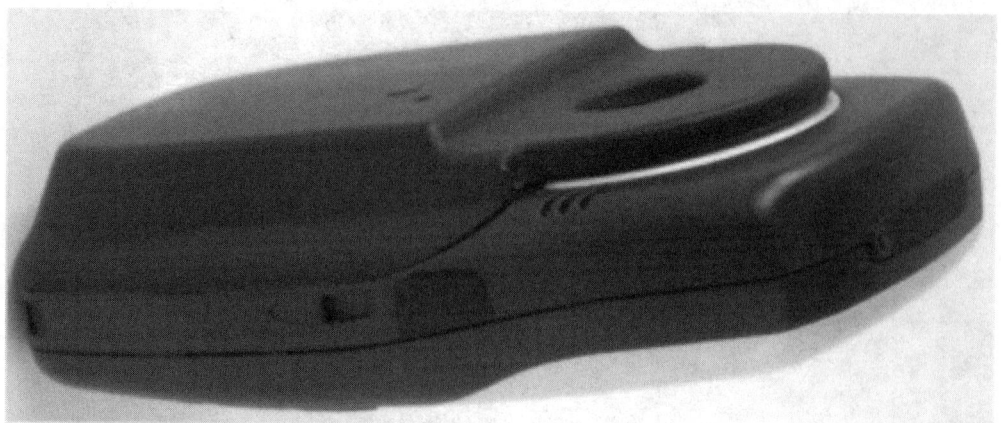

Figure 4.4 RFID reader attached to the back of a Nokia 6600 phone.

reader is on, and the button is used to turn the reader off to minimize consumption while the reader is not used. When the Bluetooth is connected, tags can be read without any button presses or actions, except the touching gesture made with the phone.

A tag is read simply by bringing the antenna of the reader close to the tag. In the prototype, the antenna is located around the lens of the phone's camera. In this case, when the tag is almost touched with the camera of the phone, the reader gives a short sound to notify that the tag is read, and the action indicated by the tag is immediately launched. In the commercial version of the reader, user's might want to be sure of the action the tag initiates and the phone should prompt the user, especially if tags offer commercial services user have to pay for, for example buying a soda from a vending machine.

Several scenarios were explored with the RFID reader device. Web page loading is perhaps the most used scenario used in the literature. We used a tag with a URL to launch the Web browser of the mobile phone. In addition, mobile phone-centric services were implemented: a tag attached to a business card or a picture launched phone call to the person in question. Text messaging service was used to get local weather information. In this case, the tag contained the message and phone number needed to get a weather forecast in a return message. A tag was also used to open other communication methods such as Bluetooth. This scenario was used to send text messages from the phone to the laptop for backup (Figure 4.5.).

Some scenarios with commercial elements were also demonstrated. These scenarios enhanced already available services by augmenting them with a tag. Background image of a mobile phone was changed by touching picture in a magazine. This scenario used already available text messaging service, where the action initiated by the tag replaced the inconvenience of writing and sending the message. In another commercial scenario, a soft drink was bought by touching the tag attached to the vending machine. The machine already had a possibility to buy a drink by calling a certain number, and the tag replaced the need for making the call manually. Both of these scenarios were easy to use and especially in the background image loading scenario, the need for a user prompt before making the purchase was evident because the service could activate accidentally if the phone were placed on top of the magazine.

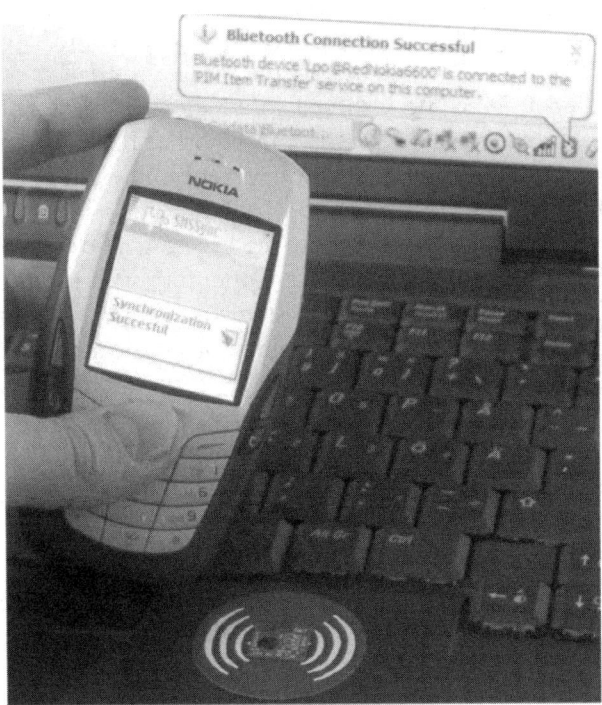

Figure 4.5 Text messages are sent to a laptop for backup.

4.9 Conclusion

In this chapter, we have discussed the concept of physical browsing, a user interaction paradigm for associating physical objects and digital information related to them. As economic and practical tagging systems are emerging, and being integrated with ubiquitous mobile terminals—mobile phones and PDAs—it will be possible to create ambient intelligence settings in grander scale than ever before and truly connect the worlds that have for a long time been largely separate, the physical world of our everyday environment and the digital world of the World Wide Web and other information systems.

We have discussed the terminology for physical browsing: the user uses physical selection to select a link in a physical object, and the mobile terminal reads an information tag attached to the object and launches some action. Physical selection can be initiated by the user, which is the case in PointMe, TouchMe and ScanMe selection methods, or it can be initiated by the tag, as is the case in the NotifyMe method. We have also described implementation technologies for physical browsing: RFID, visual tagging and infrared technologies. Our two physical browsing demonstrations are described, one that implements all three user-initiated selection methods by emulating near-future RFID tags, and another genuine RFID demonstration, which utilizes the TouchMe selection method.

Physical browsing opens many research challenges, ranging from the user interaction in physical selection to how to visualize the links in the environment. In this chapter our point of view has been user interaction, but several interesting questions can be found from other viewpoints. The contents of the tags have great impact on how and where they can be interpreted, for example if the tag contains only an identifier number it has to be resolved somewhere into a form the mobile terminal can put to use, for example a URL. If, on the other hand, the tag content is usable as is, using the tag requires only local connectivity unless it contains a link to the outside world. Optimally, standardization will provide at least some solutions to these infrastructure questions. The content of the tag also affects the way the tags can be displayed in the terminal of the user. Other important issues that will have impact on the design of physical browsing systems and infrastructures are privacy and security, which will affect the acceptability of applications utilizing tags and physical browsing.

4.10 Acknowledgments

Timo Tuomisto implemented the PointMe, TouchMe, ScanMe demonstration and provided comments for this chapter. We also thank Ilkka Korhonen and Eija Kaasinen for their helpful comments and ideas.

References

[1] Holmquist, L. E., Redström, J. and Ljungstrand, P., "Token-Based Access to Digital Information," *Proceedings of the 1st International Symposium on Handheld and Ubiquitous Computing*, Karlsruhe, Germany, 1999, pp. 234–245.

[2] Kindberg, T., Barton, J., Morgan, J., Becker, G., Caswell, D., Debaty, P., Gopal, G., Frid, M., Krishnan, V., Morris, H., Schettino, J., Serra, B. and Spasojevic, M., "People, Places, Things: Web Presence for the Real World" Mobile *Networks and Applications*, Volume 7, Issue 5, October 2002, pp. 365–376.

[3] Kindberg, T. Implementing Physical Hyperlinks Using Ubiquitous Identifier Resolution. Proceedings of the Eleventh International Conference on World Wide Web, Honolulu, Hawaii, USA. ACM Press, New York NY, USA, 2002. 191–199.

[4] Rekimoto, J. and Ayatsuka, Y. CyberCode: Designing Augmented Reality Environments with Visual Tags. *Proceedings of DARE 2000 on Designing Augmented Reality Environments*. Elsinore, Denmark, 2000, pp. 1–10.

[5] Want, R., Fishkin, K. P., Gujar, A. and Harrison, B. L. Bridging Physical and Virtual Worlds with Electronic Tags. *Proceedings of CHI 99*, Pittsburgh, PA, USA. 1999, pp. 370–377.

[6] Weiser, M. The Computer for the 21st Century. 1991. *Scientific American* Vol. 265, No. 3. 1991, pp. 94–104.

[7] Ljungstrand, P. and Holmquist, L.E. WebStickers: Using Physical Tokens to Access, Manage and Share Bookmarks to the Web. *Proceedings of DARE 2000 on Designing Augmented Reality Environments*, Elsinore, Denmark. ACM Press, New York, NY, USA, 2000. 23–31.

[8] Wellner, P. Interacting with Paper on the DigitalDesk. *Communications of the ACM*, July 1993. ACM Press, New York NY, USA.

[9] Ishii, H. and Ullmer, B. Tangible Bits: Toward Seamless Interfaces between People, Bits and Atoms. *Proceedings of CHI '97*, ACM, Atlanta, GA, 1997.

[10] Ullmer, B., Ishii, H. and Glas, D. mediaBlocks: Physical Containers, Transports, and Controls for Online Media. *Proceedings of SIGGRAPH' 98*. ACM Press, New York NY, USA, 1998.

[11] Ullmer, B. and Ishii, H. Emerging Frameworks for Tangible User Interfaces. *In Human-Computer Interaction in the New Millennium* (ed. J. M. Carrroll). Addison-Wesley, 2001. 579–601.

[12] Fishkin, K. A Taxonomy for and Analysis of Tangible Interfaces. *Personal and Ubiquitous Computing*. Springer-Verlag London, 2004. 347–358.

[13] Want, R., Weiser, M. and Mynatt, E. Activating Everyday Objects. *Proceedings of the 1998 DARPA/NIST Smart Spaces Workshop*. 1998. pp. 7–140 to 7–143.

[14] Ljungstrand, P. and Holmquist, L.E. WebStickers: Using Physical Objects as WWW Bookmarks. *Extended Abstracts of ACM Computer-Human Interaction* (CHI) '99, ACM Press, New York NY, USA, 1999.

[15] Toye, E., Madhavapedy, A., Sharp, R., Scott, D., Blackwell, A. and Upton, E. *Using Camera-Phones to Interact with Context-Aware Mobile Services*. Technical report, UCAM-CL-TR-609, University of Cambridge, 2004.

[16] Swindells, C., Inkpen, K. M., Dill, J. C. and Tory, Me. That one there! Pointing to Establish Device Identity. *Proceedings of the 15th Annual ACM Symposium on User Interface Software and Technology*. 2002. 151–160.

[17] ECMA. Near Field Communication White Paper, ECMA/TC32-TG19/2004/1, 2004.

[18] Philips Semiconductors. Philips Near Field Communication WWW page, , February 21st, 2005.

[19] Välkkynen, P., Korhonen, I., Plomp, J., Tuomisto, T., Cluitmans, L., Ailisto, H. and Seppä, H. A user interaction paradigm for physical browsing and near-object control based on tags. *Proceedings of Physical Interaction Workshop on Real World User Interfaces*, Udine, Italy. University of Udine, HCI Lab., Department of Mathematics and Computer Science, 2003. 31–34.

[20] Siegemund, F. and Flörkemeier, C. Interaction in Pervasive Computing Settings using Bluetooth-Enabled Active Tags and Passive RFID Technology together with Mobile Phones. *Proceedings of the First IEEE International Conference on Pervasive Computing and Communications* (PerCom'03). 2003.

[21] Dey, A. K. Understanding and Using Context. *Personal and Ubiquitous Computing* 5:4–7, 2001.

[22] Pradhan, S. Semantic Location. Springer-Verlag, *Personal Technologies* (2000) 4:213–216.

[23] Schmidt, A. and Van Laerhoven, K. How to Build Smart Appliances. *IEEE Personal Communications*, August 2001.

[24] Kaasinen, E., Rentto, K., Ikonen, V. and Välkkynen, P. MIMOSA Initial Usage Scenarios. 2004. Available at http://www.mimosa-fp6.com.

[25] Plain-Jones, C. Data Matrix Identification, *Sensor Review* 15, 1 (1995), 12–15.

[26] Finkenzeller, K. *RFID Handbook, Radio-Frequency Identification Fundamentals and Applications*. John Wiley & Son Ltd, England, 1999.

[27] Extending RFID's Reach in Europe. *RFID Journal* March 10, 2002. Available at

[28] *Near-Field Magnetic Communication Properties*, White Paper Aura Communications Inc. 2003.

[29] *nanoNET, Chirp-based Wireless Networks*, White Paper Version/Release number: 1.02 2004.

[30] Ma, H. and Paradiso, J. A. The FindIT Flashlight: Responsive Tagging Based on Optically Triggered Microprocessor Wakeup, in *UbiComp* 2002, LNCS 2498, 160–167.

[31] Strömmer, E., and Suojanen, M. Micropower IR Tag—A New Technology for Ad-Hoc Interconnections between Hand-Held Terminals and *Smart Objects, in Smart Objects Conference sOc'2003* (Grenoble, France, May 2003).

[32] Tuulari, E. and Ylisaukko-oja, A. SoapBox: A Platform for Ubiquitous Computing Research and Applications. Lecture Notes in Computer Science 2414: Pervasive Computing. Zürich, Switzerland, August 26-28, 2002.

CHAPTER 5
Ambient Interfaces for Distributed Workgroups: Design Challenges and Recommendations

Tom Gross

5.1 Introduction

In many situations people want and need to interact with others over distance. For instance, in industry an increasing number of global companies have employees at different sites who do joint planning, organization, and controlling; in academia scientists at various universities and research laboratories do joint research and write publications together; and in the private life people often have family and friends at various places and want to keep in touch, and plan and organize holidays, birthday parties, and trips.

There are various ways of coordinating with remote parties; however, most approaches have some unwanted side effects—having prearranged meetings can be too rigid for real-life spontaneous actions and reactions; sending emails can leave it unclear to the sender when to expect a reply; reaching somebody by phone, even if the person has a mobile phone, strongly depends on the availability and interruptibility of the callee; and sending short messages via mobile phone or instant messages via an instant messaging system can be challenging to the recipient, because of the potentially high number of messages. So, besides various technical means there is a first challenge and a tension between the need of a caller to spontaneously approach a callee over distance and the requirement to avoid disrupting the callee.

Furthermore, many knowledge workers in industry and academia are working a considerable amount of time with their computers, in order to exchange email, produce documents, do spreadsheet calculations, and so on. Additionally, they are working with real artefacts, such as printouts of long documents, books, and mail. So, as a second challenge users have to flexibly change between the electronic world and the physical world.

A number of authors have presented approaches addressing the first challenge by providing coworkers with awareness information—information about each other's presence, availability, and tasks in order to facilitate their coordination. In early approaches the information was captured and presented within one single application [1, 2]. This first generation can be called *proprietary awareness systems*. In a second generation toolkits that contained components for presenting awareness information were developed. These *awareness widget toolkits* facilitated the development of applications, because they offered complex, yet easy to handle widgets as building blocks for software developers [3, 4]). In a third generation the focus moved beyond single software applications. *Awareness information environments* made it possible to capture information from various applications and other sources and to present the information in generic representations such as with tickertapes of pop-up windows on computer desktops [5, 6].

We suggest *ambient interface systems* as a fourth generation in order to address both of the above-mentioned challenges—the coordination among remote co-workers, in both the electronic and the physical world. In fact, ambient interfaces allow to capture information in the electronic as well as in the physical world and also present the information in the electronic as well as in the physical world.

In this chapter we introduce an awareness information environment—the Theatre of Work Enabling Relationships (TOWER) environment—that provides a framework for co-orientation in distributed workgroups by capturing information about users, and providing this information to remote colleagues. This environment has been augmented by ambient interfaces. The ambient interfaces were conceptualized and development in a participatory design approach, and evaluated in field trials, which led to considerable insight on the design of ambient interfaces for distributed workgroups.

In the following we give a quick overview of related work concerning ambient displays and ambient interfaces. We introduce the TOWER environment, including the design and implementation of its ambient interfaces, and report on the participatory design and field trials. Finally, we elaborate recommendations for the design and development of ambient interfaces.

5.2 Ambient Displays and Ambient Interfaces

In this section we introduce the notions of ambient displays and ambient interfaces. We then present some prototypes and systems and bring them into a systematic classification.

Ambient interfaces basically use the whole environment of the user for interaction between the user and the system. In Webster's ambient is defined as "Latin ambient-, ambiens, present participle of ambire…to go around, from ambi- + ire to go" [7]. Wisneski [8] and colleagues at the MIT Media Lab have built a range of ambient displays; they point out that:

> Ambient displays take a broader view of display than the conventional GUI, making use of the entire physical environment as an interface to digital information. Instead of various information sources competing against each other for a relatively small amount of real estate on the screen, information is moved off the screen into the physical environment, manifesting itself as subtle changes in form, movement, sound, color, smell, temperature, or light. Ambient displays are well suited as a means to keep users aware of people or general states of large systems, like network traffic and weather.

Ambient displays use the physical environment of the user for presenting information. Ambient interfaces go one step further—they use the physical environment of the user both for presenting information and capturing information. Ambient interfaces can be seen in the vein of Weiser and Brown and their notion of calm technology [9]. Since the beginning of the 1990s, Weiser has been arguing for ubiquitous computing and calm technology, where he argues that each person will have an increasing number of computers in the walls, chairs, clothing, light switches, cars, and so forth. Consequently, there will be the danger that the computers consume a lot of our attention. A calm technology or calm device works in the background and stays in the periphery of the user's attention until the user needs it. When either the user needs the device or the device needs the user's attention for a decision and so forth, the device should be able to move to the center of the user's attention easily.

Several ambient display prototypes and systems with properties similar to the ambient interfaces of TOWER have been built. We subsequently introduce some prominent examples. The ambientROOM system [8] is a small room with special equipment for the peripheral presentation of information. For instance, water ripples on the ceiling show the intensity of activities at a remote site, where smaller and faster ripples indicate more action; ambient sounds represents activities on a remote whiteboard; and audible soundtracks plays sounds from nature such as birds and rainfalls to inform the user about the state of the email inbox. Abstract Representation Of presence supporting Mutual Awareness (AROMA) [10] is generic system architecture with mechanisms to capture, abstract, synthesize, and display presence awareness. The information is presented as an audio soundscape, an electro-mechanical merry-go-round, an electro-mechanical vibrator in the chair, thermo-electric devices to control the temperature of the hand-rest, and so forth. The Information Percolator system [11] presents information in transparent tubes. It consists of 32 transparent tubes filled with water. The tubes cover an area of 1,4m by 1,2m. Small bursts of air bubbles are released either in single tubes or in several tubes to represent more complex information (e.g., a clock, or words). The Environmental Audio Reminder (EAR) system [12] uses sound to complement visual displays, and to inform users about ongoing events and remind users about upcoming events (e.g., meetings, emails). The Audio Aura system [13] uses background auditory queues to represent serendipitous information. The auditory queues are organized in sonic ecologies—that is, sounds relating to a certain context (e.g., sound from a beach such as seagull cries, waves, and wind), which represent related events (e.g., information related to email).

These systems can be compared in various respects: We use criteria referring to the users, the information presented, and the technology used. For the user the sense(s) involved, the attention, the degree of interactivity, and the initiative of the interaction are analysed. For the information the contents (information relating to people, artefacts, or events and meetings), the context (the relation of the information to the current situation of the user), the continuity of the information, and the environment, in which the information is presented are distinguished. And, for the technology, the hardware, software, network infrastructure (existing, or new), and the integration with other technology are checked. Table 5.1 provides an overview of the ambient prototypes and systems introduced previously, and compares them according to the mentioned criteria.

Table 5.1 Overview and Evaluation of the Systems

Main criterion	Sub criteria	Systems				
		Ambient ROOM	AROMA	Information Percolator	EAR	Audio Aura
User						
	Sense	see/hear/feel	see/hear/feel	see/hear	hear	hear
	Attention	background	background	background	background	background
	Interactivity	presentation/ user input	presentation	presentation/play	presentation	presentation
	Initiative	system	system	system	system	system
Information						
	Contents	persons/ artefacts/ events	persons/ artefacts/ events	persons/ artefacts/ events	persons/ events	persons/ events/ artefacts
	Context	unrelated	unrelated	unrelated	unrelated	related
	Continuity	continual/ history	continual (e.g., slow change of temperature)	continual	continual	updates
	Environment	private	private	private	private/ public	public, but headset
Technology						
	Hardware	new	existing	new	existing	existing
	Software	new	new	new	new	new
	Network	existing	existing	existing	existing	new
	Integration	integrated	integrated	integrated	stand-alone	stand-alone

5.3 Ambient Interfaces in TOWER

In this section we give a short introduction to the TOWER environment, and present a range of ambient interfaces that have been integrated into TOWER.

5.3.1 TOWER

The TOWER environment aims to support distributed workgroups or virtual communities with mutual awareness in their current work context. It provides an infrastructure for facilitating chance encounters and spontaneous conversations among remote users. TOWER has several strengths as follows:

- *Application-independence.* TOWER is based on a client-server infrastructure—the so-called Event and Notification Infrastructure (ENI)—with fast and reliable socket-communication between the clients and the server. The server can pull information from the clients (e.g., information on user logins). Furthermore, the environment offers lightweight common gateway interfaces via HTTP. Since most office and network applications offer HTTP, TOWER can exchange data with a broad range of applications [14].

- *Sensors.* They capture user activities within the TOWER environment (e.g., logins, logouts), user activities on Win* platforms (e.g., changes to files, sharing of folders and files, starting of applications, opening of documents), user activities in shared workspaces (e.g., a sensor for the Basic Support for Cooperative Work (BSCW) system [15] records all activities in the shared workspaces such as user logins and logouts, folder creation, invitations users to shared folders, document uploads), access to Web servers [14].
- *Indicators.* A broad variety of indicators present the awareness information. Examples are lightweight indicators, such as pop-up windows with pure text or tickertapes displaying messages about the other users and shared artefacts; AwarenessMaps, which provide awareness information in the context of shared workspaces [16]; and the TowerWorld, which presents shared artifacts and users in a 3D multiuser environment [17].
- *Persistent storage.* The TOWER server receives all events captured by the sensors and puts them into a persistent storage. The storage offers various types of queries in order to retrieve the stored information [14].
- *Context models.* They allow the structuring of the huge amounts of events that are captured and stored. Context models allow to categorize events at their origin, and to map them to specific context of origin. Additionally, they allow to analyze the current situation of a user, based on all the events a user is producing in a certain situation. As a result TOWER can preselect information for users and provide them with information that is adapted to the situation [18, 19].
- *Extensibility.* TOWER provides application developers with two easy ways for extending and customizing: either by adding functionality to the ENI client via plug-ins or by communicating directly with the ENI server via HTTP [20].

With these properties TOWER offers very flexible support for facilitating co-orientation in distributed workgroups. Nevertheless, some basic challenges remain. In fact, for most users the real estate on their monitors is a scarce resource and they do not appreciate the fact that even more information is presented there. Furthermore, users are not working with their computers permanently—that is, they might miss the information presented in the 3D world and in the tickertapes while they are reading printed material or performing other activities without their computers. Still another challenge with the existing types of indicators is that users who are sitting in front of their computers are often highly concentrated and do not want to be disturbed all too often (e.g., when they are writing a paper, when they are programming, etc.).

5.3.2 Ambient Indicators

The main goal of the ambient interfaces of TOWER is to provide users with up-to-the-moment awareness information outside the computer monitor and to complement the other awareness indicators of TOWER. Therefore, key requirements for ambient interfaces in TOWER are: presentation of awareness information outside of the computer monitor; usage of the whole (office) environment of the user for presenting awareness information; and application of subtle cues in order to make use of the users' peripheral attention and capturing of information.

The TOWER project offers a broad variety of ambient interfaces as follows:

- A water tank with plastic fish releases bubbles upon the arrival of events (e.g., bubbles can be released anytime the corporate Web pages are accessed).
- A fan can start rotating and slightly blow air into the user's face while some actions last (e.g., while net traffic is low).
- Various robots can lift their hats, rotate their bodies, or raise their arms (e.g., lift the hat when a remote colleague logs in).
- The PResence, AVailability, and Task Awareness (PRAVTA) prototype provides awareness information for mobile and nomadic systems based on the wireless application protocol (WAP).
- For the ambient interfaces, which are described in the following text, mostly HTTP communication with the server was used.

5.3.2.1 Water Tank

In this approach we used a simple relaisboard that is available from electronics stores. The relaisboard we used is equipped with eight relais that can be addressed individually or together. The relaisboard can be connected to PCs via the PC parallel port. A little application—RealIndicator—was developed that allows the following: start and stop individual relais; start individual relais for a certain period of time; start and stop all eight relais; start all eight relais for a certain period of time. Now, any kind of additional hardware can be connected to the relaisboard and switched on and off upon arrival of TOWER events that match the specified event patterns [20].

Figure 5.1, shows a water tank with plastic fish that was connected to the ENI system via the relaisboard and the RealIndicator software. Upon arrival of matching events the tank starts or stops releasing bubbles. For instance, a coffee room is equipped with a WebCam and a tank. Any time somebody from outside the laboratory takes a picture, the tank starts bubbling for a few seconds.

The fish tank offers several features that are interesting for presenting awareness information in ambient interfaces and goes beyond simply bubbling or not bubbling. Depending on the frequency of the releases of the bubbles, the bubbling starts to aggregate—that is, when many events match in short distances, there are still bubbles from the former releases and the bubbling on the whole gets rougher indicating that a lot is currently going on. Additionally, the bubbling cannot only be seen, but it produces some subtle sounds of the motor and the bubbles. So, the awareness information can also be heard. The fish tank was used in a coffee room; it was mainly used to indicate who was in the coffee room when the WebCam in the coffeeroom was active. People reported that this information about being monitored was interesting and that the subtle presentation with the water tank was very good.

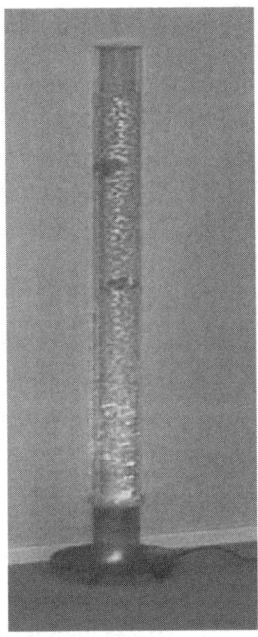

Figure 5.1 Water tank with plastic fish.

5.3.2.2 Fan

The fan uses the same approach with the relaisboard as the water tank. Figure 5.2 shows the fan.

The fan can be switched on and off depending on incoming events. We typically position the fan on the user's desktop, subtly blowing air into the user's face. This setting was used in the personal office of a user and proofed to be highly usable.

Figure 5.2 The fan.

5.3.2.3 RoboDeNiro

The Robotics Invention System is a package from LEGO that allows building robots [21]. It includes a programmable LEGO block with three plugs for sensors and three plugs for indicators (the RCX), an IR-tower for communication with a PC via infrared, two touch sensors, one light sensor, and motors. With this package it is easy to construct indicators and sensors. For accessing the motors and the sensors of the LEGO Mindstorm RCX we developed a plug-in for the ENI client called Ambient Controller. This Ambient Controller provides the user interface and functionality for specifying preferences and the functionality for communicating with the RCX via infrared [20].

We call the robots and other gadgets that we built for presenting awareness AwareBots. The RoboDeNiro AwareBot implements a thread with an event-listener that continually queries the ENI server for specific events. When a specific event occurs, the RoboDeNiro indicates it by starting or stopping a motor. We ported TclRCX to the Mac OS and developed a new ENI client that provides a bridge between the ENI server and the RCX. Figure 5.3 shows RoboDeNiro in action.

As a possible scenario, take a user who waits for a colleague to log in. When the colleague logs in, an ENI event is generated and sent to the ENI server; the RoboDeNiro application then gets this event from the ENI server, and starts the motor that lifts the hat of the RoboDeNiro. Similarly, an upload of a document that the user is waiting for can make RoboDeNiro rotate its body. The arm of the AwareBot is used for logging into the ENI system. When the user touches the arm of the AwareBot, an event is sent to the RoboDeNiro application, and the RoboDeNiro application generates an ENI event and sends it to the ENI server. This configuration was used in the personal office of a user. Since the user who was actually using RoboDeNiro also built it, he was highly satisfied and enjoyed using RoboDeNiro.

Figure 5.3 The RoboDeNiro AwareBot in action.

5.3.2.4 EyeBot

The EyeBot is another AwareBot based on the same technological infrastructure as RoboDeNiro. Figure 5.4 shows the EyeBot: On the left there is the display in the user's environment; in the middle the flag in the back signalizes that the user is currently available; and on the right the "do not disturb flag" in front indicates that the user is unavailable.

This ambient interface provides a means for indicating and sensing. The rotating eyes and dangling nose are used as indicators; the flag is used as a binary sensor. A typical application of the EyeBot is the following: The rotating eyes are connected to events produced by read activities. Examples are events produced by remote users who read a document that has been created by the user in a shared workspace. Another application is the indication of accesses to a WebCam that is located in the vicinity of the ambient display. The dangling nose is connected to events produced by communication activities. This can be a chat between project members or emails distributed via a team distribution list. A binary sensor is connected to the flag. If the user places the flag in front of the display such that it almost hides it then the user indicates that he does not want to be disturbed. This can be indicated to other users and the ENI client interprets it such that it does not display or indicate any events. If the flag is turned to the back of the display, the user indicates his availability to others, and the ENI client will also start to indicate events according to the users preferences. This configuration was also used in the personal office of the user who designed and created the EyeBot. This user used EyeBot in combination with a second PC. On the second PC several TOWER indicators were running, such as various tickertapes, the 3D multiuser world, and so forth. The user reported that this combination was very useful for him—the EyeBot gave some general and basic hints and according to personal interest, the user would then look for details on the other indicators.

5.3.2.5 EventorBot

The EventorBot can receive and react to information about actions in shared workspaces, in the email inbox, and chat requests. It uses the same technological infrastructure as RoboDeNiro.

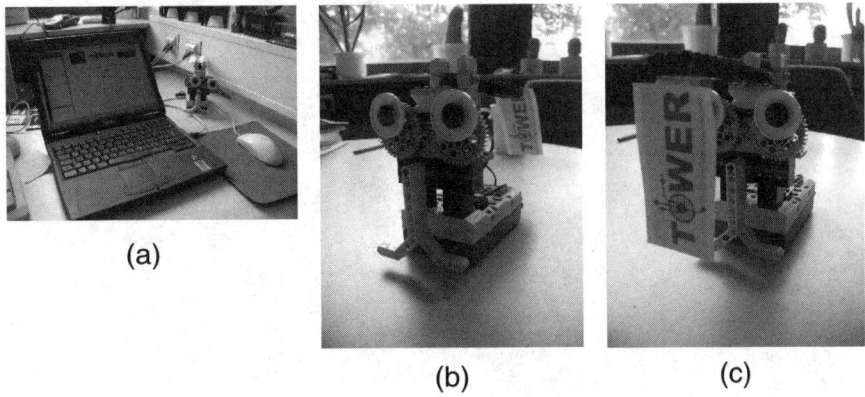

Figure 5.4 EyeBot.

Figure 5.5 shows the EventorBot and its arm. When new data is placed into a shared workspace (e.g., a document is uploaded or a schedule is put in the calendar), the EventorBot turns his left cage-arm to his back. Then it takes one LEGO brick out of the LEGO brick box by rotating its upper body to the left. After that it turns the arm back again and thereby places the LEGO brick into the collecting loop in front of it. Thus, the number of LEGO bricks in front of the EventorBot is equivalent to the number of new events in your BSCW workspace.

The EventorBot indicates read events in a shared workspace by knocking down the little tyre tree standing right beside it with its right push-arm, while turning right (see Figure 5.6). Thus, a knocked-down tree is a hint that data has been read (touched).

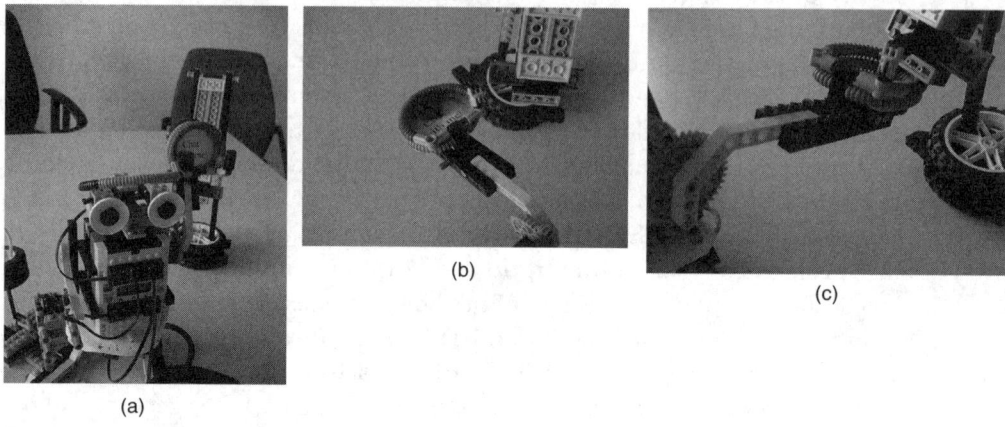

Figure 5.5 EventorBot overview and arm.

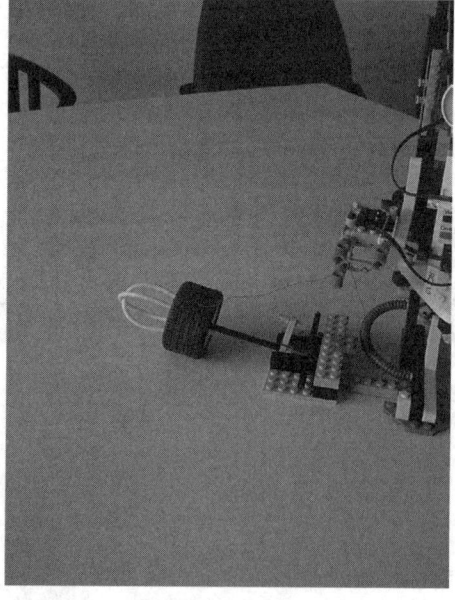

Figure 5.6 EventorBot showing touched or read data.

When users navigate to the shared workspace and acknowledge the changes (e.g., via the Catch-up function in the BSCW shared workspace system), the EventorBot lifts up the tyre tree again by turning to the left and thereby pulling the tree up with a string fixed at its push-arm. Afterward, the LEGO bricks can to be put back into the box manually. Shaking hands with the EventorBot (i.e., squeesing the push-arm) signals the robot that all bricks are put back.

The EventorBot shows the number of new emails in an email folder on a display integrated in its body. With a pointer fixed on the upper part of the body its position shows the respective number of new mails on a scale on the lower part of the body (see Figure 5.7).

If someone raises a chat request, the EventorBot turns up the cage-arm, which holds a Chat with me! sign in its cage. After turning it up, the robot moves the arm back and forth like waving with the Chat with me! Sign (see Figure 5.8).

The EventorBot is a sophisticated AwareBot. It displays action in shared workspaces, email events, and chat requests at the same time. While the users on the one hand appreciated this integration of various sources of events, this integration also entails some challenges. For instance, some users said that the meaning of the different gestures and states of the EventorBot was not as intuitively clear and easy to remember than for the other AwareBots. Furthermore, the installation of the hardware of the EventorBot is more difficult than for the other AwareBots: It has to be positioned and adjusted very precisely, in order to be able to perform all the different movements. Unless positioned and adjusted precisely it often fails when it has to display events of different types in a short sequence of time.

5.3.2.6 The PRAVTA Client

The PResence AVailability and Task Awareness (PRAVTA) client is a lightweight and mobile supplement to the other indicators and ambient interfaces that are very

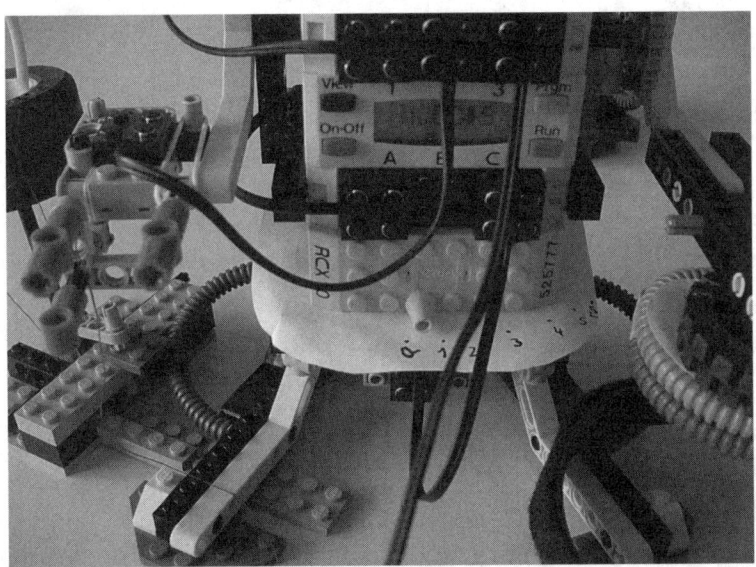

Figure 5.7 EventorBot showing incoming email.

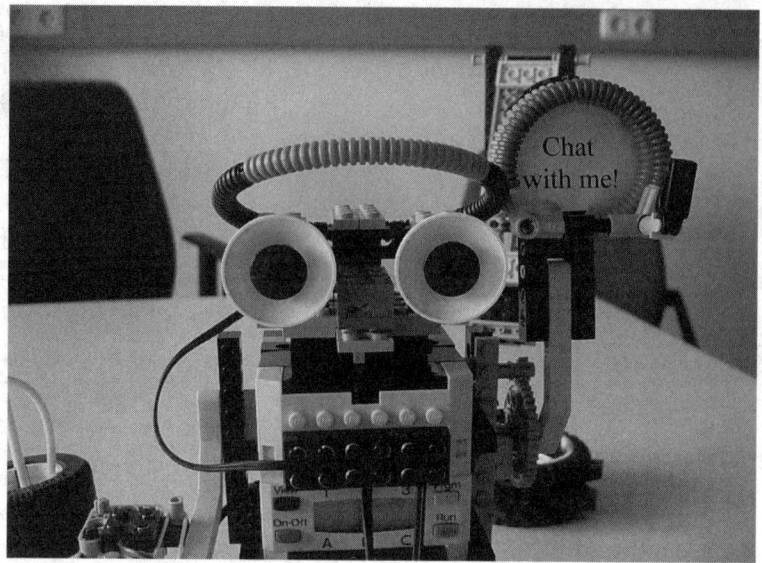

Figure 5.8 EventorBot showing chat request.

powerful and highly personalizable, but stationary [22]. In fact, PRAVTA allows users to send messages to the tickertape of the online users, to actively query various types of awareness information, to enter and update awareness information about themselves anytime and anywhere. Based on WAP [23], the PRAVTA prototype can be accessed from any mobile device that supports WAP, such as mobile phones, palmtops, and SmartPhones.

After the users have logged in, they can query information about the presence of other users, resulting in a table with all online users. They can check the availability of others and get a table with the current availability. Figure 5.9 shows a mobile phone and the login window and result of Who is online?, Check availability, and Check tasks.

For users who are in their everyday work environment the TOWER sensors can capture information about their presence, availability, activities, and so forth. Since PRAVTA can be used in any surrounding, users have the possibility to manually update their status.

5.4 User Involvement

The whole TOWER environment including the ambient interfaces was designed and developed in a participatory design fashion, and later evaluated in continuous field trials. In this section we report on the settings and methodology used.

The users of two very heterogeneous companies—Aixonix, and WSAtkins—were involved. The heterogeneity was intentional since we wanted to find out how generalizable the concepts, designs, and implementations were.

Aixonix is a small German company with a staff of 25 employees. The core business is mainly consulting and communication of complex information on the WWW, in intranets and extranets. Web-based systems for the transfer of technical

5.4 User Involvement

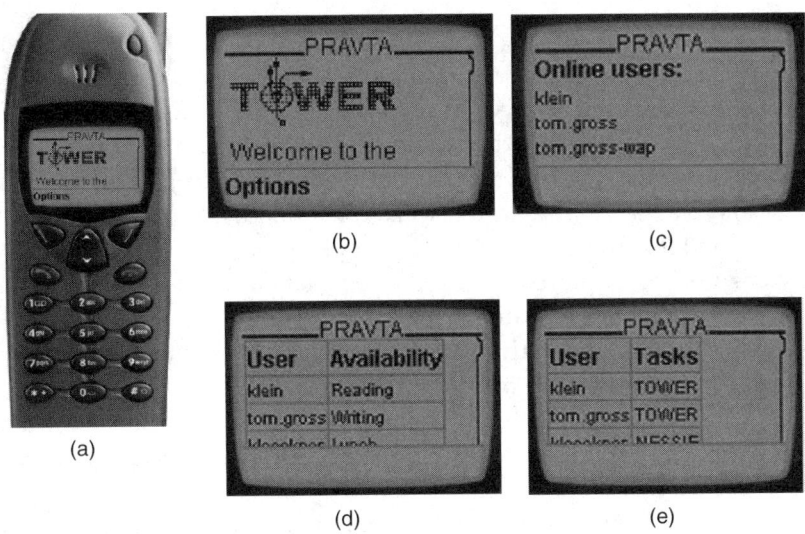

Figure 5.9 PRAVTA on a mobile phone: login window and the presentation of awareness information.

and scientific information are developed, and operated. The motivation to use the TOWER environment and the ambient interfaces was primarily to facilitate the cooperation with a major partner in the U.S. The management of Aixonix wanted to provide the employees at the German site with information about the German site; to provide the employees at the German site and at the partner's site with information about each other; and to augment its Web site with additional information for customers.

WSAtkins is a global company with several thousand employees in more than 90 offices throughout the United Kingdom, and many other offices in more than 25 countries. WSAtkins provides professional, technology-based consulting, and support services as a market leader. The motivation to use the TOWER environment and its ambient interfaces came from the wish to establish a new company culture of a one company—one team concept, after a worldwide reorganization. The technology should be used to facilitate this process by informing employees from different sites about each other, thereby improving group cohesion.

The users were involved in two levels: the outer cycle, and the inner cycle. In the outer cycle the future users at the two companies were involved, primarily in a user requirements analysis, and in a system evaluation as field trial. In the inner cycle members of the design team were involved, in the design and redesign of the system.

In the *outer cycle* the requirements were gathered in export workshops at the Aixonix company in Aachen, Germany, and at WSAtkins' main office in Epson, United Kingdom. Prospective users were interviewed about their work practices and gave feedback on early design ideas, mock-ups, and first ambient interface prototypes. Additionally, tours through the buildings and offices gave the interviewers impressions on the real work settings. Throughout the rest of the project the environment and the ambient interfaces were deployed incrementally and discussed in regular user workshops. These discussions did not only bring feedback on the presented designs, but typically also yielded additional requirements.

In the *inner circle* the design team itself started using the TOWER environment within the project. Whenever new features and ambient interfaces were ready to use, they were rapidly introduced at the different sites of the design team, for pre-testing. Additionally, several members of the design team had their own LEGO Mindstorms Robotics Invention System packages and produced their own AwareBots and defined their own mappings between the input and output of the AwareBots and the represented information and actions. Various application scenarios were developed and tried out to explore the usability of the features and to check whether they provide sufficient benefits for users to justify their introduction into work practice. Upon acceptance the features were demonstrated and discussed in the user workshops. Furthermore, we produced and analysed log files for the email correspondence (only with email headers such as sender, recipient, subject, date, and time), for the activities in shared workspaces, and for logins and logouts to and from the TOWER environment.

On a whole we primarily used qualitative measures of participatory design and evaluation, since they are often more adequate for empirical studies in cooperative settings; the evaluation of the log files was the only quantitative analysis.

5.5 Recommendations for the Design of Ambient Interfaces

In this section we will describe recommendations for the design of ambient interfaces. These recommendations originate from various sources, as follows.

- We started with some usability goals and usability principles for GUI-based systems from Preece [24], combined them with our experiences with the design, development, and use of ambient interfaces and specified some recommendations accordingly. Usability goals and usability principles can be clearly described, operationalized, and tested. Although these usability goals and usability principles cannot be applied per se, they can be tailored to our purposes.
- We were, furthermore, inspired by interaction design for consumer products from Bergman [25]. These design considerations have the advantage that they often aim at multimodal interfaces; compared with pure GUI-based systems consumer products offer many more facets of interaction than GUI-based systems and, therefore, fit very well to the challenges of ambient interfaces.
- We took some guidelines for the design of WAP applications for mobile devices from Weiss [26], and adapted it to the special requirements for a mobile client for the TOWER environment.
- We took some user experience goals from Preece [24], which are very general and are, therefore, suitable for any system with which users interact. In general, user experience goals describe how the interaction with the system feels like to the users. They are very important, but hard to operationalize and measure.

In the following text, we will present a range of recommendations for the design, development, and use of ambient interfaces.

5.5 Recommendations for the Design of Ambient Interfaces

Ambient interfaces should be effective. The effectiveness of a system refers to the quality of ambient interfaces in terms of how good they are doing what they are supposed to do. For instance, one basic goal for all ambient interfaces was to make use of users' peripheral awareness, not disturbing the user from performing the task the user has to perform. An ambient interface that is very noisy (e.g., produces loud sounds, produces abrupt and frightening movements) is, therefore, not effective.

Ambient interfaces should be efficient. Efficiency refers to the way an ambient interface supports users in carrying out their tasks. Ambient interfaces are, in general, designed to support the input of users and the presentation of information. So, the input should be very easy for the user, involving as little effort as possible. For instance, pressing the arm of RoboDeNiro logs a user into the TOWER system. This is a very easy action; in fact, the user does not have to type user name and password—so, the input is efficient. Efficiency of the output can, for instance, mean that the system only presents the amount of information that is needed to understand the underlying TOWER event. For instance, sophisticated movements of the EventorBot to simply tell the user that new email has arrived would not be efficient.

Ambient interfaces should be safe. Safe systems aim to protect the user from dangerous conditions and undesirable situations. This is a very important criterion for the success of any ambient interface. Due to the fact that ambient interfaces are, in general, multimodal—addressing various human senses at the same time—and that ambient interfaces are a relatively new area of research and development in computer science, the designers and developers of ambient interfaces have to be particularly careful with respect to safety. Several related ambient display prototypes have been described previously. For instance, the AROMA system, which produces heat close to the computer keyboard in order to let the user feel the awareness information, is a good example of a safe interface prototype: The heat is enough to be perceived by the user, but does not harm the user. Safety also includes unintended system behavior. So, for the AROMA system, it is very important that under no circumstances whatsoever should the heat get too high. Safe systems, furthermore, try to prevent users from making mistakes, and provide help in case the users have made errors.

Ambient interfaces should have good utility. Utility refers to the fact that the system provides the right kind of functionality so that users can do what they want and need to do. One particular challenge for the design of ambient interfaces is the fact that ambient interfaces are a new and fascinating area of computer science, and that, therefore, ambient interfaces are designed, developed, and used for purposes, for which they are not suitable. Ambient interfaces are best at complementing the PC and the GUI; they should not replace it. That is, ambient interfaces should be used as a means for easy input for simple actions or as subtle presentation for simple and unambiguous information. In general, ambient interfaces are not well suited for presenting complex information. The TOWER 3D multiuser world shows a complex picture of the group setting; it cannot and should not be replaced by ambient interfaces.

Ambient interfaces should be easy to learn and remember. Learnability refers to the fact that it should be easy for the users to learn a system; memorability refers to the fact that it should be easy for users to remember how to interact with the system. For ambient interfaces learnability and memorability are particular challenges for several reasons: (1) ambient interfaces are a novel way of input and output, so the users do not have experience yet; (2) ambient interfaces are often based on symbols for the input and output of information, these symbols have to be learned and memorized by the user; and (3) ambient interfaces do not (yet) offer help systems such as GUI-based systems (e.g., online help, tool tips). Basically, some help information can be printed and offered next to the ambient interface. However, the goal should be to design intuitively clear ways of interaction between the user and the ambient interface. For the TOWER users, often the same use of GUI-based indicators and ambient interfaces was very helpful in understanding the symbols of the ambient interfaces. For instance, a user working with an email program and at the same time using the EventorBot can see both new email in the inbox of the email application and the EventorBot rotating its body accordingly thereby adjusting the pointer on the number scale.

The functionality of ambient interfaces should be visible. Visibility means that at any time the system should clearly communicate to the user which choice he or she has and what the system is expecting. For instance, the arm of RoboDeNiro, with which users can perform login and logout operations, was designed and developed in a way that makes it obvious for the user that it can be pressed, in which directions it should be pressed and how long and how strong it should be pressed.

Ambient interfaces should give the users adequate feedback. Feedback refers to the information that the system presents to the users in order to tell them that the input from the user was received and analyzed properly, and that the corresponding actions have been performed or will be performed. For instance, the PRAVTA client provides the user with feedback after each action of a user: after the user has performed a login, the system writes: Logged in as <username>! [Home]; after the user has changed her activity and context, the system writes: Update successful! User does <action> in <context>. [Home]. The feedback for the login function of the RoboDeNiro confused some users: since pressing the arm can mean a login or a logout and as the RoboDeNiro does not provide information about the online-status of the respective user, some users complained that they were confused about their current state and about the action performed be RoboDeNiro upon pressing the arm.

Ambient interfaces should provide constraints. Constraining refers to a system behavior, where the system is aware of the current situation of the user and the possible next steps and appropriate actions of a user and consequently only offers options and functions that make sense in this specific situation. Constraints are an important criterion for huge and complex ambient displays and ambient interfaces (e.g., the ambientROOM described previously). For the ambient interfaces built in the TOWER project, constraints play a minor role, because the functionality of input and output is quite simple and easy to handle and does not require complex interaction or multistep interaction.

Ambient interfaces should provide an adequate mapping between controls and their effects. Mapping between controls and their effects is a particular challenge for a multimodal, distributed multiclient system such as TOWER. Ambient interfaces are, in general, used as pure input and output devices; and the effect of the input made via ambient interfaces often cannot be seen on the ambient interface per se, but rather on other clients of the TOWER system. For instance, a user logging in with the RoboDeNiro is immediately visible as avatar in the 3D multi-user TOWER world. So, the mapping often takes place between the controls on ambient interfaces and effects on other clients. For this reason, accurate mapping is very important. In fact, mapping was an important reason for using humanoid LEGO robots—users recognized their possibilities for interacting with the robots quite easily and could interpret the movements and gestures of the robots easily.

Ambient interfaces should provide consistent functionality. Consistency refers to the fact that the interfaces should offer similar operations and that similar control elements should be used for achieving similar tasks. Consistency was a challenge for the design and development of the ambient interfaces in TOWER. Since the ambient interfaces were developed in parallel at various sites, consistency was often based on good and similar intuition of the designers and developers. In order to provide consistent functionality for ambient interfaces, it is vital to take a top-down approach—that is, to plan the types of ambient interfaces, the functionality they should provide for input, the symbols they use for output, and the possible interaction in advance. Besides the general recommendations for ambient interfaces that are presented here, project-specific and detailed recommendations should be specified and followed.

Ambient interfaces should be adequate for the target domain. The target domain refers to the environment, in which the ambient interfaces are installed; the users, who will use the system; and the tasks that will be performed on the ambient interfaces. While the input and output for the PC have been standardized in a way so that a broad public can use them, several challenges remain for ambient interfaces. They are often related to the design recommendations described so far. Some examples are: the adequacy of the type of sound and its volume is highly dependent on the cultural backgrounds of the users and their ages; the unambiguity of the mapping of the controls to the respective action of the system is often dependent on the cultural backgrounds of the users; the overall acceptance of ambient interfaces is often dependent on many factors of the prospective users (e.g., some users may appreciate the use of heat, others may not; some users may appreciate wind in their face, others may not); and so forth.

Ambient interfaces should be designed and developed in a participatory environment. A participatory environment allows and stimulates users to contribute to the design of the ambient interfaces at very early stages. Participatory design has two positive consequences: Users can contribute their knowledge and ideas for the actual design of the ambient interfaces, and users have the subjective feeling that they can contribute and that their recommendations, ideas, and wishes are being heard. The experience from the TOWER project has clearly shown that users somehow liked the ambient interfaces designed and developed by others, but that they had much more fun when developing their own ambient interfaces. This is particularly the case when toolkits (e.g., LEGO) are used, with which users can create their own ambient interfaces easily and rapidly.

Ambient interfaces should provide a user experience that is satisfying, enjoyable, fun, entertaining, helpful, motivating, aesthetically pleasing, supportive of creativity, rewarding, and emotionally fulfilling.

5.6 Conclusions

A fundamental challenge for all awareness infrastructures lies in the fact that on the one hand users want to be well informed and want to have up-to-the-moment information and on the other hand users to not want to be constantly disrupted from their primary tasks. Furthermore, users want to get the information from the awareness infrastructure, but do not want to manually enter all the data. Ambient interfaces can help to reduce these trade-offs. Awareness information can be provided by indicators that are in the physical environment of the user and can be noticed by the user, but do not interrupt the user. And, sensors that are in the physical environment of the user such as movement sensors, noise sensors, and so forth, can capture awareness information. On a whole, ambient interfaces try to help the user bridge the gap between the electronic space in which they are working and the physical space they are really in. Thus, ambient interfaces will become an important aspect for the design of intelligent workplaces.

On a whole the prototypes described and the recommendations provided in this chapter only reflect on experiences made in the TOWER project. The whole area of ambient interfaces is rather new and to some extent still in its infancy. The results and recommendations provided should, therefore, be interpreted as a first attempt in combining knowledge from literature on design and usability of computing artifacts and initial experiences with the design, development, and use of actual ambient interfaces.

Further, new challenges concerning ambient interfaces for distributed workgroups arise from at least two perspectives: from the computer-supported cooperative work perspective challenges, such as privacy, lurking, and so forth, and from the human-centered design perspective challenges, such as novel metaphors, mapping, multimodality [27], hardware design, and so forth. Many research questions related to the design, development, use, and evaluation of ambient interfaces remain unanswered.

5.7 Acknowledgments

The research presented here was carried out and financed by the IST-10846 project TOWER, partly funded by the EC. I would like to thank all my colleagues from the TOWER team, and particularly the colleagues at Fraunhofer FIT—I had the pleasure and honor of working with them from 1999 to 2004. Furthermore, I would like to thank all colleagues, and the students from RWTH Aachen, Germany, and everybody inside and outside the TOWER project, who contributed to the design, development, and evaluation of the ambient interfaces.

References

[1] Beaudouin-Lafon, M. and Karsenty, A. "Transparency and Awareness in a Real-time Groupware System," *Proceedings of the ACM Symposium on User Interface Software and Technology—UIST'92* (Nov. 15–18, Monterey, CA). ACM, N.Y., 1992. pp. 171–180.

[2] Ishii, H., "TeamWorkStation: Towards a Seamless Shared Workspace," *Proceedings of the Conference on Computer-Supported Cooperative Work—CSCW'90* (Oct. 7–10, Los Angeles, CA). ACM, N.Y., 1990. pp. 13–26.

[3] Ackerman, M.S. and Starr, B., "Social Activity Indicators for Groupware," *IEEE Computer* Vol. 29, No. 6 (June 1996). pp. 37–42.

[4] Gutwin, C., Greenberg, S. and Roseman, M., "Workspace Awareness in Real-Time Distributed Groupware: Framework, Widgets, and Evaluation," *Proceedings of the Conference on Human-Computer Interaction: People and Computers—HCI'96* (Aug. 20–23, London, UK). Springer-Verlag, Heidelberg, 1996. pp. 281–298.

[5] Fitzpatrick, G., Mansfield, T., Kaplan, S., Arnold, D., Phelps, T. and Segall, B., "Augmenting the Workaday World with Elvin," *Proceedings of the Sixth European Conference on Computer-Supported Cooperative Work—ECSCW'99* (Sept. 12-16, Copenhagen, Denmark). Kluwer Academic Publishers, Dortrecht, NL, 1999. pp. 431–450.

[6] Patterson, J.F., Day, M. and Kucan, J., "Notification Servers for Synchronous Groupware," *Proceedings of the ACM 1996 Conference on Computer-Supported Cooperative Work—CSCW'96* (Nov. 16–20, Boston, MA). ACM, N.Y., 1996. pp. 122–129.

[7] Merriam-Webster, I. Merriam-Webster Online. http://www.m-w.com/, 2004. (Accessed 7/2/2005).

[8] Wisneski, C., et al., "Ambient Displays: Turning Architectural Space into an Interface between People and Digital Information," *Proceedings of the First International Workshop on Cooperative Buildings: Integrating Information, Organization, and Architecture Workshop—CoBuild'98* (Feb. 25–26, Darmstadt, Germany). Springer-Verlag, Heidelberg, 1998.

[9] Weiser, M. and Brown, J.S., "Designing Calm Technology," http://www.ubiq.com/hypertext/weiser/ calmtech/calmtech.htm, 1996. (Accessed 17/2/2005).

[10] Pedersen, E.R. and Sokoler, T., "AROMA: Abstract Representation of Presence Supporting Mutual Awareness," *Proceedings of the Conference on Human Factors in Computing Systems—CHI'97* (Mar. 22–27, Atlanta, GA). ACM, N.Y., 1997. pp. 51–58.

[11] Heiner, J.M., Hudson, S.E. and Tanaka, K., "The Information Percolator: Ambient Information Display in a Decorative Object," *Proceedings of the ACM Symposium on User Interface Software and Technology—UIST'99* (Nov. 7–12, Asheville, NC). ACM, N.Y., 1999. pp. 141–148.

[12] Gaver, W., "Sound Support for Collaboration," *Proceedings of the Second European Conference on Computer-Supported Cooperative Work—ECSCW'91* (Sept. 24–27, Amsterdam, NL). Kluwer Academic Publishers, Dortrecht, NL, 1991. pp. 293–308.

[13] Mynatt, E.D., Back, M. and Want, R., "Designing Audio Aura," *Proceedings of the Conference on Human Factors in Computing Systems—CHI'98* (Apr. 18–23, Los Angeles, CA). ACM, N.Y., 1998. pp. 566–573.

[14] Prinz, W., "NESSIE: An Awareness Environment for Cooperative Settings," *Proceedings of the Sixth European Conference on Computer-Supported Cooperative Work—ECSCW'99* (Sept. 12–16, Copenhagen, Denmark). Kluwer Academic Publishers, Dortrecht, NL, 1999. pp. 391–410.

[15] Bentley, R., et al., "Basic Support for Cooperative Work on the World-Wide Web" *International Journal of Human Computer Studies: Special Issue on Novel Applications of the WWW* (Spring 1997).

[16] Gross, T., Wirsam, W. and Graether, W., "AwarenessMaps: Visualising Awareness in Shared Workspaces," *Extended Abstracts of the Conference on Human Factors in Computing Systems—CHI 2003* (Apr. 5–10, Fort Lauderdale, Florida). ACM, N.Y., 2003. pp. 784–785.

[17] Prinz, W., et al., "Presenting Activity Information in an Inhabited Information Space," *Inhabited Information Spaces: Living with Your Data,*. Springer-Verlag, N.Y., 2003. pp. 181–208.

[18] Gross, T. and Klemke, R., "Context Modelling for Information Retrieval: Requirements and Approaches," *IADIS International Journal on WWW/Internet*, Vol. 1, No. 1 (June 2003). pp. 29–42.

[19] Gross, T. and Prinz, W., "Modelling Shared Contexts in Cooperative Environments: Concept, Implementation, and Evaluation," *Computer Supported Cooperative Work: The Journal of Collaborative Computing*, Vol. 13, Nos.3, 4 (Aug. 2004). pp. 283–303.

[20] Gross, T., "Ambient Interfaces in a Web-Based Theatre of Work," *Proceedings of the Tenth Euromicro Workshop on Parallel, Distributed, and Network-Based Processing—PDP 2002* (Jan. 9–11, Gran Canaria, Spain). IEEE Computer Society Press, Los Alamitos, CA, 2002. pp. 55–62.

[21] The LEGO Group. LEGO MINDSTORMS. http://mindstorms.lego.com/, 2005. (Accessed 17/2/2005).

[22] Gross, T., "Towards Ubiquitous Awareness: The PRAVTA Prototype," *Ninth Euromicro Workshop on Parallel and Distributed Processing—PDP 2001* (Feb. 7–9, Mantova, Italy). IEEE Computer Society Press, Los Alamitos, CA, 2001. pp. 139–146.

[23] Open Mobile Alliance, "OMA Technical Section—Affiliates—Wireless Application Protocol Downloads," http://www.openmobilealliance.org/tech/affiliates/ wap/wapindex.html, 2005. (Accessed 17/2/2005).

[24] Preece, J., Rogers, Y. and Sharp, H., *Interaction Design: Beyond Human-Computer Interaction*, Wiley, N.Y., 2002.

[25] Bergman, E., *Information Appliances and Beyond: Interaction Design for Customer Products*, Academic Press, London, UK, 2000.

[26] Weiss, S., *Handheld Usability*, Wiley, N.Y., 2002.

[27] Gross, T., "Universal Access to Groupware with Multimodal Interfaces," *Proceedings of the Second International Conference on Universal Access in Human-Computer Interaction: Inclusive Design in the Information Society* (June 22–27, Crete, Greece). Lawrence Erlbaum, Hillsdale, NJ, 2003. pp. 1108–1112.

CHAPTER 6
Expanding the Role of Wearable Computing in Business Transformation and Living

Chandra Narayanaswami

6.1 Introduction

A wearable computer is a computing device that is always available to the user and is typically worn on the body or integrated into or attached to clothes. The user must be able to get information to and from the device instantaneously and in a diverse set of situations. Back in the 1980s, researchers in wearable computing realized that the potential benefit of having a powerful computing device all the time instead of just at one's desk was huge. Early wearable computers were hardly wearable—they were bulky and weighed several pounds and drew derisive looks. They were typically worn on the user's back and included large head-mounted displays. A few brave and dedicated pioneers, however, marched along and advanced the field to where it is now. Today, more than a billion people carry a cell phone with them all the time, typically in their pocket or handbag. Not many realize that their multifunction cell phone is indeed a wearable computer with a processor whose capabilities exceed even those of a personal computer from 1995. Some of them even have a 640 x 480 pixel display, a 4-GB disk drive, and a digital camera, while weighing less than half a pound.

Early wearable computers did not include any wireless communication means. However, this did not preclude the community from developing several potential applications that cell phone manufacturers have recently adopted. For example, use of accelerometers for activity detection and user interfaces in some new cell phones are a direct consequence of research performed by the wearable computing community. In today's landscape, wireless access is available in most places. Recently, several airlines such as Lufthansa, ANA, JAL, Korean Air, and so on, provide wireless access on transatlantic and transpacific flights. With ubiquitous connectivity and the development of the World Wide Web, wearable computers are no longer islands and therefore are able to present the user with up-to-date information on a large variety of subjects.

Over time, wearable computers will not be referred to as computers, just as the early Sony Walkman was not called a wearable electric motor. By virtue of being always available and connected, wearable computers have significant potential to improve business processes and quality of life. As a proof-point, early adopters of wearable computing have included armed forces personnel, repair technicians, and people with medical needs. As described in an earlier chapter, in an ambient intelligent environment people are surrounded by networks of embedded intelligent devices that provide ubiquitous information. The term ambient intelligence[1] (or AmI) reflects this tendency, gathering best results from three key technologies: ubiquitous computing, ubiquitous communication, and intelligent user friendly interfaces. The emphasis of AmI is on greater user-friendliness, more efficient services support, user empowerment and support for human interactions. Clearly wearable computers can derive value from the intelligence in the environment and make it available to the user. In this chapter, we will explore the issues that need to be addressed in wearable computing to impact a broad audience and the progress that is being made.

6.2 Wearable Computers—History and Present Status

People have been using several special-purpose wearable electronic devices for at least two decades now. One of the earliest is the hearing aid. Pacemakers have been implanted in hundreds of thousands of people to stabilize irregular heartbeats. Pedometers can be bought for less than five dollars at a supermarket these days. Most recently, noise-canceling headsets have become popular among frequent fliers and among operators of lawn machinery, such as mowers, tractors and snow blowers. These devices started more simply as specialized electronic circuits. More recently, these devices have incorporated standard computer architecture, thanks to cheap availability of PIC microcontrollers and processors. Bluetooth headsets are the latest in this trend. Some of these devices do not conform to the commonly understood definition of a wearable computer since they perform only one task.

Since wearable computers have size limitations, innovative displays and keyboards have been developed. Miniature eyepiece displays which can be attached to a pair of glasses, come in full color with a resolution of 640 x 480 pixels or higher. With such displays, content does not have to be designed specially for the wearable computer. For example, a MicroOptical MD-6 eyeglass-mounted display allows cardiovascular surgeons to read the patients vital signs in real time while manipulating and lifting a patient's heart during surgery. Soldiers can view maps and other information without carrying bulky flat panel displays. Construction workers and outdoor technicians can read these displays in bright sunlight, thus saving trips to flat panel displays housed in trucks. Mitsubishi's Scopo wearable LCD display can be attached to cell phones and PDAs to view high-resolution content. In contrast to some wearable displays, the Scopo is positioned below the eye and does not obstruct normal vision. Virtual retinal displays which directly scan an image on the viewer's retina using tiny laser beams are also available. Keyboards such as the Twiddler allow users to wear a keyboard on one hand and type text while on the move. Glove- or ring-based data entry mechanisms have also been proposed but have had limited success in the mass market because they are rather inconvenient for daily use. Voice recognition technology has improved to the point where it can be acceptable for many tasks. However, one drawback of voice based input is that errors are generally hard to correct. Other drawbacks include the fact that voice-based input is not socially acceptable in public places and also limits the user's privacy.

Several academic and industrial institutions and defense organization have prototyped many generations of wearable computers over the last fifteen years. Some notable efforts include those at Carnegie Mellon University [2–4], Oregon Graduate Institute [5, 6], MIT [7–9], ETH Zurich [10, 11], Hewlett Packard Labs [12], and the IBM Research Division [13]. A brief but informative history of wearable computing is provided in [14]. We now describe a few of these efforts in more detail.

One of the earliest wearable computers was built by Ed Thorpe and Claude Shannon in 1961 at MIT to predict roulette wheels [15]. This analog computer included four pushbuttons and was about the size of a cigarette pack. This computer delivered tones via radio to the bettor's hearing aid in response to pushbutton sequences that indicated the speed of the spinning roulette wheel. The first computer-based head-mounted display was created by Sutherland in 1966 using CRTs that were mounted beside the ears and silvered mirrors to reflect the images to the user's eyes. Though the system was rather bulky it demonstrated what was possible in future.

The CMU wearable computing group has built several wearable computers since the early 1990s [2–4]. One of their early wearable computers, called VuMan 3, consisted of a small display monitor that is worn on the head and connects to a dial and selection buttons worn on the waist. VuMan 3 was designed for marines as they inspected amphibious tractors. Jobs that needed two people, with one performing the job and another reading out instructions, could now be done with one person. The small size of the wearable computer allowed a marine to be able to squeeze into tighter spots and inspect equipment. As the marine follows a checklist on the monitor (positioned just below the eye), he or she turns the dial and pushes buttons to report the equipment's status. It also avoided the extra step of transcribing the equipment status into a computer later. By making the entire process more efficient and by eliminating one marine in the task, the VuMan 3 reduced inspection time by 70 %. Other wearable computers built at CMU were addressed toward users in the ship building industry, mobile workers, and medical personnel. They measured and factored the amount of attention the user had to pay toward his task versus using the wearable computer. For example, the input dial on the VuMan was redesigned several times to reduce the amount of user attention required. Their efforts also studied various input and output modalities for wearable computers, with special attention to speech-based interfaces and power consumption. One of their recent efforts, Spot, is extensively instrumented for measuring power consumption.

Researchers at the Oregon Graduate Institute have built a series of wearable computers, including NetMan and MediWear [5, 6]. NetMan is a wearable groupware system designed to enhance the communication and cooperation of highly mobile network technicians. It provides technicians in the field with the capabilities for real-time audioconferencing, transmission of video images back to the office, and context-sensitive access to a shared notebook. An infrared location tracking device allows for the automatic retrieval of notebook entries depending on the user's current location. MediWear is a mobile computer that is taken home by a patient in order to monitor vital functions that could otherwise not be monitored outside a hospital. It can also inform paramedics outside the hospital. The paramedic wearable combines different functions that help the paramedic to accomplish typical tasks more rapidly.

The MIThril hardware platform [7] combines body-worn computation, sensing, and networking, that can be integrated into clothes. The MIThril body bus connects several peripherals and sensors (devices that do not require Ethernet networking) to the BSEV core. The MIThril Body Network and MIThril Body Bus simplify the physical on-body networking problem by providing a reliable single-cable power and data connection among all devices on the body. Some sensors that attach to the MIThril Body Bus include a three-axis accelerometer, an IR Active tag reader, a USB microphone and headphone, and a USB CCD camera. More recently, the MIThril researchers have started using the system for proactive healthcare monitoring—LiveNet [8].

The WearARM [11 computer from ETH Zurich and MIT is a high-performance system that uses general-purpose CPUs, DSPs, and custom circuitry. It consists of four modules namely, the core, basic, periphery and vision. The system is heterogeneous and provides user reconfigurable system architecture with advanced power management features. Mechanically flexible electronic packaging technology is used to make it comfortable to wear. The core module includes the CPU and memory as well as a graphics interface. The basic module supplies power to the core module and some hardware to allow compact flash interfaces. The periphery module includes networking interfaces and audio circuitry. The vision module is designed for image analysis and gesture recognition. The Q Belt-Integrated Computer (QBIC) [], also from ETH Zurich, uses a belt form factor to house the hardware components. A belt buckle has sufficient volume to house a powerful computer with room for expansion along the belt. The core of the system is an Intel XScale family processor: the PXA263B1C400. Main features of this processor include scalable clock frequency up to 400 MHz, dynamic core voltage scheduling and several power-save modes. A total of 256 MB of additional RAM is included in the main board for faster audio and video processing. The ETH team is using their low-power modular core and QBIC for context-aware computing. One such application is the student tracker that monitors the daily activities of a student. This includes the recognition of social interaction (e.g., discussions, lecture, meetings, and conversation), physical activity, eating and drinking habits, and operating machines in a coffee corner (See Figure 6.1).

Built in 1998, the VisionPad wearable computer prototype from IBM Research, shown in Figure 6.2, included a 233 Mhz Intel processor, 64 M of RAM and a 340 MB IBM MicroDrive. It ran Windows 98 and a set of applications including IBM ViaVoice 1998. Its lithium-ion battery lasted about 1.5-2 hours. It included USB, IrDA, and CF type II interfaces. The whole unit weighed only 299g. The head mounted display had a resolution of 320 x 240 pixels with 256 gray levels. It included an ear phone and weighed 50g. The device was controlled by a TrackPoint and a microphone.

Also around 1998, a PDA-sized flexible research platform, called Itsy [12], was developed at Compaq Computer Corporation's Western Research Laboratory (WRL), now part of HP Labs. Its aim was to enable hardware and software research

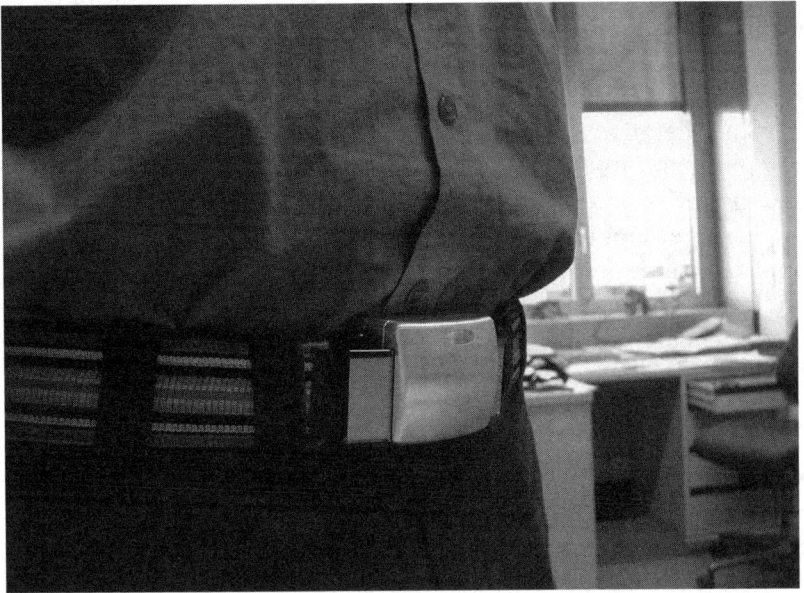

Figure 6.1 QBIC wearable computer prototype from ETH, 2002. (Image Courtesy of Wearable Computing Lab, ETH Zurich.)

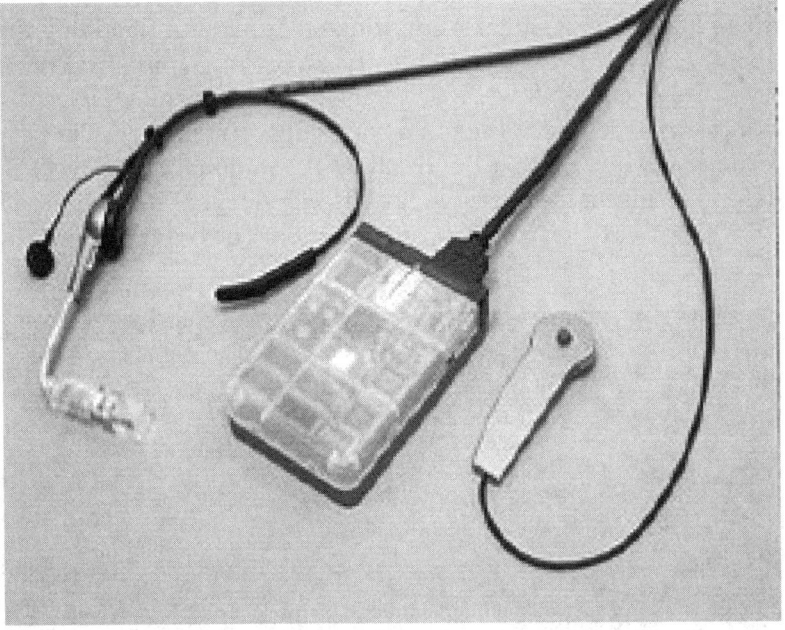

Figure 6.2 IBM research wearable computer prototype, 1998.

in pocket computing, including low-power hardware, power management, operating systems, wireless networking, user interfaces, and applications. It ran the Linux operating system and had enough computing power to run a Java Virtual Machine, show MPEG-1 videos and perform continuous speech recognition. It used a Strong-ARM SA-1100 processor whose frequency ranged from 59 MHz-191 MHz and incorporated 32 MB of Flash memory and 32 MB of RAM and had a 320 x 200 pixel display with 15 gray levels. It also showed the potential of using accelerometers for interacting with the device.

The MetaPad [16] project at IBM Research realized a new way to partition the computer and its peripherals. The MetaPad core contains the processor, memory, and disk drive and includes several connectors. A standard PC operating system runs on this core. This core can attach to a portable display sleeve which makes it ideal as a portable unit that can run standard PC applications. The display is touch sensitive and can be operated with a stylus. The core can also be attached to a docking station which can provide PC peripherals such as a full-size keyboard and display. This configuration can be utilized when the user is not mobile. A nice feature of the MetaPad approach is that the user's computing session, including running applications and windows can be preserved across the different attachment points. Some companies such as Antelope Technologies have introduced a modular computing core (MCC) that uses the same principles. The Sony Vaio U70, shown in Figure 6.3, is an extremely portable computer that belongs in the same category.

The SoulPad [17] prototype from IBM Research uses a three-layer software stack to enable a paradigm of mobile computing, where a user can suspend the computing environment on one PC and resume it on another PC that he may have never seen before. The PC boots an autoconfiguring operating system from the SoulPad, starts a virtual machine monitor, and resumes a suspended virtual machine that has the user's entire personal computing environment, including the user's files, user's operating system, installed applications, desktop configuration, all running applications and open windows. Essentially, SoulPad enables a user to hibernate a PC session to a pocket form-factor device and carry the device to some other PC and resume a session on that PC. SoulPad has minimal dependencies on PCs that can be used to resume a user session. Specifically, PCs are not required to be network connected, nor have any preinstalled software. The only requirement is the support of a

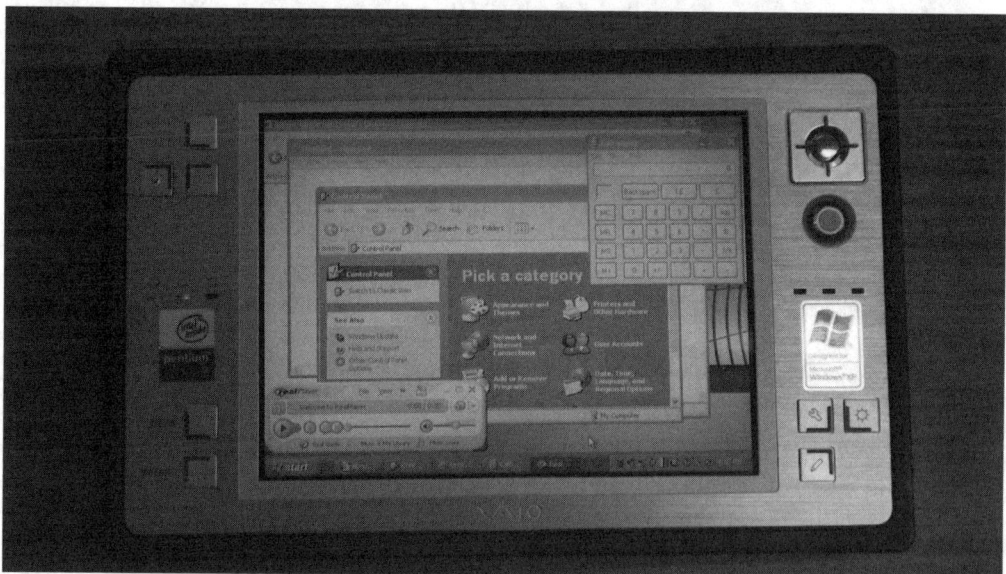

Figure 6.3 Sony Vaio U71P portable computer.

high-speed local connection to a SoulPad device for an acceptable suspend and resume times and acceptable runtime performance.

The IBM Linux Watch (or, WatchPad) [18–20] is a wrist mounted computer with short range wireless connectivity. Three versions of watches were prototyped. The first version, introduced in August of 2000, included an 18 MHz ARM7 processor, 8 MB of flash memory and 8 MB of DRAM. Short-range wireless connectivity based on IrDA and Bluetooth™ was also included. A 120 × 96 pixel touch-sensitive display and a jog dial were used to interact with the watch. The watch ran the Linux 2.2.1 kernel and included X11 graphics. The second version (shown in Figure 6.4) introduced an organic LED display, jointly developed with eMagin Corporation, with a resolution of 740 pixels per inch. This display included individual control for turning on each pixel, thereby allowing the application designer to reduce the power consumed by the display by turning on fewer pixels. This version did not have the Bluetooth hardware. The third version was jointly developed with Citizen Watch Company and had 16 MB of flash memory instead on 8 MB in the earlier versions. The user interface is composed of a 320 × 240 touch-sensitive LCD display, a jog dial, and three buttons. A fingerprint sensor, a two-axis accelerometer, and a vibration device provide additional flexibility for input and output. Several companies such as Seiko, Casio, Timex, Fossil, and Samsung, and so on, have built multifunction watches over the last decade.

The personal mobile hub (PMH, see) prototype [21] from IBM serves as a gateway between body-worn sensors, wristwatch computers, and the Internet. First demonstrated in 2002, the IBM PMH is built from the Linux watch platform, and includes both long, and short, range wireless connectivity. This three-tier configuration allows the sensor designers to focus on ergonomics, aesthetics, and data accuracy instead of attempting to pack end-to-end functionality in the sensor unit.

Figure 6.4 IBM's Linux watch prototype—2001. (Display jointly developed with eMagin Corporation.)

Figure 6.5 IBM's personal mobile hub prototype.

Clearly, the personal mobile hub can be shared between multiple body-worn devices and allows the user to add more devices to his constellation over time. The PMH also includes storage and software to collect data and analyze it to detect patterns. The rules for triggering alerts and notifications can be upgraded even after the device is purchased. One possible use for the PMH is shown in the IBM mHealth toolkit [21] which allows body-worn biometric sensors to relay information to healthcare providers.

Xybernaut Corporation has sold several wearable computers to the transportation and aerospace, retail, healthcare, education, telecommunications and media, hospitality and government and military sectors.

Some shoes have recently started incorporating computers to change the stiffness of the sole. For example, a shoe sold by Adidas uses a sensor, a microprocessor, and a motorized cable system to automatically adjust its cushioning. The sensor under the heel measures compression and decides whether the shoe is too soft or firm. That data is sent to the microprocessor and, while the shoe is in the air, the cable adjusts the heel cushion.

In summary, wearable computers have been built in several form factors. Each form factor may have some intrinsic advantages and disadvantages. The right computer for a specialized task depends on the details of the task. We believe popular form factors for wearable computers will include a cell phone, a wristwatch or a bracelet, a pendant, jewelry, a device that can be worn on a belt, a unit that can be placed into eyeglasses or earphones, shoes, and helmets.

6.3 Wearable Computing Applications

Applications of wearable computing span several domains, including defense, healthcare, field force automation, customer relationship management, and entertainment. We now briefly describe a subset of applications from these domains.

Defense and security personnel were early adopters of wearable computers, since they provided them with real-time information of the targets, the enemy, terrain maps, locations of other soldiers, and information about first-aid procedures. In addition, they allowed soldiers to capture images and relay them, and so on. Sophisticated infrared heat-sensing night-vision goggles and head-mounted displays provide valuable imagery at night and around corners. The health, location, and physical condition of soldiers can be constantly monitored and relayed to command centers. The evolution of army wearable computers is described in [22].

Advances in wearable computing technology and sensors are beginning to impact administration of healthcare in both inpatient and outpatient settings. The increasing costs of inpatient hospital stays combined with the growing evidence that faster recovery and immunity from infections is possible when the patient is back at home and ambulating, is also motivating the deployment of patient monitoring applications.

Pacemakers have been used for several decades to adjust heart rhythms and help patients live a more satisfying life. Several wearable devices that can measure and track pulse rates, EKGs, and so on, are also available now at modest costs for casual use. Some of them can be connected to fitness equipment and to personal computers. More recently, efforts are being addressed at not only measuring health signs but also on analyzing the measured data and then relaying the measured data to healthcare personnel for immediate consultation or action. Typical areas include glucose monitoring, blood pressure monitoring, coronary event prediction, pill regimen compliance, and so on. Ingestible wearable devices that include tiny cameras and wireless communication means have been developed for imaging of the digestive track. Another area where wearable computing technology is being used is in drug trials. By being able to monitor the effects of the drug on the patient in real time and in less controlled settings, researchers are able to improve the process and recruit more subjects for trials while reducing the cost of the trials. In addition, wearable monitors embedded in clothes allow more direct and accurate measurement of some parameters that have not been possible earlier. Wearable RFID tags for patient identification and storing electronic health records are now technically feasible. In some countries, implantable radio tags are already being used for identification. Other applications in the medical area include wearable pendants for nurses that include wireless notification, messaging, and voice communication. For example, the Vocera Communications Badge is a wearable device that can be clipped to a hospital garment or worn around the neck. The Vocera Communications Badge weighs less than two ounces and is controlled using natural spoken language. Its voice-based interfaces enable instant two-way voice conversation without the need to remember a phone number or manipulate a handset. Text messages and alerts can be sent to the LCD screen. In the near future, the wearable units could show images of patients and further improve the efficiency of healthcare personnel. The medical community was one of the earliest to adopt pager technology when it was invented. The same is likely with wearable computing. Drishti [23], a prototype wearable system, which includes a GPS and audio output, helps blind people navigate streets by correlating the GPS data with spatial databases that include walkways, benches, poles, and other obstacles and providing just in time information to the user. Brain-computer interfaces which allow people with certain disabilities to operate computers via brain signals are being prototyped.

The nature of medical applications dictates that they need to be highly reliable. Rebooting a wearable device in the middle of a monitoring application may be life threatening. This implies that applications must be resilient against loss of connectivity, failure to deliver messages, loss of battery, and so on. Fallback options have to be designed from the beginning.

Repair technicians in certain domains have been early adopters of wearable computing. Bell Canada, for example, reported that its utility workers were able to save about 50 minutes per day on average by being able to avoid going down from poles to trucks to use computers. Typically, workers are able to bring up repair manuals while they are still in front of complex equipment; thus, are able to better correlate the data from the manuals with the real equipment. Wearable computers have been used in the aircraft maintenance industry as well [24]. Workers at warehouses typically wear bar code scanners to help manage inventory more efficiently, because the fingers are left free, the scanner is exceptionally well suited for package handling intensive activities such as order picking, stock pulling, and put-away. Police officers and armed forces personnel were early adopters of wearable computers.

An application of wearable computing in customer relationship management is denoted as queue busting. Rental car companies have used wearable computers effectively to reduce the time it takes to drop off a rental car at the airport. Some airlines use wearable computers to check in passengers and have taken the check-in counter to where the passenger is. Personnel in warehouses use wearable computers integrated with barcode scanners to take inventory and fetch orders for customers. Employees at UPS, FedEx, and Pitney Bowes have used wearable computers to help sort, track, and deliver packages.

The remembrance agent [9, 25] is another application of wearable computing. The user's wearable computer provides information that is relevant to what the user may be reading or writing—such as the names of relevant papers to cite, email from reviewers, and so on. An often quoted extension is the following—the agent helps the user with the name of the person the user is talking to, the date and time of the last conversation with that person, and so on.

Wearable computers in the form of portable media players are already extremely popular and expected to grow when services such as television broadcasts are available on portable players. Efforts such as Digital Multimedia Broadcasting (DMB) and WiBro (Portable Internet) in Korea could serve as early test-beds.

Researchers have prototyped wearable computers that provide real-time language translation, interpret signs in foreign languages by capturing images and performing image matching, recognize leaf families while on a trek, identify bird species on a hike, and so on. A class of applications that brings wearable computing to leisure activities is emerging.

6.4 Factors Limiting the Impact of Wearable Computers

We now look at some of the factors that have limited the growth of wearable computing and how they may be addressed to extend the reach of wearable computing.

Wearable computers have typically been built in limited quantities and have not been able to exploit the benefits of scale. Typically, when devices are produced in small quantities, the nonrecurring costs have to be amortized over fewer devices. For example, building special molds for devices add significant costs for runs that are less than several thousand units. As a result, today a wearable computer is significantly more expensive than a desktop PC. The cost of a Sony Vaio U71P is almost three times that of a desktop PC. At such prices, only the researchers and then early adopters were able to enjoy the benefits of wearable computing. Only recently, cell phones have become true wearable computers and have been able to bring wearable computing to larger audiences. We believe that explosive growth of personal media players, digital cameras, and gaming devices will also increase the need for better displays at affordable prices. This in turn will make wearable computers more affordable and begin the creation of a virtuous cycle, where lower costs lead to increased adoption and increasing market size leading to creation of more applications and vice versa.

Though cell phones are powerful and ubiquitous, they generally tend to have a closed platform model. The service providers want to provide all the functions and services that are available from the phone. The device manufacturers are typically relegated to just producing hardware. Cell phone service providers are typically not able to provide a rich set of applications that a more open ecosystem will be able to provide. Cellular service providers typically do not have content and data for driving the applications. For example, though cellular service providers may be able to locate a user's device, they do not have many of the functions that Google provides. On the other hand, content and information providers do not have access to a user's physical location, since the cellular service provider guards that information carefully. As a result of this battle, several services that are technically deployable are not being made available to the end user. For example, though cellular phones include high-resolution digital cameras, cellular service providers typically want users to use their networks to transfer the pictures from the camera to other locations. The cell phone ecosystem is in sharp contrast with the PC ecosystem, where the device manufacturers, base software providers, third-party application writers and content providers are all able to create profitable businesses. The introduction of city-wide wireless networks through WiMAX, Wi-Fi, and so on, may change the balance in this ecosystem and make it lean closer to the PC ecosystem.

The power consumption and power management problem have also limited the impact of wearable computers. Users typically want to operate in 8-hour, 24-hour, or weekly patterns and expect the battery to last for such durations. In spite of advances in hardware and software technologies, the battery in a typical laptop computer today lasts only a few hours and seldom meets the expectations of users. Achieving a consistent eight hour battery life will cover large segments of users. For example, students carrying all their course material, text books, and so on, on their laptop computer can leave the power adapter behind for the day. Ubiquitous access to AC power is an interim and pragmatic solution and is being provided in newer planes, trains, automobiles, conference rooms, airport lounges, class rooms, and auditoriums. Depending on how the dynamics of these factors evolve, users of wearable computers may use the batteries only when they are physically mobile, if AC power is available at a large fraction of sites where they operate during the course of a day.

Traditional wearable computers have been bulky and are often perceived as being geeky. Head- and eye-mounted displays are not perceived as desirable gadgets, although Bluetooth headsets and music players are already being worn in large numbers and are not considered geeky. The advent of portable media players may make eye-mounted displays more popular. In order to be successful, we think that a wearable computer must be able to switch from a less socially acceptable mode to a more socially acceptable mode very easily. For example, when a user does not want to use an eye-mounted display, the wearable computer must be able to switch to a direct-view display instantaneously. Similarly, when the user wants a higher-resolution display, the applications should be able to adjust to varying display resolutions. Modular wearable computers may allow the user to switch a form factor easily. The MetaPad and SoulPad prototypes from IBM Research allow a user to transfer a session from one form factor to another while preserving the complete session and not requiring a reboot. Thus, a user can use a notepad-size wearable computer at one time and switch to a more socially acceptable pocket-size wearable computer at other times.

As the capabilities of wearable computers increase, the complexity of interacting with the wearable computer also increases. User interface paradigms that are popular in traditional desktop systems do not map well to wearable computers. The wearable computer, as well as its user interface, needs to be carefully designed, taking into consideration the types of activities the user engage in and the kinds of services the wearable will provide to the user.

The flexibility of wearable computers comes at a price. The size constraints on a truly wearable device permit the inclusion of only a small set of user interface controls. Users also find it difficult to remember which combinations or sequence of button pushes are required to accomplish the functions they are interested in. Other input modes, such as voice, may not be appropriate in many situations for wearable computers, since the user is more likely to be in a social setting. The user may also be on the move and the user's environment may be too noisy for voice input. Size limitations on wearable computers also translate to a limited amount of display area. Any information that is displayed on the small display needs to take into account the user's visual acuity. Though a user may have the option of positioning the device closer or farther away from his or her eye to make reading the display easier, such positioning may not always be desirable or even possible depending on the task the user is engaged in. Information presented on the display of the wearable computer needs to be designed differently depending on whether the user seeks to obtain that information at a glance or is willing to devote more attention and willing to consciously look at the display for a longer duration (at a stare).

Security issues related to wearable computing span several areas. For example, wearable computers are likely to be used in public spaces over wireless networks. Therefore, encryption of data transmissions is mandatory for business usage. Since mobile devices can also store several tens of gigabytes of data, it is essential to encrypt all data on the device and to protect against loss or theft of the device. Wearable computers are also more likely to execute software services that are delivered dynamically. Safeguards that test the origin and authenticity of such software services become critical. Since wearable computers either offer small direct-view displays or private head-mounted displays, the risk of unauthorized screen capture is less of an issue.

Some challenges of wearable computing are also covered in [26–28] and may be of interest to the reader.

6.5 Factors Providing Positive Feedback Loop

Several recent developments are encouraging for wearable computing. They include system on a chip architectures, ubiquitous 802.11 connectivity, portable storage, imaging sensors, ultrawide band wireless technology, global positioning systems, voice over Internet protocol, low-cost power-efficient sensors, organic displays, energy harvesting techniques, early middleware for basic operations, virtualization software, and flexible displays.

For an inordinately long period of time, processor architecture was geared toward higher performance at the expense of power consumption. For example, speculative execution of instruction streams improved performance but resulted in an increase in power consumption. Architects also paid more attention to active power consumption—that is, when the system was operating full throttle and paid less attention to conserving power when the system was more lightly loaded or idle. More recently, it has become clear [29] that passive power consumption is important and needs to be addressed. Hardware technologies such as multiple voltage islands, and so on, allow a balance between active and passive power. Processors with multiple processing cores and hierarchical cores are being developed to address the power consumption problem [30–33]. Hierarchical processing cores with differing power performance characteristics allow the use of the core that is right for the task. However, at this stage, software technologies that are necessary to exploit such hardware features are not widely available.

Since wireless subsystems consume a large fraction of total power consumed by a connected wearable computer, a series of wireless subsystems comprizing of RFID, Bluetooth, ZigBee, Wi-Fi, WiMAX, and a mechanism that powers up the right wireless subsystem are being investigated. For example, systems may include mechanisms that wake up more capable wireless subsystems after less capable wireless subsystems receive beacons indicating the need for the more capable wireless subsystem [34]. Newer wireless technologies such as ultrawide band promise to offer data transfers rates of megabytes per second while consuming only a few milliwatts of power.

Work is underway to create batteries based on fuel cell technology to provide an eight hour battery life for laptop computers. Further work is necessary in this area that will allow better power management, more rapid recharging of batteries, hot-swapping of batteries, and better and ubiquitous access to AC power.

Upcoming display technologies such as bistable LCD displays, organic LED displays, and so on, are providing more flexibility for system designers to reduce power consumption and to build more compact form-factor devices. Bistable LCD displays allow the display to retain data that has been written to it even after the display has been powered off. Organic LED displays [29] are emissive and allow finer control over the amount of power consumed by displays, since only active pixels on the display consume power. Thus, screens that have only a few pixels turned on consume very little power. The brightness of OLED displays can also vary considerably by controlling the amount of current that is supplied. Thus, such displays can consume less power when used indoors and can be used in brighter viewing conditions such as outdoors by increasing the energy supplied to the display. Flexible display technology is an active area of investigation [35]. Early prototypes appear promising. At present the bend radius of flexible displays is around 0.75 inch, which represents the radius of the smallest cylinder the flexible display can be rolled into. Projection-based wearable displays may be feasible in the future. Symbol's Laser Projection Display prototype works by scanning a laser beam along two orthogonal directions. The compact low-power display engine creates an image on any surface, at any distance, without refocusing.

Portable storage devices that can include a few to several dozens of gigabytes of storage space are very affordable. Such devices can be used to instantly provide data to wearable computers and thus change their personalities, depending on the user's need. For example, data pertaining to maps, restaurants, tourist spots, facilities, and other services for an entire city can be provided in a portable storage device including an SD card. Users visiting several cities on a vacation trip can carry several such storage devices and use the right one as their vacation progresses.

The availability of autoconfiguring operating systems, such as Knoppix (a Live Linux CD based on Debian GNU/Linux) further allows users to personalize devices. Client virtualization technologies such as VMWare Workstation, Xen, and Connectix Virtual PC, make it possible to move computing environments across client devices. Using portable storage, autoconfiguring operating systems, and client virtualization, a user can buy a new device and transfer his computing environment to the new device [17].

Accurate positioning technologies are now becoming available [36]. The GPS system works well in outdoor areas where visibility to satellites is good. In other areas a variety of technologies based on 3G, 802.11, ultrawide band, and so on., have been explored to provide location data that is accurate to a few meters and in some cases even more accurate. The availability of such data allows the wearable computer to discover services that pertain to the user's current location. By providing a list of only relevant services instead of all services that the device has seen, the usability of the system can be improved.

Energy harvesting techniques that capture energy from the environment and human activity are being developed [37, 38]. Researchers are working on several forms of harvesting—vibration, solar, thermal, and so on, to supply power. While these technologies yield small amounts of power, there is a possibility that such energy sources may be able to power the standby mode of devices and thus allow the user to worry about charging the batteries only when the device is used actively.

Vibration-based and tactile user interfaces are especially useful in a wearable computing scenario [39]. For example, calls between family and colleagues at work can be distinguished by using different vibration patterns. Accelerometers can be used [12, 40] to detect gestures and interact with wearable computers.

Security enhancements based on biometrics, Trusted Computing Group, and others, are becoming available to enhance the security of wearable computers. Biometric authentication is available on several USB flash storage devices and some USB storage drives. In some systems the biometric verification is performed on the host PC, and in some cases the verification is performed on the external storage device itself. Trusted Computing Group is a hardware-based solution for security. The goal is to provide a mechanism for establishing trust between two machines. At the heart of the Trusted Computing Group, is the idea of attestation. Attestation builds a certificate chain, from a trusted hardware component all the way up to the operating system, to identify each component of the software stack. This chain begins with hardware, whose private key is permanently embedded in a tamper-resistant chip, built specifically for this purpose and signed by the vendor providing the machine. The tamper-resistant hardware chip certifies the system firmware (e.g., BIOS). The firmware certifies the boot loader, which certifies the operating system, which in turn certifies the applications running on top of it. In essence, each layer certifies the layer above it. Overall, the Trusted Computing Group platform provides a verification chain that allows only execution of software that passes the sequence of tests.

6.6 Middleware Components for Accelerating Transformation

The availability of several pieces of middleware for core functionality can accelerate the adoption of wearable computing for business transformation. Given that we now have a reasonable understanding of some of the challenges faced by wearable computers and some of the popular applications, we are in a position to look at middleware components that can help accelerate their reach.

6.6.1 Context Sensing

Since a wearable computer may be used by business professionals in many locations other than the office, the context in which it is used is rather varied. Context could include the user's location, physical activity, intellectual activity, emotional state, future activity, and information about his environment—such as names and pictures of customers and colleagues and what they are doing. The context information can be used to adapt user interfaces, route messages, provide additional business information, and so forth.

Several methods are available to obtain location, such as GPS, cell phone, 802.11 and ultrawide band signal strength, image based [41], and more. A context middleware layer can insulate the higher levels of software from the actual physical method used to determine location. The middleware function can specify the location in some absolute or abstract way, the accuracy of the information, and the method that was used. In cases where location data is stale, the middleware function could specifies additional location datum that specify the last reliable location of the user.

The user's physical activity can be deduced either from direct analysis of sensors worn by the user or on the user's mobile devices. Several researchers have used accelerometers to reliably determine whether the user is still, walking, or running. Images from cameras in the environment near the user may also be utilized to determine the user's physical activity. Middleware functions that report the user's physical activity should specify the confidence level with which this data is reported. The user's activity can also be derived from indirect means, such as looking up the user's calendar, or based on the user's current location.

A good estimate of the user's intellectual activity can be obtained by analyzing what the user is working on—namely reading, writing, or speaking. For some users their calendar may accurately represent their activities. If the intellectual activity needs to be deduced, it can be done by analyzing the text that is being composed by the user or the words being spoken, looking at the most active and recent processes running on the user's computer. The user's future activity can sometimes be inferred from this analysis and by looking at the user's calendar. The middleware function that reports the user's mental activity should specify how the activity was determined. Researchers have also investigated how to utilize various context sources to determine whether a person can be interrupted [42].

The user's environment can be inferred in many ways. For example, the user's location may give an indication of whether the user is outdoors or indoors, in a warm or cold area, the user's time zone, and so on. The user's wearable computer can record voices of people around the user and arrive at how many people are in the user's immediate vicinity at a given time. Images from the user's camera may also give an indication of the number of people around the user.

XML representations and Web services interfaces for these types of context will make them extensible and programming language neutral. For example, they will allow a machine to parse the user's context and take appropriate action. Concepts such as a context toolkit [43] can be adapted to the field of wearable computing.

While knowing the user's current context can be very useful, it can be even more helpful if the user's future context can be predicted. A variety of techniques can be employed in this pursuit. For example, if the user is on a train traveling from place X to Y, the future location of the user is known assuming that the train schedule is available. Yet again, the user's calendar is another rich source for this information and works if the calendar is accurate. Typically, calendars for more than 50% of the users are inaccurate because of the difficulty of keeping calendars up-to-date. If the user's wearable knows that the user is going to run into person P in the corridor a minute from now, the system can quickly check to see if any messages need to be communicated when person P actually meets the user.

6.6.2 Sensor Interfaces

Several sensors may be attached to wearable devices. Low-power hardware interfaces and data formats for a variety of sensors are necessary to get some of the context information described previously. By standardizing the hardware and software interfaces, the sensor manufacturers can focus on designing the best sensor instead of worrying about the complete software and hardware stack. The IBM Personal Mobile Hub [21] is an example that uses this principle. Such a partition has already worked well in the PC industry. For example, a blood pressure sensor designer make a design more compact by not having to build long-range wireless communication capabilities into the blood pressure unit; the unit instead relies on short-range wireless which consumes less power.

Wireless means are clearly preferred for communicating between the sensors and the wearable unit. For example a sensor in the shoe may transmit a signal indicating whether the user is walking, running, or stationary. Running a wire from the shoe to the user's wearable unit, such as a cell phone, is rather inconvenient. Researchers [42] have placed accelerometers in various locations on the human body to determine the user's physical activity more accurately. Some clothes now have moisture, temperature, and ambient light sensors. The user's wearable unit can periodically communicate with all the body-worn sensors to determine the user's context.

Data from the sensors can be used to enhance the user experience. For example, imagine that a music player with a disk drive can be told ahead of time that the user is going to run. The player may be able to avoid disk accesses during the jogging period by copying the songs in the playlist that the user if going to use while running to RAM or flash memory which is less susceptible to read errors while in motion. Ambient light sensors can be used to modify the colors and brightness levels of the user interface or switch to another modality, such as voice, if appropriate.

Sensors that we can expect to communicate with the user's wearable unit include temperature, ambient light, moisture, location, and physical activity. A common event model that delivers a piece of information consistently and reliably would significantly help in streamlining data transfer and delivery.

6.6.3 Data Logging and Analysis

Wearable devices are capable of holding several tens of gigabytes of storage today. An Apple iPod with a 60 GB disk drive is already available. Cell phones have started to incorporate 4 GB drives. We can only expect this number to go up over time. One of the consequences of this is that the user can carry a lot of information on the wearable device itself.

The information stored on the wearable unit could be captured from attached sensors or from other traditional sources such as databases and Web servers. Middleware components that will be useful are embedded versions of databases that can be synchronized with larger versions on a server. For example, an athlete can log details about a particular run, such as time of day, length of run, period of run, number of paces taken, instantaneous heart, rate and so on. Such data can be stored in databases for further analysis either locally or in a more global fashion.

Data analysis on captured data can trigger further actions. In the previous jogger example, the events can be examined to look for patterns. Middleware to specify rules for patterns becomes essential. Statistical packages, event correlation engines, neural networks, and so on, can be used to launch software agents that take further action [44, 45]. For example, a simple rule might say that if the instantaneous heartbeat exceeds 170 beats per minute for 30 minutes, and the temperature is above 80°F, the user should be notified and asked to slow down. The rule engine should allow specification of simple rules and a rule composition language that allows several rules to be compounded to derive a complex rule. More complex rules could look for patterns over longer periods of time and at multiple streams of data. For example, if the user is attending a conference he might specify a rule requesting his wearable to notify him whenever there is a talk that cites one of the user's technical papers and the user's schedule is free for that time. Since rules can change over time, the middleware must allow addition of new rules and refinement of existing rules. Stream data processing is an emerging field [46–48] with particular applicability to wearable computing, since multiple sensors can stream data to the user.

When local data analysis on the wearable unit will not suffice, the collected data can be sent to remote locations for further analysis, perhaps with larger groups of users. For example, if a rule is triggered and detects a suspicious medical event, the data supporting the triggering of the rule can be sent to an expert for further analysis. The expert may have a larger collection of rules or data samples and be able to better analyze the data than the user's wearable unit. Standard methods to synchronize different data types are evolving. Technologies such as SyncML [49] may prove to be useful.

6.6.4 Energy Management and Awareness

Energy is an important consideration in wearable computing. Energy management middleware will allow the device to sense the energy status and respond to it. An example function would allow the applications to get the status of the battery powering the device. This middleware will take the user's battery recharging strategy into account as well. For example, if the user typically charges the wearable computer in the train when going home, the system does not go into the lowest power mode if it detects a relatively low battery level in the evening.

Energy management middleware can help application writers determine the amount of energy that will be consumed for a task. The developer may specify in some high-level terms, the nature of the task, and the middleware will determine an energy consumption profile for the task. This estimate may be verified after the application has been developed and can help the developer in additional tuning.

The user experience can also be tuned using energy management middleware. For example, if the battery level is low and the user is not expected to recharge the battery any time soon, the user interface can be modified to perhaps be a bit less friendly but more frugal with energy consumption.

6.6.5 Suspend, Resume, and Session Migration Capabilities

While wearable computers provide the flexibility of being available all the time, they come with several disadvantages compared with stationary counterparts. For example, a wearable computer may have a Twiddler keyboard, and the user may have slower data input rate on such a keyboard. Wearable computers also have smaller displays and may have weaker processors and less memory than stationary PCs. Wearable computers also have limited energy sources. Therefore, users of wearable computers may want to switch their computing session from the wearable to a stationary computer. Essentially, the user should be provided the ability to suspend the computing state on the wearable and then transfer it to a more convenient machine and resume the computation with out losing the context (i.e., the tasks that were running, the windows that were open, etc.) Such function is provided across PCs by the SoulPad approach. Users may want to switch device form factors depending on the nature of the task. For example, a tablet PC may be more suitable for certain tasks while a wearable computer with a head-mounted display is more suitable for others. Just as human beings select between walking, driving, riding a train, and using a plane for travel, depending on several factors, we envision that users would like to switch device formfactors depending on the nature of the task they are performing.

Another useful technology is the Sessions Initiation Protocol (SIP). Authors in [50] have used SIP to control media sessions to wearable computers. SIP separates the data path for a piece of communication from the control path. In a sense it is analogous to using pointers to refer to a data object. This technology allows, for example, the user to receive a call on the wearable device that he can subsequently be transferred to a stationary device nearby. The wearable device can either remain in the control path or drop off. A user may seamlessly transfer an ongoing conversation from his office phone at the end of the day to the cell phone while on the way to the car and then to the car phone and finally to the home phone without dropping the call.

6.6.7 Device Symbiosis

Symbiotic relationships [51] between wearable and stationary devices can alleviate several limitations of wearable computers while exploiting their advantages. Wearable devices are constrained by their input mechanisms, display capabilities, and battery life, but they are always available, and they can be personalized to meet the owner's needs. Environmental devices, on the other hand, typically have better input mechanisms, substantially higher media capabilities, and fewer energy limitations, but they are not portable and are typically shared in public spaces. Symbiotic use of devices to mitigate inherent shortcomings of each class of devices poses several issues that are beginning to be addressed. They include device and service discovery, authentication, privacy, and security.

6.6.8 Privacy and Security

Wearable computers can indeed collect a significant amount of data and analyze them. A natural question that arises is one of privacy. Who should have access to information about the user's current physical activity? How do you ensure that data from your heart rate monitor does not fall into the wrong hands? How should privacy permissions be maintained and propagated? For example, if you have agreed to release your activity information to a colleague, is he or she at liberty to reveal your activity to others?

Let us briefly examine data security before we suggest possible approaches to handle the privacy issues. Since data may be received from body-worn sensors, it is necessary to protect such data during transfer. The sensors could have a shared secret with the wearable computer and use it to encrypt data transfers. Recently, it has been shown that SSL can be implemented even on resource-constrained devices [52]. Since body-worn sensors typically have short ranges, the security issues can generally be met without requiring significant innovations.

Privacy management middleware could help define the level of privacy for each data element. For example a user could specify that his calendar entries are viewable by colleagues between 9 a.m. and 12:00 noon the week of the seventh. Another example that is coupled to rules is that if the user's heart rate is above 160 beats per minute, then person X can access the user's location. By being able to place conditions under which some information is revealed, the user gets the privacy he or she desires most of the time, while allowing for divulgation of data in case of emergencies.

The relaying and sharing of medical information brings some unique challenges in terms of security and privacy. In many parts of the world, people want to keep their health issues private for a variety of reasons and want tight control over who has access to that data. In fact, several regulations such as Health Insurance Portability and Accountability Act (HIPAA) specify how medical information can be stored, shared, and disseminated. Thus, wireless security, theft of wearable devices, and unauthorized access to medical databases are some of the issues that need to be handled.

One of the big challenges in specifying the privacy properties for any data is the usability of the system. If the method for entering the rules is complex, most users will take the default privacy levels that the system offers. The privacy middleware must have good support for cloning rules defined earlier and for editing them. A system than can learn the user's predilections for privacy will be of immense value. When such systems become available, privacy tags can be assigned automatically by the system.

6.7 Conclusions

Wearable computing is an emerging field with significant potential to change the way business is conducted and the way people entertain themselves by providing timely and relevant information and data. Over the last several years, the wearable computing community has worked towards addressing some of the fundamental challenges that face this emerging field. The broad functionality provided by modern cell phones demonstrates that the basic challenges have largely been addressed in some areas. We are now at an exciting point where increasingly larger segments of the general population can benefit from wearable computing technologies.

The combination of several technologies, including wireless communication, powerful multicore processors, voice over internet protocol, portable storage devices, client virtualization technologies, portable and trusted computing platforms, synchronization techniques, location and context sensors, high-resolution cameras, flexible and projection-based displays, energy harvesting, multimodal interfaces, embedded databases, rule engines and stream data processing bodes well for the future of wearable computing.

References

[1] Vasilakos, A., W. Pedrycz, "Ambient Inttelligence: Visions and Technologies", *Ambient Intelligence, WirelessNetworks, Ubiquitous Computing*, 2006, Artech House, Norwood MA.

[2] Bass, L. J., et al., "The Design of a Wearable Computer," *Proc. of Conference on Human Factors in Computing Systems (CHI)*, pp. 139–146, 1997.

[3] Bass, L., et al., "Constructing Wearable Computers for Maintenance Applications," *Fundamentals of Wearable Computers and Augmented Reality*, 2001, LEA Publishers.

[4] Smailagic, A., D. Siewiorek, and L. Luo, "A System Design and Rapid Prototyping of Wearable Computers," *Course 2003 International Conference on Microelectronics Systems Education (MSE'03)*, June 1–2, 2003.

[5] Kortuem, G., M. Bauer, Z. Segall, "NETMAN: The Design of a Collaborative Wearable Computer System.," *MONET*, Vol. 4, No. 1, pp. 49–58, 1999.

[6] Kortuem, G., Z. Segall, M. Bauer, "Context-Aware, Adaptive Wearable Computers as Remote Interfaces to 'Intelligent' Environments," *Proc. IEEE International Symposium on Wearable Computers (ISWC)*, pp. 58–65, 1998.

[7] DeVaul, R. W., et al., "MIThril 2003: Applications and Architecture," *Proc. IEEE International Symposium on Wearable Computers (ISWC)*, pp. 4–11, 2003.

[8] Pentland, A., "Human Design: Wearable Computers for Human Networking," *Proc. 23rd International Conference on Distributed Computing Systems (ICDCS)*, pp. 264–265, 2003.

[9] Rhodes, B. J., "The Wearable Remembrance Agent: A System for Augmented Memory," *Personal and Ubiquitous Computing*, Vol. 1, No. 3, 1997.

[10] Amft, O., et al., "Design of the QBIC Wearable Computing Platform," *Proc. 15th IEEE International Conference on Application-Specific Systems, Architectures, and Processors (ASAP)*, pp. 398–410, 2004.

[11] Lukowicz, P., et al., "The WearARM Modular, Low-Power Computing Core," *IEEE Micro*, Vol. 21, No. 3, pp. 16–28, 2001.

[12] Hamburgen, W. R., et al., "Itsy: Stretching the Bounds of Mobile Computing," *IEEE Computer*, Vol. 34, No. 4, pp. 28–36 2001.

[13] Narayanaswami, C., "Form Factors for Mobile Computing and Device Symbiosis," *Proc. Eighth International Conference on Document Analysis and Recognition (ICDAR)*, pp.335–339, 2005.

[14] Wearable Computing Timeline, http://www.media.mit.edu/wearables/lizzy/timeline.htm.

[15] Thorpe, E. O., "The invention of the first wearable computer," *Proc. Second International Symposium on Wearable Computers (ISWC)*, pp. 4–8, 1998.

[16] IBM MetaPad http://researchweb.watson.ibm.com/thinkresearch/pages/2002/20020207_metapad.shtml

[17] Caceres, R., et al., "Reincarnating PCs using Portable Soulpads," *Proc. ACM/USENIX Conference on Mobile Systems, Applications, and Services (MobiSys)*, pp. 65–78, 2005. http://www.research.ibm.com/WearableComputing/SoulPad/soulpad.html.

[18] Narayanaswami C., et al., "IBM's Linux Watch: The Challenge of Miniaturization," *IEEE Computer*, Vol. 35, No. 1, pp. 33–41, Jan 2002.

[19] Narayanaswami, C., et al., "What Would You Do with a Hundred MIPS on Your Wrist?", *IBM Research Report*, RC22057, January 2001.

[20] Raghunath, M. T., C. Narayanaswami, "User Interfaces for Applications on a Wrist Watch," *Personal and Ubiquitous Computing*, Vol. 6, No. pp. 17–30, 2002.

[21] Husemann, D., C. Narayanaswami, M. Nidd, "Personal Mobile Hub," *Proc. of Eighth International Symposium on Wearable Computers (ISWC)*, pp. 85–91, 2004.

[22] Zieniewicz, M. J., et al., "The Evolution of Army Wearable Computers," *IEEE Pervasive Computing*, Vol.1, No. 4., pp. 30–40, 2002

[23] Ran, L., S. Helal, S. Moore, Drishti, "An Integrated Indoor/Outdoor Blind Navigation System and Service," *Proc. Second IEEE International Conference on Pervasive Computing and Communications (PerCom)*, pp.23–32, 2004.

[24] Nicolai, T., T.Sindt, H. Kenn and H. Witt, "Case Study of Wearable Computing for Aircraft Maintenance," *2nd International Forum on Applied Wearable Computing (IFAWC) 2005*, pp. 97–110, VDE Verlag, March 2005.

[25] Rhodes, B. J., "Using Physical Context for Just-in-Time Information Retrieval," *IEEE Trans. Computers*, Vol. 52, No. 8, pp. 1011–1014, 2003.

[26] Starner, T., Y. Maguire, "A Heat Dissipation Tutorial for Wearable Computers," *Proc. IEEE International Symposium on Wearable Computers (ISWC)*, pp. 140–148, 1998.

[27] Starner, T., "The Challenges of Wearable Computing: Part 1," *IEEE Micro*, Vol. 21, No. 4, pp. 44–52, 2001.

[28] Starner, T., "The Challenges of Wearable Computing: Part 2," *IEEE Micro*, Vol. 21, No. 4, pp. 54–67, 2001.

[29] Kamijoh, N., et al., "Energy Trade-offs in the IBM Wristwatch Computer," *Proc. of the Fifth IEEE International Symposium on Wearable Computers (ISWC)*, pp. 133–140, 2001.

[30] Hofstee, P., M. Day, "Hardware and Software Architectures for the CELL processor," *Tutorials of Intl conf on Hardware/Software Codesign and System Synthesis*, September 18, 2005.

[31] Kumar, R., et al., "A Multi-Core Approach to Addressing the Energy-Complexity Problem in Microprocessors," *Proc. Workshop on Complexity-Effective Design (WCED)*, 2003.

[32] Kumar, R., et al., "Single-ISA Heterogeneous Multi-Core Architectures: The Potential for Processor Power Reduction," *Proc. IEEE/ACM International Symposium on Microarchitecture (MICRO)*, pp. 81–92, 2003.

[33] Olsen, C. M., L. Alex Morrow, "Multi-processor Computer System Having Low Power Consumption," *Proc. Second International Workshop on Power-Aware Computer Systems, (PACS)*, pp. 53–67, 2002.

[34] Agarwal, Y., R. Gupta, C. Schurgers, "Dynamic Power Management Using on Demand Paging for Networked Embedded Systems," *Proc. Design Automation Conference*, 2005. Proceedings of the ASP-DAC 2005. Asia and South Pacific, pp. 755–759, 2005.

[35] Narayanaswami, C., M. T. Raghunath, "Unraveling Flexible OLED Displays for Wearable Computing," *IBM Research Report RC23622*, June 2005.

[36] Borriello, G., et al., "Delivering Real-world Ubiquitous Location Systems," *Commun. ACM*, Vol. 48, No. 3, pp. 36–41, 2005.

[37] Paradiso J. A., and T. Starner, "Energy Scavenging for Mobile and Wireless Electronics," *IEEE Pervasive Computing*, Vol. 4, No. 1, February 2005, pp. 18–27

[38] Starner, T., "Human-Powered Wearable Computing," *IBM Systems Journal*, Vol. 35 No. 3/4, pp. 618-629, 1996.

[39] Gemperle, F., N. Ota, D. Siewiorek, "The design of a wearable tactile display," *Proc. Fifth International Symposium on Wearable Computers (ISWC)*, pp 5–12, 2001

[40] Rekimoto, J., "GestureWrist and GesturePad: Unobtrusive Wearable Interaction Devices," *Proc. IEEE International Symposium on Wearable Computers (ISWC)*, pp. 21–28, 2001.

[41] Sim, R., G. Dudek, "Comparing image-based localization methods," *Proc. Eighteenth International Joint Conference on Artificial Intelligence (IJCAI)*, pp. 1560–1562, 2003.

[42] Kern, N., B. Schiele, Albrecht Schmidt, "Multi-sensor Activity Context Detection for Wearable Computing," *Proc. of European Symposium on Ambient Intelligence (EUSAI)*, pp. 220–232, 2003.

[43] Dey, A. K., "Providing Architectural Support for Building Context-Aware Applications," PhD thesis, College of Computing, Georgia Institute of Technology, December 2000.

[44] Bigus, J. P., et al., "ABLE: a toolkit for building multiagent autonomic systems—Agent Building and Learning Environment," *IBM Systems Journal*, Vol. 41, No. 3, pp. 350–371, Sept 2002.

[45] Yemini, S. A., S. Kliger, E. Mozes, Y. Yemini, and D. Ohsie, "High speed and robust event correlation," *IEEE CommunicationsMagazine*, Vol. 34, No. 5, pp.82–90, 1996.

[46] Carney, D., et al., "Monitoring Streams—A New Class of Data Management Applications," *Proc. of Very Large Databases (VLDB)*, pp. 215–226, 2002.

[47] Chandrasekaran S., and M. J. Franklin, "Streaming Queries over Streaming Data", *Proc of Very Large Databases (VLDB)*, pp.203–214, 2002.

[48] Madden, S., et al., "Continuously adaptive continuous queries over streams," *Proc. ACM International Conference on Management of Data (SIGMOD)*, pp. 49–60, 2002.

[49] Hansmann, U., et al., *SYNCML: Synchronizing Your Mobile Data*, Prentice Hall Professional, September 2002.

[50] Acharya, A., S. Berger, C. Narayanaswami, "Unleashing the power of wearable devices in a SIP infrastructure," *Proc. of IEEE International Conference on Pervasive Computing and Communications IEEE (PerCom)*, pp. 159–168, 2005.

[51] Raghunath, M., C. Narayanaswami, C. Pinhanez, "Fostering a Symbiotic Handheld Environment," *IEEE Computer*, Vol. 36, No. 9, pp. 55–65, 2003.

[52] Gupta, V., M. Millard, S. Fung, Y. Zhu, N. Gura, N. Eberle, S. C. Shantz, "Sizzle: A Standards-Based End-to-End Security Architecture for the Embedded Internet," *Proc. Third IEEE International Conference on Pervasive Computing and Communications (PerCom)*, pp. 247–256, 2005.

CHAPTER 7
Grids for Ubiquitous Computing and Ambient Intelligence

Mario Cannataro, Domenico Talia, and Paolo Trunfio

7.1 Introduction

The grid extends the network paradigm offered by the Internet and offers a novel way to think and use computing resources and services. grids are distributed, heterogeneous, and dynamic. Therefore, grids can be effectively integrated with computing models such as ubiquitous systems and ambient intelligence. Ambient intelligence envisions a future scenario where humans will be surrounded by sensitive and responsive computing environments. Ambient intelligence technologies combine concepts of ubiquitous computing and intelligent systems serving people in their daily life. In doing this, ambient intelligence represents an advancement of concepts and environments that have been developed for use today such as mobile systems, ubiquitous computing, and ambient computing [1].

The concept of ambient computing implies that the computing environment is always present and available in an even manner. The concept of ubiquitous computing implies that the computing environment is available everywhere and is into everything. The concept of mobile computing implies that the end-user device may be connected even when on the move. Ambient intelligence integrates to those concepts intelligent systems solutions, which provide learning algorithms and pattern matchers, metadata and semantic models, speech recognition and language translators, and gesture classification and situation assessment [1].

Computational grids enable the sharing, selection, and aggregation of a wide variety of geographically distributed computing resources and present them as a single, unified resource for solving large-scale compute- and data-intensive applications. Due to its nature, a grid can offer effective support for applications running on geographically dispersed computing devices providing services for user authentication, security, data privacy, and decentralized cooperation. Although today grids include computing resources connected by wired communication media, in a near future wireless grids will be implemented and deployed. This is the first step toward future ubiquitous grids and ambient intelligence grids.

On the one hand, grids might represent the backbone for the implementation of ambient intelligence applications that cover large geographical areas of local settings where intelligent services must be provided to people. On the other hand, grids might represent the computing paradigm that can be used in self-organizing and self-configuring ambient Intelligence systems which can be arranged through the coordination and cooperation of dispersed computing resources such as PDAs, sensors, mobile devices, RDIF tags, data repositories, intelligent interfaces, and so on. Figure 7.1 shows a possible evolution of grids from today's grids, which are essentially wired grids of large grained computers, up to a long term scenario of ambient intelligence grids that will integrate wireless and ubiquitous grids to provide knowledge-based services in a pervasive way. In particular, the basic features of the four elements in Figure 7.1 are as follows:

- *Wired grids*: Current off-the-shelf grids coordinating large-grained computing elements connected through wired networks;
- *Wireless grids*: grids that are able to dynamically aggregate computing devices connected through wireless networks by combined mobile technology with wired grid middleware;
- *Ubiquitous grids*: Next-generation grids featuring high mobility and high embeddedness where sensor networks, mobile devices, and so on, can be dynamically connected to extend the scope of the grid itself;
- *Ambient intelligence grids (AmInt grids)*: Future grids able to support requirements of ubiquitous systems leveraging ambient Intelligence techniques (e.g., intuitive interfaces, context-awareness, knowledge discovery, and ambient semantics) and novel grids middleware.

The structure of the chapter follows the picture shown in Figure 7.1 that identifies a roadmap and the main layers of a grid infrastructure for supporting ambient Intelligence applications. Section 7.2 introduces grid computing principles, systems, and applications. Section 7.3 describes the evolution of grids towards future generation grids. Section 7.4 discusses a scenario for ubiquitous grids and outlines the basic features of grids for supporting ambient intelligence applications and how grids can integrate different technologies to implement ambient Intelligence scenarios. Section 7.5 concludes the chapter.

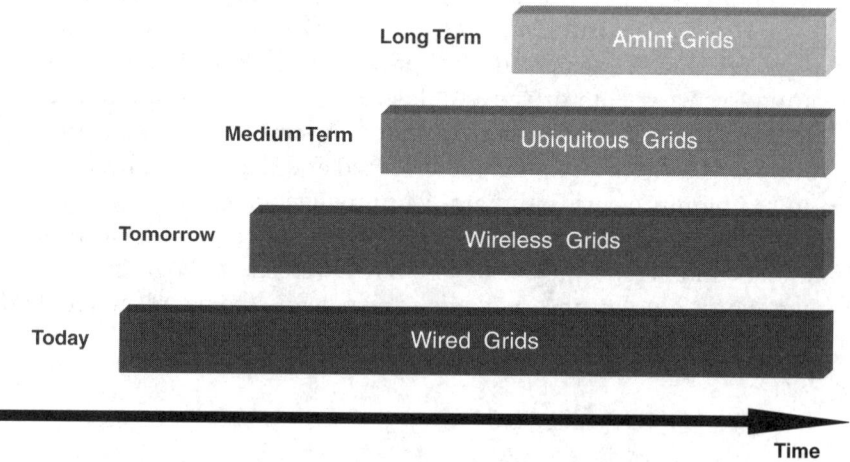

Figure 7.1 Evolution of grids—from wired grids to ambient intelligence grids.

7.2 Grid Computing

The grid computing paradigm is today broadly applied in many scientific and engineering application fields, and is attracting a growing interest from business and industry. grids provide coordinated access to widespread resources, including computers, data repositories, sensors, and scientific instruments, to enable the development of innovative classes of distributed systems and applications.

Grid computing emerged in the last decade as an effective model for harnessing the power of many distributed computing resources to solve problems requiring a large number of processing cycles and involving a huge amount of data. The term *computational grid* was adopted in the mid 1990s to denote a proposed distributed computing infrastructure for advanced science and engineering. More recently, the computational grid model evolved to include many kinds of advanced distributed computing fields, comprising commercial and business applications.

The term grid comes from an analogy to the electric power grid. As the electric power grid provides universal and standardized access to electric power to individuals and industries, the goal of computational grids is to provide users and applications ubiquitous, standardized and reliable access to computing power and resources. Foster and Kesselman in 1998 proposed a first formal definition: a computational grid is a hardware and software *infrastructure* that provides *dependable*, *consistent*, *pervasive*, and *inexpensive* access to high-end computational capabilities [2]. This definition refers to an infrastructure because a grid is mainly concerned with large-scale pooling of heterogeneous resources, including processors, sensors, data, and people. This requires hardware infrastructure to achieve the necessary interconnections among resources and a significant software infrastructure to control the resulting ensemble.

Some years later, Foster, Kesselman and Tuecke refined the previous definition stating that grid computing is concerned with coordinated resource sharing and problem solving in dynamic, multiinstitutional virtual organizations [3]. Resource sharing concerns primarily direct access to computers, software, data, and other resources, as is required by a variety of collaborative problem solving strategies in science, engineering and industry. This sharing requires the definition of sharing rules that specify clearly what is shared, who is allowed to share, and the conditions under which sharing occurs. A set of individuals and institutions defined by such sharing rules forms a so-called virtual organization (VO).

As stated by Schopf and Nitzberg [4], grids are not a new idea. The concept of using multiple distributed resources to cooperatively work on a single application has been explored by older research fields, such as network operating systems, distributed operating systems, heterogeneous computing, and metacomputing. However, there are three main differences between grid computing and older works on distributed computing:

- *Grids focus on site autonomy*. One of the underlying principles of the grid is that a given site must have local control over its resources (e.g., which users can have an account, usage policies, etc.)
- *Grids involve heterogeneity*. Instead of making every administrative domain adhere to software and hardware homogeneity, work on the grid is attempt-

ing to define standard interfaces so that any resource speaking a defined set of protocols can be used.

- *Grids focus on the user.* Previous systems were developed for and by the resource owner in order to maximize utilization and throughput. In grid computing, the specific machines that are used to execute an application are chosen from the user's point of view, maximizing the performance of that application, regardless of the effect on the system as a whole.

7.2.1 Grid Environments

The success of the grid computing model is motivated, in a significant way, by the development and availability of grid environments that provide a variety of services, ranging from basic grid mechanisms to high-level application programming interfaces. Important grid environments are Globus Toolkit [5], UNICORE [6], Legion [7], and Condor [8]. Here we discuss the main features of those environments.

7.2.1.1 Globus Toolkit

Globus Toolkit is a community-based, open-source set of services and software libraries that support the development of grid systems and applications [5, 9]. The toolkit includes software for security, information infrastructure, resource management, data management, communication, fault detection, and portability. Globus Toolkit is designed as a layered architecture in which high-level global services are built upon low-level local services.

Globus Toolkit has evolved over the years from the original version 1 (GT1), to version 2 (GT2) and version 3 (GT3), which is the reference implementations of the Open grid Services Architecture (OGSA) standard, discussed later. The protocols and services that Globus Toolkit provided changed as it evolved, still maintaining the same general approach. The main services offered by Globus include the following:

- *Grid security infrastructure (GSI).* It enables secure authentication and communication over an open network providing a number of services, including mutual authentication and single sign-on authentication, with support for local control over access rights, and mapping from global to local user identities.
- *Monitoring and discovery service (MDS).* It provides a framework for publishing and accessing information about grid resources. It is based on the combination of local and global services, giving a coherent system image that can be explored or searched by grid applications.
- *Globus resource allocation manager (GRAM).* It provides facilities for resource allocation and process creation, monitoring, and management. It simplifies the use of remote systems by providing a single standard interface for requesting and using remote system resources for the execution of jobs.
- *Grid file transfer protocol (GridFTP).* It implements a high-performance, secure data transfer mechanism based on an extension of the FTP protocol

that allows parallel data transfer, partial file transfer, and third-party (server-to-server) data transfer, using GSI for authentication.

7.2.1.2 UNICORE

UNIform Interface to COmputer REsources (UNICORE) is a project funded by the German Ministry of Education and Research [6, 9]. The design goals of UNICORE include an open architecture based on the concept of an abstract job, a consistent security architecture, minimal interference with local administrative procedures, and exploitation of existing and emerging technologies through standard Java and Web technologies. UNICORE provides an interface for job preparation and secure submission to distributed computer resources.

Distributed applications within UNICORE are defined as multipart applications, in which the different parts may run on different computer systems asynchronously or sequentially synchronized. A UNICORE job contains a multipart application augmented by the information about destination systems, the resource requirements, and the dependencies between the different parts. A UNICORE job is a recursive object containing job groups and tasks. Jobs and job groups carry the information of the destination system for the included tasks. A task is the unit that boils down to a batch job for the destination system.

The main components of UNICORE are: the Job preparation agent (JPA); the Job monitor controller (JMC); the UNICORE https server (also called the gateway); the Network job supervisor (NJS); and a Java applet-based graphical user interface (GUI). The UNICORE GUI enables the user to create, submit, and control jobs from any computer on the Internet. The client connects to a UNICORE gateway, which authenticates both the client and the user, before contacting the UNICORE servers, which in turn manage the submitted UNICORE jobs. Tasks destined for local hosts are executed via the native batch subsystem. Tasks to be run at a remote site are transferred to peer gateways. All necessary data transfers and synchronizations are performed by the servers, which also retain status information and job output, passing them to the client upon user request.

7.2.1.3 Legion

Legion is an object-based grid system that enables efficient, effective, and secure sharing of data, applications and computing power [7, 10]. Started as a research project at the University of Virginia, Legion is today a commercial product by Avaki Corporation. Legion addresses the technical and administrative challenges faced by organizations, such as research, development, and engineering groups with computing resources in disparate locations, on heterogeneous platforms and under multiple administrative domains. Legion enables these diverse, distributed resources to be treated as a single virtual operating environment with a single file structure, significantly reducing the overhead of sharing data, executing applications, and utilizing available computing power regardless of location or platform.

In Legion all things of interests to the system—for example, files, applications, application instances, users and groups—are encapsulated as objects. All objects have a name, and an interface. Each object belongs to a class and each class is by itself a Legion object. All objects export a common set of object-mandatory member functions, which are necessary to implement core Legion services. A sizable amount of what is usually considered system-level responsibility is delegated to user-level class objects. For example, classes are responsible for creating and locating their instances and for selecting appropriate security and object placement policies. Legion core objects provide mechanisms that allow user-level classes to implement chosen policies and algorithms. Legion encapsulates also system-level policy in extensible, replaceable class objects, supported by a set of primitive operations exported by the Legion core objects.

7.2.1.4 Condor

Finally, Condor is a resource management system for compute-intensive jobs [8, 9]. Users submit their jobs to Condor, and Condor consequently chooses when and where to run them based upon a policy, monitors their progress, and ultimately informs the user upon job completion. While providing functionalities similar to those of more traditional batch queuing systems, the novel mechanisms of Condor allow it to perform well in areas such as high-throughput computing and opportunistic computing. High-throughput computing aims to provide large amounts of fault-tolerant computational power over prolonged periods of time by effectively utilizing all resources available to the network. The goal of opportunistic computing is to utilize resources whenever and wherever they are available, without requiring 100% availability.

Some of the enabling mechanisms of Condor are as follows:

- *ClassAds*. A mechanism that provides a flexible framework for matching resource requests (e.g., jobs) with resource offers (e.g., machines).
- *Job checkpoint* and *migration*. Condor can transparently record a checkpoint and subsequent resume the application from the checkpoint file. A periodic checkpoint safeguards the accumulated computation time of a job, and also permits a job to migrate from one machine to another machine, enabling Condor to perform low-penalty preemptive-resume scheduling.
- *Remote system calls*. A mechanism for redirecting all of a job's I/O-related system calls back to the machine that submitted the job. As a result, users do not need to make data files available on remote workstations before Condor executes their programs there, even in the absence of a shared file system.

7.2.2 The Open Grid Services Architecture

Ever more, grid applications address collaboration, data sharing, cycle sharing, and other modes of interaction that involve distributed resources and services [11]. The need for integration and interoperability among this increasing number of applications has led the grid community to the design of the open grid services architecture (OGSA), which offers an extensible set of services that virtual organizations can aggregate in various ways [12].

OGSA aligns grid technologies with Web services technologies to take advantage of important Web services properties, such as service description and discovery, automatic generation of client and service code from service description, compatibility with emerging higher-level open standards and tools, and broad commercial support [13]. To achieve this goal, OGSA defines a uniform exposed service semantics, the so-called grid service, based on principles from both the grid computing and Web services technologies.

OGSA defines standard mechanisms for creating, naming, and discovering transient grid service instances; provides location transparency and multiple protocol bindings for service instances; and supports integration with underlying native platform facilities. OGSA also defines, in terms of the Web services description language (WSDL) [14], mechanisms required for creating and composing sophisticated distributed systems, including lifetime management and notification.

In OGSA all resources (e.g., processors, processes, disks, file systems) are treated as services, i.e., network-enabled entities that provide some capabilities though the exchange of messages. This service-oriented view addresses the need for standard interface definition mechanisms, local and remote transparency, adaptation to local operating system services, and uniform service semantics. A first specification of the concepts and mechanisms defined in the OGSA is provided by the open grid services infrastructure (OGSI) [15], of which the open source Globus Toolkit 3 (GT3) is the reference implementation.

The research and industry communities, under the guidance of the global grid forum (GGF), are contributing both to the implementation of OGSA-compliant services, and to evolve OGSA toward new standards and mechanisms. As a significant result of this process, the Web services resource framework (WSRF) [16] was proposed in 2004 as a refactoring and evolution of OGSI aimed at exploiting new Web services standards and at evolving OGSI on the base of early implementation and application experiences.

WSRF provides the means to express state as stateful resources and codifies the relationship between Web services and stateful resources in terms of the implied resource pattern, which is a set of conventions on Web services technologies, in particular XML, WSDL, and WS-Addressing [17]. A stateful resource that participates in the implied resource pattern is termed a WS-resource. The framework describes the WS-Resource definition and association with the description of a Web Service interface, and describes how to make the properties of a WS-Resource accessible through a Web Service interface. Despite the fact that OGSI and WSRF model stateful resources differently—as a grid Service and a WS-Resource, respectively—both provide essentially equivalent functionalities. Both grid services and WS-resources, in fact, can be created, addressed, and destroyed, in essentially the same ways [18].

7.3 Towards Future Grids

Since their birth, computational grids [9] have traversed different phases or generations. In the early 1990s, first-generation grids enabled the interconnection of large supercomputing centers to obtain an aggregate computational power greater than that offered by participating sites, or to execute distributed computations over thousands of workstations introducing the first real implementation of a metacomputing model. Second-generation grids were characterized by the adoption of standards (such as HTTP, LDAP, PKI, etc.) that enabled the deployment of a global-scale computing infrastructure linking remote resources. Second-generation grids started to use a grid middleware as glue between heterogeneous distributed systems, resources, users, and local policies. grid middleware targeted technical challenges such as communication, scheduling, security, information, data access, and fault detection. Main representatives of second-generation grids were systems such as Globus, UNICORE, and Legion (see Section 7.2.1). A milestone between second- and third-generation grids was posed by Foster and colleagues in [3], where they defined the grid as a flexible, secure, coordinated resource sharing among dynamic collections of individuals, institutions, and resources—what we refer to as virtual organizations. The motivation for third-generation grids was to simplify and structure the systematic building of grid applications through the composition and reuse of software components and the development of knowledge-based services and tools. Therefore, following the trend that has emerged in the Web community, the service-oriented model has been proposed (see Section 7.2.2).

A major factor that will further drive the evolution of the grid is the necessity to face the enormous amount of data that any field of human activity is producing at a rate never seen before. The obstacle is not the technology to store and access data, but perhaps what is lacking is the ability to transform data stores into useful information, extracting knowledge from them [19].

In our vision, the architecture of future generation grids, will be based on the convergence of technologies, methodologies, and solutions that are emerging in many computer science fields, apparently far away from and unaware of grids, such as mobile and pervasive computing, agent-based systems, ontology-based reasoning, peer-to-peer, and knowledge management. An example of a infrastructural methodology that is becoming common practice in many of the previous fields is the systematic adoption of metadata to describe resources, services, and data sources, and to enhance, and possibly automate, strategic processes such as service discovery and negotiation, application composition, information extraction, and knowledge discovery.

Although the ongoing convergence between grids, Web services, and the semantic Web represents a milestone towards a service-oriented grid architecture, which has the potential to face important issues such as business modeling and application programming, many other issues need research and development efforts. To be effectively adopted in different application domains, future-generation grids need to address different issues such as: an increasing complexity and distribution of applications; different goals, skills, and habits of grid users; availability of different programming and deployment models; heterogeneous capabilities, and performances of access networks and devices.

More specifically, the great availability of data and information at the different layers of grids, the maturity of data exploration techniques able to extract and synthesize knowledge, such as data mining, text summarization, semantic modeling, and knowledge management, and the demand for intelligent services in different phases of application life cycle, are the driving forces toward novel knowledge-based grid services.

Scientific and commercial applications, as well as grid middleware, will more and more produce an overwhelming quantity of application and usage data that needs to be exploited. The way how such data, at different levels of the grid, can be effectively acquired, represented, exchanged, integrated, and converted into useful knowledge is an emerging research field known as grid intelligence [20, 21]. In particular, ontologies [22] and metadata [23] are the basic elements through which grid Intelligence services can be developed. Using ontologies, grids may offer semantic modeling of user's tasks/needs, available services, and data sources, to support high-level services such as dynamic services finding and composition [24]. Semantic modeling of grid resources, services, and data, is the enabling factor to support some important emerging grid services, such as dynamic service finding, composition, and scheduling, that can be used to enhance tools for workflow composition and enactment.

Moreover, data mining and knowledge management techniques will enable intelligent services based on the exploration and summarization of stored data. Such services could be employed both at the operation layer, where grid management could gain from information hidden into usage data, and at the application layer, where users could be able to exploit distributed data repository, using the grid not only for high-performance access, movement and processing of data, but also to apply key analysis tools and instruments. Examples of knowledge discovery and knowledge management grid services are grid-aware information systems (e.g., document management and knowledge extraction on grid), and knowledge-based middleware services (e.g., scheduling and resource allocation based on resource usage patterns). Finally, context-awareness and autonomous behavior, often required in ambient intelligence applications, can be implemented by combining novel analysis of grid usage data with well-established computational and interaction models such as those offered by intelligent agents [25].

7.3.1 Requirements and Services for Future-Generation Grids

In summary, the main requirements of future-generation grids are [26]:

- *Semantic modeling applied extensively to each component of the grid* including grid services, applications, data sources, computing devices (from ambient sensors to high-performance computers);
- *Modeling of uses' tasks/needs* to enable adaptation and personalization not only in content delivery, but also in performance and quality offered to services and applications;
- *Advanced forms of collaboration* through dynamic formation and negotiation of virtual organizations and agent technologies;
- *Self-configuration* autonomic management, dynamic resource discovery and fault tolerance, (e.g. to support the seamless integration of emerging wireless grids.)

In a service oriented architecture, some novel knowledge-based grid services that will take care of those requirements, include the following:

- *Knowledge discovery and knowledge management services*, for both users' needs (e.g., intelligent exploration of data) and system management (e.g., intelligent use of resources);
- *High-level services for dynamic services finding, composition, and scheduling*;
- *Services to support pervasive and ubiquitous computing*, through environment/context awareness and adaptation;
- *Autonomous behavior*, especially when there is a lack of complete information (e.g., about environment, user goals, etc.).

7.3.2 Architecture of Future-Generation Grids

Since 2003, the European Commission recognized the importance of establishing research directions in grid computing. In the first half of 2003, a group of experts pioneered the vision of the invisible grid, (i.e. whereby the complexity of the grid is fully hidden to users and developers through the complete virtualization of resources). In 2004, the group of experts was enlarged and reconvened to continue that strategic work, and identify new additional requirements [27]. The main result of that study is that applications in the grid environment require a greater range of services than can be provided by the combination of currently evolving grid middleware and existing operating systems. In particular, the study fostered an architectural stack for next-generation grids where the development of applications is supported by three abstraction layers: grids service middleware, grids foundations middleware and operating system (from the higher to the lower abstract level). That report, as well as this chapter, also foresees the need for semantically rich knowledge-based services in both grids foundations middleware and grids services middleware.

The rationale behind such an architectural stack is the following. The application requirements provide a specification for the required services in the grids Service Middleware layer, which provides services that can be used to develop applications. The presence of such a layer of high-level grid services is justified since current grid middleware does not provide the required support to enable easy and reliable application construction and execution in a grid environment.

The grids foundations middleware, that correspond to current grid middleware such as the Globus Toolkit and associated software, has the role to elevate the interface of each operating system to that required for the grids service middleware. In a first scenario, future generation grids comprise a grids service middleware layer that provides services for supporting application building, and the coupling of a grids foundations middleware layer and an operating system layer. Furthermore, the coupling of the grids foundations middleware and operating system layers will lead to a grid operating system that will directly supports the grids service middleware layer.

In summary, semantic modeling of grid resources is recognized as one of the enabling factor to build knowledge-based services able to add intelligence to the grid. Along this research direction, some recent works focused on using ontologies to model grid resources and services and to enhance data sources description and use [28–30]. Other works discussed the provision of knowledge discovery applications on the grid [31–33]. Other than the pioneering semantic grid [34] that first has attempted to apply semantic Web results to the grid, there is now a general consensus about the need to do not reinvent the wheel [35], and to apply to the grid mature methodologies and services developed in these last decades in many research communities, such as databases, artificial intelligence, machine learning, and so on. On the other hand, there is the convincement that peculiar characteristics of grids, such as persistency, security, statefulness, and so on, can be conversely applied to some not yet mature or not yet largely adopted technologies, such as Intelligent Agents and Ubiquitous Computing. In other words, more than forcing technologies against each others, the main challenge for building future generation grids is perhaps building bridges between different research areas, busting myths and prejudices, as recently reported by Goble and De Roure [36].

7.4 Grids for Ubiquitous Computing and Ambient Intelligence

There is a twofold relationship between grids and ubiquitous computing. On the one hand, grid applications often require ubiquitous access to geographically distributed resources, including computing elements, data repositories, sensor networks, and mobile devices. We can envision such ubiquitous grids as composed of a relatively static infrastructure, to which sensor networks, mobile devices, and so on, can be dynamically connected to extend the scope of the grid itself.

From another point of view, large-scale ubiquitous applications may involve typical grid issues, such as authentication, authorization, cryptography, service description, resource discovery, and resource management, and can sometimes require the computational power of grids to accomplish their goals. grid technologies and models can thus provide the needed computational infrastructure and services to address these challenges to build ubiquitous grid environments (see Figure 7.2).

In analyzing the first scenario, the main issue to be faced is the dynamic joining of devices to the grid infrastructure. In current grids, adopted solutions take into consideration the fact that resources are usually statically connected to the network with well known IP address, reliable network connection, and have enough computational power to host heavy-weight grid middleware. Moreover, current grid systems, such as the Globus Toolkit, are designed to build organization-based environments, where a set of enrolment procedures has to be accomplished when a node joins a grid. Furthermore, grid functions are often based on hierarchical models and services (e.g., resource publishing and discovery) that assume trust relationships among resources.

In ubiquitous grids it could not be feasible to guarantee trust about resources, since devices cannot be in general univocally identified, and they could appear and disappear in different places and environments (e.g., a PDA moving across organizations). Moreover, as stated before, current grid middleware is usually huge, so it is not well suited to run on small devices typically involved in ubiquitous systems. Analogously to operating systems or software platforms that often are scaled down to run on small devices (e.g., Windows CE, Java ME), grid middleware and inner procedures should be rethought to be applicable to such devices. Moreover, strict security and interaction models should be adapted to enable intermittent participation of trusted and untrusted devices.

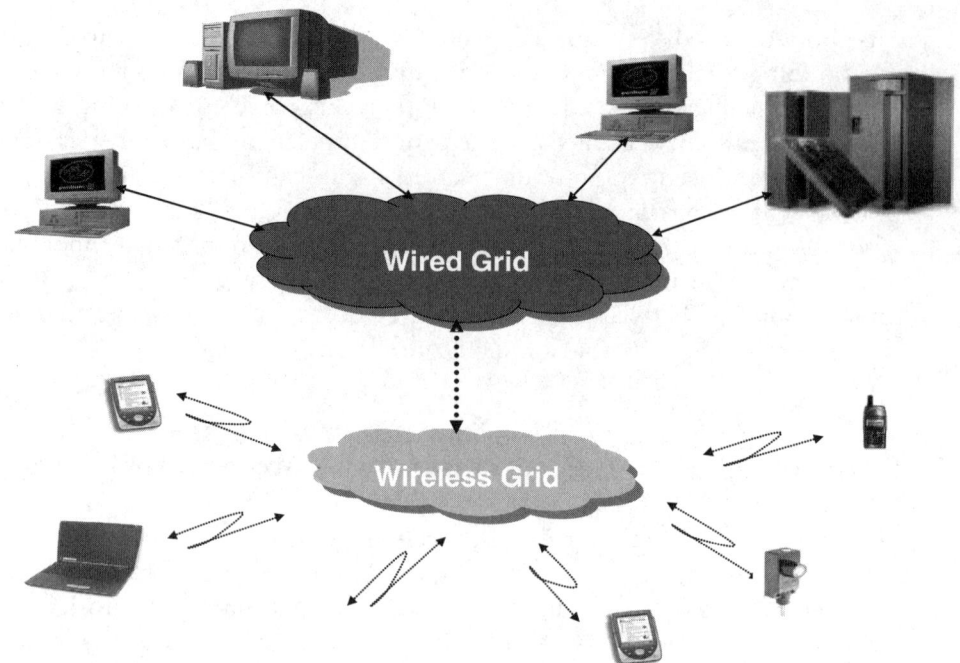

Figure 7.2 Wired and wireless grids integration to support ubiquitous computing and ambient intelligence.

When considering the second perspective, the main issue to be faced is to define appropriate models and approaches to the dynamic, on-demand, self-configuring aggregation of groups of devices able to act as a grid to accomplish possibly complex tasks. A scenario of interest is a user (e.g. his or her PDA device) who requires for a peak of computational power that can be satisfied by aggregating the power of devices available into the surrounding environment. Other than using vertically integrated solutions, which are not scalable and reusable, a possible solution is joining a ubiquitous grid for accessing its services and exploiting the available computing power.

Some feasible approaches to enable ubiquitous grids are as follows:

- Provision of grid access nodes able to support dynamic joining and leaving of intermittent devices. Such brokers should be equipped in such a way that they can support protocols and networks of ubiquitous systems, and update the grid state.
- Provision of small-footprint grid middleware to be run, possibly in a dynamic on-demand way, on devices of ubiquitous systems. Such grid middleware should take into account the characteristics of ubiquitous systems described so far, and could be based on nonhierarchical peer-to-peer protocols, technologies supporting code movement, such as agents and of course on small-footprint operating systems.
- Definition of an ubiquitous grid model through which devices of a ubiquitous system can behave as a grid, (e.g., taking into consideration well-established mobile communications and computing models, such as GSM.)

In summary, the relationships between grids and ubiquitous computing systems can be exploited in two directions. According to the first one, the grid represents a backbone for ubiquitous computing services and applications; in the second one, a self-organization of ubiquitous computing devices can be exploited for composing dynamic ubiquitous grids.

7.4.1 Grids for Ambient Intelligence

Ambient intelligence implementation needs the exploitation of advanced distributed computing systems and technologies that will support the development and delivery of intelligent services in geographically distributed environments where humans operate. Wired grids together with wireless and ubiquitous grids can be integrated to provide the distributed computing infrastructure for delivering intelligent services invisible to final users.

Grid-based ambient intelligence involves a seamless environment of computing, advanced communication technology, semantic-rich description of resources and services and intelligent interfaces [1]. Since an ambient intelligent system is aware of the particular characteristics of human presence and roles, it must take care of needs and is capable of responding intelligently to human requests, it even can engage in intelligent dialog. All this process requires the use of artificial intelligence techniques and systems that often need efficient computing resources and high-bandwidth communications to support advanced intelligent services. Today, available grids and future grid infrastructures can be the best candidates to provide ubiquitous support to run complex software systems and efficient communication protocols that are needed by ambient intelligence systems.

Ambient intelligence should also be unobtrusive and often invisible. Users should use it where needed without seeing its infrastructure all the time. Moreover, interaction with ambient intelligent systems should be comforting and agreeable for humans. These properties require the availability of high-level hardware and software interfaces that represent the front end of intelligent applications. Those interfaces also require high-performance and large data management functionality that can be provided by grids.

Grid-based ambient intelligence requires the integrated use of several advanced technologies for supporting high-performance decentralized computing and for providing pervasive data and knowledge access and management. Here, we list some of those key research areas that will provide solutions and systems that can be part of ambient intelligence systems:

- *Intelligent agents*. Agents are autonomous problem solvers that can act flexibly in uncertain and dynamic environments; there is a convergence of interests, with agent systems requiring robust infrastructure and grids requiring autonomous, flexible behaviors. Embodying intelligence in multiagent systems requires a high-performance support that can be offered by grids; at the same time, intelligent agents represent a key model for ambient intelligence systems implementations [25].
- *Peer-to-peer computing*. When a large number of dispersed nodes must cooperate, peer-to-peer networks can be used to implement scalable systems

without centralized points; recent convergence efforts between grids and peer-to-peer models can provide scalable architectural solutions [37].
- *Self-organizing systems.* Applications composed of a large number of computing elements distributed in a geographic area can benefit from self-organization features that according to emergent behavior computing, build the system functionality by integrating the single node operations.
- *Sensor networks.* Sensors are devices largely used in ambient computing frameworks to receive inputs from the environment and to register its state; sensor networks are part of the network infrastructure supporting ambient intelligence and grids can be effectively used to provide a data storage and computing backbone for sensor networks.
- *Metadata and ontologies for ambient intelligence.* Decentralized systems involve multiple entities that cooperate to ask for and deliver services. Meaningful interactions are difficult to achieve in any open system because different players typically have distinct information models. Advances are required in such areas as metadata definition, common ontology definition, schema mediation, and semantic mediation for ambient intelligence applications.

All the listed technologies are investigated both in ambient intelligence and in grid computing; therefore, advancements in those areas will be embodied in grid-based systems. This process can be a further step toward grid-based services for ubiquitous computing and ambient intelligence.

7.5 Conclusion

The grid is a new computing infrastructure that allows user to access remote computing resources and facilities that are not available in a single site. By linking user, computers, databases, sensors, and other devices, the grid extends the network paradigm offered by the Internet and the Web and offers a novel way to think and use computing. grids are distributed, heterogeneous, and dynamic; therefore, they can be effectively integrated with similar computing models such as ubiquitous systems and ambient computing. The grid can represent the backbone for implementing ubiquitous computing services and therefore, if ambient intelligence systems are implemented on top of grids, the end user can access anytime, anyhow, anywhere computation resources, data, and knowledge. This chapter discussed how current and emerging technologies can be effectively used to implement future information technology infrastructures for ubiquitous computing and ambient intelligence.

7.6 Acknowledgments

This work has been partially supported by the Italian MIUR project grid.it (RBNE01KNFP). The research work of D. Talia and P. Trunfio is also carried out under the FP6 Network of Excellence CoreGRID funded by the European Commission (Contract IST-2002-004265).

References

[1] Jeffery, K., "GRIDs and Ambient Computing: The Next Generation." *Proc. of the Fifth International Workshop on Engineering Federated Information Systems*, Coventry, UK 17–18 July, 2003.

[2] Foster, I., and C. Kesselman, "Computational grids." In: I. Foster and C. Kesselman (Eds.), *The grid: Blueprint for a New Computing Infrastructure*, pp. 15–52, Morgan Kaufmann Publishers, 1998.

[3] Foster, I., C. Kesselman, and S. Tuecke, "The Anatomy of the grid: Enabling Scalable Virtual Organizations." *Int. Journal of Supercomputing Applications*, Vol. 15, No. 3, pp. 200–222, 2001.

[4] Schopf, J.M., and B. Nitzberg, "Grids: Top Ten Questions." *Scientific Programming*, Vol. 10, No. 2, pp. 103–111, 2002.

[5] Foster, I., and C. Kesselman, "Globus: a metacomputing infrastructure toolkit." *Int. Journal of Supercomputing Applications*, Vol. 11, No. 2, pp. 115–128, 1997.

[6] Erwin, D., and D. Snelling, "Unicore: A grid Computing Environment." *Proc. European Conference on Parallel Computing (EuroPar 2001)*, LNCS Vol. 2150, 2001.

[7] Grimshaw, A.S., and W.A. Wulf, "The Legion vision of a worldwide virtual computer." *CACM*, Vol. 40, No. 1, pp. 39–45, 1997.

[8] Litzkow, M., and M. Livny, "Experience with the Condor distributed batch system." *Proc. IEEE Workshop on Experimental Distributed Systems*, 1990.

[9] De Roure, D., M.A. Baker, N.R. Jennings, and N.R. Shadbolt, "The evolution of the grid." In: F. Berman, G. Fox, and A. Hey (Eds.), grid *Computing: Making the Global Infrastructure a Reality*, pp. 65–100, Wiley, 2003.

[10] Grimshaw, A.S., A. Natrajan, M.A. Humphrey, M.J. Lewis, A. Nguyen-Tuong, J.F. Karpovich, M.M. Morgan, and A.J. Ferrari, "From Legion to Avaki: the persistence of vision." In: F. Berman, G. Fox, and A. Hey (Eds.), grid *Computing: Making the Global Infrastructure a Reality*, pp. 265–298, Wiley, 2003.

[11] Foster, I., C. Kesselman, J.M. Nick, and S. Tuecke, "Grid Services for Distributed System Integration." *IEEE Computer*, Vol. 35, No. 6, pp. 37–46, 2002.

[12] Talia, D., "The Open grid Services Architecture: Where the grid Meets the Web." *IEEE Internet Computing*, Vol. 6, No. 6, pp. 67–71, 2002.

[13] Foster, I., C. Kesselman, J. Nick, and S. Tuecke, "The physiology of the grid." In: F. Berman, G. Fox, and A. Hey (Eds.), grid *Computing: Making the Global Insfrastructure a Reality*, pp. 217–249, Wiley, 2003.

[14] Christensen, E., F. Curbera, G. Meredith, and S. Weerawarana, "Web Services Description Language (WSDL) 1.1." W3C Note, 2001. http://www.w3.org/TR/2001/NOTE-wsdl-20010315.

[15] Tuecke, S., K. Czajkowski, I. Foster, J. Frey, S. Graham, C. Kesselman, T. Maquire, T. Sandholm, D. Snelling, and P. Vanderbilt, "Open grid Services Infrastructure (OGSI) Version 1.0." 2003. http://www-unix.globus.org/toolkit/draft-ggf-ogsi-gridservice-33 2003-06-27.pdf.

[16] Czajkowski, K., D. Ferguson, I. Foster, J. Frey, S. Graham, I. Sedukhin, D. Snelling, S. Tuecke, and W. Vambenepe, "The WS Resource Framework Version 1.0." 2004. developerworks/library/ws-resource/ws-wsrf.pdf.

[17] Box, D., et al., "Web Services Addressing (WS-Addressing)." *W3C Member Submission*, 2004. http://www.w3.org/Submission/2004/SUBM-ws-addressing-20040810.

[18] Czajkowski et al., "From Open grid Services Infrastructure to WS-Resource Framework: Refactoring & Evolution." http://www-106.ibm.com/developerworks/library/ws-resource/ogsi_to_wsrf_1.0.pdf.

[19] Evolving Data Mining into Solutions for Insights (Special issue on), *CACM*, Vol. 45, No. 8, 2002.

[20] Workshop on "Knowledge grid and grid Intelligence" (KGGI 2003), October 13, 2003, Halifax, Canada.

[21] Zhong, N., and J. Liu (Eds.), *Intelligent Technologies for Information Analysis*, Springer-Verlag, 2004.

[22] Gruber, T.R., "A Translation Approach to Portable Ontologies." *Knowledge Acquisition*, Vol. 5, No. 2, pp. 199–220. http://ksl-web.stanford.edu.

[23] Keenoy, K., A. Poulovassilis, and V. Christophides (Eds.), *Metadata Management in grid and P2P Systems (MMGPS 2003)*, London, December 2003. http://www.ceur-ws.org.

[24] Cannataro, M., and C. Comito, "A Data Mining Ontology for grid Programming." *Proc. Semantics in Peer-to-Peer and grid Computing (SemPGrid 2003)*, pp. 113–134, Budapest, May 2003. http://www.isi.edu/stefan/SemPGRID.

[25] Foster, I., N.R. Jennings, and C. Kesselman, "Brain Meets Brawn: Why grid and Agents Need Each Other." *AAMAS 2004*, July 19–23, 2004, New York, USA.

[26] Cannataro, M., and D. Talia, "Semantic and Knowledge grids: Building the Next-Generation grid." *IEEE Intelligent Systems—Special Issue on e-Science*, Vol. 19, No. 1, pp. 56–63, 2004.

[27] European Commission—Information Society NGG2 Group, "Next-Generation grids 2—Requirements and Options for European grids." http://www.cordis.lu/ist/grids.

[28] Kerschberg, L., M. Chowdhury, A. Damiano, H. Jeong, S. Mitchell, J. Si, and S. Smith, "Knowledge Sifter: Agent-Based Ontology-Driven Search over Heterogeneous Databases using Semantic Web Services." *Proc. ICSNW 2004*, June 17–19, 2004, Paris, France.

[29] Liu, D.T., and M.J. Franklin, "GridDB: Data-Centric Services in Scientific grids." *Proc. SemPGrid 2004*, May 18, 2004, New York, USA.

[30] Howe, B., K. Tanna, P. Turner, and D. Maier. "Semantics: Towards Self-Organizing Scientific Metadata." *Proc. ICSNW 2004*, June 17–19, 2004, Paris, France.

[31] Cannataro, M., and D. Talia, "KNOWLEDGE grid: An Architecture for Distributed Knowledge Discovery." *CACM*, Vol. 46, No. 1, pp. 89–93, 2003.

[32] Brezany, P., J. Hofer, A.M. Min Tjoa, and A. Whrer, "GridMiner: An Infrastructure for Data Mining on Computational grids." *Proc. APAC 2003*, September 29, October 2, 2003, Queensland, Australia.

[33] Krishnaswamy, S., and S. Wai Loke, "Towards Datacentric grid Services: Cost Estimable Skeleton-Based Data Intensive Computations Executed Using Mobile Agents." *Proc. KGGI 2003*, October 13, 2003, Halifax, Canada.

[34] De Roure, D., N.R. Jennings, and N. Shadbolt, "The Semantic Grid: A future e-Science Infrastructure." In: F. Berman, G. Fox, and A. Hey (Eds.), grid *Computing: Making the Global Infrastructure a Reality*, pp. 437–470, Wiley, 2003.

[35] Geldof, M., "The Semantic grid: Will Semantic Web and grid Go Hand in Hand?" European Commission, DG Information Society, Unit "Grid Technologies," June 2004.

[36] Goble, C., and D. De Roure, "The Semantic grid: Myth Busting and Bridge Building." *Proc. ECAI 2004*, August 22–27, 2004, Valencia, Spain.

[37] Talia D., and P. Trunfio, "Toward a Synergy Between P2P and grids." *IEEE Internet Computing*, Vol. 7, No. 4, pp. 94–96, 2003.

CHAPTER 8
Peer-to-Peer Networks—Promises and Challenges

Marius Portmann, Sebastien Ardon, and Patrick Senac

8.1 Introduction

The vision of ambient intelligence [1] is an environment that intelligently and unobtrusively responds to the presence and interactions of human users. The typical ambient intelligence computing environment is made up of a large range of proactive computing devices embedded in all types of objects. These smart devices need to be able to communicate and interact with each other in order to create an impression of ambient intelligence for the user. For such a highly distributed and heterogeneous environment, the classical, centralized client-server computing model seems to be a bad match due to its inherent limitations in terms of scalability, fault tolerance and ability to handle highly dynamic environments. The peer-to-peer computing paradigm presents a promising alternative for the ubiquitous computing environments of ambient intelligence. This chapter gives an overview of peer-to-peer computing and its potential and challenges.

The last few years have seen a tremendous surge in the interest in peer-to-peer (P2P) systems, originally sparked by the popularity of applications such as Napster [2]. Napster allowed users to easily share music files stored on their own computers. During its short lifetime, Napster was a phenomenal success and managed to attract millions of users within a very short period of time. Even though it was forced to shut down not long after its launch due to problems relating to copyright infringements, Napster demonstrated the power of the P2P paradigm and has since sparked a flurry of commercial and research activity.

But what exactly is P2P? It is a somewhat elusive and amorphous term and there seems to be a lot of confusion about what it really constitutes. There have been numerous attempts to define the term P2P, but none seems to be able to completely and accurately capture its meaning.

One of the most widely accepted definitions is the following by C. Shirkey [3].

P2P is a class of applications that takes advantage of resources (storage, CPU, human presence) at the edges of the Internet. Because accessing these decentralized resources means operating in an environment of unstable connectivity and unpredictable IP addresses, P2P nodes must operate outside the DNS and have significant or total autonomy from central servers.

Another, slightly more concise definition is given in [4].

The term peer-to-peer (P2P) refers to a class of systems and applications that employ distributed resources to perform a critical function in a decentralized manner.

Rather than attempting to come up with yet another definition for P2P, we believe its meaning can be best captured with a set of key characteristics that are distinctive of P2P systems.

- *Symmetrical role of participants:* Probably the single most important characteristic of P2P systems is the symmetrical role of the participants (nodes). All nodes are equal, (i.e. peers), and they act as service consumers (clients) and service providers (servers) at the same time. The lack of separation of roles stands in contrast to the classical client/server model.
- *Decentralized organization:* P2P systems are typically decentralized and do not rely on any central infrastructure. Services and functions are implemented in a distributed fashion, through the collaboration of peers.
- *Self-organizing nature, adaptability:* Through an automatic discovery process, nodes autonomously organize themselves into a network with a topology determined by the P2P system. No manual intervention or administration is required. This discovery process is ongoing and allows the system to dynamically adapt to changes such as the arrival and departure of nodes.
- *Resource sharing and aggregation:* Since P2P systems are infrastructure-less, the participating peers need to contribute resources that can be shared with other nodes and can be aggregated to implement the required services collaboratively. These resources are typically storage space, CPU cycles, network connectivity, and bandwidth.
- *Fault tolerance:* P2P systems have the inherent feature of fault tolerance, i.e. they can cope with the failure or misbehavior of a fraction of all nodes. This is mainly due to the fact that P2P systems typically do not rely on central entities which represent a single point of failure. Since individual nodes in a P2P system are not assumed to be reliable, replication is used to achieve fault tolerance of the entire system.

These characteristics only serve as a guideline to define and identify P2P systems and should not be considered absolute. A number of practical P2P architectures involve some degree of compromise and do not fully meet all of these criteria.

For example, SETI [5] is a P2P project that aggregates unused computing power of thousands of personal computers to analyze radio telescope data in the search for extra terrestrial intelligence. Even though the computation is performed in a distributed fashion via the peer nodes, SETI relies on a central entity to coordinate the computations and collect the results, thereby violating the first two of the above mentioned P2P criteria. Other P2P systems, e.g. Napster, rely on centralized entities for other services such as searching.

If the definition of P2P were limited to merely the criterion of symmetrical roles of nodes, as suggested by some authors, a wide range of older and well-established systems would qualify as P2P systems as well. For example, the Internet itself could be considered as being P2P at the network layer, since each host has an equal role and has the ability to communicate with any other host. Under this limited definition, even the public switched telephone network would have to be deemed a P2P system.

The aim of this chapter is to give an overview of P2P computing and discuss its potential as a platform for a wide range of large-scale distributed applications. We further look at some of the key challenges that are the focus of current research and that still need to be overcome in order for P2P systems to find widespread acceptance beyond file-sharing applications.

P2P computing is a relatively young research field, with the first significant publications appearing after the year 2000. At the time of writing, Citeseer [6], a popular indexing service for scientific publications, lists more than 1,700 publications on P2P computing. A similar service recently launched by Google [7] lists over 5,000 publications with the term "peer-to-peer" between the year 2000 and 2005.

Given the intensity of research activity, it is not possible to provide a comprehensive survey of the research in the area of P2P computing. We will focus on what we consider the key areas of research and the most significant contributions.

This chapter is organized as follows: Section 8.2 provides a taxonomy of P2P systems and forms the basis for the following discussions. Sections 8.3 and 8.4 outline the promises and potential of P2P computing as well as some of its general research challenges. In Sections 8.5 and 8.6, we describe a number of prominent P2P systems and discuss their strengths, and shortcomings and identify the major research challenges. Section 8.5 specifically focuses on unstructured systems and further provides a discussion of the topology of different types of unstructured P2P systems. Section 8.6 discusses the more recently proposed structured P2P systems. Finally, Section 8.7 concludes the chapter.

8.2 Taxonomy of P2P Systems

P2P systems are implemented as virtual networks of nodes and logical links built on top of an existing network, typically the Internet. These virtual networks are called overlay networks or simply overlays.

An impressive range of P2P overlays has been proposed in the last few years and it has become increasingly difficult to keep track of all the different proposals. To provide some structure to the vast field of P2P computing, we present a simple taxonomy of P2P systems, illustrated in Figure 8.1.

P2P systems can be classified into structured and unstructured types according to the way in which the overlay topology is built.

In structured systems, the overlay topology is strictly defined and controlled by the P2P system. In unstructured systems, there is no explicit system-defined control over the topology. In these unstructured systems, individual nodes can, at least to a certain degree, autonomously determine their position in the overlay network.

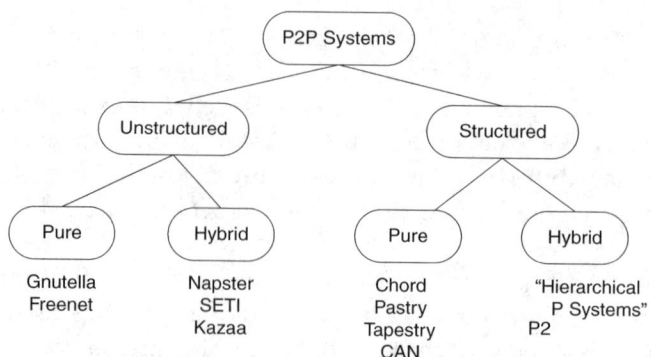

Figure 8.1 Taxonomy of P2P systems.

We can further classify P2P systems into pure and hybrid types. Similar to Schollmeier [8], we define a P2P system as pure if it strictly implements the symmetry of roles property, with all nodes providing equal functionality. The original version of Gnutella [9] is a well-known example of a pure P2P system.

In contrast, hybrid P2P systems assign special roles to a subset of all nodes. For example, a special node can act as a centralized index to locate objects such as in the case of Napster. Kazaa [10] is another example of an unstructured hybrid system. In Kazaa, specially selected nodes act as proxy or gateway nodes for more resource constrained nodes.

In structured P2P systems such as Chord [11] or Pastry [12] nodes cannot freely choose their point of attachment or their neighbors in the P2P network. These systems impose a specific structure of the overlay topology, with the main benefit of improved search efficiency. Most structured P2P systems implement the symmetry of roles property and are therefore pure P2P systems according to our classification. We will discuss structured P2P systems in more detail in Section 8.5.

The proposal by Garcés-Erice et al. [13] is one of the rare examples of hybrid structured P2P system. The authors propose a hierarchical P2P system consisting of multiple levels of pure structured P2P networks. Special nodes called super-peers act as gateways between the different hierarchy levels. At the time of writing, we are not aware of an actual implementation of such a system.

Finally, Structella is a proposal by Castro et al. [14], which does not fit in any of the four categories in Figure 8.1. Structella builds an unstructured P2P system on top of a structured system, and therefore falls somewhere between those two categories. As mentioned in [14], Structella can be implemented as either a pure or a hybrid system. Similar ideas are explored by Loo and coauthors. in [15].

Figure 8.1 shows prominent examples of actual P2P systems with their corresponding classification according to our taxonomy. We will discuss some of these systems in more detail in the following sections.

8.3 P2P—The Promises

P2P has been labelled a disruptive technology [3], with the potential to transform the way people use the Internet. *Forbes* magazine named P2P as one of the four key technologies to shape the future of the Internet. Furthermore, due to their typical characteristics discussed in this section, P2P systems can provide an essential building block for realizing the vision of ambient intelligence.

What are the qualities of P2P systems that make them such a promising technology? The most important of these positive properties are outlined below. As we will see, these properties follow more or less directly from the fundamental characteristics of P2P systems mentioned in Section 8.1.

- *Cost effectiveness:* Due to resource sharing and aggregation, P2P systems can implement services and infrastructure without the need for expensive servers and their maintenance. By tapping into vast amounts of unused resources available on today's computing devices, P2P systems can provide computation and communication services extremely cost effectively. The SETI [5] project is a remarkable example of this, since it managed to build the world's most powerful supercomputer using P2P technology at 1% of the cost of comparable systems [3].
- *Scalability:* One of the main benefits of P2P technology over the traditional client/server model is its inherent scalability, which is a direct consequence of the symmetry of roles, resource sharing and aggregation characteristic of P2P systems. Each node in a P2P system is a consumer as well as a provider of resources at the same time. When the system grows and more nodes are added, the amount of available aggregated resource also grows automatically and therefore caters to the increased demand. In contrast, the server in a traditional client/server system can quickly become a performance bottleneck under rapid growth.
- *Robustness:* The decentralized nature of P2P systems and the symmetrical role of nodes prevent the existence of a single point of failure. This leads to an increased level of robustness. P2P systems cannot and do not assume the individual nodes to be 100% reliable and therefore tolerate the failure and misbehavior of a fraction of nodes. This is achieved by employing redundancy, (i.e. by replicating the state of nodes over a set of peer nodes.) How this is achieved efficiently and reliably for the various P2P systems is still the subject of ongoing research [16, 17].
- *Adaptability:* P2P systems assume a highly dynamic environment, with nodes continuously joining and leaving the network. P2P systems employ algorithms and protocols that automatically reconfigure the individual nodes to adapt to the constantly changing network and computing environment. The ability to adapt to transient environments makes the P2P paradigm an ideal match for a range of very dynamic and infrastructure-less environments, such as ad hoc networks, and pervasive and ubiquitous computing in general.

Due to these positive and desirable qualities, P2P systems have a great potential to serve as a basis for the implementation of a wide range of distributed applications.

Applications and services using P2P technology that have been proposed so far include distributed file systems [18], event notification [19], content distribution

[20, 21], messaging [22], SPAM filtering [23], collaboration [24], Internet-based telephony (VoIP) [25] and many more.

The P2P paradigm is also an ideal match for many ambient intelligence scenarios, where highly distributed networks of intelligent devices are used to create a sensitive and responsive environment. The robustness, scalability, and ability to adapt to a dynamic environment of P2P systems are crucial qualities in an ambient intelligence context.

The number and range of proposed applications using the P2P paradigm is remarkable. However, beyond file sharing, only very few of these applications have been successfully deployed and are currently being used on a wide scale.

The main reason for the lack of deployment of a diverse range of P2P systems lies in a number of problems and research challenges that still need to be addressed and resolved. This will be the focus of the following section.

8.4 P2P—The Challenges

The great characteristics and advantages of P2P systems outlined above come with a number of critical challenges and problems that currently prevent the great potential of P2P technology to be fully realized. These key challenges, which we are going to briefly outline below, are currently the focus of extensive research efforts in the P2P research community.

8.4.1 Security

The security requirements of P2P systems are not unlike the requirements of distributed systems in general and include traditional security goals such as confidentiality, authentication, integrity, and availability. Extensive research has been done in these areas and a general discussion is beyond the scope of this chapter.

We will focus on the security problems and solutions that are new and specific to the context of P2P systems.

There are two main aspects in which P2P systems differ from traditional distributed systems, making the implementation of security a much more challenging task.

First of all, nodes in a P2P system are much more autonomous and powerful than in conventional distributed systems. For example, in numerous P2P overlays, individual nodes are responsible for assigning their own node identifiers. This makes it extremely hard for initially unknown entities to establish some level of trust, since identities can easily be changed. Furthermore, it is possible for a malicious node to create multiple identities in order to gain control of a part of the system. This is referred to as a Sybil Attack.

Individual nodes in a P2P system form part of the overlay network infrastructure and are involved in operations such as topology maintenance, routing, and forwarding of messages. This provides malicious nodes with a range of opportunities for attack and disruption. For example, messages or routing information can be misrouted, corrupted or simply dropped. This problem has been identified and discussed specifically in the context of structured P2P systems [27–29]. The solution proposed in [27] is based on a redundant routing mechanism, with the obvious trade-off between the level of redundancy (i.e., cost) and the fraction of misbehaving nodes that can be tolerated. Furthermore, the security of the suggested mechanism relies on the secure assignment of node identifiers via a trusted certification authority to thwart potential Sybil Attacks. Such a trusted entity is typically not available in P2P systems.

Another class of attacks can occur at the application layer of a P2P system. For example, a malicious node can reply to a query with incorrect data. Fortunately, cryptographic mechanisms such as *self-certifying data* can be applied to allow the system to verify the authenticity of the requested data.

The second key characteristic of P2P systems that represents a major challenge for the implementation of security mechanism is their decentralized nature. Traditional client/server systems rely on centralized and trusted entities for security mechanisms such as identity and key management, authorization, and access control.

In an open environment with mutually distrusting and potentially malicious participants, the lack of a trusted entity makes it extremely difficult to establish a level of trust and to bootstrap security mechanisms.

There have been significant efforts to design trust and reputation systems for P2P systems in a decentralized fashion [31–33]. However, the systems proposed so far are either very complex and not very practical, or are vulnerable to collusion attacks.

Finally, security requirements can vary significantly between different types of P2P systems. The difference can be in terms of the level of security that is required but can also be in regard to the fundamental security goals. For example, some systems aim to provide strong authentication and accountability, whereas others such as Freenet [34] aim to provide the conflicting goal of anonymity.

Trust and security for completely decentralized P2P systems is an extremely challenging problem and is very much an open research issue. It is arguably the biggest impediment to the widespread adoption of P2P applications.

8.4.2 Noncooperation—Freeriders

P2P systems are based on the assumption that participating nodes contribute resources and collaborate to implement a specific infrastructure or service. One of the key challenges of P2P system is the problem of noncooperation or the so-called free rider problem. A free rider is a node that accesses services and consumes resources without contributing any. This phenomenon has been observed in a number of file-sharing applications. For example, Adar and Huberman found that 70% of the Gnutella users did not share any files and 90% did not answer to any queries from other peers [35].

Due to their high degree of autonomy, nodes in a P2P system are not guaranteed to behave in the desired manner. In general, we have to assume that participants in a P2P network act rationally rather than altruistically, which means that there needs to be a clear incentive for nodes to cooperate. In a commercial context, there might even be strong incentives not to cooperate. One could imagine a scenario where nodes might decide to simply drop customer queries directed to their competitors instead of forwarding them.

The problem of rationality and self-interest in P2P networks has been identified and addressed by a number of researchers.

Economic incentives for cooperation can be provided via some form of payment, as suggested in [36], for example. The challenge here lies in implementing a practical and cost-effective payment system.

The decision to cooperate or not in a P2P context can be modeled using game theory. In [37], the classical Prisoner's Dilemma is applied to study the problem of rational behavior in P2P systems. The solution proposed in [38] uses the game theoretical concept of distributed algorithmic mechanism design (DAMD) [39]. The idea is to design the rules of a P2P system, so that the system as a whole functions properly even if the individual nodes follow self-interested strategies. The concept is relatively complex and a number of open problems remain. At this stage, it is not clear if this approach can be successfully applied to practical P2P systems.

Even though a considerable amount of research has been undertaken to address the problem of noncooperation and self-interest in P2P networks, a lot more remains to be done to find solutions that are practical and secure.

8.4.3 Search and Resource Location

Search and resource location mechanisms are a fundamental and crucial building block of most P2P systems and determine to a large degree their efficiency and scalability.

There are fundamental differences between the different types of P2P systems in regard to the implementation of their search mechanism. Unstructured P2P networks, such as Gnutella for example, typically use different variants of flooding to distribute query messages throughout the network. This search method is very robust, flexible, and easily supports partial-match and keyword queries. However, the large volume of query traffic generated by the flooding of messages limits the scalability and efficiency of this approach. There have been a number of attempts to improve the efficiency of search for unstructured pure P2P systems.

Adamic et al. [40] proposed a directed random walk approach and showed that the cost of searching can be significantly reduced in networks with power-law characteristics [40, 41]. However, this approach does not consider load balancing and places the bulk of the search cost on highly connected nodes. A similar method based on random walks and percolation theory is presented in .

The most widely deployed unstructured P2P systems such as Kazaa (FastTrack Network) [10], Gnutella2 [43] and eDonkey (Overnet) [44] are hybrid systems that use a hierarchical approach to improve search efficiency. Highly connected nodes with sufficient spare resources, called super-peers, perform the search on behalf of more resource constraint normal peers. We refer to these hybrid systems as super-peer networks. A more detailed discussion of unstructured P2P systems, super-peer networks and their topology will be given in Section 8.5.3.

Structured P2P systems employ a fundamentally different method for searching. These systems implement a distributed index or distributed hash table (DHT) by implementing an efficient object-to-node mapping functionality using a hash function. The resulting cost of searching is logarithmic in the number of nodes in the system and therefore highly scalable. Since the location of objects is based on the exact knowledge of the corresponding name (or *key*), range queries, partial-match queries, and keyword searches are not directly supported by structured P2P systems.

As proposed in [45–47], keyword search can be implemented in structured P2P systems by distributing inverted keyword lists among the nodes. An inverted list for a keyword consists of the identifiers of all documents that match the keyword. A query with k keywords can be answered by retrieving the k lists from at most k nodes. The final result is achieved through the intersection of the inverted lists. However, it has been demonstrated [48] that this technique is not feasible for large-scale keyword search due to its excessive bandwidth demand. Efficient and scalable keyword search remains one of the key research challenges for structured P2P systems.

Freenet is a rare example of an unstructured P2P system that implements a distributed index for object location based on a hash function. It therefore shares the limitation to exact-match, single key queries with structured P2P systems. In contrast to structured P2P systems, the emphasis in Freenet is to provide anonymity and censorship protection rather than efficiency.

We will further discuss the search mechanism for unstructured and structured P2P systems in more detail in the following two sections. An extensive survey of the problem of searching in P2P systems is also given in [49].

8.5 Unstructured P2P Systems

This section provides an overview of unstructured P2P systems in general and discusses some of the most relevant practical systems in more detail. As mentioned in Section 8.4, one of the key research challenges for unstructured P2P systems is the implementation of an efficient search mechanism, which greatly depends on the nature of the overlay topology. Section 8.5.3 provides a characterization and discussion of the topologies of both pure and hybrid unstructured networks. This section offers insights into the advantages of super-peer networks over pure P2P networks in regard to search efficiency.

8.5.1 Napster

The original P2P system was the Napster application, released in 1999. Napster aimed at facilitating the sharing of audio files among its users. Since then, P2P systems have greatly evolved due to a number of legal and technical reasons.

The design of Napster is relatively simple. Since its purpose is to facilitate the exchange of music files between users, a search or browsing facility needs to be implemented to allow users to locate the desired files. The search facility in Napster is realized through the use of a central index, which has a global view of the location of all files. The Napster system architecture consists of Napster clients (peers) and a central Napster server, as illustrated in Figure 8.2.

Napster works as follows: When a node starts, it sends a list of file names that it is sharing to the central index server. When a peer performs a search, it sends the query to the same central index server. The query reply contains the name of the files matching the query, and the IP addresses of the peers hosting these files. The

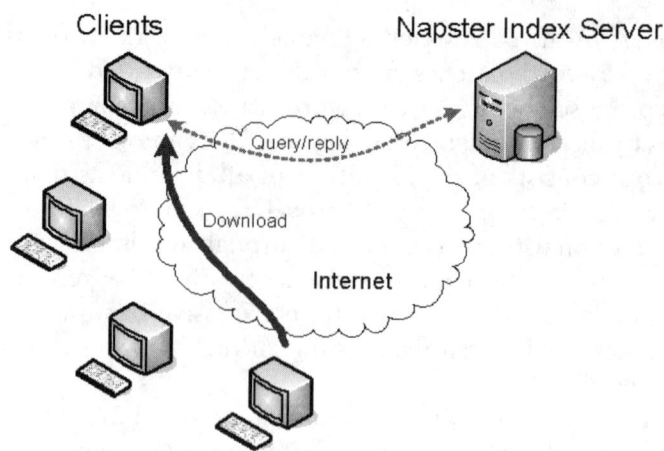

Figure 8.2 Napster architecture.

user can then decide from which peer to download the files. The file download is then performed directly between the peers (peer-to-peer) without the involvement of the index server.

The reason for Napster's failure was not technical, but legal. As the Napster network grew to a size of millions of users, the Recording Industry Association of America (RIAA) started pressuring Napster to ban the exchange of copyrighted material or to shut down the network. The RIAA eventually pressed charges and Napster was forced to shut down.

Technically, the Napster system had a number of advantages but also some shortcomings. The file availability was relatively good, since popular files were automatically replicated on the computers of a large number of users.

At its peak, Napster was able to support millions of users which shows its scalability. However, the central index server would eventually have become a scalability bottleneck if the network had grown significantly bigger. The security of the system was weak. The authenticity of the files was based on file names only and could therefore not be verified or guaranteed. Furthermore, the reliance on a central index server made the system vulnerable to denial of service attacks.

8.5.2 Gnutella

The Gnutella P2P file-sharing system [9] was developed by two employees of Nullsoft, ironically a subsidiary of AOL Time Warner, which is a member of the RIAA. The idea was to design a Napster-like system without a central point of control. This way, the network would be completely decentralized and autonomous, and would be much more robust and harder to shut down. The authors used the open-source model and released the source code and protocol specification under the GPL license to allow the system to evolve.

In Gnutella, all nodes have symmetrical roles, providing both server and client functionality. Without being able to rely on a central index server, Gnutella implements searching via controlled flooding with query messages being recursively broadcast to all neighbors in the overlay, as illustrated in Figure 8.3.

8.5 Unstructured P2P Systems

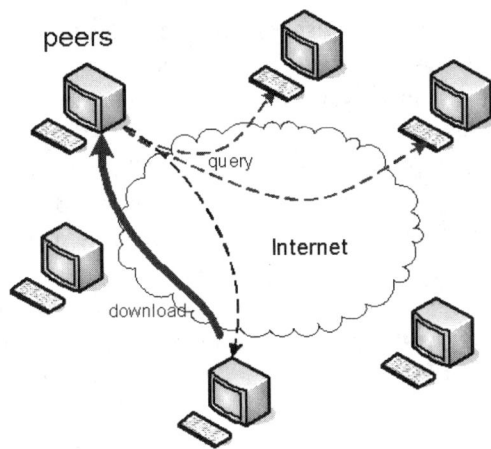

Figure 8.3 Gnutella architecture.

Query messages contain a time-to-live (TTL) field which is decremented each time it is forwarded. When the TTL value reaches zero, the message is dropped to avoid routing loops and to reduce the overhead of a search. The controlled-flooding routing algorithm is illustrated in Figure 8.4.

When a match is found, the peer replies back to the originator of the search with a list of files matching the query and its address. As in Napster, the actual file exchange is performed via a direct connection between two peers, bypassing the overlay network.

The main advantage of controlled flooding is its robustness. It can be shown for a network of 10'000 nodes that a search message with a TTL of 5 reaches on average 95% of all nodes. This is under the assumption that nodes have 4 neighbors on average and that the network exhibits power-law properties [50], which is characteristic of unstructured P2P networks.

It has been specifically shown in [40] that practical Gnutella networks have a connectivity graph that exhibits power-law characteristics.

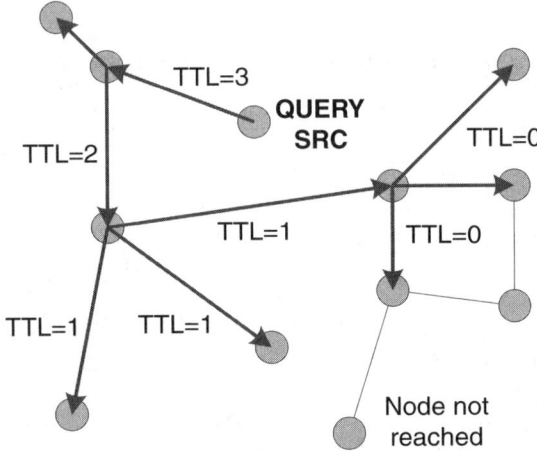

Figure 8.4 Controlled flooding in Gnutella.

Unfortunately, the robustness of unstructured P2P networks comes at the expense of scalability. For each search, the number of query messages generated in the network grows linearly with the number of nodes N and we can therefore say that in general the cost of searching is O(N).

Due to these inefficiencies, the Gnutella protocol has been replaced by the Gnutella2 [43] protocol and various proprietary protocols such as Fastrack (a.k.a. Kazaa [10]). The Gnutella2 family of protocols introduces a two-level hierarchy, consisting of a first level of interconnected peers called super-peers and a second level of so-called leaf nodes or normal peers, which are only connected to a single super-peer. An example of such a super-peer network is illustrated in Figure 8.5.

Super-peer status is presumably given to peers with a combination of good network connectivity, stability and long uptime. Super-peers maintain a temporary index of resources available in leaf nodes directly connected to them. This allows more efficient query routing since only super-peers are involved. Query routing in these systems is done in two stages. First the query message is processed at the local super-peer node and if the resource is not available locally, the message is further propagated in the network of super-peers. Queries can be aggregated and cached in this super-peer network to improve efficiency. Following the terminology introduced in Section 8.2, these networks can be classified as unstructured hybrid networks, since the role of nodes is not completely symmetrical.

As we will show in the next section, the two-level hierarchy introduces a high level of clustering of network topology graph. This can be exploited to implement more efficient search mechanisms via probabilistic routing methods such as random walks [40].

8.5.3 Topology of Unstructured Peer-to-Peer Networks

Unstructured P2P systems are composed of peers joining the overlay network according to some loose rules and without any prior knowledge of the topology. In this section, we offer a formal characterization of the network topologies for both pure and hybrid unstructured systems. In addition, we show that hybrid unstructured systems using super-peers produce a network topology with small-world properties.

One of the oldest and best studied model for network topologies is the Erdös and Rényi random graph model [51]. These networks have a nearly homogeneous

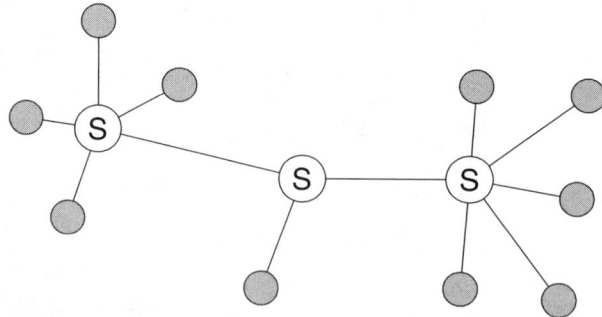

Figure 8.5 Super-peer network.

topology where each node has approximately the same average number of links. The number of links *k* of a node is also called its degree. In the Erdös and Rényi random graph model, the distribution of node degrees follows a binomial distribution which can be approximated by a Poisson distribution for networks with a large number of nodes *N*:

$$P(k) = e^{-\lambda} \lambda^k / k!, \text{ where } \lambda = N \binom{N-1}{k} p^k (1-p)^{N-1-k} \qquad (8.1)$$

However, analysis of real unstructured P2P networks has shown that the Erdös and Renyi random graph is not an accurate model for these systems. Instead, it has been demonstrated that practical unstructured P2P networks exhibit a distribution of node connectivity that follows a power law [52] (i.e., the probability density function of node degrees $P(k)$ is more heavy tailed and therefore high degree nodes are more frequent in this model than in the Erdös and Rényi model):

$$P(k) \approx k^{-\alpha} \qquad (8.2)$$

The parameter α ranges from 2.1 to 4 for practical P2P networks. This power-law property found in a large number of network topologies stems from the dynamic growth and the preferential attachment that can be observed in these networks. This is ignored by the traditional random graph model [53]. Compared with random networks, power-law networks exhibit a higher degree of resilience to random failures. However, the error tolerance of these networks comes at the price of a higher sensitivity and vulnerability to deliberate network attacks targeted toward highly connected nodes [54].

In [55] Watts and Strogatz explore the spectrum of connection topologies from completely regular ring lattices to fully random networks. They generate graphs from a ring lattice, composed of *n* vertices (nodes) and *k* edges (links), where each edge is randomly rewired with probability *p* in order to introduce an increasing degree of disorder. This rewiring process results in a new class of so called small-world networks that exhibit the interesting property of being highly clustered like regular lattices, while offering a small average path length like random graphs. Indeed, there is a broad interval of the rewiring probability *p* over which the average path length between two peers is $O(\log N)$ and the clustering coefficient is significantly higher than in the random network model.

The clustering coefficient and the network diameter are two commonly used parameters to characterize network topologies. The diameter of a network is defined as the average length of the shortest paths between any two nodes of the network. The clustering coefficient of a node is a measure of the interrelatedness of a node's neighbors. It is defined as the ratio between the number of existing links between the *k* neighbors of a node and the maximum number of links that can exist between these *k* nodes, that is $k(k-1)/2$. The clustering coefficient of a network is defined as the average clustering coefficient of all nodes.

A super-peer network enables a trade-off between the highly clustered structure of lattice networks and the short path length of random networks. The small-world structure of super-peer networks offers the interesting property of allowing fast and efficient spreading of information by means of flooding or percolation mechanisms [56].

Percolation is a form of statistical flooding in which each peer forwards a request to each of its neighbors with a probability p. Due to the similarity of this protocol with the spreading of an infectious disease, p is often referred to as the probability of infection.

Gnutella, where each node forwards messages to all its neighbors, can be considered a special case of a percolation protocol with $p = 1$.

We studied the performance of percolation-based routing for several unstructured P2P topologies. We used the BRITE topology generator [57] to generate random topologies with power-law characteristics. All the simulated network topologies are of size $N = 1000$ peers. We simulated super-peer topologies with different ratios r of super-peer nodes to the total number of nodes.

Figure 8.6 shows the spreading of information in percentage of total nodes reached as a function of the infection probability p. The simulation is based on an infect and die model where an infected node ceases to exist after it infects its neighbors. Figure 8.6 shows results for topologies with super-peer ratio $r=0.01, 0.1$ and 1. The respective clustering coefficients for these three topologies are $C=0.98, 0.83$ and 0.006. Each of these plots results from averaging 1000 simulation runs with different random topologies for the given parameters p and r.

We see that when the ratio r of super-peers decreases (i.e., when more peers are attached to super-peers and so the clustering coefficient increases) the fraction of infected nodes increases for all values of p. In addition, the figure illustrates that the critical infectiousness, i.e. the infection probability p for which half of the peers are infected, is lower for a super-peer network (i.e., around 0.7 for $r = 0.01$) than for a normal power-law network (i.e., around 0.9 for $r = 1$).

In summary, the characteristics of the topology of unstructured P2P networks can have a dramatic impact on the performance of searching and information dissemination algorithms. Super-peer networks, because of their small-world characteristics, guarantee a more efficient spreading of messages than Erdös and Renyi or

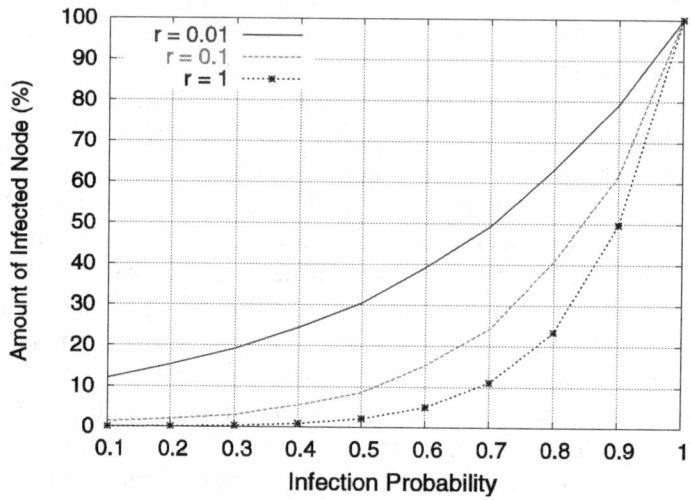

Figure 8.6 Information spreading as a function of the infection probability for various structures of super-peer networks.

power-law networks, when percolation-based protocols are used for searching or information dissemination in P2P networks.

8.6 Structured P2P Systems

8.6.1 Background

Inspired by the popularity of P2P file-sharing networks, the research community has been trying to design new applications and systems to exploit the enormous potential of unused resources of hosts connected to the Internet. However, for many applications, the performance of unstructured P2P systems is often seen as unsatisfactory and statistical performance guarantees are often not sufficient. More recently, a second generation of P2P applications has emerged. These systems are often referred to as structured P2P systems or distributed hash tables (DHTs).

In contrast to unstructured systems, nodes in structured P2P overlays are organized according to a defined structure and the location of data is also determined and managed by the system.

At the top sits the application layer which implements the actual applications. A number of such applications based on structured P2P networks have been proposed in the literature [17, 19, 21, 22].

Layer 2 implements higher layer abstractions, with distributed hash tables (DHT) being the most prominent example. DHTs are based on an idea from Plaxton [59] were originally designed to balance the load of web caches. A hash table is a data structure which provides a mapping from keys to values. Each piece of data must be associated with a key to be inserted in the hash table. For example, the content of a file (the value) is associated with the full name of the file (the key). Hash tables allow random access to pairs (key, value). A distributed hash table (DHT) is a similar data structure distributed over several nodes communicating through a network. DHTs usually provide two main primitives: get(key) and put(key, value). The put primitive is used to store a value in the hash table. It takes the data to store and the corresponding key as arguments. The get primitive is used to retrieve the data by providing the key.

The distributed object location and routing layer (DOLR) is a similar but arguably more powerful concept than a DHT. While a DHT allows storing and retrieving data, a DOLR allows locating nearby copies of objects on the network. The CAST abstraction provides scalable group communication via multicast and anycast primitives.

The key-based routing layer shown in Figure 8.7 is the underlying mechanism used to map keys onto peers (i.e., the peer responsible for a given key) and take care of the redistribution of keys in the case of nodes joining or leaving the network.

The original Plaxton DHT was designed to store a large number of Web cache objects onto a collection of hosts. It assumed a stable node population and therefore did not specify the mechanisms to manage nodes leaving and departing from the system. However, in the context of a P2P network and the Internet, nodes may join and leave the network and keys have to be managed in order to maintain the consistency of the system. Since the original Plaxton design, a number of systems have been proposed. These proposals provide different mechanisms to cope with a

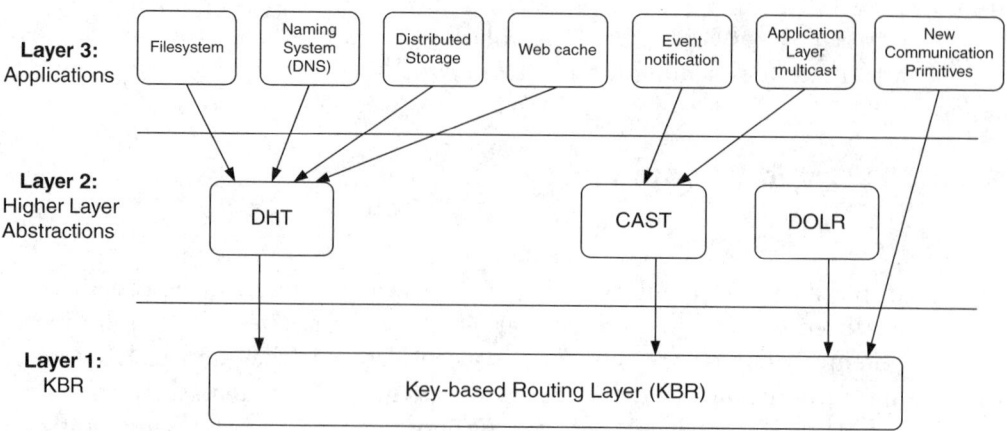

Figure 8.7 Layered view of structured P2P systems.

changing node population. Additionally, one of the key differentiators is the logical topology in which nodes are organized. We will describe some of the most significant of these proposals in the following sections.

8.6.2 The Chord Distributed Lookup Protocol

Chord is a key-based location and routing protocol designed in 2001 at MIT [11]. It provides a single operation: given a key k it maps it onto the corresponding node in the P2P overlay. Data location or any other application using this mechanism can be easily implemented on top of Chord.

Chord uses consistent hashing and proposes to use SHA-1 [60] as a hash function. The consistent hash function is used to associate each node and object with an m-bit identifier or key. Node identifiers (IDs) can be computed using the hash value of their IP address. The Node ID space is a circular ring modulo. Each node i is assigned a node ID k_i and is responsible for the objects with keys in the numerical interval $[k_{i-1}+1, k_i]$. Therefore, it is said that a key k is stored at the successor node of key k, or. By using consistent hashing to allocate keys to nodes, Chord inherently provides a load-balanced system. In a network of N nodes and K keys, each node will store $(1+\varepsilon)K/N$ keys, for small ε with a high probability.

8.6.2.1 Routing in Chord

The Chord routing algorithm is used to locate the node responsible for a given key k. To be able to find the nodes responsible for each key in the network, Chord needs to preserve the two following invariants:

1. each node maintains a correct pointer towards its successor.
2. for every key k, the node is responsible for k.

By maintaining these two invariants, any object with key k can be found by following each node's successor until the node in charge of the key is reached. This method is clearly suboptimal and would result in route lengths (i.e., the number of

8.6 Structured P2P Systems

intermediate hops before finding the key) proportional to the size of the network. To improve the routing efficiency of Chord, each node maintains a routing table, called a finger table, with at most m entries. The ith entry of the finger table of node n points to (e.g., contains the IP address of) the node that succeeds n by at least 2^{i-1} in the identifier space. This is illustrated in Figure 8.8.

To locate a key k, Chord provides a *find_successor(k)* primitive. When a node calls this function, its finger table is searched and the finger pointing to the largest node ID preceding k is returned. The algorithm is recursive and continues until the predecessor of key k is found. This node then returns its successor, which is the node responsible for k. This process is illustrated in Figure 8.9. This routing process is at the heart of Chord's scalability properties, since the average path length of lookups scales logarithmically with N. The Chord system resolves lookups via $O(\log N)$ messages to other nodes.

8.6.2.2 Node Join/Leave Operation in Chord

When a node joins the network, it computes its own unique node ID, for example, by using the hash value of its IP address. It then contacts an arbitrary node that is already a member of the network to perform the insert operation (it must know of at least one node through some external means). The node then uses Chord's *find_predecessor()* routine to find its predecessor as described previously. As stated in [11], a node joining or leaving a Chord network of N nodes will result in $O(\log^2 N)$ messages being exchanged to re-establish the Chord routing invariants and finger tables. Currently, research is being undertaken to evaluate and improve the performance of Chord under various dynamic conditions.

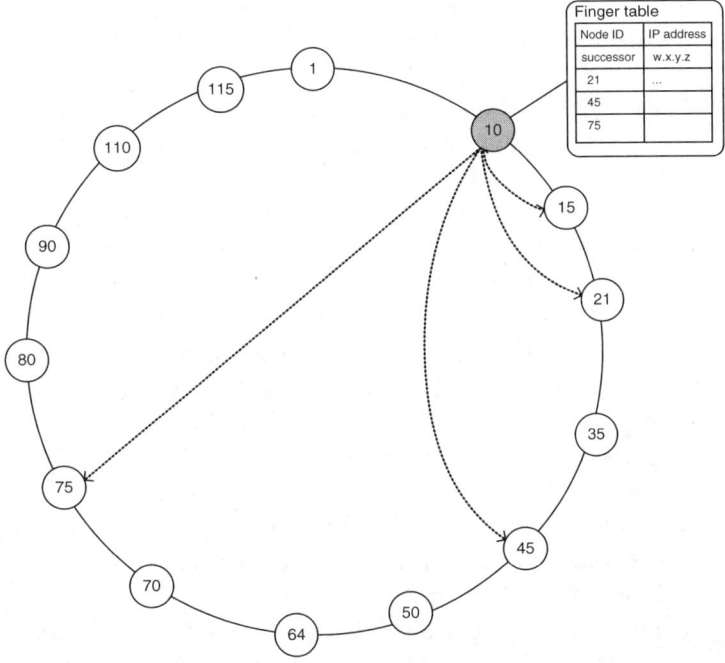

Figure 8.8 Chord routing table.

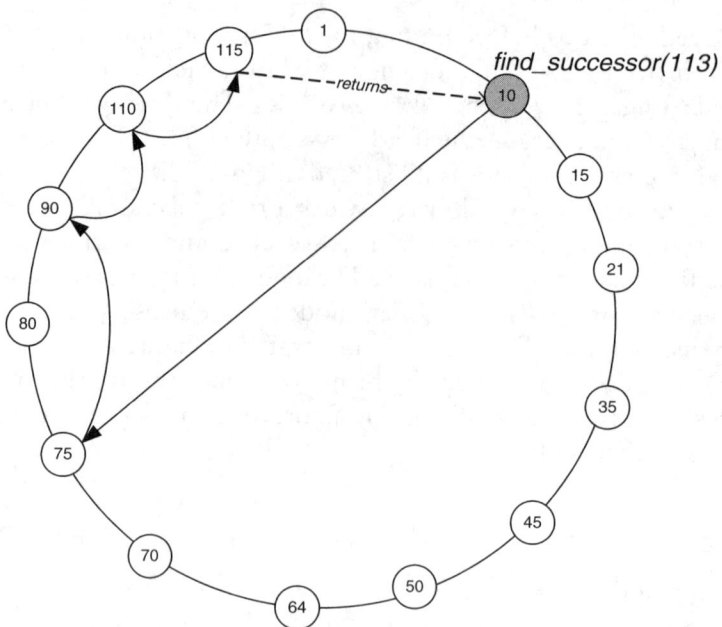

Figure 8.9 Chord routing example.

8.6.2.3 Chord-based Applications

Chord provides a scalable system to logically connect peers and locate keys amongst these peers. The scalability of the system comes from the powerful routing system which ensures that keys are reached in $O(\log N)$ steps. The following two examples illustrate how Chord is used to build practical applications:

- The cooperative file system (CFS) [61] is probably the most immediate application for a distributed hash table, since it allows us to distribute storage space evenly among peers. Peers that participate in CFS contribute some storage space and network connectivity to the system and get a redundant, decentralized storage system in return. The DHash layer introduced in CFS implements the DHT functionality and as such is responsible for retrieving and storing blocks of data, as well as maintaining replicas and providing caching. The Chord layer is responsible for scalable key location and topology maintenance.
- The Chord-based DNS [62] system attempts to eliminate many administrative problems encountered in the current DNS architecture and proposes an architecture built on top of a Chord overlay. Each host name is hashed to a key and IP addresses are stored under this key in the DHT. A main advantage of the Chord-based DNS over the existing hierarchical DNS architecture is that it does not require any centralized root server, which represents a single point of failure and is vulnerable to denial-of-service attacks. Furthermore, it does not impose any naming structure, whereas standard DNS requires host names to be hierarchically organized. Finally, while DNS is designed to map host names to IP addresses, a Chord-based system could be used to provide any name to

8.6.3 Pastry

Pastry is another pure structured P2P system. It was designed by Microsoft Research in 2001 and is also based on Key-Based Routing layer. Similarly to Chord, Pastry [12] employs consistent hashing which maps keys onto nodes IDs to achieve automatic load balancing. The routing algorithm in Pastry however differs from Chord and is based on a variant of the Plaxton distributed search technique [59], which makes use of prefix matching for routing messages between peers. Another key difference is that Pastry integrates a network proximity metric into the overlay topology construction mechanisms.

8.6.3.1 Routing in Pastry

Similar to Chord, each peer in Pastry is assigned a 128-bit node identifier on a circular space, ranging from 0 to $2^{128} - 1$. However, this node ID in Pastry is treated as a sequence of digits in base 2^b (with a typical value of $b = 4$; i.e., we consider hexadecimal digits). When forwarding a message to a destination key k, a peer will choose the peer in its routing table with the longest prefix match.

Each peer in Pastry maintains three tables: a routing table, a neighborhood set and a leaf set. A routing table is organized into $\log_{2^b} N$ rows with $2^b - 1$ entries each. It is built as follows: Entries at row n of each table refer to a node whose ID shares the present node's ID only in the first n digits. Each entry can contain the IP address of one or more nodes with an ID corresponding to this prefix, or be empty if no such node is known. The choice of b is clearly a tradeoff between the size of the routing table and the average path length. For example, with $b = 4$ and $N = 10^6$, the routing table has on average 75 entries and the average path length is 5, while for , the average path length is 7 and routing tables would contain 105 entries on average. The neighborhood set contains the ID and IP addresses of the M nodes closest to the local node according to a given proximity metric (e.g., network delay). This neighborhood set is used to maintain locality properties. The leaf set L simply contains the $|L|$ nodes with the numerically closest ID. These tables are populated when a node joins the network, through a discovery mechanism that will not be explained in detail here. The total cost of a node joining a network in terms of number of messages exchanged is $O(\log_{2^b} N)$.

The routing algorithm works as follows: To forward a message to node ID k, a node first checks if k falls within the ID range in its leaf set. If so, the message is forwarded directly to the leaf set node with the numerically closest node ID to k. If the key is not covered by the leaf set, the routing table is used and the message is forwarded to a node that shares at least one additional digit with the key k. If no such node is known, Pastry will route to a node with the same prefix but is numerically closer to the destination. Such a node must exist in the leaf set unless the message has already arrived at the destination. The algorithm always converges, since keys are always forwarded to nodes with better prefix match, or numerically closer to the destination. Using this routing algorithm, Pastry will route to the destination in less than steps.

8.6.3.2 Applications

Several applications have been designed to operate on top of Pastry. We briefly mention two examples of such applications.

SCRIBE [63] is a large-scale event notification system, which provides a publish-subscribe communication paradigm built on top of Pastry. SCRIBE can be described as a scalable application-level multicast service that uses Pastry to create groups (topics) and to build an efficient multicast tree for dissemination of events (messages) to the subscribers of the various groups.

Splitstream [20] is a high-bandwidth content distribution framework which provides a multicast communication infrastructure similar to IP multicast, but at the application layer. The key idea of SplitStream is to split the media streams into stripes, and send these through separate multicast trees, with disjoint interior nodes. Pastry is used to build and manage the tree, and split the communication load among participants.

8.7 Conclusions

In this chapter, we gave an overview of P2P computing. One of the most challenging aspects in writing such an overview is to provide a clear definition of what P2P really is. We do not believe that there is a single definition that concisely captures all aspects of P2P computing. Our approach to summarise the concept of P2P was, therefore, by giving a list of key characteristics that are typically associated with P2P systems.

We provided a simple taxonomy for P2P networks that allows us to give some structure to the large range of systems that have been proposed.

Then, we discussed the promises of P2P systems by identifying their key qualities such as inherent scalability, robustness, adaptability, and a low cost of ownership. Due to these qualities, we believe that the P2P concept has a great potential to serve as a basis for a wide range of distributed applications. For the same reasons, we also believe that the P2P paradigm is especially well suited for many ambient intelligence scenarios with networks of devices that are highly distributed and dynamic.

However, P2P is a relatively young research field and there are a number of research challenges that need to be overcome in order for P2P systems to gain widespread acceptance in different application domains.

We identified security, the problem of noncooperation, and searching as the major general research problems in the field of P2P computing.

Sections 8.4 and 8.5 presented a more detailed discussion of some of the most relevant P2P systems. According to our taxonomy, we made the distinction between unstructured and structured systems, which have very different characteristics. On the one hand, unstructured P2P networks typically use statistical routing algorithms operating on random graphs to implement searching. We have shown that the efficiency of these algorithms is increased in the case of hierarchical super-peer networks, but the cost of searching is essentially still $O(N)$, where N is the total number of nodes.

Structured P2P system, on the other hand, use the concept of consistent hashing and are able to locate objects with a cost that is only $O(\log N)$. This performance gain however does come at a price. For example, the network traffic and CPU overhead for maintaining the topology and the consistency of the hash table are significant compared with unstructured systems. Furthermore, structured P2P systems only directly support exact match searches whereas unstructured system typically have full-text and partial-match query capabilities.

P2P technology is no panacea and will never completely replace traditional client/server architectures. However, we believe that P2P is an important technology with a tremendous potential in a world that is becoming increasingly connected and decentralised. The P2P paradigm is especially well-suited for environments that are very dynamic and cannot always rely on centralized infrastructure. We therefore believe that ubiquitous computing and ambient intelligence are areas where the P2P concept will have a particularly strong impact.

References

[1] Vasilakos A., and W. Pedrycz, "Ambient Intelligence: Visions and Technologies," in *Ambient Intelligence, Wireless Networking, and Ubiquitous Computing*: Artech House, 2006.

[2] The Napster Home Page, www.napster.com, 2001.

[3] *Peer-to-Peer, Harnessing the Power of Disruptive Technologies*: O'Reilly, 2001.

[4] Milojicic, D. S., V. Kalogeraki, R. Lukose, N. Nagaranja, J. Pruyne, B. Richard, S. Rollins, and Z. Xu, "Peer-to-Peer Computing," HP Laboratories, Palo Alto HPL-2002-57, 2002.

[5] E. Korpela, D. Werthimer, D. Anderson, J. Cobb, and M. Lebofsky, "SETI@home-massively distributed computing for SETI," *Computing in Science & Engineering*, Vol. 3, pp. 78–83, 2001.

[6] Citeseer home page, citeseer.ist.psu.edu.

[7] Scholar home page, www.scholar.google.com.

[8] R. Schollmeier, "A Definition of Peer-to-Peer Networking for the Classification of Peer-to-Peer Architectures and Applications," *International Conference on Peer-to-Peer Computing*, Linköping, Sweden, 2001.

[9] The Gnutella home page, gnutella.wego.com, 2001.

[10] The Kazaa home page, www.kazaa.com, 2005.

[11] Stoica, I., R. Morris, D. Karger, F. Kaashoek, and H. Balakrishnan, "Chord: A Scalable Peer-to-peer Lookup Service for Internet Applications," *SIGCOMM*, San Diego, CA, USA, 2001.

[12] Rowstron A., and P. Druschel, "Pastry: Scalable, Distributed Object Location and Routing for Large-scale Peer-to-peer Systems," *IFIP/ACM International Conference on Distributed Systems Platforms (Middleware)*, Heidelberg, Germany, 2001.

[13] Garcés-Erice, L., E. W. Biersack, P. A. Felber, K. W. Ross, and G. Urvoy-Keller, "Hierarchical Peer-to-peer Systems," *ACM/IFIP International Conference on Parallel and Distributed Computing*, Klagenfurt, Austria, 2003.

[14] Castro, M., M. Costa, and A. Rowstron, "Should we build Gnutella on a structured overlay?," *2nd Workshop on Hot Topics in Networks*, Cambridge, MA USA, 2003.

[15] Loo, B., R. Huebsch, I. Stoica, and J. Hellerstein, "The case for a hybrid P2P search infrastructure," *The 3rd International Workshop on Peer-to-Peer Systems (IPTPS)*, San Diego, CA USA, 2004.

[16] Gopalakrishnan, V., B. Silaghi, B. Bhattacharjee, and P. Keleher, "Adaptive Replication in Peer-to-Peer Systems," *International Conference on Distributed Computing Systems (ICDCS)*, Tokyo, Japan, 2004.

[17] Cohen E., and S. Shenker, "Replication Strategies in Unstructured Peer-to-peer Networks," *SIGCOMM*, Pittsburgh, PA, USA, 2002.

[18] Kubiatowicz, J., D. Bindel, Y. Chen, S. Czerwinski, P. Eaton, D. Geels, R. Gummadi, S. Rhea, H. Weatherspoon, W. Weimer, C. Wells, and B. Zhao, "Oceanstore: An Architecture for Global-scale Persistent Storage," *International Conference on Architectural Support for Programming Languages and Operating Systems (ASPLOS)*, Cambridge, MA, USA, 2000.

[19] Cabrera, L. F., M. B. Jones, and M. Theimer, "Herald: Achieving a Global Event Notification Service," *Workshop on Hot Topics in Operating Systems*, Elmau, Germany, 2001.

[20] Castro, M., P. Druschel, A.-M. Kermarrec, A. Nandi, A. Rowstron, and A. Singh, "Splitstream: High-bandwidth Content Distribution in a Cooperative Environment," *International Workshop on Peer-to-Peer Systems (IPTPS)*, Berkeley, CA, USA., 2003.

[21] Freedman M. J., and D. Mazieres, "Sloppy Hashing and Self-organizing Clusters," *International Workshop on Peer-to-Peer Systems (IPTPS)*, Berkeley, CA, US, 2003.

[22] Mislove, A., C. Reis, A. Post, P. Willmann, P. Druschel, D. Wallach, X. Bonnaire, P. Sens, and L. A. J-M. Busca, "POST: A Secure, Resilient, Cooperative Messaging System," *IEEE Workshop on Hot Topics in Operating Systems (HotOS-IX)*, Kaui, US, 2003.

[23] Zhou, F., L. Zhuang, B. Y. Zhao, L. Huang, A. Joseph, and J. Kubiatowicz., "Approximate Object Location and Spam Filtering on Peer-to-Peer Systems," *ACM/IFIP/USENIX International Middleware Conference*, 2003.

[24] Groove Networks home page, www.groove.net, 2005.

[25] Skype home page, www.skype.com.

[26] Douceur, J. R., "The Sybil Attack," *International Workshop on Peer-to-Peer Systems (IPTPS)*, Cambridge, MA, USA, 2002.

[27] Castro, M., P. Druschel, A. Ganesh, A. Rowstron, and D. S. Wallach, "Secure Routing for Structured Peer-to-Peer Overlay Networks," *Symposium on Operating Systems Design and Implementation (OSDI)*, Boston, MA, USA, 2002.

[28] Sit, E., and R. Morris, "Security Considerations for Peer-to-Peer Distributed Hash Tables," *International Workshop on Peer-to-Peer Systems (IPTPS)*, Cambridge, MA, USA, 2002.

[29] Wallach, D. S., "A Survey of Peer-to-Peer Security Issues," *International Symposium on Software Security*, Tokyo, Japan, 2002.

[30] Fu, K., M. F. Kaashoek, and D. Mazieres, "Fast and Secure Distributed Read-only File System," *Symposium on Operating Systems Design and Implementation (OSDI)*, San Diego, CA, USA, 2000.

[31] Kamvar, S. D., M. T. Schlosser, and H. Garcia-Molina, "The EigenTrust Algorithm for Reputation Management in P2P Networks," *International World Wide Web Conference*, Budapest, Hungary, 2003.

[32] Aberer K., and Z. Despotovic, "Managing Trust in a Peer-2-Peer Information System," *Conference on Information and Knowledge Management*, Atlanta, GA, USA, 2001.

[33] Cornelli, F., E. Damiani, and S. D. Capitani, "Choosing Reputable Servents in a P2P Network," *International World Wide Web Conference*, Honolulu, Hawaii, USA, 2002.

[34] Clarke, I., T. W. Hong, O. Sandberg, and B. Wiley, "Protecting Free Expression Online with Freenet," *IEEE Internet Computing*, Vol. 6, pp. 40–49, 2002.

[35] Adar E., and B. A. Huberman, "Free Riding on Gnutella," *Technical Report, Xerox PARC*, August 2000.

[36] Kamvar, S. D., M. T. Schlosser, and H. Garcia-Molina, "Addressing the Non-Cooperation Problem in Competitive P2P Networks," *First Workshop on Economics of P2P Systems*, Berkeley, CA, USA, 2003.

[37] Lai, K., M. Feldman, J. Chuang, and I. Stoica, "Incentives for Cooperation in Peer-to-Peer Networks," *Workshop on Economics of Peer-to-Peer Systems*, Berkeley, CA, USA, 2003.

[38] Shneidman J., and D. C. Parkes, "Rationality and Self-Interest in Peer to Peer Networks," *International Workshop on Peer-to-Peer Systems (IPTPS)*, Berkeley, CA, USA, 2003.

[39] Feigenbaum J., and S. Shenker, "Distributed Algorithmic Mechanism Design: Recent Results and Future Directions," *International Workshop on Discrete Algorithms and Methods for Mobile Computing and Communications*, Atlanta, GA, USA, 2002.

[40] Adamic, L. A., R. M. Lukose, B. Huberman, and A. R. Puniyani, "Search in Power-Law Networks," *Physical Review, The American Physical Society*, Vol. 64, 2001.

[41] Gkantsidis, C., M. Mihail, and A. Saberi, "Random Walks in Peer-to-Peer Networks," INFOCOM, Hong-Kong, China, 2004.

[42] Sarshar, N., P. O. Boykin, and V. P. Roychowdhury, "Percolation Search in Power Law Networks: Making Unstructured Peer-To-Peer Networks Scalable," *Fourth International Conference on Peer-to-Peer Computing (P2P'04)*, Zürich, Switzerland, 2004.

[43] The Gnutella2 Developer Network, www.gnutella2.com, 2005.

[44] eDonkey home page, www.edonkey2000.com, 2005.

[45] Shi, S., G. Yang, D. Wang, J. Yu, S. Qu, and M. Chen, "Making Peer-to-Peer Keyword Searching Feasible Using Multi-level Partitioning," *International Workshop on Peer-to-Peer Systems (IPTPS)*, La Jolla, CA, USA, 2004.

[46] Bhattacharjee, B., S. Chawathe, V. Gopalakrishnan, P. Keleher, and B. Silaghi, "Efficient Peer-To-Peer Searches Using Result-Caching.," *International Workshop on Peer-to-Peer Systems (IPTPS)*, Berkeley, CA, USA, 2003.

[47] Harren, M., J. M. Hellerstein, R. Huebsch, B. T. Loo, S. Shenker, and I. Stoica, "Complex Queries in DHT-based Peer-to-Peer Networks," *International Workshop on Peer-to-Peer Systems (IPTPS)*, Cambridge, MA, USA, 2002.

[48] Li, J., B. T. Loo, J. M. Hellerstein, M. F. Kaashoek, D. R. Karger, and R. Morris, "On the Feasibility of Peer-to-Peer Web Indexing and Search," *International Workshop on Peer-to-Peer Systems (IPTPS)*, Berkeley, CA, USA, 2003.

[49] Risson J., and T. Moors, "Survey of Research Towards Robust Peer-to-Peer Networks: Search Methods," University of New South Wales, Technical Report UNSW-EE_P2P-1-1, 2004.

[50] Portmann, M., P. Sookavantana, S. Ardon, and A. Seneviratne, " The Cost of Peer Discovery and Searching in the Gnutella Peer-to-Peer File Sharing Protocol," *International Conference on Networks (ICON)*, Bangkok, Thailand, 2001.

[51] Erdös P., and A. Rényi, "On the Evolution of Random Graphs," *Publ. Math. Inst. Hungar. Acad. Sci*, Vol. 5, pp. 17–61, 1960.

[52] Faloutsos, M., P. Faloutsos, and C. Faloutsos, "On power-law relationships of the Internet topology," *ACM SIGCOMM*, Cambridge, MA, USA, 1999.

[53] Barabási A.-L., and R. Albert, "Emergence of scaling in random networks," *Science*, Vol. 286, pp. 509–512, 1999.

[54] Albert, R., H. Jeong, and A.-L. Barabási, "Error and attack tolerance in complex networks," *Nature*, Vol. 406, 2000.

[55] Watts, D., and S. Strogatz, "Collective dynamics of small world networks," *Nature*, Vol. 393, pp. 440–442, 1998.

[56] Newman, M. E. J., I. Jensen, and R. M. Ziff, "Percolation in a two-dimensional small world," *Physical review*, Vol. 65, 2002.

[57] Medina, A., I. Matta, and J. Byers, "On the Origin of Power Laws in Internet Topologies," *ACM Computer Communications Review*, 2000.

[58] Dabek, F., B. Zhao, P. Druschel, J. Kubiatowicz, and I. Stoica, "Towards a Common API for Structured Peer-to-Peer Overlays," *International Workshop on Peer-to-peer Systems (IPTPS)*, Berkeley, CA, USA, 2003.

[59] Plaxton, C. G., R. Rajamaran, and A. W. Richa, "Accessing nearby copies of replicated objects in a distributed environment," *ACM Symposium on Parallel Algorithms and Architectures*, Rhode Island, USA, 1997.

[60] F. 180-1, "Secure Hash Standard," *U.S. Department of Commerce/N.I.S.T., National Technical Information Service*, 1995.

[61] Dabek, F., M. F. Kaashoek, D. Karger, R. Morris, and I. Stoica, "Wide-area cooperative storage with CFS," *ACM Symposium on Operating Systems Principles (SOSP '01)*, Banff, Alberta, Canada, 2001.

[62] Cox, R., A. Muthitacharoen, and R. Morris, "Serving DNS using Chord," *International Workshop on Peer-to-Peer Systems (IPTPS)*, Cambridge, MA, USA, 2002.

[63] Rowstron, A., A.-M. Kermarrec, M. Castro, and P. Druschel, "SCRIBE: The Design of a Large-scale Event Notification Infrastructure," *International Workshop on Networked Group Communication*, London, UK, 2001.

CHAPTER 9
Comparative Analysis of Routing Protocols in Wireless Ad Hoc Sensor Networks

D. Li, H. Liu, and A. Vasilakos

A wireless ad hoc sensor network is a collection of low-cost, low-power, multifunctional sensor nodes that communicate unattended over wireless channel. Recent advancement in wireless communication and electronics has made the deployment of wireless ad hoc sensor networks a reality. Ad hoc sensor networks can be used for various application areas like surveillance, mine field, emergency rescue and response, healthcare, environmental monitoring, manufacturing, and traffic control. The main advantage is that they can be deployed in almost any kind of terrain with a hostile environment where it might be impossible to use traditional wired networks. In order to facilitate communication within the network, a routing protocol is used to not only discover routes between sensor nodes, but also collect the relative information from the network. Due to limited resources of sensor nodes, it is a challenging issue to design a correct, scalable and efficient routing protocol that results in a robust, long-lived and low-latency ad hoc sensor network. This chapter first examines routing protocols for ad hoc sensor networks and evaluates them in terms of a given set of parameters. This chapter also provides an overview of different protocols by presenting their characteristics and functionality, and then gives a comparison and discussion of their respective merits and drawbacks.

9.1 Introduction

Intelligent environments are opening unprecedented scenarios where people interact with electronic devices embedded in environments that are sensitive and responsive to the presence of people. The term ambient intelligence [30] (or, AmI) reflects this tendency, gathering best results from three key technologies: ubiquitous computing, ubiquitous communication, and intelligent user friendly interfaces. The emphasis of AmI is on greater user friendliness, more efficient services support, user empowerment and support for human interactions. People are surrounded by smart proactive devices that are embedded in all kinds of objects: an AmI environment is capable of recognizing and responding to the presence of different individuals, working in a seamless, unobtrusive, and often invisible way.

Devices are (wired or unwired) plugged into the network. The resulting system consists of multiple devices, computer equipment, and software systems that must interact among them. Some of the devices are simple sensors, other ones are actuator owning a crunch of control activity on the environment (central heating, security systems, lighting system, washing machines, refrigerator, etc.).

In recent years, advances in micro-electro-mechanical systems (MEMS) technology and digital electronics have made it economically feasible to manufacture microsensors. These tiny sensor nodes have sensing, data processing, and wireless communication capabilities. As shown in Figure. 9.1, a sensor node is made up of four basic components: a sensing unit, a processing unit, a transceiver unit and a power unit. A sensor node may also have optional application-dependent components, such as a position finding system, power generator, and mobilizer. The sensing circuitry measures ambient conditions in the environment surrounding the sensor and then transforms them into electric signals. The sensor processes such electric signals to reveal some characteristics about the phenomena located in the vicinity. The sensor sends the collected data to a sink usually via radio transmitter. If a large number of such distributed sensors collaborate and operate unattended via wireless links across a geographical area, they form a dense wireless ad hoc sensor network.

There is a wide range of applications for wireless sensor networks (WSN). Some of the application areas are military, environment, and health. Military applications are quite numerous, such as monitoring friendly forces, combat field surveillance, target field imaging, intrusion detection, and battle damage assessment. For example, the use of WSNs can limit the need for personnel involvement in the usually dangerous reconnaissance of opposing forces and terrain. WSNs can be deployed in critical terrains, and some valuable, detailed, and timely information about the opposing forces and terrain can be collected within minutes. Some environment applications of WSNs include tracking the movements of animals, forest fire detection, flood detection, pollution study, and precision agriculture. Some of the health applications for WSNs provide interfaces for the telemonitoring of human physiological data, diagnostics, drug administration in hospitals, and tracking and monitoring doctors and patients inside a hospital. A wireless sensor network can be deployed in these applications through two methods: random fashion (e.g.,

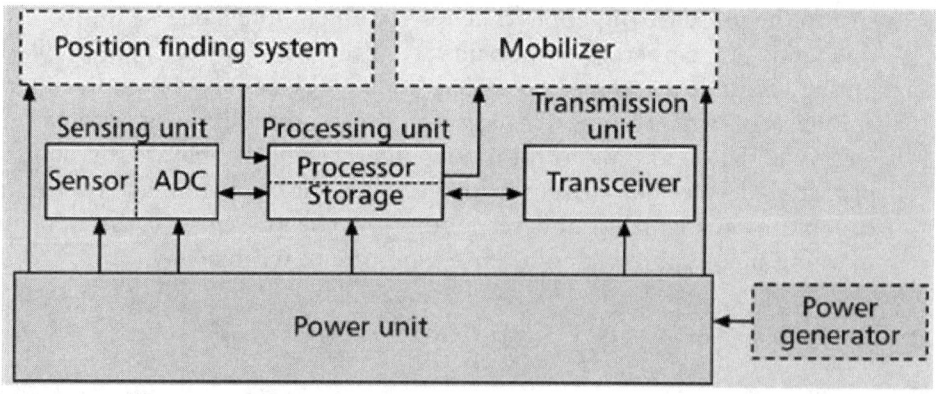

Figure 9.1 The components of a sensor node.

dropped from a helicopter in a disaster rescue application) or manual fashion (e.g., alarm sensors in forest fire detection or sensors planted underground for precision agriculture).

WSNs have recently received intensive attention in the research literature such as collaboration among sensors in data gathering and processing, and coordination and management of the sensing activity. However, sensor nodes are constrained in energy supply and communication bandwidth. Energy consumption occurs in three domains: sensing, data processing, and communications. Communication is the major consumer of energy in WSNs. Such constraints combined with a typical deployment of a large number of sensor nodes pose many challenges to the design and management of WSNs and necessitate energy-awareness at all layers of the networking protocol stack. Therefore, it is necessary to eliminate energy inefficiencies that shorten the lifetime of WSNs and to implement efficient use of the limited communication bandwidth via creative techniques. For example, during design of the network layer, the main goal is to find methods for energy-efficient route discovery and relaying of data from the sensor nodes to the sink so that the lifetime of the network is maximized.

While wireless sensor networks share many commonalities with existing ad hoc networks concepts, there are also a number of inherent characteristics and specific challenges for routing in WSNs. Some of the most important points that make the WSNs different and challenging are as follows:

- Wireless sensor networks are application-specific. It is unlikely that there will be one-size-fits-all solutions for all the potentially very different application scenarios.
- The number of sensor nodes in a sensor network can be several orders of magnitude higher than the nodes in an ad hoc network.
- Sensor nodes are densely deployed.
- Sensor nodes are prone to failures.
- Sensor nodes deployed in ad hoc manner need to be self-organizing since the as hoc deployment of these nodes requires the system to form connections and cope with the resultant nodal distribution, especially since the operation of sensor networks is unattended.
- In some applications, the topology of a sensor network may change frequently, because some sensor nodes may be allowed to move and change their location. However, in most applications, nodes in WSNs are generally stationary after deployment except for maybe a few mobile nodes.
- Sensor nodes are tightly constrained in terms of energy, computation capacities, and storage.
- Sensor nodes may not have global identification (ID), since the overhead of ID maintenance is high due to the deployment of a large number of sensor nodes.
- The networking paradigms of WSNs are data-centric architectures instead of node-centric architectures. The importance of any one particular node is considerably reduced as compared with traditional networks. More important is the data that these nodes can observe.

- There is a high probability that observed data by sensor nodes has some redundancy. Some redundancy needs to be exploited by routing protocols to improve energy and bandwidth utilization.

Due to the unique characteristics of wireless sensor networks, many new algorithms have been proposed for the routing problem in WSNs in order to meet specific design goals. These routing mechanisms have considered the inherent features of WSNs combined with the application and architecture requirements. Finding and maintaining routes in WSNs is tough, because energy limitations and sudden failures of sensor nodes cause frequent and unpredictable topological changes. Almost all the routing protocols can be classified into flooding based, gradient based, hierarchical based and location based according to data communication techniques. Flooding based protocols disseminate all the information at each node to every node in the network assuming that all nodes in the network may be interested in the information through controlled flooding. Gradient based routing protocols are query based, and all data generated by sensor nodes is named by attribute-valued pairs, which helps in eliminating many redundant transmissions. Hierarchical based protocols aim at clustering the nodes so that cluster heads can do some aggregation and reduction of data in order to save energy. Location based protocols utilize position information to relay the data to the desired regions rather than the whole network. We propose some performance metrics to evaluate wireless sensor network protocols. These performance metrics consist of energy consumption, network lifetime, latency, accuracy, and amount of data disseminated per unit of energy. Energy consumption may be measured as the total consumed energy over a period of time. Network lifetime can be measured by generic parameters such as the time until the first node die or by application-directed metrics, such as when the network stops providing the application with the desired information about the phenomena. The latency is the time from the moment a sensor node sends its sensed data out to the time the sink receives the data. The accuracy depends on whether the result of the sensor network matches what is actually happening in the environment. In this chapter, we explore these routing techniques in WSNs that have been developed in recent years, including their advantages and drawbacks, compare them via the previous performance metrics, and highlight some guidelines for improvement.

The remainder of this chapter is organized as follows. Section 9.2 describes the communication architecture of wireless sensor networks. We present the factors influencing sensor network design in Section 9.3. In the Section 9.4, we discuss these four types of routing protocols: flooding based, gradient based, hierarchical based and location based routing approaches.

9.2 Communication Architecture

Figure 9.2 shows the communication architecture of a wireless sensor network. Sensor nodes are usually scattered in a geographical area, where the sensor nodes are deployed. Sensor nodes cooperate among themselves to produce high-quality information about the physical phenomenon. Each sensor node has the capability to collect data and route data either to other sensor nodes or back to an external sink. The

9.2 Communication Architecture

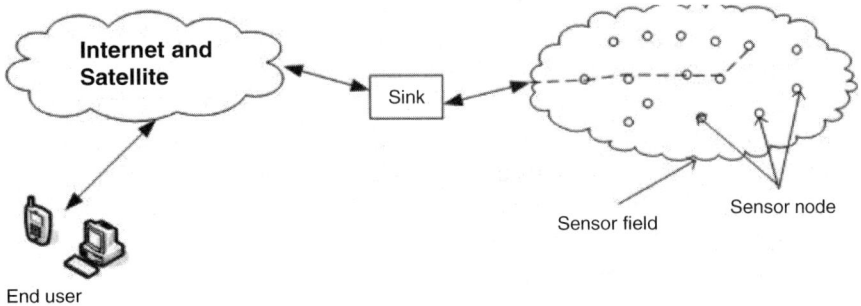

Figure 9.2 Communication architecture of a wireless sensor network.

sink may be a fixed or mobile node capable of communicating with the end user via Internet or satellite.

The inherent features of WSNs have introduced a lot of challenging issues into designing protocols in network layers. Figure 9.3 illustrates the protocol stack used by the sink and sensor nodes. The protocol stack consists of the physical layer, MAC layer, network layer, transport layer, application layer, power management plane, mobility management plane, and task management plane. The main problem of the physical layer is how to transmit energy as efficiently as possible with respect to classical radio transmission. Thus, it addresses the needs of simple but robust modulation, transmission, and receiving techniques. MAC is still one of the most active research areas for WSNs. The medium access control (MAC) protocol must be able to minimize collision with neighbors' broadcasts, and power aware, for example, it can ensure that the sensor nodes can sleep as long as possible. The network layer takes care of routing issues, and should meet the more stringent requirements regarding scalability, energy efficiency and data-centricness. The transport layer helps to maintain the flow of data if the sensor network application requires it. At the application layer, processes aim to create effective new capabilities for efficient extraction, manipulation, transport, and representation of information derived from collected data of sensor nodes.

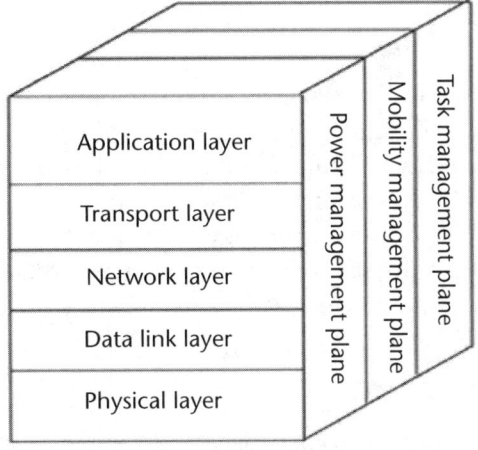

Figure 9.3 The wireless sensor networks protocol stack.

The power management plane manages how a sensor node uses its power. The mobility management plane detects and registers the movement of sensor nodes, so the sensor nodes can keep track of who their neighbor sensor nodes are, and a path back to the sink is always maintained. The task management plane balances and schedules the sensing tasks given to a specific region. The power, mobility, and task management planes monitor the power, movement, and task distribution among the sensor nodes. These planes help the sensor nodes coordinate the sensing task and lower overall power consumption.

9.3 Design Factors

The design of routing protocols in wireless sensor networks is influenced by many factors, including energy consumption, fault tolerance, scalability, and hardware constrains. These factors are important because they serve as a guideline to achieve efficient communication in wireless sensor networks. We summarize some of the design factors that affect the routing process in wireless sensor networks below.

9.3.1 Power Consumption

A sensor node, being a microelectronic device, can only be equipped with a limited battery power, and can use up its limited battery power performing computation and transmitting information in a wireless environment. Often the battery of a sensor node is not rechargeable, and the need to prolong the lifetime of a sensor node has a deep impact on the system and networking architecture. Thus, energy consumption is the most important factor to determine the life of a sensor network. Each node plays a dual role as data originator and data router in a multihop wireless ad hoc sensor network. The malfunctioning of some sensor nodes due to power failure can cause significant topological changes, and might require rerouting of packets and reorganization of the network. It is for these reasons that researchers are currently focusing on the design of power-aware protocols and algorithms for sensor networks.

9.3.2 Fault Tolerance

Sensor nodes may fail due to lack of power, physical damage, or environment inferences. It may be difficult to replace or recharge the existing sensors; the network must have the ability of fault tolerance, which is the ability to maintain sensor network functionalities without any interruption due to any failure of sensors. Fault tolerance may be achieved through reducing energy consumption or replicating data. Therefore, multiple levels of redundancy may be needed in a fault-tolerant sensor network.

9.3.3 Scalability

Scalability for sensor networks is a critical factor. The number of sensor nodes deployed in the sensing field may be on the order of hundreds or thousands, or more.

New routing schemes must be able to work this huge number of nodes. At the same time, routing schemes must also utilize the high density of the sensor networks to respond to events in the environment. In order to ensure scalability, it is important to localize interactions through hierarchy and aggregation.

9.3.4 Hardware Issues

A sensor node is expected to be of a matchbox size. The required size may be smaller than even a cubic centimeter. There are some other stringent hardware constraints for sensor nodes apart from small size. These sensor nodes must be able to implement the following items:

- Consume extremely low power;
- Operate in high volumetric densities;
- Have low production cost and be dispensable;
- Be autonomous and operate unattended;
- Be adaptive to the environment.

9.4 Routing in Wireless Sensor Networks

In this section, we survey a few well-cited routing protocols for wireless sensor networks and analyze their performance metrics. Routing in WSNs can be broken down into flooding-based routing, gradient-based routing, hierarchical-based routing and location-based routing, depending on data communication techniques. In the rest of this section we present a detailed overview of the main routing paradigms in WSNs.

9.4.1 Flooding-Based Routing

Flooding and gossiping are two classical methods to disseminate data in sensor networks without the need for any routing algorithms and topology maintenance. Each node receiving a data packet broadcasts it to all of its neighbors and this process continues until the packet arrives at the destination node or the maximum number of hops for the packet is arriving in flooding. In gossiping, the receiving sensor node sends the data packet to some randomly selected neighbors which also pick another random neighbors to forward the packet to and so on. Although flooding is very easy to implement, it has several drawbacks [2].

Such drawbacks include implosion caused by duplicated messages sent to the same node. As shown in Figure 9.4, node A starts by flooding its data to all of its neighbors. Node D gets two copies of the data, which is not necessary.

There is overlap when two nodes sensing the same region send similar packets to the same neighbor. As shown in Figure 9.5, two sensors cover an overlapping geographic region and C gets same the copy of data from these sensors. There then is resource blindness caused by consuming a large amount of energy without consideration for the energy constraints [2]. Gossiping avoids the problem of implosion by

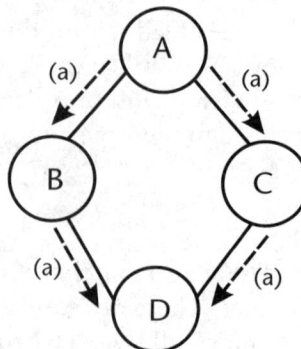

Figure 9.4 The implosion problem.

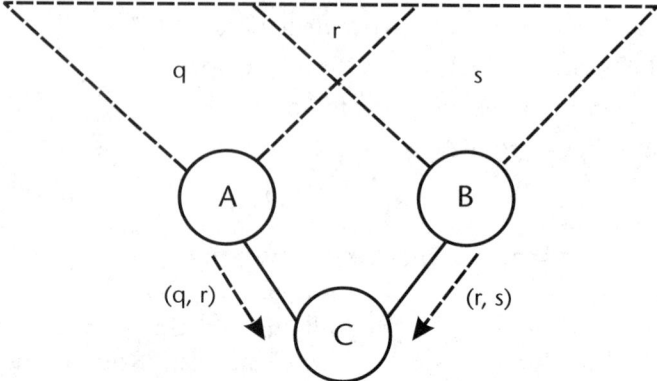

Figure 9.5 The overlap problem.

just selecting a random node to send the packet rather than broadcasting. However, this cause delays in propagation of data through the nodes.

9.4.1.1 T&D flooding

Liu et al. [3] propose a location-information-free and flooding-like routing protocol for WSNs, called throw and drowned (T&D) flooding protocol. This protocol is a multihop implicit routing protocol. Packets are propagated without knowing the network topologies or previously and establishing the route. T&D is also energy saving. It can turn the sensor node into the sleeping mode for the purpose of power saving. It is suitable for applications with dynamic topology change and scalable WSNs, and instead of pure flooding in existing WSNs, T&D has three mechanisms for power saving. First, T&D can turn sensor node into sleeping mode to save power. Second, T&D is designed as a flooding-based and implicit routing protocol. Third, T&D uses link layer relaying instead of network-layer-forwarding to propagate packets to sink, which saves energy costing on next-hop selection and route establishment. T&D uses these three schemes for packet routing and to prolong the lifetime more than 142% to the pure flooding and gossiping scheme in simulation. Knowledge of location information is necessary for most of the sensing tasks and part of WSNs routing protocol. T&D does not use the sensor node's location information. However, it can be applied to the applications whether they have the location information or not.

T&D divides time into slots and intends to propagate one packet in a slot. A node receives a packet and throws it after a random back-off time by broadcasting. Nodes receiving more than one packet will be drowned by the packet and stop throwing it. T&D does not need to maintain the neighbors table during the packet relaying stage, and it may have more than one route propagating the packet to the sink.

T&D is designed for the WSN with a dynamic topology changed environment and for the WSN with a large number of sensor nodes. Simulation results show that the T&D have good packet reachability, particularly in a WSN with more than one sinks. The average packet delay can be predicted and is directly proportional to the length of PPP. T&D can also be used instead of pure flooding in other WSNs when broadcasting management or query messages.

9.4.1.2 SPIN (SPIN-1, SPIN-2)

Sensor protocols for information via negotiation (SPIN) [2, 4] is a serial of adaptive protocols that disseminate all the information at each node to every node in the networks assuming that all nodes in the network are potential BSs. SPIN's metadata negotiation solves the classic problems of flooding, such as redundant information passing, overlapping of sensing areas and resource blindness—thus, achieving a lot of energy efficiency. There is no standard metadata format, and it is assumed to be application specific (e.g. using an application level framing.) SPIN is a three-stage protocol since sensor nodes use three types of messages, ADV, REQ, and DATA, to communicate. ADV is used to advertise new data, REQ to request data, and DATA is the actual message itself. Figure. 9.6, redrawn from [2], summarizes the steps of the SPIN protocol.

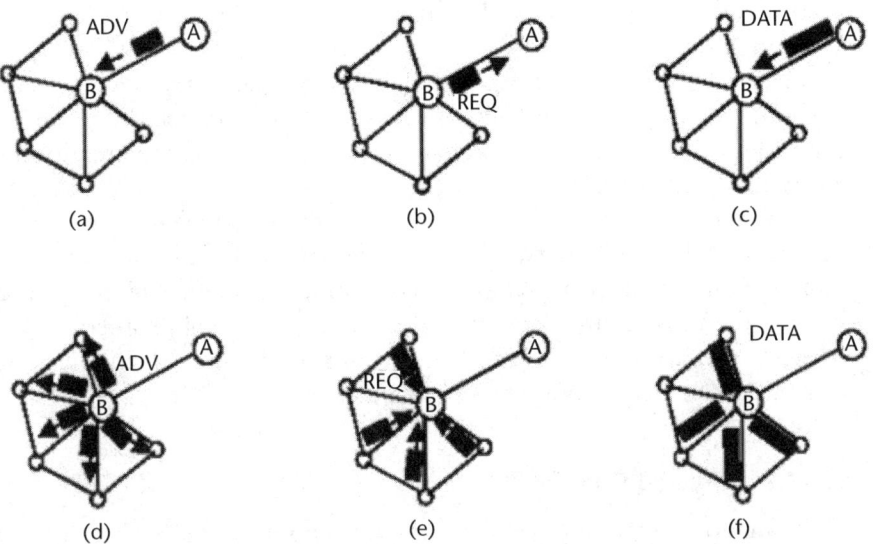

Figure 9.6 SPIN protocol (*From:* [2].)

As shown in Figure 9.6, node A starts by advertising its data to node B (a). Node B responds by sending a request to node A (b). After receiving the requested data (c), node B then sends out advertisements to its neighbors (d), who in turn send requests back to B (e, f).

One of the advantages of SPIN is that topological changes are localized since each node needs to know only its single-hop neighbors. SPIN gives a factor of 3.5 less than flooding in terms of energy dissipation and metadata negotiation almost halves the redundant data. However, SPIN's data advertisement mechanism cannot guarantee the delivery of data. For instance, if the nodes that are interested in the data are far away from the source node and the nodes between source and destination are not interested in that data, such data will not be delivered to the destination at all. Therefore, SPIN is not a good choice for applications such as intrusion detection, which require reliable delivery of data packets over regular intervals.

The SPIN family of protocols includes many protocols. The main two are called SPIN-1 and SPIN-2; the SPIN-1 protocol is a three-stage protocol, as described previously. An extension to SPIN-1 is SPIN-2, which incorporates a threshold-based resource awareness mechanism in addition to negotiation. When energy in the nodes is abundant, SPIN-2 communicates using the three-stage protocol of SPIN-1. In conclusion, SPIN-1 and SPIN-2 are simple protocols that efficiently disseminate data while maintaining no per-neighbor state. These protocols are well suited to an environment where the sensors are mobile, because they base their forwarding decisions on local neighborhood information.

9.4.2 Gradient-Based Routing

The key idea in gradient-based routing (GBR) [5] is to memorize the number of hops when the interest is diffused through the whole network. As such, each node can calculate a parameter called the height of the node, which is the minimum number of hops to reach the BS. The difference between a node's height and that of its neighbor is considered the gradient on that link. A packet is forwarded on a link with the largest gradient.

The authors aim at using some auxiliary techniques such as data aggregation and traffic spreading along with GBR in order to balance the traffic uniformly over the network. Nodes acting as a relay for multiple paths can create a data combining entity in order to aggregate data.

The data-spreading schemes strive to achieve an even distribution of the traffic throughout the whole network, which helps in balancing the load on sensor nodes and increases the network lifetime. The employed techniques for traffic load balancing and data fusion are also applicable to other routing protocols for enhanced performance. Through simulation GBR has been shown to outperform directed diffusion in terms of total communication energy.

9.4.2.1 Directed Diffusion

C. Intanagonwiwat et al. [6, 7] proposed a popular data aggregation paradigm for WSNs called directed diffusion. Directed diffusion is a data-centric (DC) and application-aware paradigm in the sense that all data generated by sensor nodes is named

9.4 Routing in Wireless Sensor Networks

by attribute-value pairs. The main idea of the DC paradigm is to combine the data coming from different sources en route (in-network aggregation) by eliminating redundancy, and minimizing the number of transmissions, thus saving network energy and prolonging its lifetime. Unlike traditional end-to-end routing, DC routing finds routes from multiple sources to a single destination that allows in-network consolidation of redundant data.

In directed diffusion, sensors measure events and create gradients of information in their respective neighborhoods. The BS requests data by broadcasting interests that describe tasks required to be done by the network. As the interest is propagated throughout the network, gradients are set up to draw data satisfying the query toward the requesting node. Each sensor that receives the interest sets up a gradient toward the sensor nodes from which it receives the interest. This process continues until gradients are set up from the sources back to the BS. More generally, a gradient specifies an attribute value and a direction. The strength of the gradient may be different toward different neighbors, resulting in different amounts of information flow. Figure 9.7 shows an example of directed diffusion (sending interests, building gradients, and data dissemination).

When interests fit gradients, paths of information flow are formed from multiple paths, and then the best paths are reinforced to prevent further flooding according to a local rule. In order to reduce communication costs, data is aggregated on the way. The goal is to find a good aggregation tree that gets the data from source nodes to the BS. The BS periodically refreshes and resends the interest when it starts to receive data from the source(s). This is necessary because interests are not reliably transmitted throughout the network.

Directed diffusion differs from SPIN in two aspects. First, directed diffusion issues data queries on demand as the BS sends queries to the sensor nodes by flooding some tasks. In SPIN, however, sensors advertise the availability of data, allowing interested nodes to query that data. Second, all communication in directed diffusion is neighbor to neighbor, with each node having the capability to perform

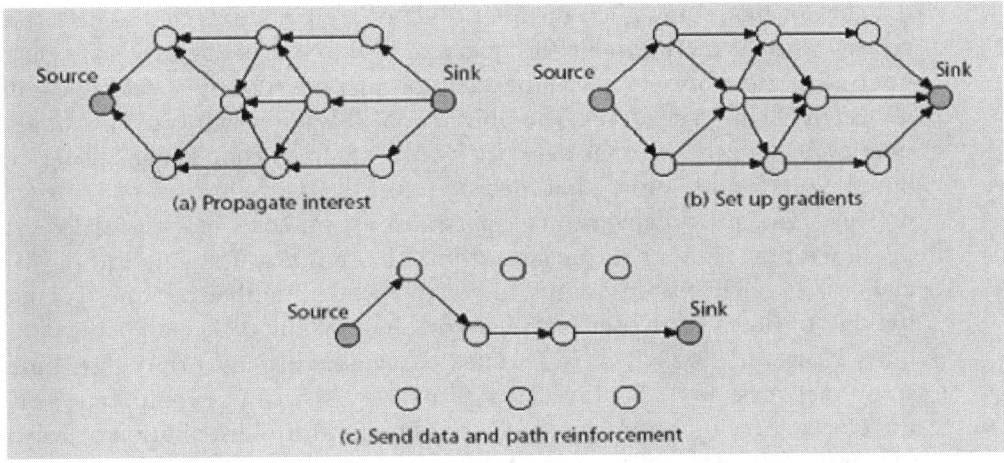

Figure 9.7 An example of interest diffusion in a sensor network.

data aggregation and caching. Unlike SPIN, there is no need to maintain global network topology in directed diffusion. However, directed diffusion may not be applied to applications (e.g., environmental monitoring) that require continuous data delivery to the BS. This is because the query-driven on-demand data model may not help in this regard. Moreover, matching data to queries might require some extra overhead at the sensor nodes.

9.4.2.2 Rumor Routing

Rumor routing [8] is a variation of directed diffusion and is mainly intended for applications where geographic routing is not feasible. The rumor routing algorithm employs long-lived packets called agents in order to flood events through the network. A node adds the event to its local table, called an events table, when it detects an event, and then generates an agent. Agents travel the network to propagate information about local events to distant nodes. The nodes that know the route may respond to the query by inspecting its event table when a node generates a query for an event. So there is no need to flood the whole network, which reduces the communication cost. Rumor routing maintains only one path between source and destination, as opposed to directed diffusion where data can be routed through multiple paths at low rates. Simulation results showed that rumor routing can achieve significant energy savings compared with event flooding and can also handle a node's failure.

9.4.2.3 Zonal Rumor Routing

Zonal rumor routing (ZRR) [9] is an extension to the rumor routing algorithm. Compared with a traditional rumor routing algorithm, the ZRR algorithm improves the percentage of query delivery and generates fewer transmissions, thus reducing the total energy consumption in a sensor network. In Zonal rumor routing, the network is partitioned into zones, where each node is a member of exactly one zone. Zones are partitions of the network, thus any node present in one zone becomes a member of that zone. Each member node of a zone stores that zone's zone-ID. In rumor routing, the agent or the query randomly selects an unvisited neighboring node as the next hop. In ZRR, the agent or query randomly selects a neighbor in an unvisited neighboring zone. The objective of this approach is to spread agents and queries to a large part of the network by moving from zone to zone using long steps. Simulation results show that decision for the next hop in ZRR significantly improves the query delivery rate and transmission costs of query delivery as compared with the rumor routing algorithm. The rumor routing algorithm [8] is a special case of ZRR corresponding to the case when number of zones is equal to the number of nodes in the sensor a network (i.e., one node is assigned to one unique one.) However, in ZRR, it is assumed that nodes are uniformly distributed in the sensor network, so that in this case optimal zone size can be computed easily. It does not consider the optimum zone size with a nonuniform distribution of sensors in the sensor field. It also does not evaluate overhead of different zone creation schemes for sensor networks.

9.4.3 Hierarchical-Based Routing

Similar to other communication networks, scalability is one of the major design attributes of sensor networks. A single-tier network can cause the gateway to overload, with an increase in sensors density. Such overload might cause latency in communication and inadequate tracking of events. In addition, the single-gateway architecture is not scalable for a larger set of sensors covering a wider area of interest since the sensors are typically not capable of long-haul communication. To allow the system to cope with additional load and to be able to cover a large area of interest without degrading the service, networking clustering has been pursued in some routing approaches.

Hierarchical or cluster-based routing methods, originally proposed in wired line networks, are well-known techniques with special advantages related to scalability and efficient communication. In a hierarchical architecture, higher-energy nodes can be used to process and send the information, while low-energy nodes can be used to perform the sensing in the proximity of the target. The creation of clusters and assigning special tasks to cluster heads can greatly contribute to overall system scalability, lifetime, and energy efficiency. Hierarchical routing is an efficient way to lower energy consumption within a cluster, performing data aggregation and fusion in order to decrease the number of transmitted messages to the BS. Hierarchical routing is mainly a two-layer routing, where one layer is used to select cluster heads and the other for routing.

9.4.3.1 LEACH

Low energy adaptive clustering hierarchy (LEACH) is a hierarchical clustering protocol, which includes distributed cluster formation. LEACH randomly selects a few sensor nodes as cluster heads (CHs) and rotates this role to evenly distribute the energy load among the sensors in the network. The operation of LEACH is separated into two phases, the setup phase and the steady-state phase. In the setup phase, the clusters are organized and CHs are selected. In the steady-state phase, the actual data transfer to the BS takes place.

LEACH assumes that all nodes can transmit with enough power to reach the BS if needed and that each node has computational power to support different MAC protocols. Therefore, it is not applicable to networks deployed in large regions. It also assumes that nodes always have data to send, and nodes located close to each other have correlated data. It is not obvious how the number of predetermined CHs (p) is going to be uniformly distributed through the network. Therefore, there is the possibility that the elected CHs will be concentrated in one part of the network; hence, some nodes will not have any CHs in their vicinity. Furthermore, the idea of dynamic clustering brings extra overhead (head changes, advertisements, etc.), which may diminish the gain in energy consumption. Finally, the protocol assumes that all nodes begin with the same amount of energy capacity in each election round, assuming that being a CH consumes approximately the same amount of energy for each node.

LEACH achieves over a factor of 7 reduction in energy dissipation compared with direct communication and a factor of 4–8 compared with the minimum transmission energy routing protocol. The nodes die randomly and dynamic clustering increases the lifetime of the system. LEACH is completely distributed and requires no global knowledge of network. However, LEACH uses single-hop routing, where each node can transmit directly to the cluster head and the sink. Therefore, it is not applicable to networks deployed in large regions. Furthermore, the idea of dynamic clustering brings extra overhead (e.g. head changes, advertisements etc.,) which may diminish the gain in energy consumption.

9.4.3.2 BCDCP

Base station controlled dynamic clustering(BCDCP) [10] is a clustering-based routing protocol. It utilizes a high-energy base station to set up clusters and routing paths, perform randomized rotation of cluster heads, and carry out other energy-intensive tasks. In BCDCP, the key ideas are the formation of balanced clusters, where each cluster head serves an approximately equal number of member nodes to avoid cluster head overload, uniform placement of cluster heads throughout the whole sensor field, and utilization of cluster-head-to-cluster-head (CH-to-CH) routing to transfer the data to the base station. As shown in the following text, BCDCP yields an improved system lifetime and better energy savings over the previously mentioned clustering-based routing protocols.

BCDCP assumes a sensor network model with the following properties:

- A fixed base station is located far away from the sensor nodes.
- The sensor nodes are energy constrained with a uniform initial energy allocation.
- The nodes are equipped with power control capabilities to vary their transmitted power.
- Each node senses the environment at a fixed rate and always has data to send to the base station.
- All sensor nodes are immobile.

The two key elements considered in the design of BCDCP are the sensor nodes and base station. The sensor nodes are geographically grouped into clusters and capable of operating in two basic modes, the cluster head mode, the sensing mode

In the sensing mode, the nodes perform sensing tasks and transmit the sensed data to the cluster head. In cluster head mode, a node gathers data from the other nodes within its cluster, performs data fusion, and routes the data to the base station through other cluster head nodes. The base station in turn performs the key tasks of cluster formation, randomized cluster head selection, and CH-to-CH routing path construction.

BCDCP utilizes the high-energy base station to perform most energy-intensive tasks. By using the base station, the sensor nodes are relieved of performing energy-intensive computational tasks, such as cluster setup, cluster head selection, routing path formation, and TDMA schedule creation.

Performance of the proposed BCDCP protocol is assessed by simulation and compared with other clustering-based protocols such as LEACH. The simulation results show that BCDCP outperforms its comparatives by uniformly placing cluster heads throughout the whole sensor field, performing balanced clustering, and using a CH-to-CH routing scheme to transfer fused data to the base station. It is also observed that the performance gain of BCDCP over its counterparts increases with the area of the sensor field. Therefore, it is concluded that BCDCP provides an energy-efficient routing scheme suitable for a vast range of sensing applications.

9.4.3.3 TTDD

Two-tier data dissemination (TTDD) [11], provides data delivery to multiple mobile BS. In TTDD, each data source proactively builds a grid structure which is used to disseminate data to the mobile sinks by assuming that sensor nodes are stationary and location aware, whereas sinks may change their locations dynamically. Once an event occurs, sensors surrounding it process the signal, and one of them becomes the source to generate data reports. Sensor nodes are aware of their mission, which will not change frequently. To build the grid structure, a data source chooses itself as the start crossing point of the grid, and sends a data announcement message to each of its four adjacent crossing points using simple greedy geographical forwarding. When the message reaches the node closest to the crossing point (specified in the message), it will stop. During this process, each intermediate node stores the source information and further forwards the message to its adjacent crossing points except the one from which the message came. This process continues until the message stops at the border of the network. The nodes that store the source information are chosen as dissemination points. After this process, the grid structure is obtained. Using the grid, a BS can flood a query, which will be forwarded to the nearest dissemination point in the local cell to receive data. Then the query is forwarded along other dissemination points upstream to the source. The requested data then flows down in the reverse path to the sink. Trajectory forwarding is employed as the BS moves in the sensor field. Although TTDD is an efficient routing approach, there are some concerns about how the algorithm obtains location information, which is required to set up the grid structure. The length of a forwarding path in TTDD is larger than the length of the shortest path. The authors of TTDD believe that the suboptimality in the path length is worth the gain in scalability. Finally, how TTDD would perform if mobile sensor nodes are allowed to move in the network is still an open question. Comparison results between TTDD and directed diffusion show that TTDD can achieve longer lifetimes and shorter data delivery delays. However, the overhead associated with maintaining and recalculating the grid as network topology changes may be high. Furthermore, TTDD assumed the availability of a very accurate positioning system which is not yet available for WSNs.

9.4.4 Location Based Routing

Location based routing is using an area instead of a node identifier as the target of a packet; any node that is positioned within the given area will be acceptable as a destination node and can receive and process a message. In the context of sensor networks, such location based routing is important to request sensor data from some region (request temperature in living room); it will also often be combined with some notion of multicast: specifically, stochastically constrained multicast.

9.4.4.1 GAF

Geographic adaptive fidelity (GAF) [12] is an energy-aware location-based routing algorithm designed primarily for mobile ad hoc networks, but it may be applicable to sensor networks as well. The network area is first divided into fixed zones and formed into a virtual grid. Inside each zone, nodes collaborate with each other to play different roles. GAF conserves energy by turning off unnecessary nodes in the network without affecting the level of routing fidelity. Each node uses its GPS-indicated location to associate itself with a point in the virtual grid. Nodes associated with the same point on the grid are considered equivalent in terms of the cost of packet routing. Such equivalence is exploited in keeping some nodes located in a particular grid area in sleeping state in order to save energy. Thus, GAF can substantially increase the network lifetime as the number of nodes increases. A sample situation is depicted in Figure. 9.8, which is redrawn from [12].

In Figure 9.8, node 1 can reach any of nodes 2, 3 and 4 and nodes 2, 3, and 4 can reach 5. Therefore nodes 2, 3 and 4 are equivalent, and two of them can sleep.

Nodes change states from sleeping to active in turn so that the load is balanced. There are three states defined in GAF. These states are discovery, for determining the neighbors in the grid; active reflecting participation in routing, and sleep, when the radio is turned off. The state transitions in GAF are depicted in Figure 9.9. Which node will sleep for how long is application dependent, and the related parameters are tuned accordingly during the routing process.

In order to handle the mobility, each node in the grid estimates it's the time of leaving grid and sends this to its neighbors. The sleeping neighbors adjust their sleeping time accordingly in order to keep the routing fidelity. Before the leaving time of the active node expires, sleeping nodes wake up and one of them becomes active. GAF is implemented both for nonmobility (GAF basic) and mobility (GAF-mobility adaptation) of nodes.

9.4.4.2 GEAR

Energy aware routing (GEAR) [11] uses energy-aware and geographically informed neighbor selection heuristics to route a packet toward the destination region. The

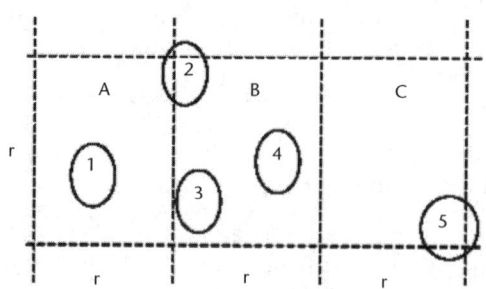

Figure 9.8 Example of a virtual grid in GAF. (*From*: [13].)

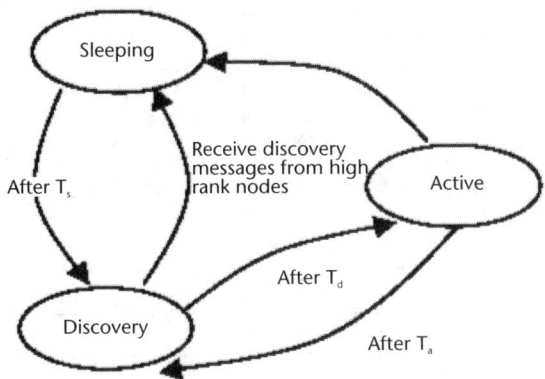

Figure 9.9 State transitions in GAF.

key idea is to restrict the number of interests in directed diffusion by only considering a certain region rather than sending the interests to the whole network. By doing this, GEAR can conserve more energy than directed diffusion.

Each node in GEAR keeps an estimated cost and a learning cost of reaching the destination through its neighbors. The estimated cost is a combination of residual energy and distance to destination. The learned cost is a refinement of the estimated cost that accounts for routing around holes in the network. A hole occurs when a node does not have any closer neighbor to the target region than itself. If there are no holes, the estimated cost is equal to the learned cost. The learned cost is propagated one hop back every time a packet reaches the destination so that route setup for the next packet will be adjusted. There are two phases in the algorithm:

1. Forwarding packets toward the target region: Upon receiving a packet, a node checks its neighbors to see if there is one neighbor that is closer to the target region than itself. If there is more than one, the nearest neighbor to the target region is selected as the next hop. If they are all farther than the node itself, this means there is a hole. In this case, one of the neighbors is picked to forward the packet based on the learned cost function. This choice can then be updated according to the convergence of the learned cost during the delivery of packets.

2. Forwarding the packets within the region: If the packet has reached the region, it can be diffused in that region by either recursive geographic forwarding or restricted flooding. Restricted flooding is good when the sensors are not densely deployed. In high-density networks, recursive geographic flooding is more energy efficient than restricted flooding. In that case, the region is divided into four subregions and four copies of the packet are created. This splitting and forwarding process continues until the regions with only one node are left. An example [14] is depicted in Figure. 9.10.

9.4.4.3 GPSR

Greedy perimeter stateless routing (GPSR) [15] is a novel routing protocol for wireless datagram networks that uses the positions of routers and a packet's destination

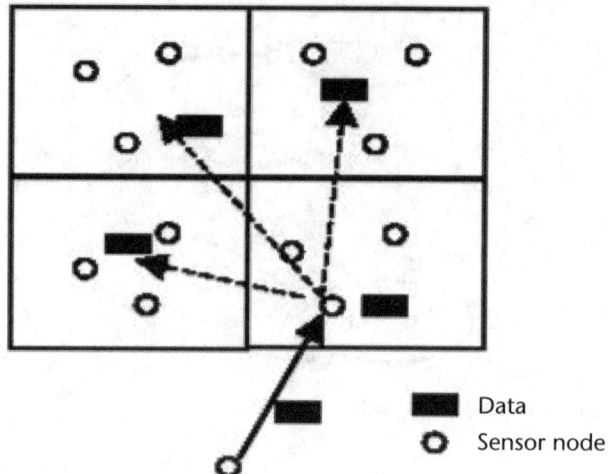

Figure 9.10 Recursive geographic forwarding in GEAR. (*From*: [14].)

to make packet forwarding decisions. GPSR makes greedy forwarding decisions using only information about a router's immediate neighbors in the network topology. When a packet reaches a region where greedy forwarding is impossible, the algorithm recovers by routing around the perimeter of the region. By keeping state only about the local topology, GPSR scales better in per-router state than shortest-path and ad hoc routing protocols, as the number of network destinations increases. Under mobility's frequent topology changes, GPSR can use local topology information to find correct new routes quickly. GPSR's benefits all stem from geographic routing's use of only immediate-neighbor information in forwarding decisions. Routing protocols that rely on end-to-end state concerning the path between a forwarding router and a packet's destination, as do source-routed, DV, and LS algorithms, face a scaling challenge as network diameter in hops and mobility increases because the product of these two factors determines the rate that end-to-end paths change. Hierarchy and caching have proven successful in scaling these algorithms. Geography, as exemplified in GPSR, represents another powerful lever for scaling routing.

Although the GPSR approach reduces the number of states a node should keep, it was designed for general mobile ad hoc networks and requires a location service to map locations and node identifiers. GEAR not only reduces energy consumption for route setup, but also performs better than GPSR in terms of packet delivery. The simulation results show that for uneven traffic distribution, GEAR delivers 70% to 80% more packets than GPSR. For uniform traffic pairs GEAR delivers 25% to 35% more packets than GPSR.

9.4.4.4 LCR

Logical coordinate routing (LCR) [16] is a novel logical coordinate framework that encodes connectivity information for routing purposes without the benefit of geographic knowledge, while retaining the constant-state advantage of geographic routing.

The main idea of LCR is for each node to maintain hop counts to a small number of landmarks. This hop count vector is the logical coordinate of the node. The difference vector between two node vectors represents the distance between them. The routing scheme forwards packets in the direction that minimizes the magnitude of the remaining difference vector. Compared with current routing protocols, this simple scheme has several advantages. First, it translates the routing problem into a different logical domain in which the state kept on each node is constant (only immediate neighbor coordinates). Moreover, it can encode a configurable amount of topological information that depends on the number of chosen landmarks, which are called configurable dimensionality, from which are observed several positive implications. Most importantly, since the logical coordinate dimensions can be arbitrarily enriched by increasing the number of landmarks, logical vectors contain inherent redundancy, which significantly improves robustness with respect to node failures and physical voids. Selecting an appropriate number of landmarks at suitable locations makes it possible to eliminate voids in the logical coordinate space despite their existence in the physical space. Compared with location-based approaches, the logical coordinates directly encode connectivity relationships among nodes, rather than physical proximity. Hence, they reflect and abstract the more relevant topological information in a simple and efficient manner. Finally, the logical coordinates of each node can be used to bound the actual hop distance between two arbitrary nodes. As a result, they can be leveraged to predict the delivery delay between nodes, which is of use in soft real-time sensor networks such as EnviroTrack [17] and Mobicast [18].

Compared with other ef?cient routing protocols that are geographical location independent [19, 20], LCR is the ?rst one to use con?gurable logical dimensionality, and directly encode a bound on hop count. Besides, simulation-based comparisons show that LCR protocol demonstrates a considerable performance improvement in delivery ratio, especially in the presence of voids.

Based on these simulation results, it can be concluded that every protocol has its desirable features. Although GPSR is the scheme that theoretically guarantees a 100% delivery ratio, it has drawbacks in sensor networks. The route discovery process GPSR can be severely degraded by localization inaccuracy. The design of LCR avoids these problems. LCR is not affected by localization errors thus guaranteeing a better delivery ratio than greedy GF and degraded GPSR. Its protocol overhead, however, is low and it exhibits exceptional ability to avoid voids in the network. In addition to ef?ciency in the absence of geographic knowledge, LCR has two important advantages: it improves robustness in the presence of voids compared with other logical coordinate frameworks, and it allows inferring bounds on route hop count from the logical coordinates of the source and destination nodes, which makes it a candidate for use in soft, real-time systems. The scheme is evaluated in simulation demonstrating the advantages of the new protocol.

9.5 IP Mobility Management in Wireless Sensor Networks

The widespread growth of mobile wireless networks, applications, and services has ushered in the era of mobile computing, where handheld computing devices (or terminals) have become the predominant choice for users [21]. At the same time, major efforts are underway to deliver applications and services to mobile nodes (MNs) over a packet-switched access network that is homogeneous with the Internet. So the trend in mobile wireless network evolution is directed toward an all-IP network. Work has already begun on such an end-to-end IP-based solution that will combine mobility with multimedia-rich content, high bit rate, and IP transport, with support for quality-of-service (QoS) management and authentication, authorization, and accounting (AAA) security [22].

9.5.1 IP Mobility Protocols

Since many sensors and other AmI devices will use IP, it is necessary to give a brief review of Mobile IP v4/6 and its ad hoc derivatives. Thus, we present an overview of a set of IP-based mobility protocols—Mobile IP, Hawaii, Cellular IP, Hierarchical MIP—that will play an important role in the forthcoming convergence of IP and legacy wireless networks. A comparative analysis with respect to system parameters such as location update, handoff latency and signaling overhead exposes their ability in managing micro/macro/global-level mobility.

9.5.1.1 Mobile IP (MIP)

Mobile IP (MIP) [23] was the proposed standard (and MIPv6 [24] is the proposed draft standard) by the Internet Engineering Task Force (IETF). MIP was originally designed to serve the needs of globally mobile users who wish to connect their MNs to the Internet and maintain connectivity as they move from one place to another, establish new links, and move away from previously established links. This is achieved by providing a care-of address (CoA) to an MN when it moves out of its home network (HN) to visit a foreign network (FN). While in an FN, the location of an MN is captured by its CoA assigned by a foreign agent (FA) in the FN. A home agent (HA) in the HN maintains a record of the current mobility binding (i.e., the association of an MN's home address with its CoA during the remaining lifetime of that association). The HA intercepts every packet addressed to the MN's home address and tunnels them to the MN at its current CoA. This is known as triangular routing.

For managing mobility on the level of the global Internet, Mobile IP offers a practical solution. However, frequent handoffs inside a relatively small geographic area tend to generate a remarkable amount of signaling overhead due to required control messages between a mobile host and home agent. Additionally, the need for obtaining a new CoA and notifying it to a possibly distant home agent results in latency and disruption to user traffic during every handoff. Smooth, fast, and transparent handoffs are impossible to do with the present basic Mobile IP [25].

Several IP micro/macro mobility protocols, such as Hawaii and Cellular IP and Hierarchical MIP (HMIP), proposed over the past several years within the IETF, complement MIP in better handling local movement (e.g., within a subnet) without much interaction with the MIP-enabled Internet. Despite many apparent differences, the operational principle of the protocols is quite similar in complementing base MIP by providing local handoff control.

9.5.1.2 Hawaii

The Hawaii protocol [26] from Lucent Technologies proposes a separate routing protocol to handle intradomain mobility on top of using MIP. Four alternative path setup schemes control handoff between access points. The appropriate path setup scheme is selected depending on the operator's priorities between eliminating packet loss, minimizing handoff latency, and maintaining packet ordering. Hawaii also uses IP multicasting to page idle mobile hosts when incoming data packets arrive at an access network and no recent routing information is available.

9.5.1.3 Cellular IP (CIP)

The Cellular IP protocol [27, 28] from Columbia University and Ericsson Research supports paging and a number of handoff techniques. Location management and handoff support are integrated with routing in CIP networks. An MN communicates to its HA with the local gateway's address as the CoA. Consequently, after intercepting the packets from a CN, the HA sends them in encapsulated form to the MN's gateway. The gateway decapsulates the packet and forwards it to the MN. To minimize control messaging, regular data packets transmitted by mobile hosts are used to refresh host location information. Cellular IP uses mobile-originated data packets to maintain reverse path routes. Nodes in a Cellular IP access network monitor (i.e., snoop) mobile originated packets and maintain a distributed, hop-by-hop location data base that is used to route packets to mobile hosts. Cellular IP uses IP addresses to identify mobile hosts. Since the CoA does not change in local mobility control, CIP eliminates the overhead for location updates in the Internet [29]. It supports a fast security model based on special session keys, where base stations (BSs) independently calculate keys. This eliminates the need for signaling in support of session key management, which would otherwise add additional delay to the handoff process.

9.5.1.4 Hierarchcal MIP (HMIP)

The Hierarchical Mobile IP protocol [30] from Ericsson and Nokia employs a hierarchy of FAs to locally handle Mobile IP registration. It is an extension of MIP. In this protocol mobile hosts send Mobile IP registration messages (with appropriate extensions) to update their respective location information. Registration messages establish tunnels between neighboring FAs along the path from the mobile host to a gateway FA (GFA). Packets addressed to the mobile host travel in this network of tunnels, which can be viewed as a separate routing network overlay on top of IP. The use of tunnels makes it possible to employ the protocol in an IP network that carries non-mobile traffic as well.

9.5.1.5 Mobile IPv6

With the huge (128 bits long) address space of MIPv6 [24], a tiny part is reserved for all current MIPv4 [23] addresses. Another tiny part is reserved for link-local addresses, which are not routable but are guaranteed to be unique on a link. Design of MIPv6 is adjusted to account for the few special needs of MNs that can perform decapsulation. A set of new destination options, called binding update and binding acknowledgment, manages the cache entries of CNs. MNs must be able to send binding updates and receive binding acknowledgments. Based on the lifetime field in the binding updates it sends, every MN must keep track of which other MNs may need to receive a new binding as a result of any recent movement by the MN [23].

9.5.2 Limitations of Mobile IP

It's important to realize that Mobile IP has certain limitations in its usefulness in a wireless environment. It was designed to handle mobility of devices, but only relatively infrequent mobility. This is due to the reconfiguration work involved with each change. It can be an issue for real-time mobility such as roaming in a wireless network, where handoff functions operating at the data link layer may be more suitable. Mobile IP was designed under the specific assumption that the attachment point would not change more than once per second. Mobile IP is intended to be used with devices that maintain a static IP configuration. Since the device needs to be able to always know the identity of its home network and normal IP address, it is much more difficult to use it with a device that obtains an IP address dynamically, using something like DHCP.

9.6 Conclusions

Wireless sensor networks have become popular due to the recent advances in sensing, communication, and computation. To make wireless sensor networks useful, we need to develop routing protocols for them that meet several unique requirements and constraints. Thus, routing in wireless sensor networks has attracted a lot of attention in the recent years. In this chapter, we have surveyed the communication architecture of wireless sensor networks, recent research results on data routing in wireless sensor networks, and classified the approaches into four main categories according to data communication techniques—namely flooding-based routing, gradient-based routing, hierarchical- and location-based routing. Moreover, we have highlighted the advantages and drawbacks of each protocol and analyzed some performance metrics of the protocol.

Flooding-based protocols disseminate all the information at each node to every node in the network, assuming that all nodes in the network may be interested in the information through controlled flooding. They do not rely on global interaction or information, and aim to achieve desired global behavior with an adaptive localized algorithm. However, they are hard to model in a dynamic environment [6].

Gradient-based routing protocols are query-based, and all data generated by sensor nodes is named by attribute-valued pairs, which helps in eliminating many redundant transmissions. However, the naming schemes, such as attributed-value pairs, might not be sufficient for complex queries and they are usually dependent on the application. Efficient standard naming schemes are one of the most interesting future research directions related to this category.

Hierarchical-based protocols aim at clustering the nodes to efficiently relay the sensed data to the sink. The cluster heads are sometimes chosen as specialized nodes that are less energy constrained. A cluster-head performs aggregation of data and sends it to the sink on behalf of the nodes within its cluster. However, the most interesting research issue related to such protocols is to find the novel techniques of network clustering so that the energy consumption and contemporary communication metrics such as latency are optimized, and the network lifetime is maximized. Furthermore, the process of data aggregation and fusion among clusters is also an interesting problem to explore.

Location-based protocols utilize position information to relay the data to the desired regions, rather than the whole network. The number of energy-aware location based protocols found in the literature is rather small. The problem of intelligent utilization of the location information in order to aid energy-efficient routing is the main research issue. The problem of estimating spatial coordinates of the node is referred to as localization. There is a need to develop efficient methods of establishing a coordinate system without relying on an existing infrastructure. Many researchers have proposed localization techniques based on recursive trilateration/multilateration techniques, which would not provide enough accuracy in wireless sensor networks.

Other possible future research for routing protocols includes the integration of sensor networks with wired networks (i.e., the Internet). Most applications in security and environmental monitoring require the data collected from sensor nodes to be transmitted to a server so that further analysis can be done. On the other hand, the requests from the user should be made to the BS through the Internet. Since the routing requirements of each environment are different, further research is necessary for handling these kinds of situations. It is also very important to propose a general mathematical framework that can be used to measure the total performance of a network protocol for wireless sensor networks.

References

[1] Vasilakos, A., W. Pedrycz, "Ambient Intelligence: Visions and Technologies," *Ambient Intelligence, Wireless Networking, and Ubiquitous Computing*, Norwood MA: Artech House, 2006.

[2] Heinzelman, W., J. Kulik, and H. Balakrishnan, "Adaptive Protocols for Information Dissemination in Wireless Sensor Networks," *Proc. 5th ACM/IEEE Mobicom*, Seattle, WA, Aug. 1999. pp. 174–85.

[3] Hung-Huan Liu, Wu, J.-L.C., Chun-Jui Wang, "A multi-hop implicit routing protocol for sensor networks," *Vehicular Technology Conference*, 2004. VTC2004-Fall. 2004 IEEE 60th Volume 4, 26–29 Sept. 2004.

[4] Kulik, J., W. R. Heinzelman, and H. Balakrishnan, "Negotiation-Based Protocols for Disseminating Information in Wireless Sensor Networks," *Wireless Networks*, Vol. 8, 2002, pp. 169–85.

[5] Schurgers C., and M.B. Srivastava, "Energy Efficient Routing in Wireless Sensor Networks," *MILCOM Proc. Commun. for Network-Centric Ops.: Creating the Info. Force*, McLean, VA, 2001.

[6] Intanagonwiwat, C., R. Govindan and D. Estrin, "Directed diffusion: A Scalable and Robust Communication Paradigm for Sensor Networks", *Proceedings of the 6th Annual ACM/IEEE International Conference on Mobile Computing and Networking (MobiCom'00)*, Boston, MA, August 2000.

[7] Intanagonwiwat, C., R. Govindan, D. Estrin, J. Heidemann, and F. Silva, "Directed Diffusion for Wireless Sensor Networking," *IEEE/ACM Trans. Networking*, Vol. 11, Feb. 2003, pp. 2–16.

[8] Braginsky D., and D. Estrin, "Rumor Routing Algorithm for Sensor Networks," *Proc. 1st Wksp. Sensor Networks and Apps.*, Atlanta, GA, Oct. 2002.

[9] Banka, T., G. Tandon, and Anura P. Jayasumana, "Zonal Rumor Routing for Wireless Sensor Networks," *Proceedings of the International Conference on Information Technology: Coding and Computing* (ITCC'05)

[10] Muruganathan, S.D., Ma, D.C.F., Bhasin, R.I., Fapojuwo, A.O., "A Centralized Energy-efficient Routing Protocol for Wireless Sensor Networks," *Communications Magazine*, IEEE, Volume 43, Issue 3, March 2005, pp. S8–13

[11] Ye, F. et al., "A Two-Tier Data Dissemination Model for Large-Scale Wireless Sensor Networks," *Proc. ACM/IEEE MOBICOM*, 2002.

[12] Xu, Y., J. Heidemann, and D. Estrin, "Geographyinformed Energy Conservation for Ad-hoc Routing," *Proc. 7th Annual ACM/IEEE Int'l. Conf. Mobile Comp. and Net.*, 2001, pp. 70–84.

[13] Xu, Y., J. Heidemann, and D. Estrin, "Geography-informed energy conservation for ad hoc routing," *Proceedings of the 7th Annual ACM/IEEE International Conference on Mobile Computing and Networking (MobiCom'01)*, Rome, Italy, July 2001.

[14] Yu, Y., D. Estrin, and R. Govindan, "Geographical and Energy-Aware Routing: A Recursive Data Dissemination Protocol for Wireless Sensor Networks," *UCLA Comp. Sci. Dept. tech. rep.*, UCLA-CSD TR-010023, May 2001.

[15] Karp, B., and H. T. Kung, "GPSR: Greedy Perimeter Stateless Routing for Wireless Sensor Networks," *Proc. MobiCom 2000*, Boston, MA, Aug. 2000.

[16] Cao, Q., Abdelzaher, T., "A scalable logical coordinates framework for routing in wireless sensor networks," Real-Time Systems Symposium, 2004. *Proceedings. 25th IEEE International* Dec. 5–8, 2004 pp. 349–358.

[17] T. A. et al. Envirotrack: Towards an Environmental Computing Paradigm for Distributed Sensor Networks, *In International Conference on Distributed Computing Systems (ICDCS)*, 2004.

[18] Huang, Q., C. Lu, and G.-C. Roman. "Spatiotemporal Multicast in Sensor Networks," *First ACM Conference on Embedded Networked Sensor Systems (SenSys'03)*, 2003.

[19] Rao, A., C. Papadimitriou, S. Shenker, and I. Stoica. "Geographic Routing Without Location Information." *In Proceedings of Mobicom*, 2003.

[20] Newsome, J., and D. Song. "Gem: Graph Embedding for Routing and Datacentric Storage in Sensor Networks Without Geographic Information," *Proceedings of the SenSys2003*, 2003.

[21] Stuckman, P., *The GSM Evolution—Mobile Packet Data Services*, Wiley, 2003.

[22] Akyíldíz, I. F., et al., "Mobility Management in Current and Future Communications Networks," *IEEE Network*, Vol. 12, July/Aug. 1998, pp. 39–49.

[23] Perkins, C., "IP Mobility Support for IPv4." RFC 3344, 2002.

[24] Johnson, D., C. Perkins, J. Arkko, "Mobility Support in IPv6." RFC 3775, 2004.

[25] Saraswady, D.; Shanmugavel, S.; "Performance Analysis of Micromobility Protocol in Mobile IP Networks," *Networking, Sensing and Control, 2004 IEEE International Conference*, Vol. 2, 2004 pp. 1406–1411.

[26] Ramjee, R. et al., "HAWAII: A Domain-Based Approach for Supporting Mobility in Wide-area Wireless Networks," *Proc. Int'l. Conf. Network Protocols*. 1999.

[27] Valko, A., "Cellular IP: A New Approach to Internet Host Mobility," *ACM SIGCOMM Comp. Commun. Rev.*, Vol. 29, No. 1, Jan. 1999, pp. 50–65.

[28] Campbell, A. T., et al., "Design and Performance of Cellular IP Access Networks," *IEEE Pers. Commun., Special Issue on IP-Based Mobile Telecommunications Networks*, K. Basu, A. T. Campbell, and A. Joseph, Guest Eds., Vol. 7, No. 4, Aug. 2000, pp. 42–49.

[29] Sidahmed, E.; Ali, H.H.; "A QoS Based Model for Supporting Multimedia Applications over Cellular IP," *System Sciences, 2005. HICSS'05. Proceedings of the 38th Annual Hawaii International Conference*, Jan. 3–6 2005, p. 305b.

[30] Gustafsson, E., A. Jonsson. and C. Perkins. "Mobile IP Regional Registration ," Internet draft, draft-ietf-mobileip-reg-tunnel-03, work in progress, July 2000.

CHAPTER 10
Pose Awareness for Augmented Reality and Mobile Devices

J.Caarls and P.P. Jonker

In this chapter, we describe in detail the self-localization algorithm for our augmented reality set-up, using a camera and inertia sensors. The idea is that the algorithms and the hardware can be used in intelligent mobile devices as well. For instance, a mobile phone could mute the sound when it knows it is in a conference room. Different applications require different accuracies, latencies and update rates. Although we developed our algorithms primary for augmented reality, we try to make the system dynamic in the sense that sensors can be removed or added by the end-user, depending on the accuracy needed.

10.1 Introduction

Wireless technologies that enable continuous connectivity for mobile devices will lead to new application domains. An important paradigm for continuously connected mobile users based on laptops, PDAs, or mobile phones, is context-awareness. Context is relevant in a mobile environment, as it is dynamic and the user interacts in a different way with an application when the context changes. Context is not limited to the physical world around the user, but also incorporates the user's behavior, his terminal and the network characteristics. As an example of a context aware application, we are developing an augmented reality system that can be connected to a roaming PDA. Our user carries a wearable terminal and a see-through display in which the user can see virtual visual information that augments reality (Figure 10.1). Augmented reality differs from virtual reality in the sense that the virtual objects are rendered on a see-through headset. As with audio headphones, where one can hear sound in private, partly in overlay with the sounds from the environment, see-through headsets can do that for visual information. The virtual objects are in overlay with the real visual world (Figure 10.2). It can also be used to place visual information on otherwize empty places, such as white parts of walls of a museum. The 3D vector of position and orientation (or heading) is referred to as pose. Knowing the pose of those walls and the pose of a person's head, visual data can be perfectly inlaid on specific spots and kept there while the head is moving. To lock the virtual objects in the scene, the head-movements must be sam-

Figure 10.1 Our augmented reality setup.

Figure 10.2 Example of augmentation of the visual reality.

pled with such a frequency and spatial accuracy that the rendering of virtual images does not cause motion sickness.

In the field of cognition, one distinguishes the personal space (usually an arm length [1m] around the body), the action space (the size of a large room; radius 15m) and the vista space (everything further away than 15m) [1]. The action space is typically covered by a person's stereo vision capability. The personal space is usually for interaction with oneself, (e.g., reading, eating, computing, etc.) The action space is for interaction with others and the vista space is too far for interactions, it only shows us and brings us to places where we can interact. We focus on the action space and focus on a system that can accurately operate in a range from 1 to 15 meters. Our system can be used for applications such as tour guiding, remote maintenance, design visualization and games. Also without the headset, the pose estimation system can be used by intelligent devices, such as UMTS cellular telephones or PDAs that can provide position / context based services such as positional reminders, auto mute in theaters, path finding, etc.

Our wearable system contains a radio link that connects the user to computing resources and services from the Internet. It can be used to answer questions from the user of a PDA that relate to his environment. For outdoor augmented reality, course location information can be determined e.g. based on differential GPS while UMTS can be used for the connection with backbone services. It can lead us, for instance, to places where we can interact, such the entrance of a building, a corridor, a room. Indoor, a presence and location server can be based on the signal strength of WiFi and/or Bluetooth access-points [2]. Based on course pose information, either from DGPS (outdoors) or from location services over the network (indoor or near buildings), a camera on the AR headset can capture the user's environment, which, fused with data from an inertia system: gyroscopes, accelerometers, and compass, can make the PDA fully aware of the user's head pose (the absolute position and orientation of the user's) with a resolution that matches his action space. This can be performed by sending camera images to the backbone and match this to a 3D description of the environment, to determine the user's head pose. A much easier approach is to use coded tags e.g. fixed at known positions on the walls. People use the same technology in daily life, in the form of street names, house and room numbers. These tags can then be used by the system as calibration points to determine the absolute pose of the user's head. Such an approach requires less communication bandwidth and backbone processing effort. Features in the moving camera images can be used to track the user's pose in between the reading of those calibration tags. Finally, for proper augmented reality, rendering latencies lower than 10 ms are necessary when one wants to reduce motion sickness to a minimum. In conventional systems with a refresh rate of 50 Hz it takes 20 ms to display a single frame. The time to render that frame will add to the total latency. It is clear that it is not possible to reach the required latency for augmented reality (<10 ms) by sequentially rendering and displaying a frame. Consequently, we developed a rendering system that renders a part of the frame ahead of the display's raster beam and has a combined rendering and display latency of 8 ms [3].

10.1.1 Problem Description

In this text we address the problem of the fusion of the sensors that are needed to obtain an accurate, fast and stable rendering system for augmented reality in a person's action space. For this we describe the fusion of data of various sensors with different update rates and accuracies, including camera and DGPS, by using Modular Kalman Filtering. The novelty in our approach is first the use of quaternion descriptions inside the indirect filter. Furthermore, we foresee a system in which all the sensors might either not always be there, or might not give sensible data for awhile. Moreover, applications with less stringent demands, such as path finding services in mobile phones or PDA's, might be able to do their job with less or less accurate sensors. Consequently, we aimed for a modular sensor framework in which sensors can be plugged in and out. For this we have split the Kalman filter up into a so called federated Kalman filter and modified this to allow hot plugging. Finally, we worked on a camera system that is able to accurately detect coded visual tags on the wall, which can be used to determine the pose of a user's head in his action space with the accuracy needed for augmented reality.

10.1.2 System Setup

To track the pose (position and orientation) of the user's head, we use a combination of sensors, which can be divided into relative sensors (angular velocity and linear acceleration) and absolute sensors (orientation and position). For the relative sensors we used three gyroscopes [4] and three accelerometers [5] combined in one board linked to a LART platform [6]. For the absolute sensors we use a Precision Navigation TCM2 compass with tilt sensor [7] and a JAI CV-S3300 camera [8], as well as a Philips DICA Smartcam [9].

The Murata Gyrostar piezoelectric vibrating gyros can measure up to 300°/s. They are inexpensive but have a large bias that varies with time up to 9°/s. Consequently, we had to correct for this bias. After correction, the noise level is around 0.2°/s when sampled at 100 Hz.

The accelerometers (ADXL202) also have a varying offset. This offset can be 0.5 m/s^2 and the residual noise level is around 0.06 m/s^2 when sampled at 100 Hz. The maximum acceleration that can be measured is $2g$ in both directions.

The TCM2-50 liquid inclinometer and compass uses a viscose fluid to measure the inclination with respect to the gravity vector with an accuracy of 0.2° (static orientation). The heading is calculated using three magnetometers with an accuracy of 0.5 to 1.5°. Because the liquid will slightly slosh when accelerations are applied, we have to cope with an error of about 20°. The update rate is 16 Hz.

The JAI CV-S3300 color camera with a resolution of 320 x 240 pixels in grayscale has a wide-angle lens with a 90° opening angle, which introduces spherical distortions. We calibrate the camera using the Zhang algorithm [10]. Images are grabbed at 15 Hz.

The Philips DICA Smart Camera equipped with an onboard 180 MHz processor and a 1280 x 1024 gray-value CMOS sensor will provide us with a better accuracy than the JAI. We use a similar lens as on the JAI, so the spherical distortions are also present. Currently we don't do processing onboard, but we will in the future.

The LART platform which is used for data acquisition and preprocessing has an 8-channel fast 16-bit AD-converter to acquire synchronous data from the accelerometers, gyros, and temperature data. The gyros and the accelerometers are analog devices, which are sampled at 100 Hz by the AD converter. The TCM2 updates at 16 Hz and is read via a serial line. Of each measurement of the sensors, the expected delay is determined. In that way the relative latencies can be compensated.

Figure 10.3 shows the software architecture we use. The idea is that every sensor has its own readout program. These programs maintain measurement noise and measurement latencies. When a measurement arrives, the information is sent to a central location that can be easily accessed. We have made a message passing system using shared memory that enables us to reduce unnecessary copying of data. In our system it is even possible that some sensor data comes from another computer in the network.

10.2 Notation

Throughout the chapter, we will use the following notations.

Ψ will be a coordinate frame. A point p expressed in a frame b will be noted as $\vec{p}^{\,b}$

A point with normal coordinates will have a non-capital designation. Homogeneous coordinates will be noted with a capital. For a 3D point or vector, the notation becomes: $\vec{P} = (\vec{p} \quad 1)^T = (x \quad y \quad z \quad 1)^Y$.

In some situations, it is necessary to specify a rotation or translation uniquely. This will be done using the notation:

$$\mathbf{H}_A^{C,B} = \begin{pmatrix} \mathbf{R}_A^{C,B} & \mathbf{T}_A^{C,B} \\ 0 & 1 \end{pmatrix}$$

This is a transformation that brings coordinate frame B to A expressed in frame C. Often a more compact notation will be used: $\mathbf{H}_A^B = \mathbf{H}_A^{B,B}$ which can also be viewed as a coordinate transformation from A to B.

10.3 Fusion Framework

For our fusion framework (Figure 10.4), we use the following coordinate systems (Figure 10.5):

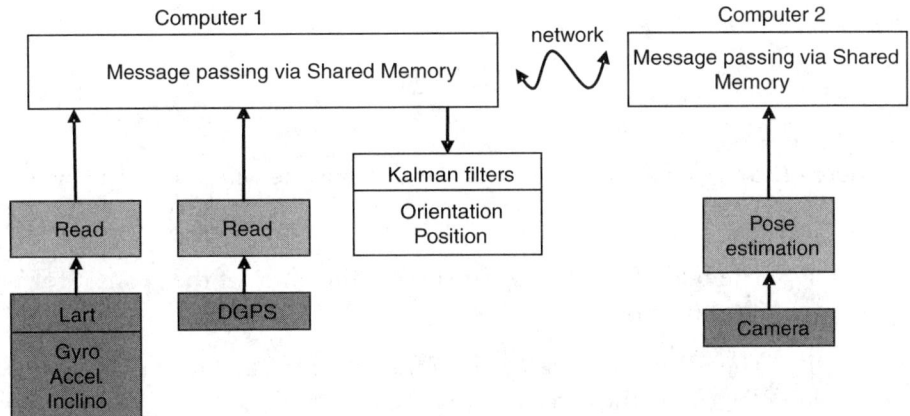

Figure 10.3 Schematic view of our software architecture for pose estimation.

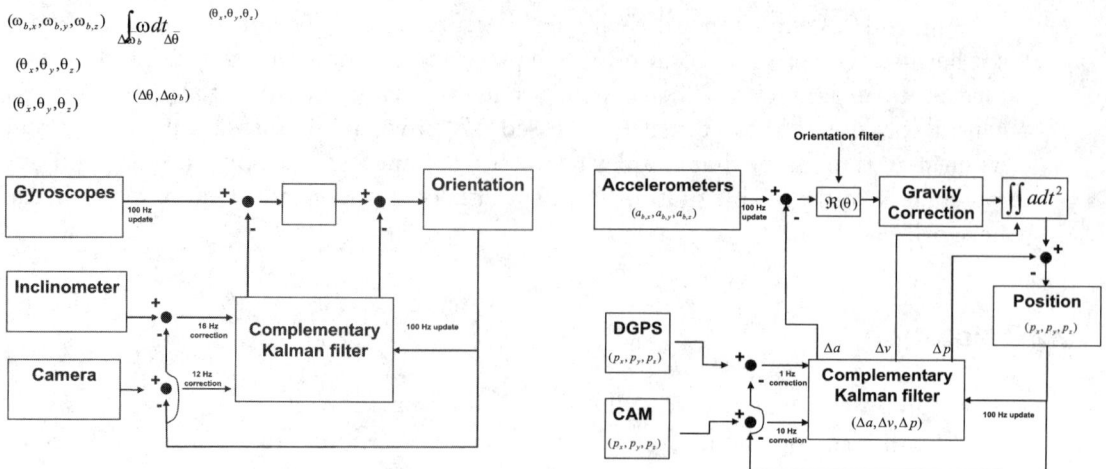

Figure 10.4 Fusion of data from the sensors for pose tracking.

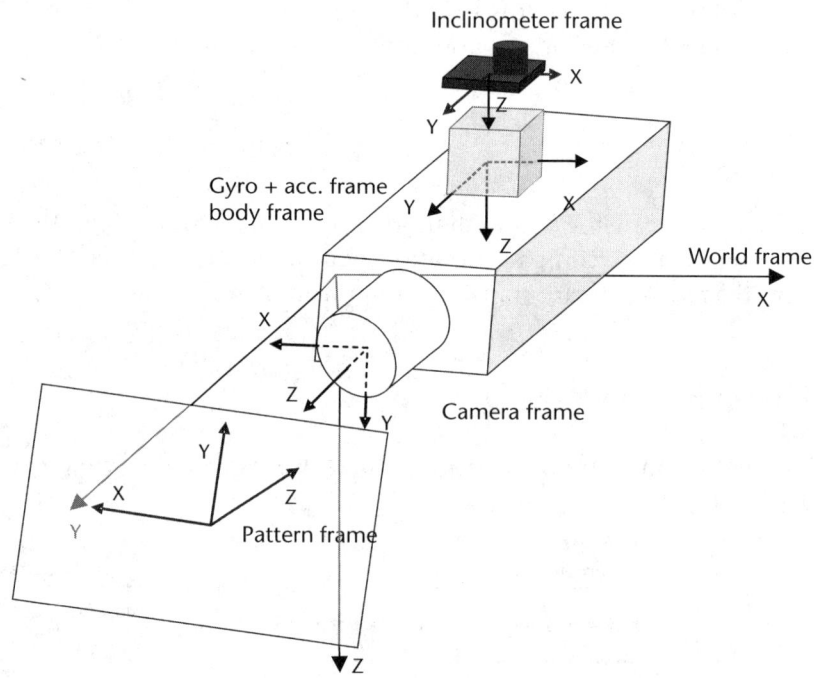

Figure 10.5 Schematic picture of a number of coordinate systems used for sensor fusion.

Ψ_b The body frame. It is attached to the body of the headset for which we need the pose.

Ψ_p The 3D pattern frame. This frame is attached to the pattern used to determine the camera's position.

Ψ_C The 3D camera frame. This frame has the optical axis as z axis.

Ψ_n The navigation frame. This frame is a rotated body frame. It has the z-axis pointing in the direction of the gravity vector, while the x-axis is pointing to the earth's North Pole.

Ψ_w The world frame. This frame is like the navigation frame, but has a fixed origin.

Sub indices like $\Psi_{b,1}$ denote a specific instance of a moving frame. To be able to combine sensor measurements with different update rates and error characteristics, we have used a Kalman filter setup [11, 12].

Due to the rotations, the Kalman equations become nonlinear, and hence we need to linearize the filter. We have chosen to use the errors in position and orientation as filter states, since then we can update the real states using nonlinear formulas to obtain a better performance. To overcome singularities when representing orientation in Euler angles, we used the quaternion notation. Quaternions can be used to represent orientations in 3D. The advantage over the use of common Euler angles is that the representation of the orientation is continuous (i.e., without jumps from 2π to 0.) This filter set-up will be used to explain our sensor fusion algorithm in the next sections, but the idea is that the filter is made modular later on. That will enable us to use accurate pose estimation only when it is needed, reducing computational load and power requirements if sensors can be shut down.

10.3.1 Quaternions

A quaternion has a scalar part and a vector part:

$$q = \begin{pmatrix} q_0 \\ \vec{q} \end{pmatrix} \text{ and } q^* = \begin{pmatrix} q_0 \\ -\vec{q} \end{pmatrix} \tag{10.1}$$

in which θ is the angle of rotation around the normalized vector $\vec{n}$, and is the complex conjugate of q^*. If q represents an orientation (i.e., a unit quaternion), then its inverse becomes:

$$q^{-1} = \frac{q^*}{\|q\|^2} = q^* \tag{10.2}$$

A quaternion that represents a rotation of a frame Ψ_A with respect to frame Ψ_B expressed in terms of frame Ψ_B is represented by q_A^B: the rotation that rotates frame Ψ_B to frame Ψ_A expressed in Ψ_B.

Let the quaternion representation of a vector $\vec{v}^a$ be:

$$q_{\vec{v}^a} = \begin{pmatrix} 0 \\ \vec{v}^a \end{pmatrix} \tag{10.3}$$

Then the rotation of this vector is obtained by a double quaternion multiplication:

$$q_{\vec{v}^a} = q_A^B \otimes q_{\vec{v}^a} \otimes q_A^{B*} = \begin{pmatrix} 0 \\ R_A^B \cdot \vec{v}^A \end{pmatrix} \tag{10.4}$$

in which the operator $\otimes$ is the quaternion multiplication. In matrix form this becomes:

$$q_1 \otimes q_2 = q_1 q_2 = \begin{pmatrix} q_{1,0} & -q_{1,x} & -q_{1,y} & -q_{1,z} \\ q_{1,x} & q_{1,0} & -q_{1,z} & q_{1,y} \\ q_{1,y} & q_{1,2} & q_{1,0} & -q_{1,x} \\ q_{1,z} & -q_{1,y} & q_{1,x} & q_{1,0} \end{pmatrix} \begin{pmatrix} q_{2,0} \\ q_{2,x} \\ q_{2,y} \\ q_{2,z} \end{pmatrix}$$

$$= \bar{q}_2 q_1 = \begin{pmatrix} q_{2,0} & -q_{2,x} & -q_{2,y} & -q_{2,z} \\ q_{2,x} & q_{2,0} & q_{2,z} & -q_{2,y} \\ q_{2,y} & -q_{2,2} & q_{2,0} & q_{2,x} \\ q_{2,z} & q_{2,y} & -q_{2,x} & q_{2,0} \end{pmatrix} \begin{pmatrix} q_{1,0} \\ q_{1,x} \\ q_{1,y} \\ q_{1,z} \end{pmatrix} \tag{10.5}$$

in which $\tilde{q}$ is the quaternion matrix and $\bar{q}$ the transmuted quaternion matrix.

The representation of angular velocity vectors using quaternions is analogous to the case of rotations of position vectors over angles.

The angular velocity of Ψ_j with respect to Ψ_i expressed in Ψ_i is given by:

$$\dot{R}_i^j = R_i^j \cdot \tilde{\omega}_i^{i,j} = R_i^j \cdot \begin{pmatrix} 0 & -\omega_x & \omega_y \\ \omega_x & 0 & -\omega_z \\ -\omega_y & \omega_z & 0 \end{pmatrix} \tag{10.6}$$

The subscripts i and j in $\tilde{\omega}_i^{i,j}$ are removed for clarity. In quaternion notation this is:

$$\dot{q}_i^j = q_i^j \otimes \tfrac{1}{2} q_{\omega_i^{i,j}} = \tfrac{1}{2} q_{\omega_i^{i,j}} \cdot q_i^j \tag{10.7}$$

As the scalar part of $q_{\omega_i^{i,j}}$ is zero, the solution to (10.7) is:

$$q_i^j(t) = e^{\tfrac{1}{2} q_{\omega_i^{i,j}} \cdot t} q_i^j(0)$$

$$= \left(I \cdot \cos\left(\tfrac{1}{2}\|\vec{\omega}_i^{i,j}\|t\right) + \sin\left(\tfrac{1}{2}\|\vec{\omega}_i^{i,j}\|t\right) \cdot \frac{q_{\omega_i^{i,j}}}{\|\omega_i^{i,j}\|} \right) q_i^j(0) \tag{10.8}$$

or:

$$q_i^j(t) = q_i^j(0) \otimes \begin{pmatrix} \cos\left(\tfrac{1}{2}\|\vec{\omega}_i^{i,j}\|t\right) \\ \sin\left(\tfrac{1}{2}\|\vec{\omega}_i^{i,j}\|t\right) \dfrac{\vec{\omega}_i^{i,j}}{\|\vec{\omega}_i^{i,j}\|} \end{pmatrix} \tag{10.9}$$

A more detailed treatment is given in [13].

10.4 Kalman Filters

To illustrate why we can't just use the measurements of the inclinometer and interpolate in between samples using the faster gyroscope, we show in Figure 10.6 the output of a static gyroscope during 20 hours. The gyroscope was powered on at the start of the measurement and the heating inside the device causes the very steep slope. In addition, the output clearly has a low-frequency component which we call drift. This drift has to be corrected, to be able to correctly interpolate between two inclinometer samples.

Another reason to use filtering is that the inclinometer output is noisy. If the inclinometer measurement is used without filtering, the resulting orientation estimate will be bumpy even in the case of a static orientation and correct interpolating.

Using a model of the orientation change in time, the Kalman filter can estimate the orientation and the change in orientation. The inclinometer measurement will be tested against the expected orientation at the time of measurement, and the two estimates (one of the filter, and one of the inclinometer) will be optimally combined using information about the accuracy of the estimate. These accuracies are maintained in covariance matrices.

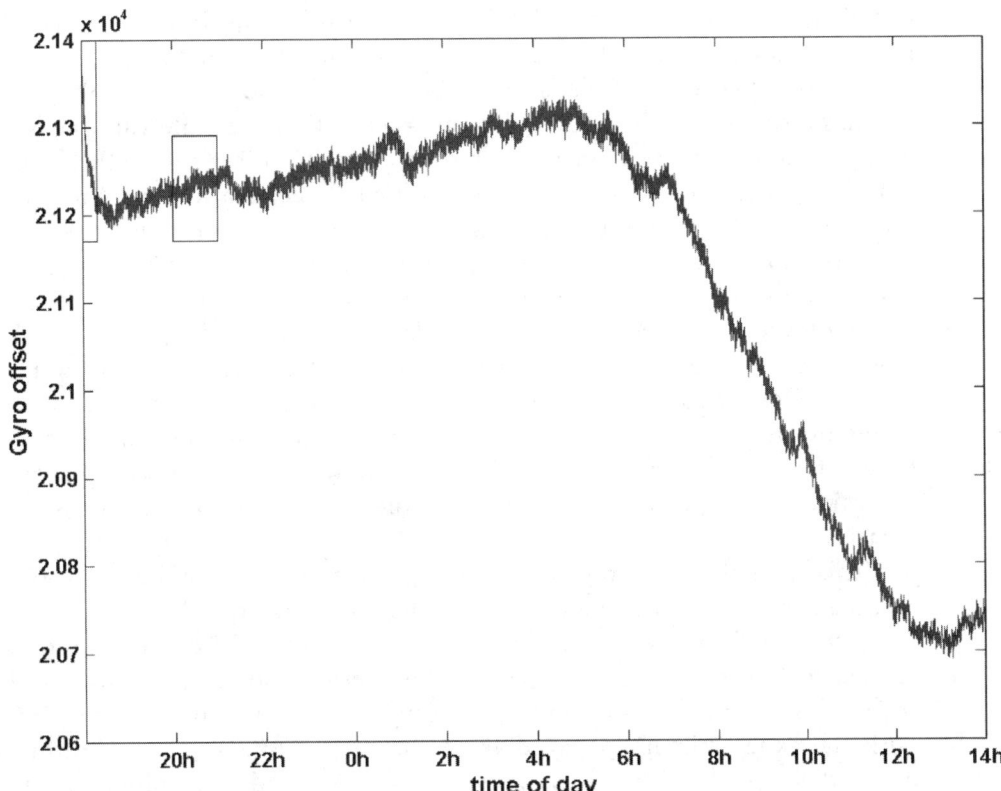

Figure 10.6 Output of a gyroscope in rest, sampled at 100 Hz. The vertical range here translates to 6 deg/s. The lowest frequency drift is probably the effect of temperature change in the room.

10.4.1 Filter Variants

The Kalman filter (KF) is a set of mathematical equations that predict the optimal (in statistical sense) estimate from an assumed process model and new incoming measurements. The filter is useful for sensor fusion since it can deal with sensors with different sampling rates and latencies, and because of its simple recursive form it is suited for real-time processing.

The KF in its standard form is known as the discrete KF. The counter part of the discrete KF is the continuous KF. Although the continuous KF, is not useful in practice it has theoretical purposes. The discrete Kalman filter is not an approximation, as in many other discrete theories. It is exact, provided that the difference equations are correct. However, physical processes are usually not discrete but continuous. A KF with continuous process and measurement model is called a continuous KF. The continuous KF is not useful for implementation, but it gives insight in the theoretical concepts of the KF.

In the indirect KF, the term indirect means that the state vector contains error states instead of the normal states. The introduction of error states is necessary, because sometimes it is not possible to model the process with the actual states. Take for instance an accelerometer with sensor output containing temperature drift. In this case it is not correct to model the sensor with acceleration and bias as states, because the state transition for the acceleration is not known; i.e., we don't have a reasonable guess what the acceleration will be one step in the future. Then why not just make use of a constant model? This is good as long as the situation is stationary, but it will fail as soon as the accelerometer is moving. Then, the acceleration will be registered as bias by the KF. Error states can help to solve this problem. If we take the bias error as a state variable, we can assume that this state variable is constant. The right acceleration can be recovered by setting the bias error to 0. Then, after the bias error is corrected by the KF, the estimated acceleration is retrieved form the sensor output by subtracting this estimated bias. Since the indirect KF uses error state variables, it also needs error measurements. For this purpose we need another sensor. As error measurement, we can use the difference of the measurements of another sensor with the calculated, i.e., estimated, corresponding variable using only accelerometer measurements. Sticking to the example of the accelerometer, a camera can be used as a second sensor. When the states of the accelerometer are extended with velocity error and position error (the state transition is fortunately known), the difference between the position measurement of the camera and position estimation of the accelerometer can be used as error measurement.

This example shows the beginning of sensor fusion; with the use of the indirect KF, the camera is correcting the drift of the accelerometer.

Until now it was assumed that the process model and the measurement model are linear. However, in most cases these models are nonlinear. A way to deal with the nonlinear problem is to linearize about a trajectory that is continually updated with the state estimates derived from the measurements. The resulting filter is called the extended KF (EKF). The EKF is often used in combination with error states, because usually the linear first approximation of the error state is good enough.

10.4.2 General Setup

First, we will give the well-known formulas for the Kalman filter. The time update for a Kalman filter without control inputs is given by:

$$\hat{x}_{\bar{k}} = \Theta_{t_k,t_{k-1}} \hat{x}_{\overline{k-1}}$$
$$P_{\bar{k}} = \Theta_{t_k,t_{k-1}} P_{k-1} \Theta_{t_k,t_{k-1}} + Q(t_k,t_{k-1})$$
$$Q(t_k,t_{k-1}) = \int_{t_{k-1}}^{t_k} \Theta_{t_k,s} \cdot Q(s) \Theta_{t_k,s}^T ds \quad (10.10)$$
$$\approx \Theta_{t_k,t_{k-1}} \cdot Q(t_{k-1}) \cdot \Theta_{t_k,t_{k-1}} \cdot (t_k - t_{k-1})$$
$$(Q)t = \text{cov}(w(t),w(\tau))$$

in which $\hat{x}_{\bar{k}}$ is the current apriori estimate of the state x_k, Θ is the state transition matrix that projects a state into the future, and $w(t)$ is the white noise process, which represents not modeled changes in the state. We don't use inputs in our filter, but the transition matrix is in our case dependent on the inertia sensor output. When an observation is done, the aposteriori estimate of the state can be calculated from the estimate of the observation:

$$y_{\bar{k}} = H\hat{x}_{\bar{k}}$$
$$\hat{x}_k = \hat{x}_{\bar{k}} + K(y_k - y_{\bar{k}})$$
$$P_k = (I - KH)P_{\bar{k}} \quad (10.11)$$
$$K = P_{\bar{k}} H^T \left(H P_{\bar{k}} H^T + R \right)^{-1}$$
$$R = \text{cov}(v(t),v(\tau))$$

in which $y_{\bar{k}}$ is the predicted observation, y_k is the real observation, H is the observation matrix, K is the Kalman gain, and $v(t)$ is the white noise that models measurement noise in y_k.

In the sequel we use for the estimate of the real states (aposteriori and apriori) and for the error states of the Kalman filter. The state vector contains orientation, angular velocity, position, linear velocity, linear acceleration, current gyro, and current accelerometer bias:

$$X = \left(q_b^n, \vec{\omega}_b^{b,n}, \vec{p}_b^n, \vec{v}_b^n, \vec{a}_b^b, \vec{b}_{gyro}^b, \vec{b}_{acc}^b \right)^T \quad (10.12)$$

To simplify the system we use two Kalman filters. One for the orientation, with state:

$$dX_{orient} = \left(dq_b^n, d\vec{b}_{gyro}^b \right)^T \quad (10.13)$$

being the error in orientation and the drift in the gyroscopes, and one for the position containing the error in position, linear velocity, and accelerometer drift:

$$dX_{pos} = \left(d\vec{p}_b^n, d\vec{v}_b^n, d\vec{b}_{acc}^b\right)^T \tag{10.14}$$

The accelerometer bias and its error state are expressed in body coordinates, and therefore the position filter depends on the orientation. So in this setup it is essential to have an accurate orientation estimate, as its error will be propagated with t^2 into the position.

10.4.3 Time Update

Each time a measurement is obtained from the inertial sensors, the estimate of the actual state X is updated, using (10.9) and:

$$\begin{aligned}
{}^{r_n}_{v_b}(\Delta t) &= {}^{r_n}_{v_b}(0) + \left(R_b^n(0) \cdot {}^{r_b}_{a_b}(0) - {}^{r_n}_{g}\right)\Delta t \\
{}^{r_n}_{p_b}(\Delta t) &= {}^{r_n}_{p_b}(0) + \frac{1}{2}\left({}^{r_n}_{v_b}(0) + {}^{r_n}_{v_b}(\Delta t)\right)\Delta t
\end{aligned} \tag{10.15}$$

Note, that we do not take the rotational speed into account, which results in an error in $\vec{a}_b^b(t)$ of about $3 \cdot 10^{-3}$ m/s^2 when rotating at 200 °s and $\Delta t = 1/100$s, in a direction perpendicular to the gravity vector.

The state update for the position Kalman filter is based on the formulas and , and we neglect rotations errors in R_b^n. The change in rotation should be small due to the high update rate. The state transition is now given by:

$$dX_{pos}(t_k) = \Phi dX_{pos}(t_{k-1}) \Rightarrow$$
$$\begin{pmatrix} d\vec{p}_n^b \\ d\vec{v}_n^b \\ d\vec{b}_{acc}^b \end{pmatrix}_{(t_k)} = \begin{pmatrix} I & I \cdot \Delta t & \frac{1}{2}R_n^b(t_k) \cdot \Delta t^2 \\ 0 & I & R_n^b(t_k) \cdot \Delta t \\ 0 & 0 & I \end{pmatrix} \begin{pmatrix} d\vec{p}_n^b \\ d\vec{v}_n^b \\ d\vec{b}_{acc}^b \end{pmatrix}_{(t_{k-1})} \tag{10.16}$$

The state update for the orientation Kalman filter is more complicated, because we use the orientation difference in quaternion notation:

$$q_b^n = q_b^{n-} \otimes dq_b^n \Leftrightarrow dq_b^n = q_b^{n-*} \otimes q_b^n \tag{10.17}$$

in which q_b^{n-} is the estimated orientation by integration using (10.9) and q_b^n is the real state. Using:

$$\begin{aligned}
q_b^{n-*} \otimes q_b^{n-} &= 0 \Rightarrow \dot{q}_b^{n-*} \otimes q_b^{n-} + q_b^{n-*} \otimes \dot{q}_b^{n-} = 0 \Rightarrow \\
\dot{q}_b^{n-*} &= -q_b^{n-*} \otimes \dot{q}_b^{n-} \otimes q_b^{n-*}
\end{aligned} \tag{10.18}$$

The time derivate of the estimate of dq_b^n becomes:

10.4 Kalman Filters

$$d\dot{q}_b^n = \dot{q}_b^{n-b} \otimes q_b^n + q_b^{n-*} \otimes q_b^n$$
$$d\dot{q}_b^n = -\dot{q}_b^{n-*} \otimes q_b^{n-} \otimes q_b^n + q_b^{n-*} \otimes \left(q_b^n \otimes q_{\omega_b^{b,n}}\right) + \tfrac{1}{2} q_b^{n-*}$$
$$d\dot{q}_b^n = -q_b^{n-*} \otimes \left(\tfrac{1}{2} q_b^{n-} \otimes q_{\omega_b^{b,n-}}\right) \otimes dq_b^n + \tfrac{1}{2} dq_b^n \otimes q_{\omega_b^{b,n}} \quad (10.19)$$
$$d\dot{q}_b^n = \tfrac{1}{2} q_{\omega_b^{b,n-}} \otimes dq_b^n + \tfrac{1}{2} dq_b^n \otimes q_{\omega_b^{b,n}}$$
$$d\dot{q}_b^n = \tfrac{1}{2} q_{\omega_b^{b,n-}} \cdot dq_b^n + \tfrac{1}{2} q_{\omega_b^{b,n}} \cdot dq_b^n$$
$$d\dot{q}_b^n = \tfrac{1}{2} \left(q_{\omega_b^{b,n}} - q_{\omega_b^{b,n-}}\right) \cdot dq_b^n$$

We can calculate the aposteriori estimate $\omega_b^{b,n}$ from its apriori estimate $\omega_b^{b,n-}$ with:

$$\omega_b^{b,n} = \omega_b^{b,n-} - d\vec{b}_{gyro}^b \quad (10.20)$$

Then (10.19) becomes, after some elaboration:

$$dq_b^n = \tfrac{1}{2}(q_\omega - q_{\omega-}) \cdot dq = \tfrac{1}{2}\begin{pmatrix} -d\vec{b} \cdot d\vec{q} \\ dq_0 \cdot d\vec{b} - d\vec{b} \times d\vec{q} + 2\vec{w} \times d\vec{q} \end{pmatrix} \quad (10.21)$$

Now this can be linearized around the state dx_{orient} at time $t = 0$ (the previous estimate). In the indirect filter setup, the error states are reset after every observation update. This means that $d\vec{b}$ will be assumed constant and 0, $d\vec{q}$ will be small, and dq_0 will be approximately 1. Our linearization of the time derivative becomes:

$$\begin{pmatrix} d\dot{q}_0 \\ d\dot{\vec{q}} \\ d\dot{\vec{b}} \end{pmatrix} = \begin{pmatrix} 0 & 0 & 0 \\ 0 & w & \tfrac{1}{2} \\ 0 & 0 & 0 \end{pmatrix} \begin{pmatrix} dq_0 \\ d\vec{q} \\ d\vec{b} \end{pmatrix} \quad (10.22)$$

In this case $\tilde{w}$ is assumed constant (zero order hold), we can find the state transition to be:

$$dX_{orient}(t_k) = \Phi dX_{orient}(t_{k-1}) \Leftrightarrow$$
$$\begin{pmatrix} dq_0 \\ d\vec{q} \\ d\vec{b} \end{pmatrix}_{(t_k)} = e^{\begin{pmatrix} 0 & 0 & 0 \\ 0 & w_{(t_k)} & \tfrac{1}{2} \\ 0 & 0 & 0 \end{pmatrix} \Delta t} \begin{pmatrix} dq_0 \\ d\vec{q} \\ d\vec{b} \end{pmatrix}_{(t_{k-1})} \quad (10.23)$$

Of course, after this update, dq should be normalized to unity again. The algebraic solution to this exponent of a matrix was found using a mathematical software package.

10.4.4 Observation Update

When an inclinometer measurement becomes available, it is converted to quaternion notation. Now it becomes easy to determine the observation estimate of the error in orientation:

$$dq^n_{b,obs\_est} = q^{n-*}_b \otimes q^n_{b,inclino} \tag{10.24}$$

This observation estimate replaces y_k in (10.11).

Because both estimates are in quaternions, the output matrix H becomes:

$$y_{\bar{k}} = H\hat{x}_{\bar{k}} = (I \quad 0)\begin{pmatrix} d\vec{q} \\ d\vec{b} \end{pmatrix} \tag{10.25}$$

The Kalman measurement noise $\vec{v}$ looks complicated, it is dependent on both the measurement and the estimated real state. However, if we remember that we are measuring here the deviation from the estimate, and not the deviation from the truth, it is clear that we only have to take into account the measurement noise of the sensor. To find the covariance matrix R, we determined the linearized matrix G that relates the noise of the deviation in quaternion representation, dq^n_{b,obs_est}, to the noise expressed in Euler angles $\delta\vec{\theta}$:

$$
\begin{aligned}
dq^n_{b,obs}\left(\vec{\theta}_{inclino} + \delta\vec{\theta}\right) &\approx dq^n_{b,obs}\left(\vec{\theta}_{inclino}\right) + \frac{\partial\, dq^n_{b,obs}\left(\vec{\theta}_{inclino}\right)}{\partial\vec{\theta}} \delta\vec{\theta} \\
\delta dq^n_{b,obs} &= \frac{\partial\, dq^n_{b,obs}\left(\vec{\theta}_{inclino}\right)}{\partial\vec{\theta}} \delta\vec{\theta} = G \cdot \delta\vec{\theta} \\
\vec{v} &= G \cdot \vec{w} \\
R &= \text{cov}(\vec{v}(t),\vec{v}(\tau)) = G \cdot \text{cov}(\vec{w}(t),\vec{w}(\tau)) \cdot G^T
\end{aligned}
\tag{10.26}
$$

in which $\vec{w}(t)$ is the presumed white noise in the inclinometer measurement.

In the position update, the estimated position error is:

$$d\vec{p}^n_{b,ops\_est} = \vec{p}^{n-}_b - \vec{p}^n_{b,observation} \tag{10.27}$$

and using y_k for $d\vec{p}^n_{b,obs\_est}$ we get (10.11):

$$y_{\bar{k}} = H\hat{x}_{\bar{k}} = (I \quad 0 \quad 0)\begin{pmatrix} d\vec{p} \\ d\vec{v} \\ d\vec{b} \end{pmatrix} \tag{10.28}$$

Since the position difference measurement is in the same coordinate system as the state, there is no need for a matrix G to calculate the measurement noise.

10.4.5 Coping with Lag

The inclinometer output is delayed with about 0.375s (or 6 samples @ 16 Hz). When we ignore this delay, the orientation Kalman filter will assume an error in orientation and will adjust the current error and bias estimate. After the rotation, the filter needs some time to recover. A bigger problem is that the position filter uses the orientation, and this delay will hence introduce a nonexistent acceleration.

In our method we store all the observations of the sensors, as well as the Kalman states and matrixes, at every step and keep a history of 30 steps.

When an inclinometer measurement finally arrives, we step back to the position in time of that measurement, and do the filtering in the Kalman state that belongs to that point in time (see Figure 10.7.) From here on all the other measurements (such as gyro, camera, GPS) are processed again up to the current time. In this way the best estimate at the current time is achieved. We are aware that this method is computationally intensive, but for now it poses no problem as the CPU load is almost 0% on a Pentium 3 GHz. Some possible methods to address that problem of out-of-sequence measurements are given in [14–17].

10.4.6 Divergence Problems

Problems in Kalman filtering are addressed as divergence problems. Two common sources of divergence are round-off errors and modeling errors. Round-off errors can slowly affect the error covariance, such that it becomes asymmetric or non-positive definite. In both cases the KF algorithm becomes unstable. One should be cautious with round-off errors, but they can easily be dealt with. Modeling errors, however, form a bigger problem, because they are less noticeable. Typical modeling errors include the following:

- Modeling of systematic errors as white noise. In this case the KF filter can't give the optimal estimate. The estimated states will have a systematic error as well. An example would be the inclinometer which has a systematic error on acceleration.

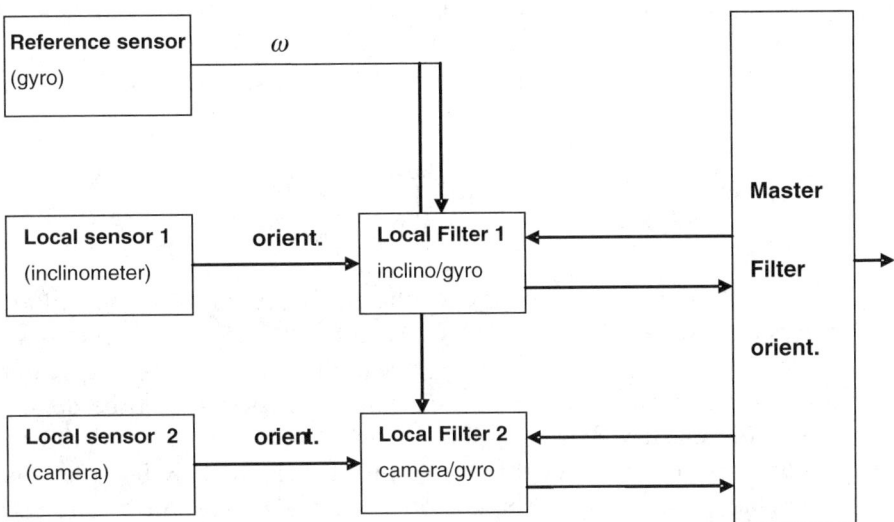

Figure 10.7 Schematic picture of a Federated Kalman Filter for the orientation.

- Modeling of a dynamic process as a static process. The KF simply tries to fit the wrong curve through the measurements. That would be the case if we don't estimate the change in bias of the inertia sensors.
- Taking the process noise too small. In this case, after a while, the error covariance becomes so small that the filter doesn't believe the measurements anymore.

The EKF is linearized around the estimated states. This is a little risky, since the estimated states are just the statistical optimal states, and by no means are they for certain the actual states. In combination with modeling errors, the EKF can have serious problems.

10.4.7 The Decentralized KF

In a decentralized KF, the sensor outputs are first processed in local filters and then the result of the local filters are combined in a master filter. Compared with the centralized filter (one filter for all sensors) the decentralized structure needs less computational power and has a greater fault tolerance when one of the sensors fails [18]. The most popular decentralized design is the federated filter of Carlson [19]. The idea is that measurement updates are processed in local filters, and the estimates from the local filters are combined in a master filter to give the best estimate. The federated filter is equivalent to the centralized filter, provided that all filters have the same states, the time-update formulas and process noise are the same for all local filters, and the best estimate has to be fed back to the local filters (zero reset or full feedback). The general setup is illustrated in Figure 10.7.

For orientation estimation the reference sensors would be the gyros. One local filter combines gyros and inclinometer, and the other combines gyros and the camera orientation. All filters have the orientation as state, and the local filters also have the gyro bias errors as states. The local state $\mathbf{x}_i$ and its corresponding error covariance $\mathbf{P}_i$ of local filter i is defined as:

$$\mathbf{x}_i = \begin{pmatrix} \mathbf{x}_{ci} \\ \mathbf{x}_{di} \end{pmatrix} \qquad (10.29)$$

$$\mathbf{P}_i = \begin{pmatrix} \mathbf{P}_{cci} & \mathbf{P}_{cdi} \\ \mathbf{P}_{dci} & \mathbf{P}_{ddi} \end{pmatrix} \qquad (10.30)$$

with $\mathbf{x}_{ci}$ the common states and $\mathbf{x}_{di}$ the drift states of the ith local filter. The two local filters are exactly the same as the central filter from the previous section. When a camera measurement comes in, the local filter updates its estimate of the orientation and the bias errors. To ensure optimality, a fusion should be done immediately in which the master filter combines the estimates of both local filters and resets the common states of the local filters to the best estimate. When the fusion is done at a later stage, the result will be less optimal. The master filter can be seen as a global filter with augmented state vector:

10.4 Kalman Filters

$$\mathbf{x} = \begin{pmatrix} \mathbf{x}_{ci} \\ \vdots \\ \mathbf{x}_{cN} \end{pmatrix} \quad (10.31)$$

with N the number of local sensors. The corresponding error covariance is:

$$\mathbf{P}_i = \begin{pmatrix} \mathbf{P}_{11} & \cdots & \mathbf{P}_{1N} \\ \vdots & \ddots & \\ \mathbf{P}_{N1} & & \mathbf{P}_{NN} \end{pmatrix} \quad (10.32)$$

Given a set of local state estimates $\hat{\mathbf{x}}_i$, the globally best estimate $\hat{\mathbf{x}}_m$ is the one that minimizes the weighted least squares cost function:

$$\sum_{j=1}^{N} \sum_{i=1}^{N} (\hat{\mathbf{x}}_j - \mathbf{x}_i)^T P_{ij}^{-1} (\hat{\mathbf{x}}_j - \mathbf{x}_j) \quad (10.33)$$

When we assume that the cross variances in (10.32) are 0, the solution is simple:

$$\mathbf{P}_m = \left(\mathbf{P}_{11}^{-1} + \cdots + \mathbf{P}_{NN}^{-1} \right)^{-1} \quad (10.34)$$

$$\hat{\mathbf{x}}_m = \mathbf{P}_m \left(\mathbf{P}_{11}^{-1} \hat{\mathbf{x}}_{c1} + \cdots + \mathbf{P}_{NN}^{-1} \hat{\mathbf{x}}_{cN} \right)^{-1} \quad (10.35)$$

The last assumption actually means that the local filter states are treated uncorrelated. As can be seen from (10.34) the more independent estimates for the common states, the smaller the covariance matrix. After the fusion however, all filters take over the estimate for the common states, as well as the covariance for those states. That means that all filters become correlated. They stay correlated because the local filters share the same time-update sensor and process noise. This problem is solved by feeding back $\gamma_i \mathbf{P}_m$ instead of $\mathbf{P}_m$, with $\sum_{i=1}^{N} \frac{1}{\gamma_i} = 1$. If (10.34) is applied immediately after feedback, the original covariance matrix is recovered. So in this way the local filters can be treated uncorrelated. To make sure that the local filters can be treated uncorrelated after time updates, the process noise for the common states should also be multiplied with γ_i in the time-update stage of the local filters. Usually all the γ_i are taken the same; for N local filters this translates to $\gamma=N$. The fusion formulas for the local filters are found by requiring that the result of the federated filter should be the same as the result for the centralized filter. These formulas can be proven to be:

$$\mathbf{x}_i = \begin{pmatrix} \hat{\mathbf{x}}_m \\ \hat{\mathbf{x}}_{di} + \mathbf{P}_{dci} \gamma \mathbf{P}_{cci}^{-1} (\hat{\mathbf{x}}_m - \hat{\mathbf{x}}_{ci}) \end{pmatrix} \quad (10.36)$$

$$\mathbf{P}_i = \begin{pmatrix} \gamma \mathbf{P}_m & \mathbf{P}_m \gamma_i \mathbf{P}_{cci}^{-1} \mathbf{P}_{cdi} \\ \mathbf{P}_{dci} \gamma \mathbf{P}_{cci}^{-1} \mathbf{P}_m & \mathbf{P}_{ddi} - \mathbf{P}_{dci} \gamma_i \mathbf{P}_{cci}^{-1} \left(I - \mathbf{P}_m \gamma_i \mathbf{P}_{cci}^{-1} \right) \mathbf{P}_{cdi} \end{pmatrix} \quad (10.37)$$

When using the indirect filter in federated mode, the error states should be reset only after the fusion step. This means that a time update should correctly update the nonzero error states found after a measurement update. The federated filter works optimally when all observations by the local filters are immediately fused with the other filters. The measurements should also be ordered in time. In our case we have delays in the sensors, and we use the same trick as in the central filter to incorporate these measurements. This means that if a measurement is done, all filters go back in time to the time of the observation. The local filter does an observation update and fuses its estimate with the rest of the filters. Then, all filters redo the measurements in proper time order until the current time. Normally the computational load of the federated filter is less, because each local filter only has the common states and their own bias states, whereas the central filter should include all local bias states. But in our case, the computational load is a lot higher, for more filters are used in parallel, and all have to redo the measurements. For now, that poses no problem since we have a fast computer to do the estimation (see Figure 10.8.)

10.4.8 A Modular Kalman Filter

The federated Kalman filter is a suitable filter for sensor fusion, but in its original form, the FKF cannot serve as a modular filter since the reference sensors have a far too dominant role for that. In the FKF all the local filters contain the reference sensor states. If another sensor replaces this reference sensor, all local filters have to be changed to accommodate other reference sensor states, and that is not the modularity we require.

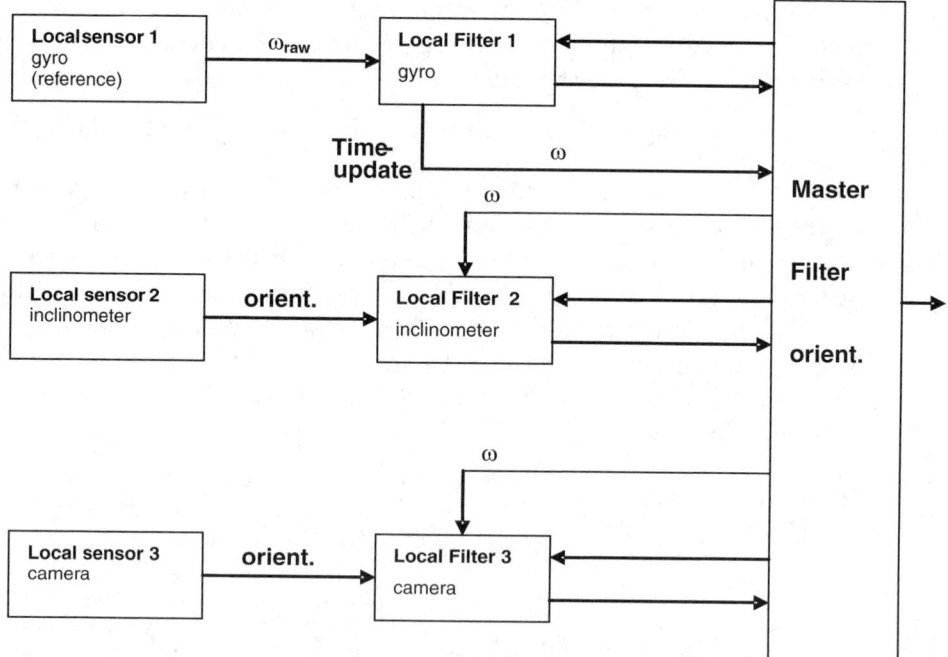

Figure 10.8 Schematic picture of our PKF implementation for orientation.

We designed a pluggable Kalman filter (PKF). The main difference with the federated filter is that every sensor now has a local KF. The local filters are still indirect KFs, but instead of the reference sensor states (gyro/accelerometer) they only have the common states, and their own bias states. However, we still need the reference sensors for the time update of the local filters. The reference sensor will be chosen by the master filter, and at every measurement of the reference sensor, the bias corrected version will be sent to all local filters.

The new setup is suitable to serve as a pluggable filter because the sensor states are not dependant on other sensor bias states. The change of the local filter structure we designed has one serious consequence: the increase in the covariance matrix in the time-update step is not the same for all filters. The process noise is still the same for all local filters, but only the one with the reference sensors will have an extra increase because of uncertainty in the bias of these sensors. The effect is that the estimate of the reference local filter is a little bit less trusted with respect to the original federated filter.

The implemented filter actually consists of two filters: one for orientation and one for position. The orientation filter's structure is depicted schematically in . The position filter structure has a similar architecture. Figure 10.8 shows the information flow between sensors, local filters, and the master filter. In addition, the local filters for position also need the master estimate of orientation as in.

The local filters do the actual filtering of the sensor output. With the aid of the local filters it is tried to model and correct for systematic errors in its own sensors (e.g., drift). Some sensors errors (e.g. errors in camera and compass output) are more difficult to model. In these cases, we only tried to diminish the influence of those sensor estimates to the master filter estimate, by varying the measurement noise.

Finally, some problems concerning filter design and software design need to be solved.

1. The federated filter has a big problem when the reference sensor is unplugged, because all local sensors use the dynamics and bias errors of the reference sensor. Our PKF does not have this problem, but still the reference sensors are used to update the local filters. When such an updating sensor is removed, its updating function could be taken over by another sensor. In our system, it is required that all updating sensors measure the same variable, for instance angular velocity. Preferably, this new updating sensor has a periodic output, a high sampling rate and a small lag. When the sampling rate is too low, more time updates can be generated with a constant value, but with an increasing process noise. If no other sensor is found, a dummy sensor can be used with zero value and a very high process noise. Aside from the question of whether it is desirable or not to remove the accelerometer, the important point is that the filter keeps functioning properly.

2. The PKF software needs to solve the problems of recognizing sensors that are plugged in and out and managing their filter data. To deal with this, every local sensor module needs to send the necessary information to the master filter at its initialization. When running, the modules should also send a heartbeat to the master filter indicating that they are still alive. In order to

make the data accessible for all modules, we use shared memory. The PKF loop can be summarized by the following actions: The sensors post their data on the location of the sensor data in the shared memory. The local filters wait until there is new data for them. When data from the reference sensor has arrived, they perform the time update. On local sensor data they do a measurement update and request a fusion step to the master filter. The master filter notices that new data has arrived in the local filters' locations and performs the fusion algorithm, after which it posts the result in the master filter's location, so the local filters can perform the fusion update step. With sensors that have delays, the fusion step should be done back in time. Redoing all measurements is no problem when all measurements are stored centrally, but when the filter is split up in separate processes, the synchronization of the local filters to do all measurements in proper observation time order is not easy.

10.5 Camera Positioning

10.5.1 Problem Specification

A camera provides a tremendous amount of information about the world. Ideally, the location of the camera can be found just by looking at the surroundings, as people do.

That means that natural landmarks should be used for localization, but that still proves to be difficult. Although it is our intention to incorporate natural landmarks in the future, we currently use only man made patterns.

We set the following criteria for the pattern:

- The pattern should easily be recognized.
- It should not be too obtrusive.
- The pattern template should have enough unique different instantiations.
- Its ID should be fast to read.
- The pattern should provide enough information to determine the position.
- To be usable in an office, the pattern-camera combination should provide estimates of positions of at least up to 5 meter with a lens that has a 90° opening angle.
- The less patterns in a room the better.

Because we want to minimize the number of patterns in the environment, we don't want to use the circular fiducials of Naimark and colleagues [14]. Although these fiducials can be detected in a fast way and provide good localization, at least three are needed to provide a full pose (6D).

We started with a 3 × 5 checkerboard pattern. It has 6 saddle points, and with a minimum of 4 points needed to estimate its position, there is some redundancy. In [20] we describe how that saddle-point pattern was recognized. The problem, however, is that we cannot attach an ID to that marker. We found, in the augmented reality toolkit [21], that a rectangular pattern with a big black border is easy to recognize, and provides enough space in the inner part to distinguish many different codes.

10.5.2 Marker Layout

The marker that we use now is depicted in Figure 10.9. The inner part consists of a 2D grid of 5 x 3 blocks. The blocks can be white or black. The color of the four blocks in the corners is chosen such that we can always determine the correct orientation of the marker. The other 11 blocks are used to determine the ID of the marker, and that means that there are 2,048 different IDs possible. The number of codes is of course much lower than the circular fiducials [14], but it is not always necessary that all patterns are unique. Combinations of patterns and other clues can be used to determine an absolute position. We settled for an A4 size pattern, which is not really big and is easy to produce on a normal printer. Next, we will describe how we detect these markers. Figure 10.10 gives the operations schematically.

10.5.3 Canny Edge Detection

A good method to find edges or contours is the Canny edge detector [22].

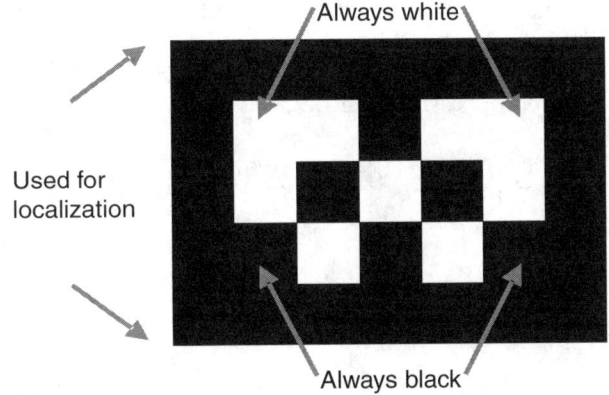

Figure 10.9 Layout of the pattern used for self-localization.

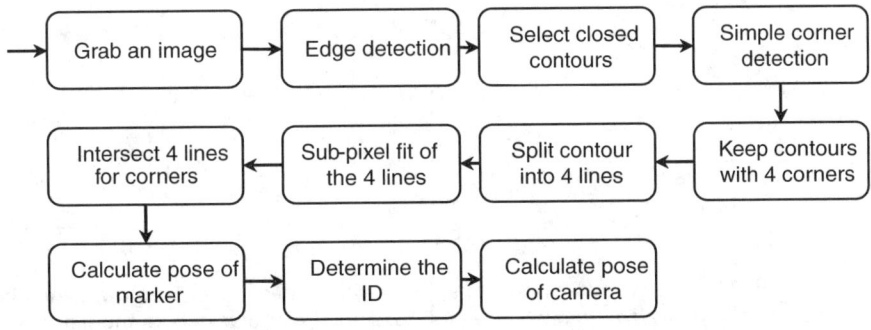

Figure 10.10 Schematic representation of the marker detection algorithms.

Canny uses the fact that a point on the edge has a gradient magnitude that is a maximum perpendicular to the edge. This is depicted in Figure 10.11.

Since the derivative of an edge usually is modeled as a Gaussian, a Gaussian derivative filter should be used to calculate the gradient image. Because we are interested in small details, we use a small sigma of 1.5. A proper filter should use a [7 x 7] neighborhood with filter coefficients in floating point.

For performance reasons we simplified the gradient calculation. We found that under the perceived noise levels and optical characteristics the derivatives or gradients in the image can be approximated using the following mask, which can be implemented using integer math:

$$g_u = \begin{bmatrix} -1 & 0 & 1 \end{bmatrix}, g_v = \begin{bmatrix} -1 & 0 & 1 \end{bmatrix}^T, \vec{g} = \begin{pmatrix} g_u \\ g_v \end{pmatrix} \quad (10.38)$$

Note that no smoothing is applied in the perpendicular direction. When we need to use the gradients to determine the precise edge location in Section 10.5.7 we need a better approximation of the Gaussian derivative for a better accuracy (see Section 10.9.1). Now a useful approximation was found as:

$$gu = \begin{bmatrix} -7 & -19 & 0 & 19 & 7 \end{bmatrix} \quad (10.39)$$

All gradients are now scaled by some factor, but that has no influence on the rest of the algorithm.

With this filter, the gradient orientation and magnitude can be calculated, and non-maximum suppression can be done:

$$nms(\vec{r}) = \begin{cases} 1 & if \|\vec{g}(\vec{r})\| \text{ is a maximum in the gradient direction} \\ 0 & otherwise \end{cases} \quad (10.40)$$

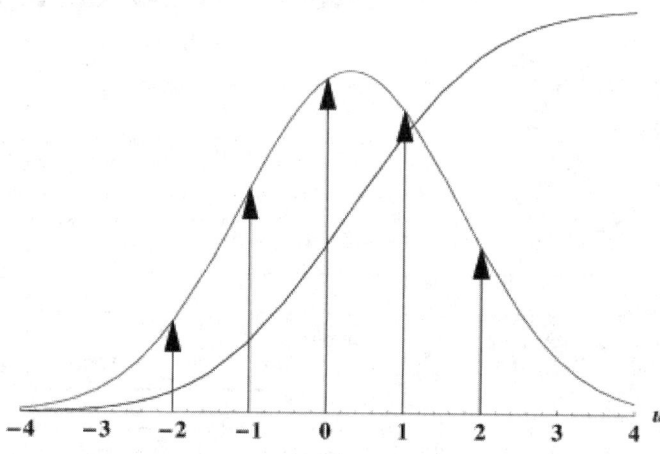

Figure 10.11 Intensity on a horizontal edge, with its derivative. The edge is modeled as an error function. The derivative then becomes a Gaussian. The arrows denote the samples of the derivative in the image. The Canny algorithm selects the sample at u = 0 as an edge point.

10.5 Camera Positioning

Not all maximum points are edges, as noise also generates maximum points. We may also want to disregard very weak edges. Canny proposed to do hysteresis thresholding, shown in Figure 10.12.

The number of strong edges is defined as:

$$\text{\# strong edges} = P_{high} \cdot N \qquad (10.41)$$

with p_{strong} a user-specified proportion of all maximum points N with the largest gradient magnitudes. To determine the corresponding threshold t_{strong}, a histogram h is made for the gradient magnitude of the maximal points. We use a histogram with 65,536 bins in which bin 65,535 holds the maximum gradient found in the image. Now the threshold can be determined as:

$$\sum_{i=t_{strong}}^{65535} h(i) = P_{strong} \cdot N \qquad (10.42)$$

in which i is the bin number in the histogram. By definition the threshold for weak edges is:

$$t_{weak} = p_{weak} \cdot t_{strong} \qquad (10.43)$$

in which p_{weak} is a user-specified proportion. Now points with gradient magnitudes between t_{weak} and t_{high} belong to weak edges, while points with a gradient magnitude greater than t_{high} are classified as strong edge points. After this a maximum point is classified as an edge, only if it belongs to a strong edge, or if it belongs to a weak edge that is connected—directly or via other weak edge points—to a strong edge point. What we obtain are thin borders on the presumed edges in the image.

10.5.4 Contour Detection

We define a contour as a set of all connected edge points. For further processing we want to distinguish different kinds of contours. For this we classify each edge point looking at the number of neighboring edge points n. This is shown in Figure 10.13.

In case of our marker, we are only interested in contours without branch or end points. To reduce the amount of data we make a list of contours. Of each contour all special points are stored ($n \neq 2$). In case there are no special points, one edge point is chosen. For each stored point, a list is made that stores the 8-connected direction of

Figure 10.12 Black part consists of strong edge points. The gray part consists of weak edge points. The hysteresis thresholding selects the strong edge points, and all weak edge points connected to them. In this case the weak edge points at the right side are disregarded as they don't connect to strong edge points.

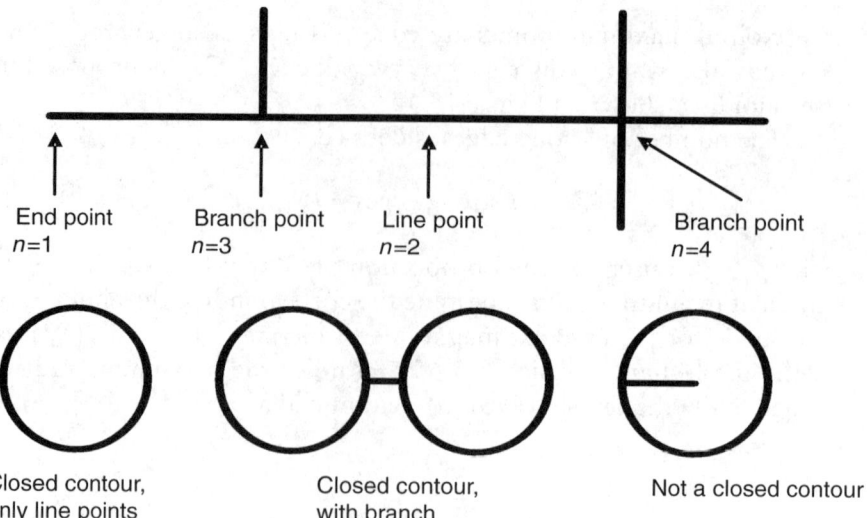

Figure 10.13 Overview of different edge points. Bottom: Some contour examples.

the neighboring edge points, as well as a link to the special point that can be found following that direction,

In normal images we found that the Canny edge is sometimes more than one pixel thick. This means that the above method of classifying edge points is not valid. Therefore we apply a simple edge thinning algorithm that removes points with $n > 1$, without splitting a contour.

This is done in a 3 x 3 neighborhood, and can be implemented quickly using a lookup table. The result can now be called a skeleton.

Now we have a set of contours that includes the contours of our markers.

10.5.5 Rejecting Unwanted Contours

To be the outer edge of a marker, a contour has a few restrictions. The contour should be closed, so no end points or branch points should be present. From we know that the outer border is black, and thus darker than the surroundings. This means that the gradient on the outer edge is pointing outward. Only one random point has to be checked to verify this. Contours not satisfying these restrictions are not considered further.

10.5.6 Corner Detection

If we want to have 4 borderlines the closed contour has to be split in 4. The splitting should be done on the corner positions. So first we approximate the corner positions by applying a corner detector on the contour. It is similar to the Harris corner detector [23]. We found that although this detector is useful to find corners, it is not accurate enough to be used for our position estimation.

From the distribution of gradients around a corner or a line in Figure 10.14, we can see that if an ellipse is fit around all the gradients with (0,0) as center, the length of the main axes say something about how well the data describes a corner or a line.

10.5 Camera Positioning

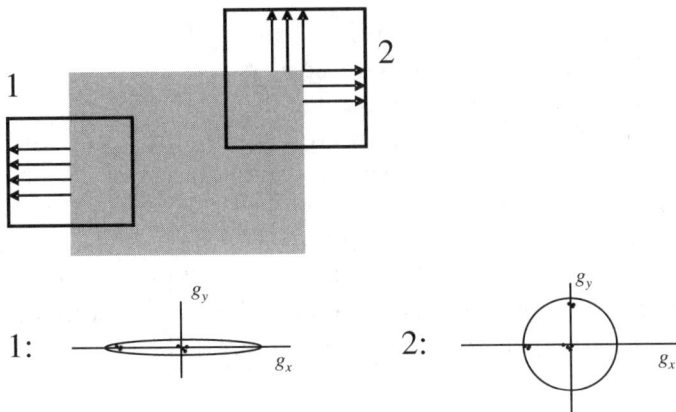

Figure 10.14 Gradient directions on the edge of a gray patch. Region 1 is a line and region 2 is a corner. Bottom: The dots show the gradients in region 1 and 2. A centered ellipse is fit around these gradients.

To find the ellipse we use the covariance matrix of the gradients, g_x g_y, in a 5 x 5 neighborhood.

This covariance matrix is also referred to as the Gradient Structure Tensor:

$$GST = \sum_{5x5} \begin{pmatrix} g_x \cdot g_x & g_y \cdot g_x \\ g_x \cdot g_y & g_y \cdot g_y \end{pmatrix} \quad (10.44)$$

The eigenvalues of this matrix give the length of the two main axes. When a 90° corner is present, the ellipse would be circular, which means that the two eigenvalues are equal. A measure can be constructed that is 0 on a line, and 1 on a circle. This q-value [24] is given by:

$$q = 2\frac{\sqrt{\lambda_1 \lambda_2}}{\lambda_1 + \lambda_2} \quad (10.45)$$

A problem is that due to noise, this q value can also approach 1 in a homogeneous region. Therefore, the eigenvalues have to be threshold before using the q value. In our case, we are always on an edge, so we only have to threshold the q value. To have a robust count of corners, we do a non-maximum suppression of this q value along the contour. The remaining corner candidates are threshold using:

$$t_{corner} = p_{corner} \cdot \max(q) \quad \text{per contour} \quad (10.46)$$

in which p_{corner} is selected by the user. If we're not left with 4 corners, the contour is rejected as a marker.

10.5.7 Fitting Lines

Now the four corner positions are known we can determine the 4 lines. The contour can now be broken into 4 separate lines. Because we model the lines as straight lines, we don't want any influence of the corners. Thus, the lines will be eroded a bit, so they will be disconnected from the corners.

To find the best fitting line through the contour points, we use a least squares fit. The more edge-points used, the more accurate the result will be. For speed reasons the number of edge points used is set to a maximum of 15. The best fit is found if the edge positions are subpixel accurate, and if the line to be fitted is perfectly straight.

We know that our images suffer from large lens distortions, so the camera was calibrated to be able to correct for that. The sub pixel edge points found will be corrected before the fit is performed. From section it is assumed that perpendicular to an edge, the gradients follow a Gaussian profile:

$$g(x) = a \cdot e^{\left(\frac{x-\mu}{\sigma}\right)^2} \tag{10.47}$$

in which g is the gradient magnitude, x is the distance from a Canny edge point, μ is the estimated true sub-pixel position of the edge, and σ is a scale.

If we take the natural logarithm of the gradient, we obtain a simple model:

$$\ln(g(x)) = \left(\frac{x-\mu}{\sigma}\right)^2 + c = a'x^2 - a'2\mu x + c' \tag{10.48}$$

If we now fit a parabolic function through the gradients on a line perpendicular to the edge:

$$\ln(g(x)) = ax^2 + bx + c \tag{10.49}$$

μ can be calculated to be:

$$\mu = -\frac{b}{2a} \tag{10.50}$$

The gradient g is estimated using the filter (10.39). To do the fitting in a fast way, we only fit horizontally or vertically. Furthermore, we only use three neighboring points. In case the edge orientation is more horizontal than vertical, we find:

$$\begin{aligned} x &\in [-1,0,1] \\ y(x) &= \ln(g(c+x)) \end{aligned} \tag{10.51}$$

in which c is the position of the contour point, and x the distance to the neighboring point. After fitting a parabola, μ is the horizontal sub pixel shift from the contour point.

Because we use three constant displacements, x, an analytic solution for μ can be given:

$$\mu = \frac{y(-1) - y(1)}{(y(-1) + y(1) - 2y(0))} \tag{10.52}$$

10.5 Camera Positioning

In Section 10.9.1 the accuracy of this position μ is determined. All calculated sub pixel edge points will now be transformed to undistorted-image coordinates (Section 10.6.2) so that we can just fit a straight line using a standard least square fit. The intersection point of the two lines adjacent to the estimated corner, give the sub-pixel accurate corner in undistorted image coordinates.

10.5.8 Determining the ID of the Marker

If the position of the camera with respect to the marker is known (see Section 10.6.3), we can try to find its ID. For that we want to know the intensity of the blocks within the pattern. We know in pattern-coordinates where the midpoints of the blocks are:

$$\vec{m}_{i,j}^{P} = \begin{pmatrix} (i-2)\frac{width}{7} \\ (j-1)\frac{height}{5} \\ 0 \end{pmatrix}, i \in [0,4], j \in [0,2] \tag{10.53}$$

Thus, we can calculate the midpoint positions in image-coordinates using the estimate of the camera position in marker coordinates using the formulas in Section 10.6.2. For those 15 points, the intensity in the image is determined:

$$y_{i,j} = I(\vec{m}_{i,j}^{i}) \tag{10.54}$$

The inner part of the marker consists of black and white patches. Both colors are always present, so we determine the threshold as:

$$t_{block} = \frac{\min(y_{i,j}) + \max(y_{i,j})}{2} \tag{10.55}$$

$$b_{i,j} = \begin{cases} 1 & y_{i,j} > t_{block} \\ 0 & otherwise \end{cases} \tag{10.56}$$

At this point, it is still possible that the candidate marker is not valid. For instance, a black computer monitor also has four corners and a dark inner part. To invalidate such cases we also reject a candidate if the separation between black and white intensity is too low:

$$reject\ if\ \left(\max(y_{i,j}) - \min(y_{i,j})\right) < t_{sep} \tag{10.57}$$

Now the ID can be determined as:

$$bit = 5j + i, b_{bit} = b_{i,j}$$
$$ID = \sum_{bit=0}^{14}(2_{bit} \cdot b_{bit}) \tag{10.58}$$

This is also explained in Figure 10.15.

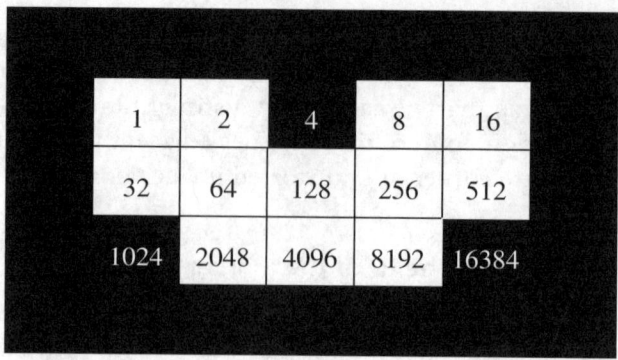

Figure 10.15 Marker with ID = 4 + 1024 + 16,384 = 17,412.

Using this method, we theoretically only need 1 pixel per block, so the inner part could be as small as 5 by 3 pixels. However, in our case the border will then also be 1 pixel wide, so the marker won't be detected

10.6 Determining the Pose from a Marker's Feature Points

10.6.1 Coordinate Systems

For pose estimation, we use the following coordinate systems:

Ψ_M The 2D model frame, or the x,y-plane of the pattern frame

Ψ_P The 3D pattern frame. This frame is attached to the pattern that is used to determine the camera's position.

Ψ_C The 3D camera frame. This frame has the camera's optical axis as z axis.

Ψ_U The 2D undistorted image frame. The pinhole projection of the camera frame.

Ψ_D The 2D distorted image frame. Distortions of the lens applied to the undistorted image frame.

Ψ_I The 2D image frame, pixel coordinates, includes distortions from the sampling grid of the sensor.

Some of these frames are depicted in Figure 10.5.

10.6.2 Camera Model

As is mostly done, we transform pixel coordinates such that a simple pinhole camera model can be used. The steps to go from this model to pixel coordinates will be explained now. Please remember that we use capitals to denote homogeneous coordinates, and superscripts to specify which coordinate system is used. We start from a point in homogeneous camera coordinates:

$$\vec{P}^C = \begin{pmatrix} p_x^C & p_y^C & p_z^C & 1 \end{pmatrix}^T = \begin{pmatrix} \vec{p}^C & 1 \end{pmatrix}^T \quad (10.59)$$

10.6 Determining the Pose from a Marker's Feature Points

in which T is the transpose operator. The camera coordinate system is a right-handed coordinate system with the optical point as center, and the optical axis as z axis.

This point is imaged on a fictive sensor plane which we call the undistorted image plane U, at distance 1 from the optical point:

$$\vec{P}^U = \frac{1}{p_z^C} \vec{p}^C \tag{10.60}$$

Because our lens has a very wide opening angle of 90°, we use a distortion model with the second and fourth-order spherical distortions, more orders did not yield more accurate results:

$$\vec{p}^D = \left(1 + k_1 r_u^2 + k_2 r_u^4\right)\vec{p}^U = c \cdot \vec{p}^U, \text{ with } r_u = \left\|\vec{p}^U\right\| \tag{10.61}$$

with D short for distorted image plane. The distorted image plane can be seen as a scaled version of the real sensor plane. We go directly to pixel coordinates using the formula:

$$\vec{p}^P = (u \quad v) = \begin{pmatrix} S_u & 0 & u_{offset} \\ S_{sk} & S_v & v_{offset} \end{pmatrix} \vec{p}^U = A\vec{P}^U \tag{10.62}$$

This means that we expect different sampling distances in x and y as well as a skewing of the v axis.

Note that we use a model description going from camera to pixel coordinates. To go from pixels to camera coordinates these formulas have to be inverted. The only formula that poses a problem comes from the distortion model (10.61). The formula can be rewritten in the form:

$$f(r) = r_d - \left(1 + k_1 r^2 + k_2 r^4\right) r = 0 \tag{10.63}$$

in which the root of f, within the image's interval, is the sought r_u. The root is found iteratively with the Newton method with r_d as initial guess.

10.6.3 Estimating the Pose

A candidate marker has 4 corners which are expressed in undistorted-image coordinates. Below, the superscript P will be used to specify the pattern coordinate system. The relation between a point expressed in pattern coordinates and in undistorted image plane coordinates is given by:

$$s\vec{P}^U = s\begin{pmatrix} u \\ v \\ 1 \end{pmatrix} = \vec{p}^C = \begin{pmatrix} \mathbf{R}_P^C & \mathbf{T}_P^C \end{pmatrix}\begin{pmatrix} \vec{P}^P \end{pmatrix}, s = p_z^C \tag{10.64}$$

in which **T** is a translation and **R** is a 3 x 3 rotation matrix to go from pattern coordinates to camera coordinates. This formula can be simplified by the fact that we defined $p_z^P = 0$ because a corner always lies in the x,y plane of the pattern frame. When we remove the z coordinate from formula (10.64) we obtain:

$$s\vec{P}^U = s\begin{pmatrix} u \\ v \\ 1 \end{pmatrix} = (\vec{r}_1 \quad \vec{r}_2 \quad \mathbf{T}) \begin{pmatrix} p_x^P \\ p_y^P \\ 1 \end{pmatrix} = \mathbf{H}\vec{P}^M \qquad (10.65)$$

This can be rewritten as:

$$s\begin{pmatrix} u \\ v \\ 1 \end{pmatrix} = \begin{pmatrix} H_1 \\ H_2 \\ H_3 \end{pmatrix} \vec{P}^M \qquad (10.66)$$

This set of equations can be reordered to:

$$\begin{aligned} s &= H_3 \vec{P}^M \\ H_1 \vec{P}^M - u H_3 \vec{P}^M &= 0 \\ H_2 \vec{P}^M - v H_3 \vec{P}^M &= 0, or \\ \begin{pmatrix} \vec{P}^{MT} & 0 & -u\vec{P}^{MT} \\ 0 & \vec{P}^{MT} & -v\vec{P}^{MT} \end{pmatrix} \begin{pmatrix} H_1^T \\ H_2^T \\ H_3^T \end{pmatrix} &= Lx = 0 \end{aligned} \qquad (10.67)$$

in which P^{MT} is the transpose of P^M. The matrix L can be extended downwards for all 4 points, and the solution for **x** is found using singular value decomposition as the right singular vector of L, associated with the smallest singular value. To get L numerically well conditioned, data normalization can be used. From **x** we can reconstruct **H**. To complete the rotation matrix **R** we use:

$$\vec{r}_3 = \vec{r}_1 \times \vec{r}_2 \qquad (10.68)$$

Because **R** is estimated, the matrix is not orthonormal. We can find the best orthonormal matrix using singular value decomposition as well:

$$\mathbf{R}_{estimate} = \mathbf{UDV}^T \Rightarrow \mathbf{R}_{orthonormal} = \mathbf{UIV}^T \qquad (10.69)$$

The resulting camera pose is given by the homogeneous transformation matrix:

$$\mathbf{H}_P^C = \begin{pmatrix} \mathbf{R} & \mathbf{T} \\ 0 & 1 \end{pmatrix} \qquad (10.70)$$

We found that the resulting camera pose was not accurate enough, so we applied a Levenberg-Marqardt algorithm that optimizes the pose by maximizing a fitness criterion. This criterion as given by:

$$\begin{pmatrix} \vec{p}_c^{u'} \\ 1 \end{pmatrix} = \frac{1}{s}(\mathbf{R} \quad \mathbf{T})\vec{p}_c^l > \textit{fitness} = -\frac{1}{4}\sum_{c=1}^{4}\left(\vec{p}_c^U - \vec{p}_c^{U'}\right)^2 \qquad (10.71)$$

in which $\vec{p}_c^U$ is the measured position of corner c. This fitness value says something about how good the 4 corner points can be described by our model of the marker and the camera. If we only take into account the model of the marker, we can say that if the fitness is low, that the 4 corners were not part of our marker. So the final step is to threshold the fitness value to get rid of candidate markers that do not match well enough.

Note that we now have the camera's position, but the camera frame itself is not known yet. For sensor fusion all estimates should be given for a common frame, and be expressed in the same frame of reference. Expressing the estimated pose is possible by looking up the pose of the marker in world coordinates using its ID. Generating a pose estimate for the common body frame is only possible if the transformation of our camera frame is known with respect to the body frame:

$$\mathbf{H}_p^b = \mathbf{H}_C^b \mathbf{H}_P^C \qquad (10.72)$$

The calibration to find is described in Section 10.8.3.

10.7 First Experiment

A simple experiment was done to see the effect of sensor fusion using our pluggable Kalman filter setup. We put all the sensors on a mobile robot. We had gyroscopes, inclinometer, and a JAI camera for the orientation estimation, and accelerometers, odometer, and the JAI camera for position estimation. The odometer of the robot counts wheel rotations, and we used it to calculate the current speed of the robot. The robot was pushed in a square rectangle and the output of the different sensors, as well as the fused sensor estimate, was recorded. Outside the square two markers were put for the camera pose estimation. Figure 10.16 shows the result.

These results show the capability of the fusion setup to recover from large errors, and the ability to use knowledge about the expected accuracy of a sensor. The camera output for instance is not accurate at large distances. Its inaccuracy is modeled as an increase of the measurement noise as a function of distance to the marker. In effect the filter doesn't believe the camera until the marker is very near. At that moment any error accumulated on the way (left lower corner) is corrected (left upper corner). On the left side something clearly is more wrong than the camera estimation alone. We have determined that at that point a large magnetic disturbance is present beneath the floor. The compass readings there have an error of 5° to 10°.

It can be seen from Figure 10.16 that the camera only gives reasonable estimates for distance smaller than 1.5 m. With one marker in view the JAI camera is clearly not good enough for usage in office buildings. Consequently, we want to extend the range of a camera by increasing the resolution of the sensor. The DICA camera has a resolution of 1,280 x 1,024 instead of 320 x 240, so we expect a 4 times higher range. In Section (10.9) we look at the accuracies of the JAI and DICA cameras. The

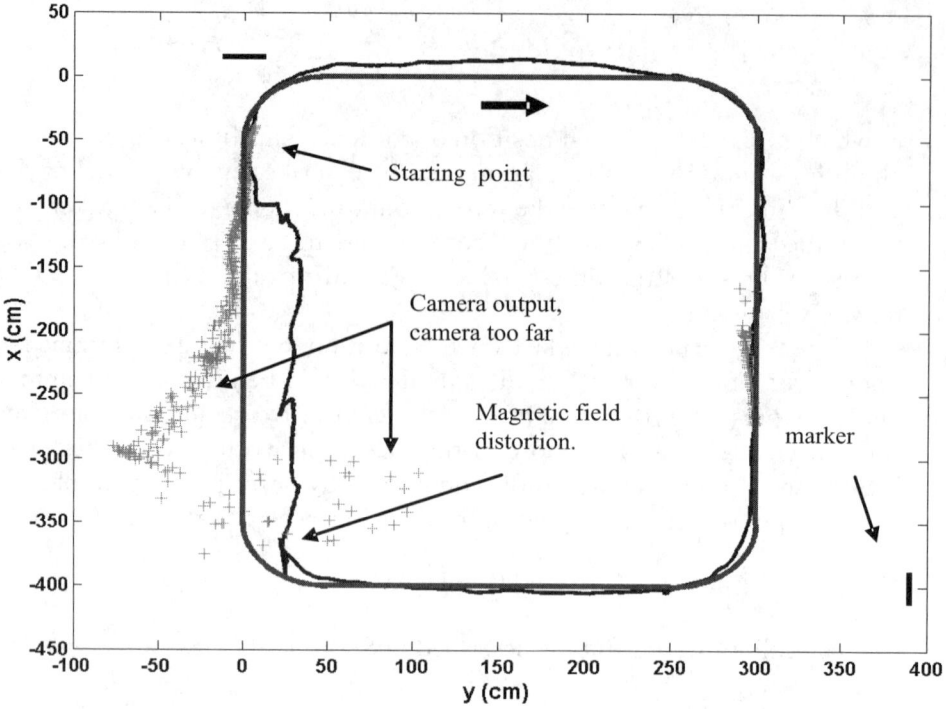

Figure 10.16 Experiment in which the robot was moved clockwise on the red line. The Kalman output in shown in blue, and the camera position in green.

results could be used to make a better model of the measurement noise of the camera. In Figure 10.16 one may notice that the camera pose estimation is quite good at 2m, but the filter thinks the pose is bad.

10.8 Calibration

10.8.1 Camera Calibration

Because of the large lens distortions for a 90° opening angle lens, we had to at least correct for the fourth-order radial distortion. The Zhang algorithm [10] estimates the second- and fourth-order radial distortion parameters. Six views of a big checkerboard pattern were used, and the extracted corners (saddle points) were used in the algorithm. We found that adding more distortion parameters did not show a significant decrease in the residual error.

10.8.2 Pattern Pose

Normally the pose of a pattern in world coordinates should be known in advance. With mobile demonstrations however that is not the case. If the pose of the body frame is known, then the pose of the pattern is obtained by:

$$H_P^W = H_b^W H_C^b H_P^C \qquad (10.73)$$

The problem here is that H_C^b has to be calibrated as well, so we estimate H_P^W directly using 3 vectors, expressed in world coordinates as well as in pattern coordinates. If a linear combination of these vectors spans the whole 3D space, the rotation matrix can be found as:

$$\begin{pmatrix} \vec{v}_1^W & \vec{v}_2^W & \vec{v}_3^W \end{pmatrix} = R_P^W \begin{pmatrix} \vec{v}_1^P & \vec{v}_2^P & \vec{v}_3^P \end{pmatrix}$$
$$R_P^W \begin{pmatrix} \vec{v}_1^W & \vec{v}_2^W & \vec{v}_3^W \end{pmatrix} \begin{pmatrix} \vec{v}_1^P & \vec{v}_2^P & \vec{v}_3^P \end{pmatrix}^{-1}$$
(10.74)

A good method would be to translate the body in these three directions, but we cannot measure these vectors in world coordinates. Another method is to view those vectors as rotation axes. The body can now be rotated around each of these axes, measuring starting orientation and ending orientation. The rotation axis in world coordinates can be found by using the quaternion notation of the inclinometer output:

$$\begin{pmatrix} 1 \\ \vec{v}^W \end{pmatrix} \cdot \begin{pmatrix} \cos(\varphi/2) \\ \sin(\varphi/2) \end{pmatrix} = q_{b,end}^W \otimes q_{b,start}^W - 1$$
(10.75)

and a similar formula in pattern coordinates; noise on the measurements should be averaged to get a good result. Now only the orientation of the pattern is known. The position of the pattern is dependent on the position of the origin. If the origin can be chosen freely, the position of the marker can be chosen freely.

10.8.3 Camera Frame to Body Frame

Rewriting (10.73) gives:

$$H_b^W = H_p^W H_C^P H_b^C$$
$$H_b^C = \left(H_p^W H_C^P \right)^{-1} H_b^W$$
(10.76)

We have the pose of the camera in pattern coordinates. Hence, we have derived the pose of the pattern in world coordinates. We have the orientation of the body frame in world coordinates. Only the true position of the body frame is not known yet. The rotation of the body frame with respect to the camera frame can be calculated any time as the rotation is fixed. By rotating the body around known axes (in body frame coordinates) the translation from camera origin to body origin can be determined as well.

10.9 Measurements

10.9.1 Performance of the Subpixel Edge Detector

Edge position estimators use interpolation to find the subpixel location. Normally a quadratic surface is fit around a maximum of the gradient. We want to validate this approach by looking at the response of the camera on a vertical step edge.

A step edge was printed on a A4 paper Figure 10.17 (a) and placed at 2m on a micro stage. The stage is able to move the edge horizontally by a single micro meter. We moved the stage linearly in 100 steps. In case of the JAI camera we used 400-μm steps, and for the DICA camera we used 200-μm steps.

In b the response of 3 single pixels in the edge neighborhood is shown. One can observe that the edge is imaged very sharply; also the noise in the values is very low (3σ error bars).

For all 50 frames of the 100 stage positions, the edge position was determined using the simplest edge detector from section Because no ground truth was available, we estimated the real edge position by fitting a straight line through the 100 * 50 estimations. The mean and sigma of the deviation are plotted in Figure 10.18.

The remarkable sine-like error in the mean deviation can be explained by the mismatch between the presumed parabolic behavior of the gradient on an edge, and the actual behavior of the gradient, which is more Gaussian-like. From the figure, an

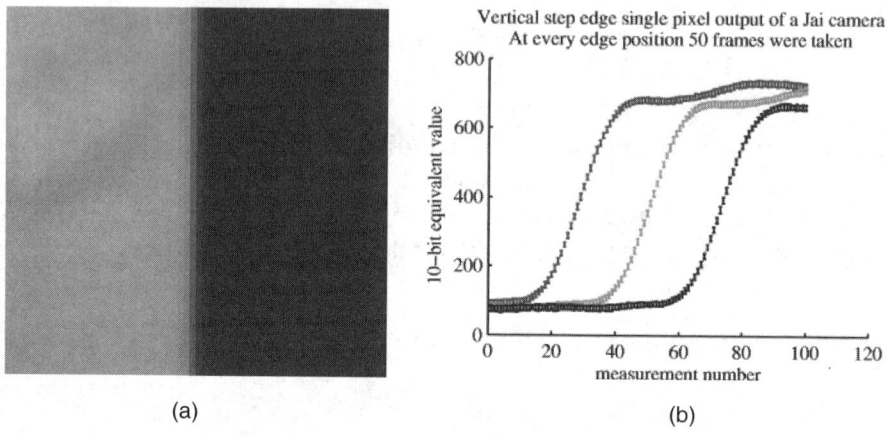

(a) (b)

Figure 10.17 (a) One image of the step edge, (b) Intensity of pixels on a single row in columns 26,28,30.

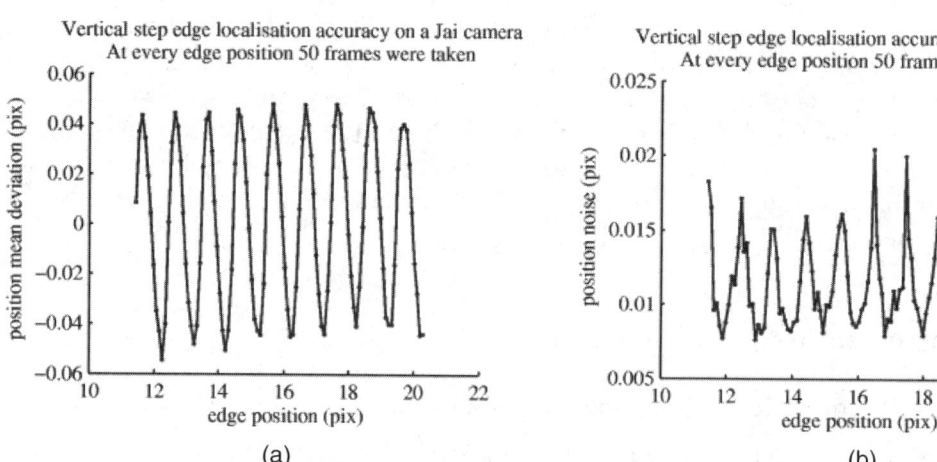

(a) (b)

Figure10.18 Station from estimated truth and noise of a simple edge detector.

accuracy of about one 25th of a pixel can be seen. If we correct for the Gaussian profile by taking the logarithm of the derivatives before fitting the parabola, we obtain Figure 10.19.

Consequently, the accuracy is much better, about one 50th of a pixel, but at the expense of some processing time. With a simple Gaussian derivative the mean deviation can be made a little bit better still, but the processing time is again longer than the simple gradient method.

10.9.2 The Dependence of the Pose Accuracy on the Viewing Angle

To test the accuracy of the positioning algorithm we placed a marker on a pan-tilt unit. A camera was placed at various distances, with the marker in the middle of the image.

Using the pan-tilt unit we rotated the marker, and at every orientation, we grabbed 50 images. In we show with a blue '+' the direction of the marker. With red dots the estimated orientation for all frames is depicted (50 per orientation). Clearly, the direction of the noise seems dependant on the angle—in pan, tilt coordinates. In the right figure, the orientation of the noise is graphed with respect to the pan, tilt angles. The sigma of the noise in that direction did not show dependence on the angle.

The ground truth is the settings for the pan-tilt unit, but the camera could not be fixed with a known orientation. Therefore, we calibrated the pose output of the algorithm to match best with the ground truth. The formula is given in (10.77). All matrices are rotation matrices, and the 6 parameters for the left-hand and right-hand correction matrices are estimated.

$$R_{calib} = R_1 R_{meas} R_r \qquad (10.77)$$

In Table 10.1 and Table 10.2, some numbers are given for the noise in the measurements. With the sigma of the noise we mean the variation around the mean of the estimate. With the sigma of the deviation we mean the variation around the

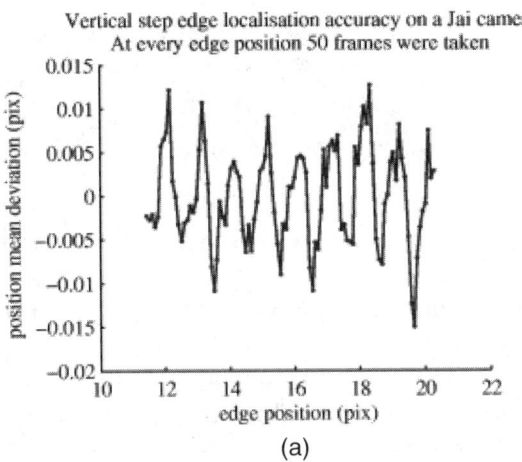
(a)

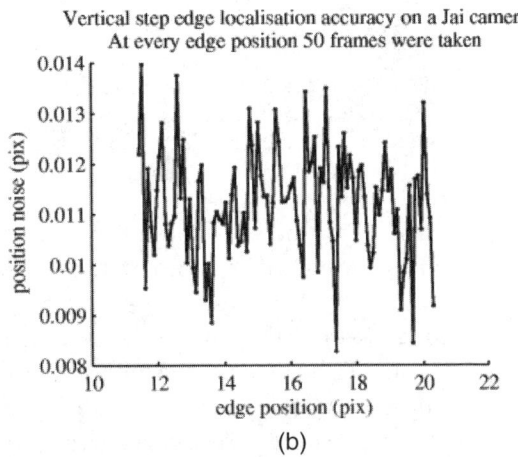
(b)

Figure 10.19 Deviation from estimated truth and noise of the log-intensity edge detector.

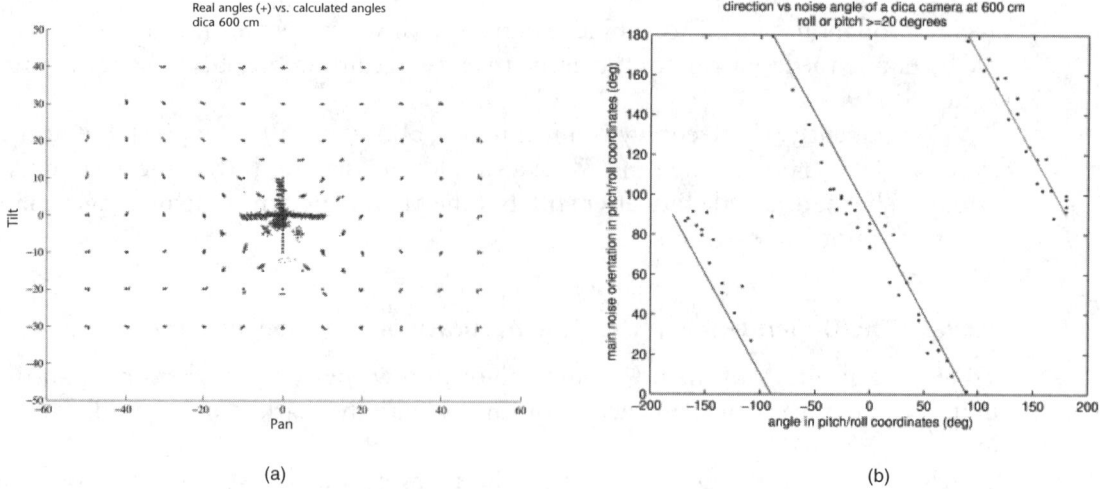

Figure 10.20 Measurements of the orientation of the marker on a pan-tilt-unit

Table 10.1 Orientation Accuracy with Roll or Pitch Greater than 20 Degrees. (The 95% best data is used and angles are in degrees. Left: DICA camera. Right: JAI camera.)

distance	noise		deviation	
	sigma	max	sigma	max
200 cm	0.032	0.069	0.35	0.59
300 cm	0.047	0.10	0.49	0.89
400 cm	0.12	0.27	0.76	1.4
500 cm	0.13	0.27	0.54	0.94
600 cm	0.18	0.48	0.62	1.1
650 cm	6.1	30	12	49

distance	noise		deviation	
	sigma	max	sigma	max
140 cm	0.11	0.24	0.66	1.23
200 cm	0.24	0.54	0.75	1.31
300 cm	0.90	3.69	2.14	4.47

Table 10.2 Orientation Accuracy with Roll and Pitch Lower than 20 Degrees. (The 95% best data is used and angles are in degrees. Left: DICA camera. Right: JAI camera.)

distance	noise		deviation	
	sigma	max	sigma	max
200 cm	0.11	0.25	0.64	1.4
300 cm	0.17	0.44	0.89	2.4
400 cm	0.69	2.3	2.9	6.9
500 cm	0.65	2.4	2.4	9.5
600 cm	0.53	1.5	1.3	5.2
650 cm	0.65	2.0	2.9	11

distance	noise		deviation	
	sigma	max	sigma	max
140 cm	0.31	0.68	1.3	2.0
200 cm	1.1	3.3	2.7	6.8
300 cm	2.7	7.9	4.6	16

ground truth, and that number could be used in a Kalman filter. The roll and pitch angles are used to estimate the noise and deviation.

From these results we first can deduce that the accuracy of the orientation is better when the marker is viewed under an angle (here 20°s). Second is that the sigma of the deviation from the ground truth is much higher than just the noise in the measurements. This suggests that there are not modeled influences on the orientation or a not entirely correct calibration of the camera.

To estimate the accuracy of the position, we first estimated the ground truth. The midpoint of the marker will not be on the joint of the pan-tilt unit, so we need to estimate 6 parameters: The vector to the joint and the vector from the joint to the center of the marker. Again we fit a correction for the unknown direction of the pan-tilt unit, thus in total there are 9 parameters to calibrate. The measurements are tested against the corrected set points:

$$\vec{r}_{est.truth} = \vec{r}_{joint} + R_{corr} R_{set} \cdot \vec{r}_{center} \tag{10.78}$$

In Table 10.3 and Table 10.4 numbers for the noise and deviation of the three coordinates are given.

Please note that the deviation is generally much larger than the noise and that only the accuracy in z is really dependent on the distance. Even at 6.5m the accuracy in x and y coordinates is extremely good. From the camera's point of view this means that the center of the marker in image coordinates is very accurate, this in turn means that the angle to the center of the pattern is very accurate. That of course is one of the reasons that people use multiple markers reasonable far apart to determine the position of a camera.

10.9.3 The Dependence of the Pose Accuracy on the Location in the Image

We have looked at the accuracy of the pose estimation with the marker in the middle of the image at various angles. That means that lens distortions are not really important. In this experiment, we looked at the accuracy if the marker is imaged at various positions in the image. This was accomplished by putting the camera on a pan-tilt unit instead of the marker. As seen in the previous experiment, the pose estimation is best when the marker is viewed under an angle. So we chose to put the marker under 30°s (pan direction). In Figure 10.21 an example can be seen of the JAI camera, with pan and tilt at −30°s.

Table 10.3 Position Accuracy of the DICA Camera. (The 95% best data is used and numbers are in cm.)

	distance	x		y		z	
		sigma	max	sigma	max	sigma	max
noise	200 cm	0.0033	0.0084	0.0015	0.0036	0.025	0.068
	300 cm	0.0025	0.0067	0.0022	0.0053	0.052	0.14
	400 cm	0.0054	0.015	0.0064	0.016	0.17	0.47
	500 cm	0.0096	0.025	0.0075	0.019	0.23	0.60
	600 cm	0.0053	0.014	0.018	0.046	0.36	0.97
	650 cm	0.015	0.055	0.025	0.067	0.49	1.30
deviation	200 cm	0.027	0.070	0.012	0.024	0.18	0.55
	300 cm	0.043	0.13	0.047	0.12	0.32	0.91
	400 cm	0.021	0.069	0.030	0.059	0.52	1.4
	500 cm	0.028	0.068	0.025	0.063	0.89	2.4
	600 cm	0.034	0.07	0.059	0.14	1.3	3.9
	650 cm	0.039	0.14	0.072	0.16	1.2	3.2

Table 10.4 Position Accuracy of the JAI Camera. (The 95% best data is used and numbers are in cm.)

	distance	x		y		z	
		sigma	max	sigma	max	sigma	max
noise	140 cm	0.0047	0.012	0.0057	0.014	0.058	0.16
	200 cm	0.015	0.039	0.011	0.029	0.16	0.43
	300 cm	0.022	0.059	0.027	0.075	0.54	1.5
deviation	140 cm	0.013	0.030	0.021	0.043	0.22	0.50
	200 cm	0.040	0.11	0.029	0.069	0.40	1.1
	300 cm	0.059	0.15	0.075	0.21	1.6	4.2

In these experiments, we did not see much difference between pan/tilt values inside or outside the range of 20 degrees. Therefore, the numbers are given for the whole range. Equation (10.77) is used again to get a best fit of the data.

In Figure 10.22 the ground truth and calculated orientation are depicted. Equation (10.78) is used again, but since the camera is now rotating, the inverse of R_{set} is used.

In Figure 10.23 the deviation in x and y is depicted for each orientation. At each orientation, pan/tilt axes, a virtual coordinate system is constructed. In that coordinate system, the deviation from the ground-truth, at that orientation, is plot. The scale is such that one degree corresponds to 1 cm.

Table 10.5, Table 10.6 and Table 10.7 give the accuracies of orientation and position estimation for both the JAI and the DICA camera.

Looking at the sigma of the deviation, the position estimation is much worse than in the case of a rotating marker. In this case, a wrong estimate of the distance – z – has more influence on the estimate of x and y. This can be seen by looking at

Figure 10.21 Example picture taken by our JAI camera mounted on a pan-tilt unit.

10.9 Measurements

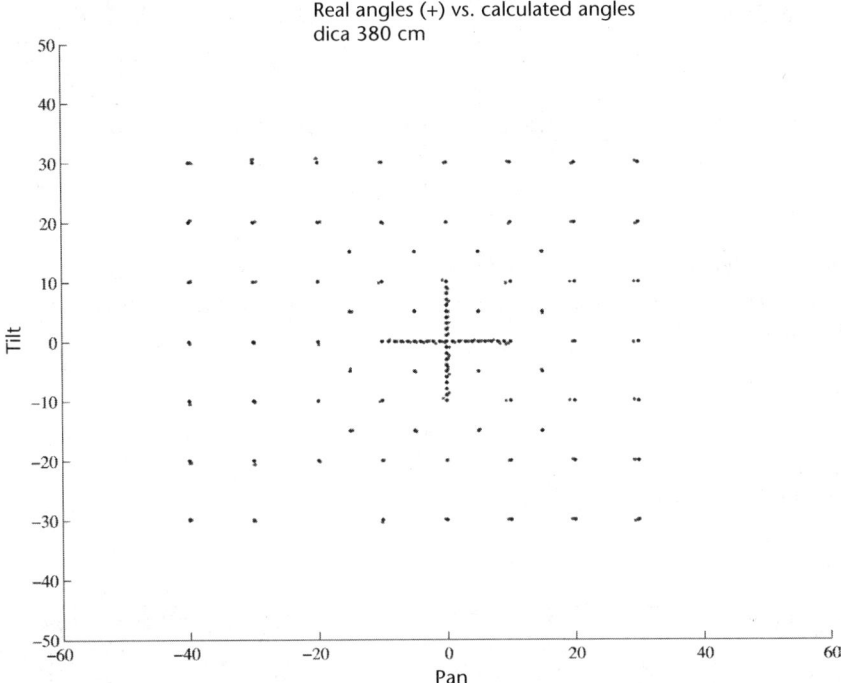

Figure 10.22 Measurement result with a DICA camera on a pan-tilt unit. The plusses are the ground truth orientations, the dots are the calculated orientations.

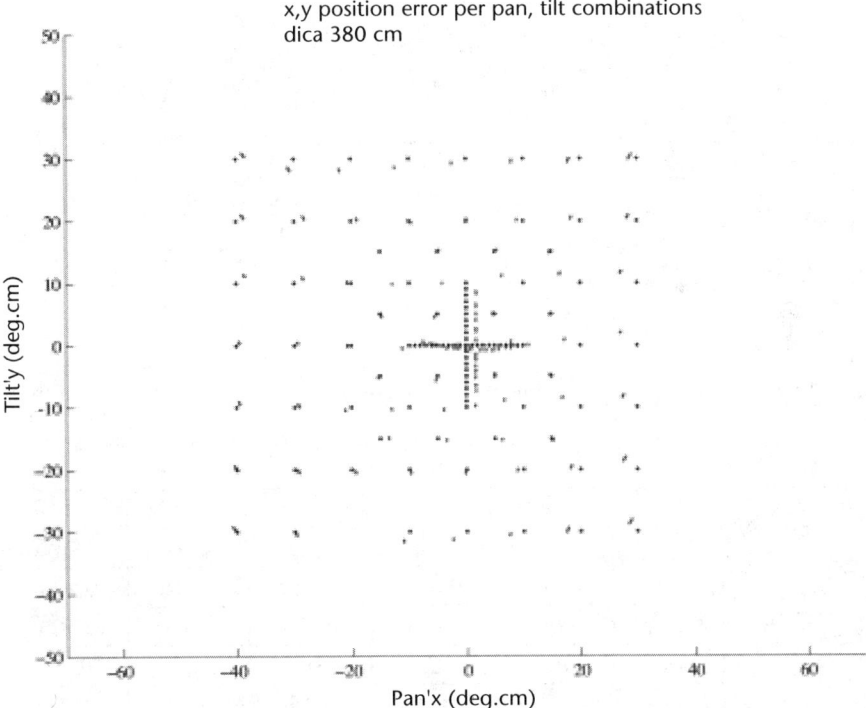

Figure 10.23 Measurement result with a camera on a pan-tilt unit. The plusses are the ground truth positions, the dots are the calculated positions (see text for detailed explanation).

the projection formula and realizing that the midpoint of the marker is not in the z-axis:

$$\frac{M_x}{M_z} = u \Leftrightarrow M_x = u \cdot M_z \qquad (10.79)$$

$$\delta M_x = u \cdot \delta M_z + \delta u \cdot M_z$$

in which u is the u-coordinate of the projection of the middle point of the marker on the undistorted image plane. If u is high—near the image border—the influence of an error in z is greatest.

Table 10.5 Orientation Accuracy. (The 95% best data is used and numbers are in degrees. Left: DICA camera. Right: JAI camera)

distance	noise sigma	noise max	deviation sigma	deviation max
180 cm	0.03	0.07	0.30	0.59
280 cm	0.03	0.07	0.44	0.79
380 cm	0.07	0.15	0.42	0.77

distance	noise sigma	noise max	deviation sigma	deviation max
120 cm	0.06	0.13	0.58	1.57
180 cm	0.90	1.86	2.27	3.84

Table 10.6 Position Accuracy of the DICA Camera. (The 95% best data is used and numbers are in cm.)

	distance	x sigma	x max	y sigma	y max	z sigma	z max
noise	180 cm	0.0040	0.016	0.0131	0.031	0.003	0.011
noise	280 cm	0.0075	0.026	0.0249	0.058	0.006	0.023
noise	380 cm	0.023	0.098	0.063	0.163	0.017	0.07
deviation	180 cm	0.51	1.4	0.18	0.45	0.20	0.5
deviation	280 cm	1.3	2.6	0.45	1.3	0.64	1.6
deviation	380 cm	1.6	3.2	0.59	1.6	0.7	1.8

Table 10.7 Position accuracy of the JAI camera. The 95% best data is used and numbers are in cm.

	distance	x sigma	x max	y sigma	y max	z sigma	z max
noise	120 cm	0.0055	0.017	0.0055	0.016	0.023	0.054
noise	180 cm	0.029	0.081	0.035	0.10	0.096	0.24
deviation	120 cm	0.11	0.28	0.099	0.27	0.20	0.39
deviation	180 cm	0.21	0.58	0.24	0.54	0.73	1.6

On the other hand, the sigma of the deviation has gone way up, which suggests that the lens model may be not good enough for the estimation to be noise limited.

10.10 Conclusions

10.10.1 Pluggable Filter

A first test (Section 10.7) showed that the filter is capable of weighing the different sensor estimates such that a reasonable position is found. Our filter structure makes it possible to implement sensors as plugins as long as they provide enough information (such as accuracies). The dominant role of the reference sensors in the original federated filter is diminished at the loss of some optimality. We have to perform more experiments to measure the performance of the different filter strategies.

10.10.2 Usability for Mobile Devices

In the current setup our system is bulky, because we really need a computer to do the image processing and to filter the measurements. Our current system relies on man-made markers, and those are usually only found indoors. The JAI camera can only be used when the device is within 2 meters of a marker. For pose estimation this means that the inertia sensors will not be updated much, as the markers are almost never recognized. The DICA camera, however, can be used within 6 meters of a marker, thus allowing more observations to calibrate the inertial sensors. Another good aspect of the DICA is that the image processing can be done almost entirely in the camera which means that the extra computer is not needed anymore.

When the device is outdoors, the modular set-up enables us to disconnect the camera and continue tracking on (D)GPS data. Experiments showed however that the DGPS had a slowly varying bias of up to 5 meters with a measurement rate of 1 Hz. This was not enough to calibrate the bias of the accelerometers, so the filter setup for the position estimation was not usable. In the future we want to use the camera to estimate the movement or velocity of the device, which will enable the use of the accelerometers again for better accuracy.

10.10.3 Usability for Augmented Reality

The only test we did was using virtual reality to test the camera positioning algorithms. As the algorithm optimizes the mismatch in the image, it may be not surprising that a virtual object on top of the marker is accurately displayed, even when the pose estimation has medium errors. With the augmented reality application we have in mind, the marker will be attached on ceilings and walls, and the virtual objects are projected at a different position. The relation between the pose in camera coordinates of the marker m and virtual object o is given by:

$$\begin{aligned} \vec{p}_o^c &= H_m^c \vec{p}_o^m = R_m^c \vec{p}_o^m + T_m^c \\ d\vec{p}_o^c &= \frac{\delta R}{\delta \vec{\theta}} d\vec{\theta} \cdot \vec{p}_o^m + dT_m^c \end{aligned} \qquad (10.80)$$

To test the applicability of our setup for augmented reality, we have to set a required accuracy in the position of the virtual object. We set this to 1% of the distance to that object, which corresponds to roughly 0.5°s error in the direction from the optical point to the object. Actually, this accuracy should be specified for the coordinates in the human eye frame, but for now we will use the camera frame.

We performed a simulation with a marker and a virtual object on the optical axis, to find out at what distance the virtual object can be displayed in accordance to our requirement, as function of the distance of the marker. The set-up is shown in Figure 10.24.

We took the best camera from our results, the DICA camera, and we presume that we do not have modeling errors. This means that we will use the relation between the noise in pose estimation and the distance to the marker (Table 10.1 and Table 10.2). Figure 10.25 gives the lower and upper bound on the displayed distance for different marker sizes.

With our current A4 sized marker, a high displayable range is only realized when the marker is within 1.4m of the camera. Outside that region, the object

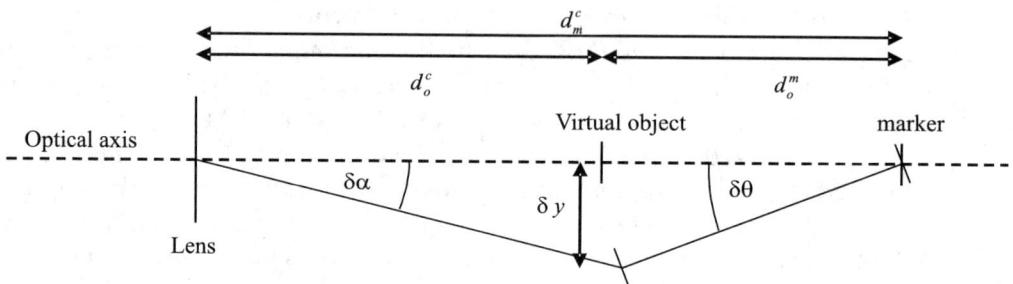

Figure 10.24 Imulation setup with object and marker on the optical axis. Our requirement says that due to should be less than 0.01. Or, should be less than 0.5 degrees.

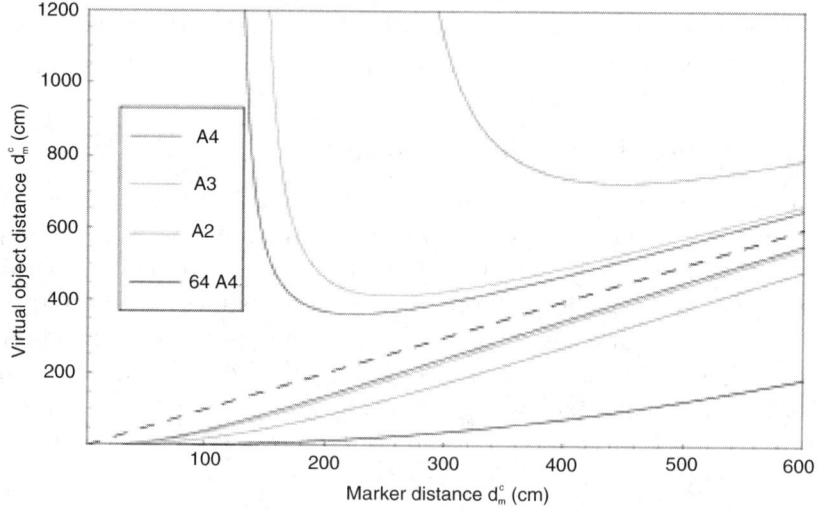

Figure 10.25 Upper and lower bounds on distance of projected virtual object vs. distance to the marker. The dashed line indicates that the virtual object is displayed on the marker.

should be within tens of centimeters from the marker. The 64 times A4 sized marker (8 x 8 A4's) is not feasible in reality, but it is used to show the effect of four normal markers at the four corners of that big marker.

As already mentioned, this figure gives an overestimate. When we use the numbers for the deviation instead of the noise, the admissible distance between marker and object can be calculated to be around 5% of the marker distance, even with the biggest simulated marker.

10.11 Conclusions

The conclusion is that with a high-resolution camera with only a single A4 sized marker, the noise and deviation properties are not good enough for full image augmented reality. Only the virtual object near the marker will have a stable position. It may well be that when the full Kalman filter is used, the required accuracy is met, provided of course that the systematic errors can be removed. Measurements need to be done to test that.

What determines the accuracy is shown to be related to the minimum size of the region of interest in the image that contains the features. When the size of the marker is increased, this region grows. When multiple markers are used, the region will be bigger as well. Both methods will increase the accuracy. However, as one of the goals was to minimize the number of markers, we think it is worth trying to combine the single marker with the use of natural landmarks, possibly augmented with a simple model of the environment. This yields a bigger region of the image that can be used for pose estimation, which again means a higher accuracy.

References

[1] Cutting, J.E., Vishton, P.M., "Perceiving Layout and Knowing Distances: The Integration, Relative Potency, and Contextual Use of Different Information about Depth," *Perception of Space and Motion*, pp. 69–118, edited by W. Epstein and S. Rogers, 1995.

[2] Jonker, P.P., Caarls, J., Eijk, R. van, Peddemors, A., Heer, J. de, Salden, A., Määttä, P., Haataja, V., "Augmented Reality Implemented on a Mobile Context Aware Application Framework," submitted to IEEE Computer (2003).

[3] Pasman, W., Schaaf, A. van der, Lagendijk, R.L., Jansen, F.W., "Information Display for Mobile Augmented Reality: Merging the Real and Virtual World," *Proceedings of the IMC'98* (Rostock, 1998). *http://www.cg.its/tudelft.nl/~wouter.*

[4] Murata Piezoelectric Vibrating Gyroscopes (GYROSTAR), http://www.murata.com.

[5] Analog Devices, ADXL202 accelerometer, http://www.analog.com/en/prod/0,2877,ADXL202,00.html.

[6] Bakker, J.D., Mouw, E., Pouwelse, J., "The LART Pages," Delft University of Technology, Faculty of Information Technology and Systems (2000). Available at *http://www.lart.tudelft.nl*.

[7] Precision Navigation, TCM2-50, http://www.pnicorp.com/index?nodeId=c36.

[8] JAI, datasheet of CV-S3300 Color CCD camera, http://www.jai.com/db_datasheet/cv-s3200_3300db.pdf.

[9] Philips Industrial Vision, DICA Fast High-Resolution camera, http:// www.apptech.philips.com/industrialvision/products.htm.

[10] Zhang, Zhengyou, "A Flexible New Technique for Camera Calibration," http://www.research.microsoft.com/~zhang/calib.

[11] Kalman, R.E., "A New Approach to Linear Filtering and Predicting Problems," *Journal of Basic Engineering*, March 1960.

[12] Brookner, E., Tracking and Kalman, *Filtering Made Easy*, John Wiley & Sons Inc. (1998)

[13] Caarls, J., "Geometric Algebra with Quaternions," Technical Report (2003). http://www.ph.tn.tudelft.nl/Publications/phreports.

[14] Naimark, L., Foxlin, E., "Circular Data Matrix Fiducial System and Robust Image Processing for a Wearable Vision-Inertial Self-Tracker," ISMAR 2002: 27-36

[15] Bar-Shalom, Y., Mallick, M., Chen, H., and Washburn, R., "One-Step Solution for the General Out-of-Sequence-Measurement Problem in Tracking," *IEEE Aerospace Conference Proceedings*, Vol. 4, 2002, pp.1551–1559.

[16] Mookerjee, P., Reifler, F., "Application of Reduced State Estimation to Multisensor Fusion with Out-of-Sequence Measurements," *IEEE Radar Conference*, 2004

[17] Nebot, E.M., Bozorg, M., Durrant-Whyte, H.F., "Decentralized Architecture for Asynchronous Sensors," *Autonomous Robots* Vol. 6, No. 2, pp. 147–164, 1999

[18] Lawrence, P.J., Jr.; Berarducci, M.P., "Comparison of Federated and Centralized Kalman Filters with Fault Detection Considerations," *IEEE Position Location and Navigation Symposium*, 1994.

[19] Carlson, N.A., "Federated Square Root Filter for Decentralized Parallel Processors," *IEEE Transactions on Aerospace and Electronic Systems*, Vol. 26, May 1990.

[20] Caarls, J., Jonker, P.P., Persa, S., "Sensor Fusion for Augmented Reality," in: Aarts, E., Collier, R., Loenen, E. van, Ruyter, E. de, (eds.), *Ambient Intelligence (Proc. 1st European Symposium EUSAI 2003*, Veldhoven, Netherlands, Nov.3–4, Lecture Notes in Computer Science, Vol. 2875, Springer Verlag, Berlin, 2003, pp.160–176.

[21] Hirokazu, K., Billinghurst, M., Augmented Reality Toolkit, http://www.hitl.washington.edu/artoolkit/.

[22] Canny, J., "A Computational Approach to Edge Detection," *IEEE Transactions on Pattern Analysis and Machine Intelligence*, Vol 8, No. 6, Nov 1986.

[23] Harris, C., and M. Stephens, "A combined corner and edge detector," *Proc. Alvey Vision Conf.*, Univ. Manchester, pp. 147–151, 1988.

[24] Haralick, R.M., L.G. Shapiro, *Computer and Robot Vision*. Addison-Wesley, 1992 and 1993.

[25] Jonker, P.P., Persa, S., Caarls, J., Jong, F. de, Lagendijk, I., "Philosophies and Technologies for Ambient Aware Devices in Wearable Computing Grids," *Computer Communications Journal*, Vol. 26, No. 11, 2003, pp. 1145–1158. www.sciencedirect.com/science/journal/01403664.

CHAPTER 11
Dynamic Synthesis of Natural Human-Machine Interfaces in Ambient Intelligence Environments

Nikolaos Georgantas, Valérie Issarny, and Christophe Cerisara

11.1 Introduction

Ambient intelligence (AmI) targets user-centric service provisioning enabling anywhere, anytime access to information and computation, compared to the conventional computer-centric approach. Systemically, this is realized as a synergistic combination of intelligent human-machine interfaces and ubiquitous computing and networking. The ubiquitous property implies a useful, pleasant and unobtrusive presence of the system everywhere—at home, en route, in public spaces, at work. Ubiquitous computing and networking provide base system support for AmI, employing a highly complex distributed software infrastructure. Service provisioning in AmI environments poses a number of fundamental requirements on the software infrastructure. First, omnipresence of the system entails its capacity to integrate diverse, heterogeneous computing and networking facilities. This calls for an open, multiplatform computing and networking environment. Interoperability shall then be established based on generic and pervasive, that is, widely accepted, software paradigms. Second, the highly dynamic character of the AmI environment, due to the intense use of the wireless medium and the mobility of the users, requires that applications be structured dynamically, incorporating loosely coupled interacting components. Third, AmI environments are mostly populated by wireless, resource-constrained devices, hence the need for lightweight software infrastructures. In [1], an in-depth analysis is presented on the requirements and challenges that AmI software infrastructures are facing.

Our work described herein was initiated as part of the effort of the IST Ozone[1] project, and is currently being pursued within the IST Amigo[2] project. In this context, we have been elaborating a generic framework for consumer-oriented AmI applications offering anywhere, anytime service provisioning. Our solution to a base software infrastructure for AmI covering the above raised requirements is the WSAMI core middleware infrastructure based on the Web services[3] paradigm, which has been released as open source software[4] (see the lower part of Figure 11.1). A Web service is defined as a software entity, which is deployed on the Web;

---

1. http://www.hitech-projects.com/euprojects/ozone/.
2. http://www.hitech-projects.com/euprojects/amigo/.
3. http://www.w3.org/2002/ws/.
4. http://www-rocq.inria.fr/arles/download/ozone.

exposes a public interface—comprising a set of operations—described in an XML-based language (namely WSDL[5]); and can interact with other software entities using an XML-based protocol (namely SOAP[6]) over standard Internet transport protocols (like HTTP). This very generic definition of Web Services makes minimal assumptions about the underlying and overlying technologies, enabling interoperation of Web services developed and deployed independently on heterogeneous platforms. This, along with the pervasiveness of the ubiquitous Web, guarantees availability of Web services in most environments, which replies to the ubiquity and interoperability requirement. Further, Web services are Web resources, developed and deployed independently on the Web, enabling dynamic, loose structuring of applications integrating a number of Web services. Building on this feature, WSAMI establishes the base for dynamic discovery and integration of mobile Web services in AmI environments [2]. Finally, WSAMI is a lightweight middleware specifically targeting Web service deployment on wireless, resource-constrained devices.

Besides the base AmI requirements covered by WSAMI, a number of advanced requirements are further posed in AmI environments. AmI requires software infrastructure support for the enactment of complex user tasks related to accessing information and computation. Thus, in a common situation, a user enters into an AmI environment most often carrying a portable device, such as a PDA or a smartphone. The user will be able to execute complex tasks within the environment, dynamically employing the currently available environment's functionalities, which come in the form of services (or applications) and resources. Thus, the software infrastructure shall provide support for dynamic task synthesis, which involves composition and reconfiguration of tasks dependent on the dynamic situation of the AmI environment. Further, AmI tasks require employing advanced human-machine interfaces. User interfaces shall be natural, to be accepted and used by nonspecialized users. This means that they shall be inspired by and mimic the communication between humans. Interfaces shall further be context-aware, dynamically adapting to a changing context, concerning, e.g., the location of the user, physical environmental conditions, etc. Traditional interfaces that assume a fixed context (e.g., a user at work) are insufficient for AmI. Finally, various, application-independent user interfaces shall be offered by the AmI software infrastructure, and be adaptable and reusable between applications. This, on the one hand, relieves application developers from the burden of developing advanced, complex human-machine interfaces for each application, and makes interfaces available in the AmI environment as any other resource. On the other hand, users may use the same, familiar interfaces across applications, not bothering adapting themselves to new interfaces for each new application. An illustrating scenario of the enactment of a user task, with focus on the employment of advanced human-machine interfaces, is presented next; it is an adapted extract from one of the Ozone demonstrator scenarios.

> The Rocquencourt city offers cybercar transportation; these are automated vehicles that do not require any driver assistance and may be booked via the Internet. Michel waits for his friend Paul at the Rocquencourt tennis club. They arranged their game

---

5. http://www.w3.org/TR/wsdl20/.
6. http://www.w3.org/TR/soap12-part0/.

3 days ago. Michel lives very close to the tennis club, Paul a little further away. Paul has already reserved a cybercar to take him to the tennis club. While waiting for the cybercar to arrive, he makes arrangements for a visit to Paris after the game. He uses his PDA to access the Ozone Transport Information System for the multimodal transport option (taxi, bus, railway). While he is checking a long train timetable on the small screen of his PDA, the cybercar arrives. He continues his transport browsing in the cybercar, benefiting from the cybercar's advanced human-machine interface facilities, such as multimodal interaction via speech and gestures on a touch screen, and combined visual and audio feedback via an animated agent and speech synthesis. At some point on the way, there is a lot of noise due to major work in the street and the stereo loud speakers of the cybercar are not powerful enough. Paul is not able to understand the audio feedback and the interaction modality is turned into textual interaction. Being indecisive, he has not yet chosen his transport modality, when he arrives at the tennis club. He asks Michel's opinion on the best way to Paris, resorting again to his PDA.

The following presents our approach toward enabling dynamic synthesis of complex user tasks incorporating advanced human-machine interfaces within the AmI environment. We elaborate a fine-grained task architectural model, which precisely models environment functionality in the form of software components, as well as functionality integration into user tasks, with special focus on synthesis of user interfaces (Section 11.2). This model is built on top of the WSAMI core middleware infrastructure, along with a related middleware service that supports dynamic, situation-sensitive composition and reconfiguration of user tasks. We further elaborate a human-machine interface functional architecture, which enables natural, multimodal, context-aware user interaction (Section 11.3). We focus on speech recognition, which is certainly the most natural modality, and associated dialog understanding. We incorporate application-independent context, including user preferences and environmental conditions, which affects the choice of the best interaction modality at any time. Further, our interface functional architecture is application-independent and reusable, supporting an abstract description language by which it may be configured and used by different applications. Then, to enable incorporation of advanced human-machine interfaces into dynamically synthesized user tasks, we structure the human-machine interface functional architecture in terms of the task architectural model, identifying dynamically composable user interface components (Section 11.4). To illustrate our approach, we apply it to the presented scenario. Finally, we position our current achievements within the context of related research efforts and indicate future perspectives of our work (Section 11.5).

11.2 A Task Architectural Model for Ambient Intelligence

The fundamental architectural element of the WSAMI infrastructure is the core middleware that provides essential functions, enabling interaction among mobile Web services. Building upon the generic core middleware, a task architectural model has been elaborated, along with a supporting middleware service. The WSAMI architecture is depicted in Figure 11.1.

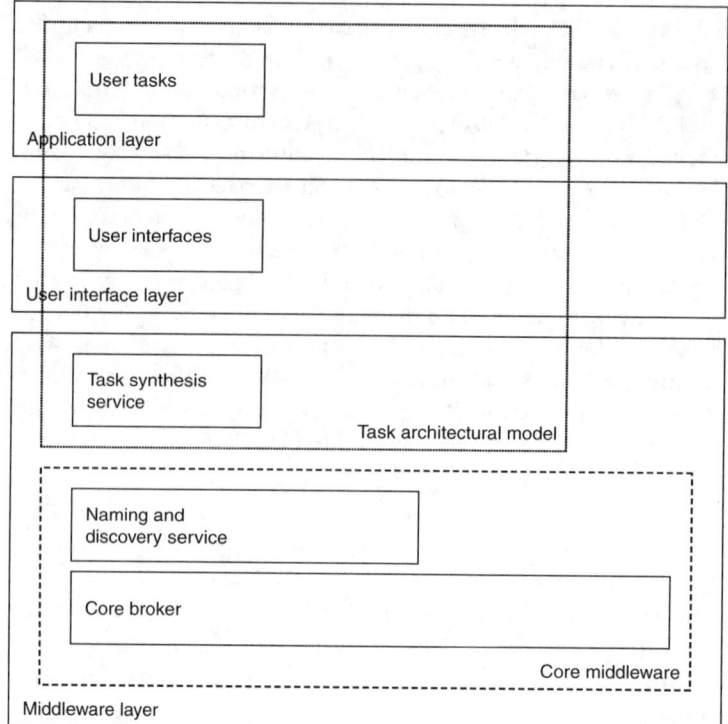

Figure 11.1 WSAMI architecture.

The core middleware [2] comprises: the WSAMI language for specifying mobile composite Web services; a lightweight, efficient middleware core broker realizing SOAP; and the naming and discovery service for dynamically locating requested services. The WSAMI language is a XML-based language enriching the WSDL-based description of Web services in order to enable specification of composite Web services. The minimal core middleware will be deployed on every node of the environment. Additional WSAMI middleware-level services provide advanced features, but do not have to be supported by all nodes.

The task architectural model enables structuring complex user tasks that are performed by employing the services and resources of the AmI environment [3]. The task model spans all three WSAMI layers. First, it specifies the structure of user tasks lying in the application layer. Further, the user interface layer hosts generic, reusable user interface components, which may be incorporated by different applications, possibly each time adaptable to the specific application; the task model specifies these components. Finally, in the middleware layer, the task synthesis service undertakes dynamic composition and reconfiguration of user tasks, based on their architectural modeling. The task model identifies a number of component classes, which enable: modeling of the environment's services; modeling of the environment's resources, for example I/O devices; reasoning on patterns for composition of services/resources, for example favoring a centralized coordination or a peer-to-peer scheme; modeling of client functionality, that is, functionality accessing the services/resources of the environment; modeling of user interface devices; and composition of user tasks from all the above functional entities. This model provides the base for managing the different forms of functionality in AmI environments, allowing for the guided composition of complex functionality. In the following, we detail the component classes of our model (Sections 11.2.1 and 11.2.2), and introduce the task synthesis service (Section 11.2.3).

11.2.1 Service/Resource Components

We model an atomic service or resource of the environment as an elementary service/resource component [4] exposing a public Web service (WS) interface, as depicted in Figure 11.2. We consider I/O devices as a special class of elementary service/resource components, which we call I/O components.

A number of elementary service/resource components may be composed to provide composite functionality. Service/resource components of the environment may follow a passive behavior model, meaning that they expect to be invoked by a controlling entity wishing to use their functionality. For service/resource components that exhibit active behavior comprising outgoing operations, such as a sensor control component, there may be an initialization phase, during which the controlling entity manifests itself to the component as the target of its outgoing behavior by suitable control operations on the component. Alternatively, a more generic behavior model may envisage peer-to-peer interactions among composed components. This requires that components be aware of the composition and contain inherent composition functionality. Since we cannot assume this for any component, we adopt a composition scheme based on centralized coordination [5]. However, this centralized coordination scheme does not exclude peer-to-peer interaction among composed service/resource components; this interaction will normally take place under the initialization and control of the coordinating entity.

Composition is realized by employing a coordination component and a computation component, as depicted in Figure 11.2. The coordination component encloses conversation functionality with each service/resource component. The

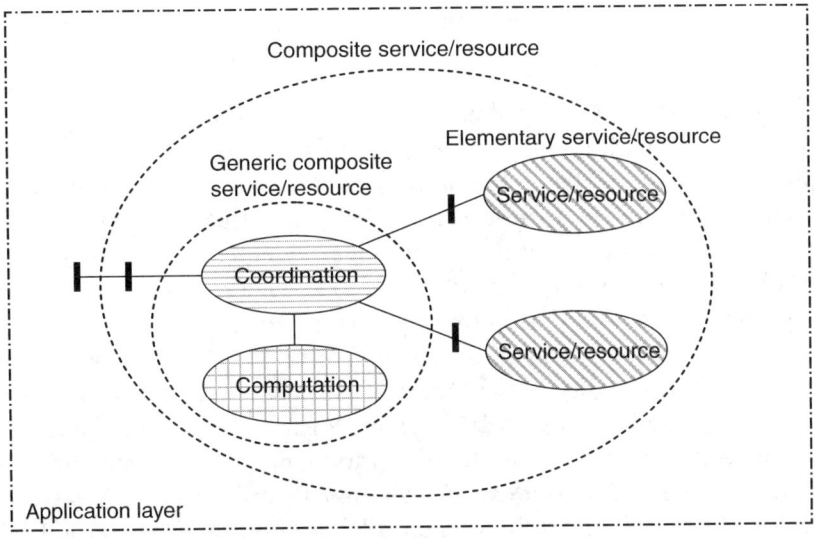

Figure 11.2 Elementary/composite service/resource component.

term *orchestration* is used in Web services to define this meaningful integration of conversations with a number of Web services in order to realize some useful function. The coordination component can, thus, be developed by using one of the proposed orchestration workflow languages like BPEL4WS [6] or the OWL-S process model [7]. It is assumed that the coordination component does not itself contain any processing functionality on the data exchanged with the service/resource components. The computation component adds processing functionality to the coordination component. Combination of a coordination and a computation component provides incomplete composite functionality. We call this a generic composite service/resource component. This component exposes two public WS interfaces: one connecting to the elementary components being composed, and one allowing access to the composite functionality. Note that a public WS interface is denoted in Figure 11.2 by a bold bar, while lack of this bar denotes an internal or non-WS interface. The result of the composition is a composite service/resource component, which exposes a public WS interface, as in the case of an elementary component. This interface is the related interface of the generic composite component. Since there is a very close relation between the coordination and the computation component forming a generic composite component, they are normally developed together. Furthermore, there is no meaning in separate discovery and dynamic composition of these two components: normally, they are collocated. To support dynamic composition, a generic composite component is enabled to bind to the service/resource components of the composition at run-time. The presented composition scheme may be applied recursively, meaning that a composite component may participate in another composition providing a new composite component. This results in nesting an arbitrary number of composition levels.

Elementary service/resource components and the components constituting a generic composite service/resource component lie in the application layer of the WSAMI architecture. I/O components may also lie in the user interface layer, if they make part of a reusable user interface, as discussed in the next section.

11.2.2 Tasks

To provide complete user task functionality, end-use functionality, that is, client application functionality shall be added to the components we have introduced so far. End use functionality is provided by an end use coordination component, a computation component and one or more user interface (UI) components, as depicted in Figure 11.3. This end-use functionality may either connect to an elementary/composite component or compose a number of elementary/composite components. The result of adding end use functionality to service/resource components is a task. We note here that the end use coordination component of a task is similar to the coordination component of a composite service/resource component. Nevertheless, coordination functionality in the first case is specific and complete, while in the second case, it is generic enough leaving space for additional end-use functionality.

Incorporation of multiple UI components allows distribution of user interaction. For example, a user executing a task from his/her PDA may use for part of the task execution, exclusively or in parallel, another user interface device: a traditional one such as a power-plugged workstation, or an advanced one such as a large inter-

11.2 A Task Architectural Model for Ambient Intelligence

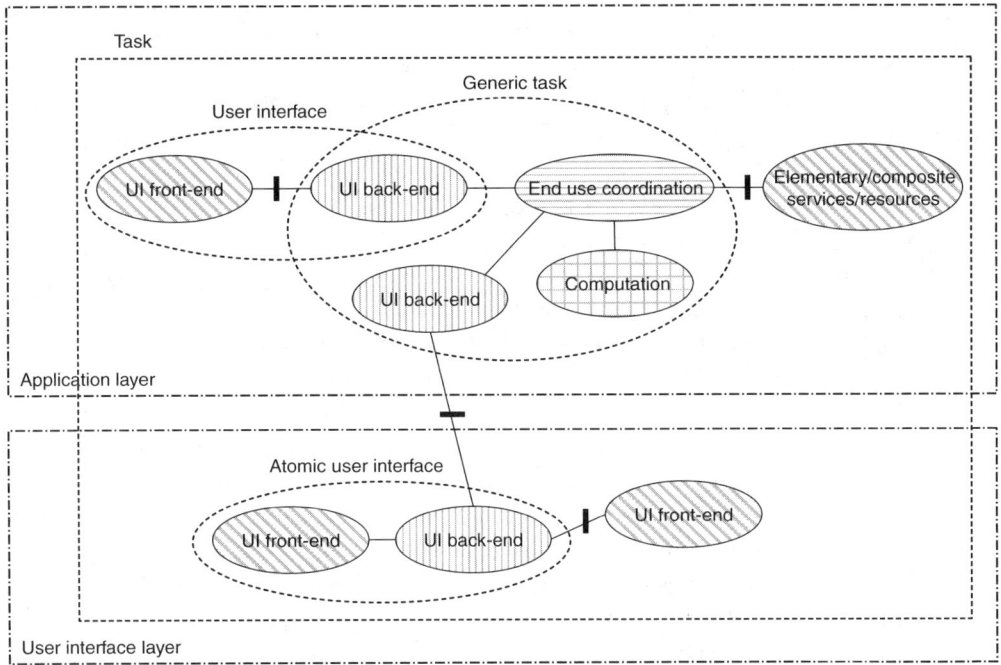

Figure 11.3 Task architectural model.

active wall screen. We decompose each UI component into a UI front end component and a UI back end component. The UI front end component is an I/O component and contains generic functionality, while the UI back end component contains more specific functionality, which can vary between absolutely task-specific functionality and reusable functionality, adaptable to different tasks. We give the following examples illustrating the distinction between the UI front end and UI back end components:

- A graphical UI may be developed as a Web application. Then, the task-specific UI back end is based on technologies like JSP and servlets, while the UI front end is a generic Web browser. The two components communicate via HTML/HTTP.
- A UI based on speech recognition may comprise a generic speech recognition front end turning speech into text, and a task-specific back end interpreting text into commands meaningful to the task. The two components communicate via WS interfaces.
- A graphical UI may be implemented in Java. We may consider as UI front end the JVM and the generic Java GUI library classes, and as UI back end the task-specific Java classes. If the two components are not collocated, mobile code over HTTP shall be employed.

Combination of an end use coordination, a computation and one or more UI back end components provides incomplete task functionality. We call this a generic task component. Since there is a very close relation between the end use coordination and the computation component included in a generic task component, they are normally developed together and are collocated. These components lie in the application layer of the WSAMI architecture.

As for the UI back end component, it may be distributed between the application and user interface layers, thus, having a task-specific part and a reusable part. The task-specific part is normally developed together with the end-use coordination and the computation component and is collocated with them. However, the reusable part may be networked; then, the task-specific part dynamically binds to the reusable part at run time, and possibly specializes it to the task in question by communicating to it task-specific information.

Further, UI front end components may lie either in the application layer, thus, provided and used by a specific application, or in the user interface layer, thus, reusable by several applications. A UI back end component may dynamically bind to a (several) UI front end component(s) at run time. Nevertheless, this may not always be possible or even required by the designer, meaning that, in this case, the UI front end component shall be collocated with the UI back end component. In this case, the UI component is considered atomic. In the Java GUI example presented above, the UI shall be atomic if the designer does not wish to integrate a mobile code capability. Since our model is based on Web services, we are interested in the case where the UI front end and the UI back end components communicate via SOAP/HTTP, allowing as well for the special case of HTML/HTTP communication between a Web application and a Web browser. If neither of these is the case, we consider the UI atomic.

A generic task component exposes a public WS interface that connects to elementary or composite service/resource components. Then, other public interfaces are exposed in the following cases: a WS interface between a task-specific and a reusable part of a distributed UI back end component; and a WS or HTML/HTTP interface between a UI back end and a UI front end component, forming a non-atomic UI component.

11.2.3 The Task Synthesis Service

The WSAMI core middleware provides the core broker (CB) and the naming and discovery (ND) service (see Figure 11.1). CB and ND are deployed on every networked node of the AmI environment. Each such node is a peer ND node. Peer ND nodes discover each other by employing the service location protocol (SLP) [8]. The WSAMI middleware layer of a networked node may be optionally enhanced with the task synthesis (TS) service, which is a base solution to dynamic composition and reconfiguration (upon execution) of user tasks. We introduce a distributed approach to task synthesis, requiring that TS be deployed on every node hosting components that integrate other components, i.e., generic composite service/resource components, generic task components, and networked UI back end components. In this way, task synthesis is performed locally for each component wishing to integrate other components. This approach allows efficient, localized control on the bindings between possibly mobile components.

A user wishing to carry out a task within the AmI environment launches the generic task component on a device of the environment, e.g., his/her PDA. The generic task component contacts the local TS, conveying its configuration policy; from then on, TS undertakes configuration management for this component. TS employs ND to retrieve from the AmI environment all the components to be integrated into the task in question. According to the configuration policy, some components may be optional, which means that their absence does not prevent execution of the task. Then, TS performs all necessary bindings. If a component is retrieved that shall further integrate other components, the same composition procedure is carried out for this component by the TS instance of the node hosting this component.

Then, two essential cases of task reconfiguration upon execution are treated: one performed upon the availability of new, desirable components and one performed upon disconnection of existing components. Task reconfiguration is carried out by: adding or removing an optional component; substituting a new component for a component currently employed. Reconfiguration is performed locally for each component integrating other components.

For the first case, ND is capable of generating an event notification to its collocated TS upon discovery of a new ND node. Upon this event, TS employs ND to retrieve all the relevant components residing on the new ND node. Then, TS may add a new optional component to the task, or substitute a new component for an existing one, following the configuration policy. Thus, the latter may state that new optional components shall be directly inserted, while only certain substitutions may take place, e.g., to exploit new powerful user interface components. Finally, TS performs binding or rebinding where necessary.

For the second case, CB is capable of generating an event notification to its collocated TS when a call is made to a disconnected component. Upon this event, TS employs ND to retrieve from the AmI environment a substitute for the disconnected component. If one is found, TS puts the new component in the place of the disconnected one, performing rebinding. Again, the configuration policy determines whether a disconnected component is optional or not, and whether its replacement is indispensable for the recovery of the task.

11.3 A Human-Machine Interface Functional Architecture for Ambient Intelligence

The elaborated task architectural model puts special emphasis on the modeling of user interfaces being integrated within tasks, since enabling rich user interaction with the system is a major requirement for AmI. User interaction within AmI environments shall further integrate advanced features prescribed by the user-centrism of AmI; specific requirements have been identified in a number of preliminary works in this domain [9, 10]. In this section, we elaborate a human-machine interface functional architecture that enables advanced user interaction. In the next section, we employ our task architectural model to structure this functional architecture from a software engineering viewpoint; our ultimate goal is to incorporate advanced user interaction into dynamically synthesized user tasks. Addressing all requirements on user interfaces within AmI environments is not possible within the timeframe of a single project. Therefore, in a first step, we focused our efforts on the most fundamental ones, namely, context awareness and naturalness of user interactions. Thus, the interface architecture supports natural communication based on different modalities, and context-awareness applied in the selection of the most appropriate modality. Furthermore, the interface architecture will form a baseline for developing applications that incorporate the provided user interfaces. Therefore, reusability of the supported functionalities by different applications and their extensibility are major concerns of our approach. The proposed functional architecture is positioned in the user interface layer of the WSAMI architecture (see Figure 11.1); it is detailed in Figure 11.4.

The roles of the different functionalities that appear in Figure 11.4 are summarized next. The context-awareness functionality collects and manages context information that concerns the modality employed in the communication with the user. It includes the following functionalities:

- The context manager centralizes every piece of contextual information. It is actually composed of several sub-functionalities, each one specialized in a certain kind of information.
- The user profile stores information about the user preferences, for example, his/her preferred interaction modalities.
- The modality advisor uses different contextual information—derived from the environment and the user preferences—to suggest the best interaction modality to use at any time.

The role of the dialog manager is to mediate between the application and the user: it interprets the user inputs and transforms them into commands that can be handled by the application. It, thus, manages the sequence of dialog acts with the user to solve ambiguous inputs, to help the user in specifying his/her will, etc. It includes the following functionalities:

- The domain model contains all the concepts relevant to the application (the application ontology). This includes, for example, the graphical objects that

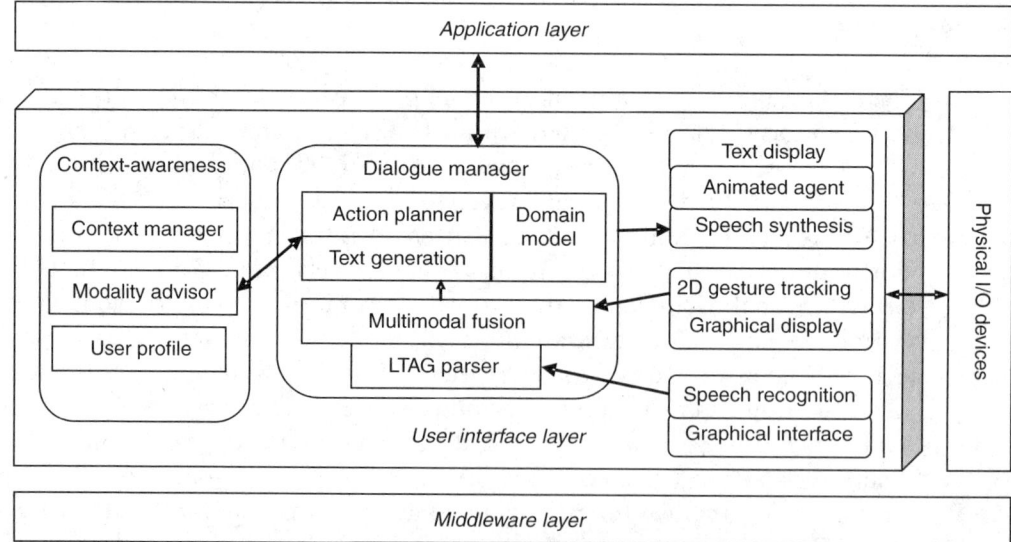

Figure 11.4 Human-machine interface functional architecture.

can be displayed or selected by the user, the vocabulary associated with these objects, etc. Domain model information for a specific application is communicated to the dialog manager by the application itself, thus specializing the whole interface architecture to the application in question.

- The action planner decides which dialog act the system shall use next in its interaction with the user. It considers that an interaction session is complete when it can send a command to the application. Currently, the action planner is based on a "template" form that needs to be filled in with pieces of information provided by the user; for example, for a train reservation application, this information concerns the destination, the departure time, etc.
- The text generation outputs sentences of the system to be displayed or synthesized in the language of the user.
- The multimodal fusion outputs a semantic representation of the user inputs, by merging two modalities: speech and gesture, the latter concerning 2D gestures made by the user, e.g., by moving an electronic pen on a touch-screen.
- The LTAG parser parses a sentence pronounced by the user into a syntactic tree based on the lexicalized tree adjoining grammar (LTAG) formalism [11].
- The interaction functionalities employed in the communication with the user are the following:
- The speech synthesis transforms text into an audio stream.
- The animated agent is a 2D cartoon-like character that can reproduce emotions visually. It is combined with the speech synthesis functionality to produce a talking avatar. Alternatively, the animated agent's speech may be displayed on a text display.
- The 2D gesture tracking captures and analyzes the 2D gestures made with the electronic pen of a tablet PC on a graphical display that is application-specific and makes part of the domain model.
- The speech recognition translates the user speech into a textual form that can be processed by the LTAG parser. It comes with a graphical interface, which gives to the user some control over the recognition process, as well as some feedback to facilitate the communication with him/her.

In the following, we focus on the three requirements raised, i.e., context-awareness (Section 11.3.1), natural interaction (Section 11.3.2) and reusability (Section 11.3.3), and discuss how the proposed human-machine interface functional architecture fulfills them.

11.3.1 Context-Awareness

The definition of context is highly variable. AmI systems address the consumer-oriented challenge, and target non-specialized and average consumers. Therefore, the most important context is certainly the user himself, and, at the second rank, his/her domestic, mobile and professional environments. The architecture that we propose focuses on these two aspects, but also develops mechanisms to progressively extend this context to new domains. The core of the context-awareness module is the context manager, which centralizes every kind of contextual information. Each such piece of information represents a specific context clue, and is represented by a sub-functionality of the context manager. Thus, the context manager is composed of a variable number of sub-functionalities, depending on which type of context is needed at a given time. The context manager can be considered as a manager of all the subfunctionalities, but it may also include reasoning functionality to merge the different kinds of context and interpret their correlation. Discussion on the concepts behind the context manager can be found in [12, 13].

In Ozone, we implemented three context-specific sub-functionalities: the interaction user profile, environment context and modality advisor. The first one models the user's preferred interaction modalities (speech input, gesture input, text input, graphical output); the second one captures and analyzes the light and noise levels in the user's environment; and the third one merges the recommendations of both the previous ones to help the dialog manager in deciding which interaction modality it should use. This merging mechanism is based on a set of rules that are semi-automatically learned using a reinforcement learning approach. Similar machine-learning algorithms are used to train the user profile recommendations. For example, the user profile may indicate that the user prefers textual rather than audio output, but, when sunlight focuses directly on a laptop screen, it may be difficult to read, and audio output will be chosen. Similarly, speech recognition may not be adequate when there is a lot of noise near the user. The best interaction modalities can certainly be configured or dynamically set by the user, but they are also automatically trained based on the dialog history: the dialog manager evaluates the success of any past interaction session, based on different criteria, such as the number of corrections from the user, the normalized duration of the session, the number of confirmations required, etc. This score is then passed by the dialog manager to the modality advisor; the latter adjusts the weights of its rules, based on this feedback, and transfers this information to the user interaction profile, which acts similarly. This training procedure is inspired from [14].

The modality selection mechanism introduces context-awareness in our interface architecture; adaptation to the specific context is further realized by a number of features of the speech recognition functionality. Thus, the user identity is communicated to the speech recognizer, which automatically loads user-adapted acoustic models whenever the user is known; this greatly improves the recognition accuracy. A specific application has also been developed that supports creation of a new set of adapted models for a new user registering with the system. Further, a basic noise robustness technique (online cepstral mean normalization [15]) is used at the speech recognition module to reduce the effect of the background noise when using a far-talking microphone in out-of-office environments. Finally, the graphical interface of the speech recognizer is scalable, and can be dynamically resized, moved, or hidden, depending on the size of the screen available, and on the display space left by the other interaction functionalities, such as the animated agent.

11.3.2 Natural Interaction

Natural interaction is highly important in AmI environments, to address users that are not necessarily familiar with traditional computers, but also to support rich interaction of any user with the system, which shall make part of the human natural behavior and be ambient, not restricted to a single conventional device. Human-machine interactions are thus considered to be natural when they are similar to human-to-human interactions. In general, this is very hard to achieve, because when humans communicate, they extensively exploit the common ground, or common knowledge, which includes a common language, common semantic concepts, a common world model, the knowledge of what their interlocutor is aware of, etc. Humans do not have such a general common ground with machines, but it is feasible to develop this common ground for very specific domains.

Natural interaction is realized in our interface architecture by a number of features. The dialog manager supports natural language. Further, multimodal user interactions are supported, specifically through speech input and 2D gestures recognition. The third important point is that a high-quality, efficient and multimodal feedback is delivered to the user, so that he/she perfectly understands and feels in control of what is going on. Finally, the user should feel comfortable when using the system, and this can only be achieved with efficient response times.

Specifically, we have developed a high-quality dialog manager, coupled with a fast and accurate speech recognizer, together enabling several advanced features. First, the linguistic analysis is based on the LTAG formalism. This allows easily deriving a semantic representation of the sentence, which can be used to resolve past references, ellipsis, etc. Then, the potential discrepancies between the speech recognition and the dialog manager have been minimized: the speech recognizer makes use of a context-free grammar that is directly derived from the LTAG grammar in order to obtain a high integration/compatibility with the dialog manager. Further, the speech recognition grammar is augmented with parallel out-of-grammar networks to still be able to interpret, as much as possible, users' incorrect sentences. This strategy allows still providing to the linguistic parser perfectly correct (in the LTAG grammar sense) sentences, despite some deviations to this grammar by the speaker. The user is thus given some freedom in the way he/she can speak, such as hesitations or repetitions. He/she can also use words that are not known to the system, as long as these words do not change the meaning of the sentence. Even when the user says something that is completely unrelated (lexically and syntactically) to the LTAG grammar, the speech recognizer can deliver a null sentence that is derived only from the out-of-grammar network. Thus, the system can "know" when it has not understood at all the user, which is very important, to choose a fallback strategy. Finally, with regard to response time, the speech recognizer implements a very fast procedure (typically 0.5s) to dynamically load new grammars at any time. The objective is to improve the recognition accuracy by exploiting the current dialog state to constrain the search space. Another benefit is the speeding up of the recognition process.

Moreover, the feedback information that is delivered to the user is of very high quality. Thus, the graphical interface of the speech recognizer assists the user when he/she talks to the system. This interface is composed of a few control buttons and sliders, and a VU meter[7] that changes its color when the system listens to the user and detects his/her voice. The recognized sentence is also displayed. Further, a contextual help functionality can be activated at any time by the user to let him/her

---

7. A Volume Unit (VU) meter displays the average volume level of an audio signal.

know how he/she can proceed from the specific point. Finally, high quality speech synthesis, developed by EPICTOID[8], one of the Ozone partners, is delivered to the user. This goes along with the animated agent, having a high graphical quality, which personalizes the system and greatly helps relax the user, by showing pleasant emotions. This avatar has also been developed by EPICTOID.

11.3.3 Reusability

The elaborated human-machine interface architecture will help developers propose and implement novel applications. We have thus paid great attention to produce a modular, reusable and extensible design of our interface architecture functionalities, so that they can easily be plugged into new applications and possibly be modified to better fit the new requirements.

To achieve this objective, we have introduced well-defined interfaces and description languages. Thus, we have elaborated the multimodal interchange language (MMIL) that is used to encode interaction between any two modules in our interface architecture. This language has been designed to represent multimodal content independently of its emitting source. This elaboration contributes to the ongoing standardizing efforts under the joint ISO-ACL/SIGSEM committee TC 37/SC 4[9]. It is based on XML, and is thus completely compatible with the SOAP protocol employed by the WSAMI infrastructure. An in-depth description of the MMIL language can be found in [16].

In the same direction, another concern of ours has been the design of functionalities that are as much independent as possible of each other, although such a design may penalize the performance of the overall system. There is often a trade off between the level of independence and the global optimization of the system. For example, speech recognition and dialog management seem at first glance relatively independent, since it is perfectly conceivable to simply transfer the recognized text from the former to the latter. However, this approach does not guarantee that the recognition and parsing grammars match, and it also prevents, for example, improving the speech recognition accuracy by exploiting the linguistic and semantic aspects coming from the dialog manager. The alternative consists in increasing the amount of information that is transferred from one module to the other, but this approach increases the complexity of the interfaces and the dependence between modules. To improve the speech recognition accuracy, incomplete word graphs (instead of simple text) may be transferred from the speech recognizer to the dialog manager. These word graphs correspond to concurrent hypothesis at different stages of the speech recognition process, and can be rescored linguistically and transferred back to the speech recognizer, which can then proceed further and complete these graphs using acoustic information.

A third challenge toward reusability has been to reduce the correlation between the dialog manager and the application. In our view, the dialog manager plays the role of a mediator between the user and the application. Our dialog manager translates the utterances of the user into a simplified set of speech acts, called high level interaction protocol primitives (HLIPs). These HLIPs are very similar to FIPAS's

---

8. http://www.epictoid.nl/.
9. http://www.tc37sc4.org/.

Agent Communication Language[10] or to the MMIL specification. As an example, suppose that the user says "Ring the bell" and a function "RING" exists in the application. The corresponding HLIP would be: ORDER(e=NEW, RING(e)). This literally means: "make the proposition that there is a new ringing event true." Note that propositions are constrained to denote events whose type is directly linked to an existing application function. We have defined the following HLIPs in our interface architecture:

- SayThat(P): The application wants to inform the user that the proposition P holds.
- AskForValueAmong(list, lambda X.P): The application wants to know which value into list makes P true according to the user.
  Example: AskForValueAmong(["Fiction," "Documentary"], lambda X.wantsToWatch [User, X]).
- AskForValue(lambda X.P): Same as the previous one but without any restriction.
- Order(P) : The application orders the user to make P true.

11.4 Synthesizing Natural Human-Machine Interfaces

The elaborated task architectural model enables fine representation of the different forms of functionality within the AmI environment, with focus on user interfaces. Thus, specialized classes of networked components have been identified, along with the task synthesis service that supports their dynamic integration into complex user tasks. Further, the elaborated human-machine interface functional architecture enables advanced context-aware, natural, multimodal user interaction, and allows its reuse by a wide variety of applications. Our ultimate objective is to enable incorporation of the advanced human-machine interfaces into dynamically synthesized user tasks. To this end, we structure, from a software engineering standpoint, the human-machine interface functional architecture in terms of the task architectural model, identifying dynamically composable user interface components (Section 11.4.1). This enables composition and reconfiguration of the human-machine interfaces according to the dynamics of the AmI environment. We demonstrate this capability by applying our approach to realize the scenario introduced in Section 11.1 (Section 11.4.2).

11.4.1 Dynamically Composable Human-Machine Interfaces

The context-awareness and dialog manager modules of Figure 11.4 realize reusable backbone functionalities within the human-machine interface functional architecture. Thus, we integrate these two modules in a reusable UI back end component, which we call reusable dialog UI back end component. The reusable dialog UI back end connects to a number of generic UI front end components, as depicted in Figure 11.5:

---

10. www.fipa.org.

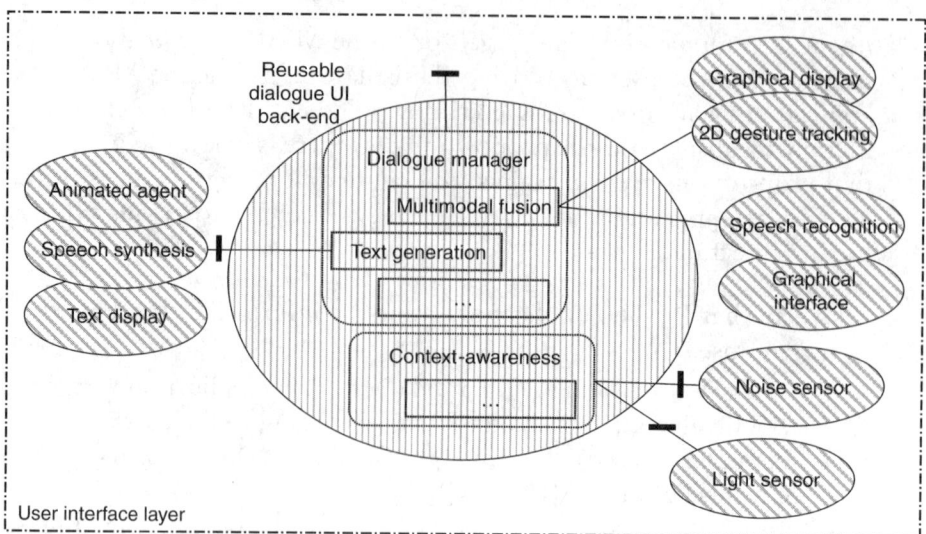

Figure 11.5 Human-machine interface component architecture.

- The speech recognition functionality forms a UI front end component interacting with the user. It is coupled with a graphical interface forming a second UI front end component. Both components connect to the multimodal fusion functionality of the dialog UI back end.
- Similarly, the 2D gesture tracking functionality and its associated graphical display form two UI front end components that also connect to the multimodal fusion functionality.
- The interrelated speech synthesis, animated agent and text display functionalities make three UI front end components that connect to the text generation functionality of the dialog UI back end.
- We add two more UI front end components with respect to Figure 11.4: a noise sensor and a light sensor connecting to the context-awareness functionality of the dialog UI back end.

The interfaces between the dialog UI back end component and each of the speech synthesis-related and sensor-related UI front end components are public Web service interfaces. Furthermore, the dialog UI back end may dynamically bind to these UI front ends at run time. However, the dialog UI back end—specifically, the dialog manager functionality—is closely interrelated with the speech recognition and 2D gesture tracking UI-front ends, namely, there is a high interdependency between the involved functionalities. We discussed this dependency for the speech recognition functionality in Section 11.3.3. Thus, most existing generic speech recognizers are not appropriate for being integrated with the dialog manager, as they do not support all the required features discussed in Section 11.3.2. Further, gesture analysis is merged with speech analysis, and both inputs are interpreted by the dialog manager, which also imposes a strong dependency on the 2D gesture tracking functionality. Thus, we consider that each of the speech recognition related and 2D gesture tracking related UI-front ends form atomic UI components with the dialog UI back end. Finally, the dialog UI back end exposes a public Web service

11.4 Synthesizing Natural Human-Machine Interfaces

interface—coming from the dialog manager functionality—for being integrated into a task.

11.4.2 Realization of the Scenario

In our scenario, Paul executes the task "accessing the Ozone Transport Information System (OTIS)" to check the different transport modalities and make a reservation. The OTIS service is a service component on the Internet offering transport information and reservation capabilities. It has been implemented as an elementary component, but it could as well be a composite one integrating several elementary components, e.g., each one serving a single transport modality. The task "accessing OTIS" is based on the generic task component OTIS client, which resides on Paul's PDA, as depicted in Figure 11.6. This component offers a classical WIMP[11] user interface, which is an atomic—combined front/back end—UI component. Via the WIMP UI, suitable for the small screen of the PDA, Paul can access timetables for the different transport modalities, and enter information such as his destination and preferred departure time. The OTIS client includes further the OTIS dialog UI back end component, which supports connecting to advanced human-machine interfaces that support dialog understanding. However, the limited resources of the PDA do not allow deploying on it such an interface. Thus, if a suitable external advanced interface is not available, the OTIS dialog UI back end component stays inactive.

To access OTIS while waiting for the cybercar, Paul launches the OTIS client on his PDA. The TS on the PDA retrieves the OTIS service, thus composing the required task. No advanced interface is retrieved; while waiting for the cybercar, Paul has to content himself with the WIMP UI of the PDA. This initial configuration of the task "accessing OTIS" is depicted in Figure 11.6.

When Paul gets into the cybercar, the TS on the PDA perceives the availability of a new component: the cybercar offers the reusable dialog UI back end component of Figure 11.5. TS retrieves this component. According to the configuration policy

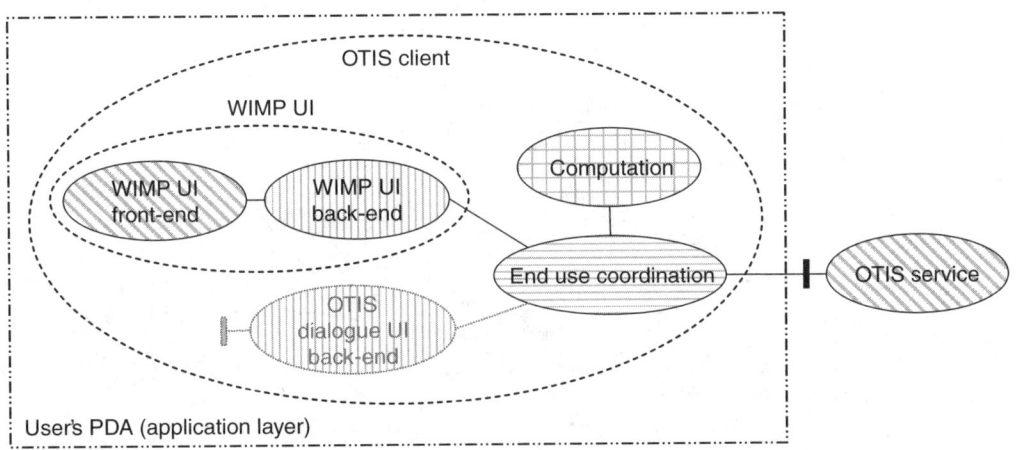

Figure 11.6 Waiting for the cybercar.

---

11. Windows, Icons, Menus and Pointing device.

of the OTIS client, a retrieved UI back end component suitable for connecting to the OTIS dialog UI back end shall directly substitute for the embedded WIMP UI, which is now deactivated. Further, the reusable dialog UI back end asks its local TS to dynamically retrieve the required UI front end components (see Figure 11.5) that are not statically attached to it (i.e., forming atomic UI components). These UI front end components are also offered by the cybercar. The new configuration of the task "accessing OTIS" is depicted in Figure 11.7. Upon binding of the new components, the OTIS dialog UI back end conveys to the reusable dialog UI back end the domain model of OTIS, thus specializing the reusable UI back end to the specific task.

Paul now enjoys advanced human-machine interaction. The OTIS client on the PDA employs the touch-screen (tablet PC) of the cybercar to display a map covering the route from Rocquencourt to Paris. Paul may use both speech, captured by a tie-microphone, and gestures, made with the electronic pen of the tablet PC on the displayed map, in the following ways:

- Specify a destination and/or an origin. This can be done either by voice ("I want to leave from Rocquencourt"), by selecting a city with the pen, or by a combination of voice and pen selection ("I want to go there"). Once both the origin and destination are known, the possible ways are highlighted on the map.
- Ask for information about a given way: the easiest solution is to use both voice and gesture ("How long does it take to go by this way?").

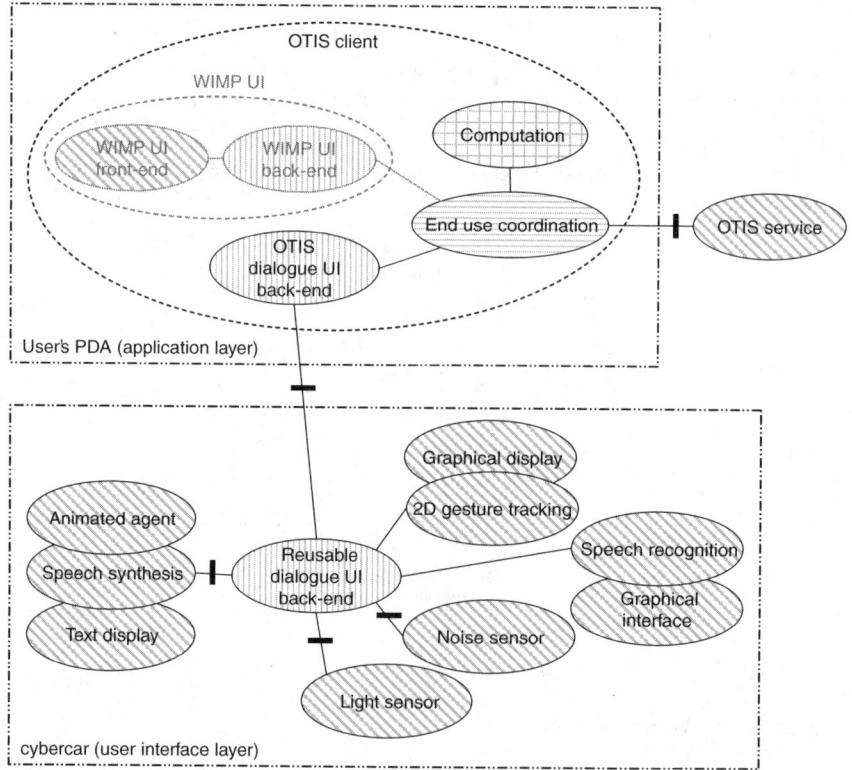

Figure 11.7 In the cybercar.

- Select a way, either by gesture only or by a combination of voice and gesture ("I choose this way").
- Specify the preferred departure time: this can be done by voice.

Screenshots of OTIS are depicted in Figure 11.8, where the animated agent and the speech recognition graphical interface are included. In the first screenshot, the user is asked for his/her identity, so that the speech recognizer can load user-adapted acoustic models.

The supported human-machine interaction is further context aware based on the inputs of the noise sensor (actually, a simple ambience microphone) and light sensor. Thus, when Paul encounters major work in the street, and his comprehension of the audio feedback is evaluated by the modality advisor as low, the animated agent's speech is displayed as text.

When Paul is out of the cybercar again, cybercar's resources disappear, and the TS on the PDA is notified of the disconnected components. TS performs a new

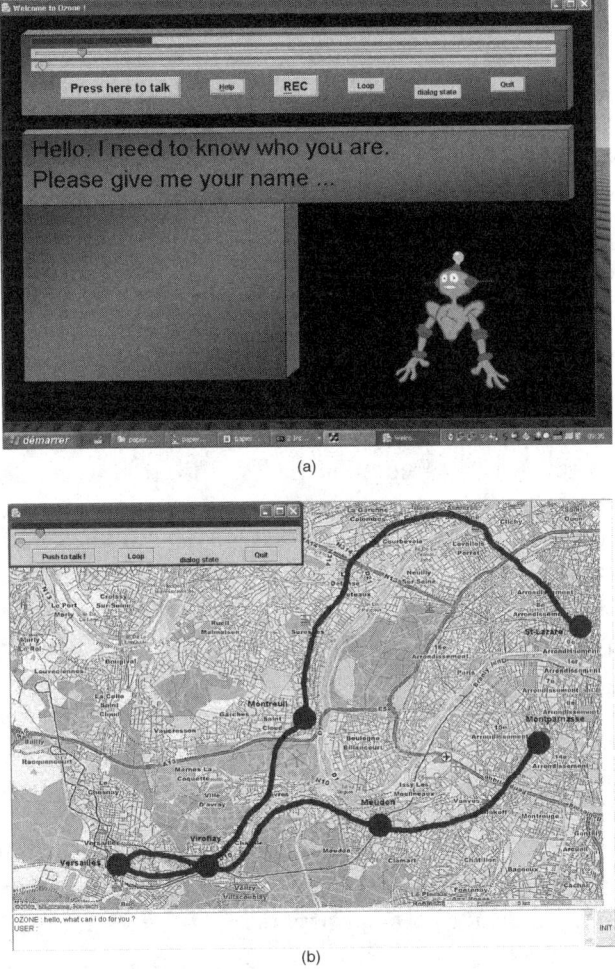

Figure 11.8 Screenshots of OTIS.

reconfiguration restoring the PDA's WIMP UI and the initial configuration of the task as depicted in Figure 11.6. Paul uses the WIMP UI to ask Michel's advice.

11.5 Current Achievements and Future Perspectives

Building upon the well-defined WSAMI middleware, based on the Web services paradigm, we have elaborated a fine-grained task architectural model and an associated task synthesis middleware service, which provide for dynamic, situation-sensitive composition and reconfiguration of complex user tasks within AmI environments. Our architectural model focuses especially on the synthesis of user interfaces, enabling multiple distributed interfaces in a task that may dynamically incorporate: specialized, however reusable, UI back end components; and generic UI front end components which may be general-purpose I/O devices. This approach effectively models user interfaces in AmI environments, where rich, distributed, dynamically composed, according to the AmI dynamics, user interaction is required. Besides, our task synthesis service provides a base solution towards task composition and reconfiguration, applying a distributed approach to the composition and reconfiguration of components that integrate other components, which is further guided by a separate configuration policy for each such component.

We have as well elaborated a generic functional architecture for advanced human-machine interfaces within AmI environments. We address three principal AmI requirements that affect the design of such interfaces: naturalness of user interactions, reusability of the interfaces and context-awareness. Concerning naturalness of interactions, we focus on speech, which is the most natural modality for conveying simple or complex communication. The speech recognition functionality has been designed to be dynamically adaptable to different users and, to some extent, different background noise. It further guarantees a high quality of integration with any linguistic parser based on the LTAG deep parsing solution, and handles the most common grammatical errors in conversations, such as hesitations, repetitions and out-of-vocabulary words. Regarding reusability of the user interfaces, we have introduced the XML-based MMIL description language, which offers a common communication scheme between every module that manipulates multimodal information. With respect to context-awareness, our approach introduces a generic mechanism, and then specializes it to a context that relates to user interaction modalities.

Aiming at dynamically synthesized human-machine interfaces, we have combined these two above elaborations. The task architectural model can adequately model the diverse functionalities integrated within the human-machine interface functional architecture, structuring them into dynamically composable user interface components. Then, by employing the task synthesis service, composition and reconfiguration of the interfaces are realized, as demonstrated via our scenario. More specifically, the scenario includes: replacing an embedded user interface with a new discovered, networked, advanced one; incorporating and specializing a reusable user interface to a specific task; and substituting for a disconnected user interface.

11.5 Current Achievements and Future Perspectives

Reporting on the current status of our work, we note that a prototype of the human-machine interface functional architecture has been implemented and deployed over the open source software prototype of WSAMI to realize a static version of our scenario, where no dynamic synthesis of the user interfaces is supported. We aim at completing the dynamic approach to synthesizing user interfaces presented herein and developing a prototype thereof. Certain aspects have to be further elaborated, including the configuration policy guiding the task synthesis service, and, certainly, a number of important issues concerning software systems reconfiguration in general, such as ensuring consistency in terms of functionality, state, and users' data [17–19].

A number of research efforts have focused on the dynamic composition of functionality in AmI or—coming from the relative domain—pervasive computing environments. In project Aura [20–22], an approach similar to our synthesis of tasks has been taken. Tasks are represented as abstract service coalitions; abstract services are implemented by wrapping existing environment's applications to conform to Aura APIs. Diverse interaction mechanisms among component services are encapsulated into connectors. Service abstraction in Aura follows a proprietary model, while our approach builds upon the widely accepted Web services standard. On the other hand, Web services define a specific interaction mechanism based on SOAP/HTTP, which is restrictive compared to Aura connectors. However, component interoperability is thus directly enabled; further, the pervasiveness of the Web and the minimal requirements that Web services pose upon software components guarantee the availability of rich conforming functionality in any environment. Further, our detailed task architectural model, identifying the different classes of functionality within the AmI environment, allows the guided composition of complex functionality.

In project Gaia [23, 24], integration of environment resources into an application is treated by a system infrastructure encompassing conventional operating system functions. Thus, a traditional application may be partitioned onto different devices and be dynamically reconfigured while executing. Gaia builds upon distributed CORBA objects. Gaia introduces an application partitioning model that extends the traditional Model-View-Controller model [25]. Gaia's building on distributed objects implies a rather strong coupling among components, which is undesirable for highly dynamic environments. Gaia's application partitioning model resembles our task architectural model; however, our concept of a task is more general than the concept of an application: A task is a user's activity that may span several applications.

The BEACH [26] software infrastructure provides support for constructing synchronous collaborative applications for active meeting rooms. BEACH supports multiple-device and multiple-user interaction, as well as multiple-user devices such as interactive walls, which may be shared by more than one user. The BEACH architecture is organized in four layers and five models separating basic concerns within each layer. Compared with our model, BEACH enables more complex, multiuser activities, however, as admitted by its designers, all active room components will have a permanent network connection and no slow CPUs, which makes the environment rather static and excludes small devices like PDAs.

At Xerox PARC, an approach called recombinant computing [27, 28] focuses on establishing interoperability among software components based on generic interfaces that any existing or future component may provide irrespectively of its specific functionality. Then, aided by the user's knowledge of the semantics of components, any components may be composed in an ad hoc manner, even if no application has been expressly written to combine these components. Specific semantic interactions among components are then enabled by leveraging mobile code, much like the Jini architecture [29], which, however, does not support the recombinant feature. Certainly, this approach requires the deployment of a specific mobile code support platform on every node.

Integration and interoperability between heterogeneous software components is a major issue in AmI environments, resulting from the requirement for ubiquity. In our approach presented herein, interoperability at software infrastructure level is based on the generic and pervasive Web services paradigm, where we assume its availability in most environments. Further, integration at the software component level is based on the assumption that the interfaces of dynamically retrieved components perfectly conform—syntactically, but also semantically—to the specification that guided their retrieval. However, these two assumptions may prove too restrictive for true AmI ubiquity. To address this issue, we have been studying solutions to integration of heterogeneous software components based on semantic modeling of software architectures, which can support reasoning on partial conformance between components and associated interoperability mechanisms at both software infrastructure and component level [30].

Further, human-machine interfaces in AmI environments have been studied in several research efforts. Natural, speech-based interaction is supported in our solution principally by the dialog manager functionality; most of the dialog manager architectures can be classified into three broad categories, according to [31]:

- Finite state-based systems, where the dialog states are encoded into a predefined graph;
- Frame-based systems, where the dialog manager aims at filling in empty slots in a template form;
- Agent-based systems, where the human-machine interaction is modeled as a communication process between two agents that can reason about their own and others' actions and context.

Our solution belongs to the second category; nevertheless, the third one is promising in handling unstructured and complex dialogs. It has only been implemented until now for very specific applications, such as in [32]. Another important feature addressed in our approach is multimodality. Combining modalities for user interactions can serve several purposes as follows:

1. Disambiguate the user's inputs. For example, speech recognition and lip reading using cameras can be combined to improve the accuracy of speech recognition [33]. More generally, it has been showed in [34] that 80% error avoidance can be achieved via mutual disambiguation across media.

2. Accommodate a wider range of users than with traditional interfaces. Multimodal interfaces provide a clear added value for disabled people, but also for specific categories of people [35]. Note that several projects have specifically addressed the challenge of building AmI interfaces for disabled people, such as EMBASSI[12], DynAMITE[13], or TéléTact[14], which has developed a multimodal (audio and tactile) white stick for blind people. The EMBASSI project further proposed an amodal[15] XML-based user interaction description language [36] to transfer information to mobile devices using different modalities (for example in Braille[16]) [37, 38].
3. Capture and merge different kinds of information that might be important for interpreting the user's inputs. For example, speech recognition can be combined with facial expression recognition in order to annotate the text stream with metadata such as sentence types (questions, orders, etc.), dialog acts, emotions, etc.
4. Improve the naturalness of AmI user interactions, since real communication between people naturally exploits several modalities [39].
5. Improve the effectiveness and quality of the way in which information and feedback are presented to the user, for example, with an emotional talking head.

With respect to context awareness, besides context that relates to user interaction modalities, which is addressed in our approach, AmI applications will more generally exploit information about the context in which they operate; to this end, they will share the same context terminology. In [40], an adaptable and extensible context ontology is proposed. This ontology is employed in application adaptation and generation of device-specific user interfaces. Another context modeling language is the context ontology language (CoOL) [41], which associates context information (e.g., a number) with an aspect (e.g., distance) and a scale (e.g., kilometers). Other such languages are CoBrA [42] and SOCAM [43]. In our future work, we aim at extending our approach to context to include other characteristics of the AmI environment. We intend to enhance at the same time the MMIL language so that it can handle contextual ontologies.

In our task architectural model, the decomposition of a user interface component into a UI front end and a UI back end component enables the incorporation of different generic I/O devices into an application. This is further related to dynamic adaptation of user interfaces to different interaction devices. In [44], an integrated user interface is proposed that automatically adapts to different consumer electronic appliances and lets the user control them with his/her PDA. Further, two major approaches are proposed to develop universal interfaces. The first one develops a standard for the discovery, selection, operation, and substitution of user interfaces and related options. This ANSI standard is called universal remote console

---

12. http://www.embassi.de.
13. http://www.dynamite-project.org.
14. http://www.limsi.fr/Individu/bellik/TeleTact/TeleTact.htm.
15. Modality-independent.
16. Braille is a code that enables blind persons to read and write.

[45]. The second one proposes a method for automatically generating interfaces for intermediate devices, such as PDAs or mobile phones, via an abstract specification of the appliance to be controlled [46].

Finally, an advanced feature of user interfaces in AmI environments, which we intend to address in our future work, concerns the so-called implicit (alternatively transparent, unobtrusive, calm) interactions. The main idea is to constantly analyze and interpret the user's behavior in order to proactively activate services that may be helpful, look for useful information, etc., without disturbing the user in his/her current tasks. To achieve this goal, the system has to share a common ground of semantic concepts with the user. The notion of shared knowledge is central to any communication between two entities [9]; however, it can be currently realized only for very limited application domains. For example, concrete implicit interaction platforms have been proposed for the World Wide Web [47]. The basic claim is that users need to be continuously aware of the presence of remote users to efficiently collaborate or talk with them. Thus, cyberwindows are proposed that let the user know, through discrete ambient sounds, what the remote users are doing in their own Web domain. The objective is to emulate, e.g., what happens in real offices, where others are typing, walking, etc. Other projects that propose innovative paradigms for implicit interactions are the Tangible User Interfaces [48], where users interact with the system by moving physical objects, the Internet Foyer [49] and the Dangling String [50].

11.6 Conclusions

In the extremely rich ambient intelligence domain, synergies are required between diverse research areas. Advanced, natural human-machine interfaces have a key position in the user-centric ambient intelligence vision. Providing them with adequate software infrastructure support for dynamic, ad hoc integration and interoperability will enable true omnipresent intelligence in our physical worlds.

References

[1] Vasilakos, A., W. Pedrycz, "Ambient Intelligence: Visions and Technologies" *Ambient Intelligence, Wireless Networking, and Ubiquitous Computing*, Artech House, Norwood MA, 2006.

[2] Issarny, V., et al., "Developing Ambient Intelligence Systems: A Solution based on Web Services," *Journal of Automated Software Engineering*, Vol. 12, No. 1, January 2005.

[3] N. Georgantas, V. Issarny, "User Activity Synthesis in Ambient Intelligence Environments," *Adjunct Proc. 2nd European Symposium on Ambient Intelligence (EUSAI '04)*, Eindhoven, The Netherlands, November 8–10, 2004.

[4] Borriello, G. and Want, R., "Embedded Computation Meets the World Wide Web," *Communications of the ACM*, Vol. 43, No. 5, 2000.

[5] Benatallah B.,et al., "Towards Patterns of Web Services Composition," *Patterns and Skeletons for Parallel and Distributed Computing*, Springer Verlag, United Kingdom, 2002.

[6] Curbera, F., et al., "Business Process Execution Language for Web Services," 2002.

[7] D. Martin, et al.,. "Bringing Semantics to Web Services: The OWL-S Approach," *Proceedings of the First International Workshop on Semantic Web Services and Web Process Composition (SWSWPC 2004)*, July 6–9, 2004, San Diego, CA, USA.

[8] E. Guttman, et al., "Service Location Protocol, ver. 2," IETF RFC 2608, June 1999.

[9] Riva, G., et al., (eds.). Ambient Intelligence. IOS Press, 2005, http://www.ambientintelligence.org.

[10] Ark W.S., and T. Selker, "A Look at Human Interaction with Pervasive Computers," *IBM Systems Journal*, Vol. 38, No. 4, pp. 504–507, 1999.

[11] Joshi A., and Y. Schabes, "Tree-Adjoining Grammars," *Handbook of Formal Languages*, Vol. 3, pp. 69–124. Springer, Berlin, 1997.

[12] Buil, V. P., et al., "Context-Aware Personal Remote Control: A Future Remote Control Concept," Nat.Lab. Technical Note 2001/533, 2002.

[13] Media Interaction Group, Philips Research Project. Window on the World of Information, Communication and Entertainment, April 2002, http://www.research.philips.com/password/archive/13/downloads/PW13_wwice.pdf

[14] Smart W., and L.Kaelbling, "Practical Reinforcement Learning in Continuous Space," *Proc. of the Seventeenth Int. Conf. on Machine Learning (ICML-2000)*, 2000.

[15] Westphal M., "The Use of Cepstral Means in Conversational Speech Recognition," *Proc. Eurospeech*, Rhodos, Greece, 1997.

[16] Landragin F., and L. Romary, "Dialogue History Modeling for Multimodal Human-Computer Interaction," *Eighth Workshop on the Semantics and Pragmatics of Dialogue (Catalog'04)*, pp.41–48, Barcelona, Spain, 2004.

[17] Oreizy, P., N. Medvidovic, and R. N. Taylor, "Architecture-Based Runtime Software Evolution," *Intl. Conf. on Software Engineering*, Kyoto, Japan, April 1998.

[18] Shang-Wen Cheng, et al., "Software Architecture-Based Adaptation for Pervasive Systems," *International Conference on Architecture of Computing Systems (ARCS'02): Trends in Network and Pervasive Computing*, April 8–11, 2002.

[19] Chetan, S., A. Ranganathan, and R. Campbell, "Towards Fault-Tolerant Pervasive Computing," *First International Workshop on Sustainable Pervasive Computing*, at Pervasive 2004, Vienna, Austria, April 20, 2004.

[20] Sousa, J.P., D. Garlan, "The Aura Software Architecture: An Infrastructure for Ubiquitous Computing," *Carnegie Mellon Univ. Technical Report CMU-CS-03-183*, 2003.

[21] Sousa, J. P., and D. Garlan. "Aura: an Architectural Framework for User Mobility in Ubiquitous Computing Environments," *Proceedings of the Working IEEE/IFIP Conference on Software Architecture*, Montreal, August 25–31 2002.

[22] Garlan, D., et al.,. "Project Aura: Towards Distraction-Free Pervasive Computing," *IEEE Pervasive Computing*, Vol. 1, No. 2, April–June, 2002.

[23] Roman, M., et al., "Gaia: A Middleware Infrastructure for Active Spaces," *IEEE Pervasive Computing*, Vol. 1, No. 4, pp. 74–83, October–December 2002.

[24] Roman, M., and R. Campbell, "A Middleware-Based Application Framework for Active Space Applications," *Proc. of ACM/IFIP/USENIX International Middleware Conference (Middleware 2003)*, Rio de Janeiro, Brazil, 2003.

[25] Krasner, G. E. and S. T. Pope, "A Description of the Model-View-Controller User Interface Paradigm in the Smalltalk-80 System," ParcPlace Systems, Inc., Mountain View 1988.

[26] Tandler, P., "Software Infrastructure for Ubiquitous Computing Environments: Supporting Synchronous Collaboration with Heterogeneous Devices" *Proceedings of Ubicomp 2001: Ubiquitous Computing*, Atlanta, Georgia, 2001.

[27] Newman, M., et al., "Designing for Serendipity: Supporting End-User Configuration of Ubiquitous Computing Environment," *Proceedings of DIS '02*, June 2002.

[28] Edwards, W.K., et al., "The Case for Recombinant Computing," Technical Report CSL-01-1, Xerox Palo Alto Research Center, Palo Alto, CA, April 20, 2001.

[29] Waldo, J., "The Jini Architecture for Network-centric Computing," *Communications of the ACM*, July 1999, pp. 76–82.

[30] Georgantas, N.,et al.,."Semantics-Aware Services for the Mobile Computing Environment," *Architecting Dependable Systems III, Lecture Notes in Computer Science*, Springer-Verlag (to appear).

[31] McTear, M., "Spoken Dialogue Technology: Enabling the Conversational User Interface," *ACM Computing Surveys*, Vol. 34, No. 1, pp. 90–169, 2002.

[32] Smith, R., and D. Hipp, *Spoken Natural Language Dialog Systems: A Practical Approach*, Oxford University Press, New York, 1994.

[33] Waibel, A., "Multimodal Human Computer-Interaction," Proc. Eurospeech, 1993.

[34] Oviatt, S.L., "Mutual Disambiguation of Recognition Errors in a Multimodal Architecture," *ACM Conference on Human Factors in Computing Systems (CHI'99)*, Pittsburgh, PA, May 15–20, pp. 576–583, 1999.

[35] Oviatt, S.L., "Multimodal Interfaces," *The Human-Computer Interaction Handbook*, 2003.

[36] Dubinko, M., et al., XForms 1.0, W3C Recommendation, World Wide Web Consortium, 2003.

[37] Herfet, T.,et al., "Embassi: Multimodal Assistance for Infotainment and Service Infrastructures," *Computers & Graphics*, Vol. 25, pp. 581–592, 2001.

[38] Richter, K., "Remote Access to Public Kiosk Systems," *Proceedings of the 1st International Conference on Universal Access in Human-Computer Interaction*, 2001.

[39] Nijholt, A., *Algorithms and Ambient Intelligence*, Kluwer Academic Publishers, Boston/Dordrecht/London, 2003.

[40] Preuveneers, D., et al., "Towards an Extensible Context Ontology for Ambient Intelligence," *Proc. EUSAI '04*, pp. 148–159, 2004.

[41] Strang, T., etal., "CoOL: A Context Ontology Language to enable Contextual Interoperability," LNCS Vol. 2893: *Proc. of the 4th IFIP WG 6.1 International Conference on Distributed Applications and Interoperable Systems (DAIS 2003)*, Paris, France, Springer Verlag, pp. 236–247, 2003.

[42] Chen, H.,et al.,. "An Ontology for Context-Aware Pervasive Computing Environments," *Special Issue on Ontologies for Distributed Systems, Knowledge Engineering Review*, 2003.

[43] Gu, T.,et al., "An Ontology-Based Context Model in Intelligent Environments," *Proceedings of Communication Networks and Distributed Systems Modeling and Simulation Conference*, San Diego, California, USA, 2004.

[44] Ali, A., and S. Nazari, "A Novel Interaction Metaphor for Personal Environment Control: Direct Manipulation of Physical Environment Based on 3D visualization," *Computers & Graphics*, Vol. 28, No. 5, pp. 667–675, October 2004.

[45] Zimmermann, G., et al. "Toward a Unified Universal Remote Console Standard," *CHI Extended Abstract*, pp. 874, 2003.

[46] Nichols, J., "Automatically Generating User Interfaces for Appliances," *Advances of Pervasive Computing*, pp. 105–110, April 18, 2004.

[47] Liechti, O., et al.,." A Non-Obtrusive User Interface for Increasing Social Awareness on the World Wide Web," *Personal Technologies*, Vol. 3, pp. 22–32, 1999.

[48] Ishii, H., Ulmer, B., "Tangible Bits: Towards Seamless Interfaces between People, Bits and Atoms," *Proceedings of CHI '97*, ACM Press, 1997.

[49] Benford, S., et al., "Shared Spaces: Transportation, Artificiality and Spatiality," *Proc. of CSCW*, Cambridge, USA, ACM Press, 1996.

[50] Weiser, M., and J.S. Brown, "Designing Calm Technology," *PowerGrid Journal*, Version 1.01, July 1996, http://powergrid.electriciti.com/

CHAPTER 12
Emotional Interfaces with Ambient Intelligence

G. Andreoni, M. Anisetti, B. Apolloni, V. Bellandi, S. Balzarotti, F. Beverina, M. R. Ciceri, P. Colombo, F. Fumagalli, G. Palmas, and L. Piccini

12.1 Introduction

Intelligence performs also through the heart. AmI can be defined as a pervasive and unobtrusive intelligence in the surrounding environment supporting the activities and interactions of the users [1] in order to bring computation into the real, physical world and to allow people to interact with computational systems in the same way they would with other people. In this chapter, we will deepen the role of nonverbal communication in the interaction between the artificial intelligence of the ambient and the natural one of the human. This kind of communication is based on signals not focused to transmit an explicit—symbolic—message, but rather analogical messages directly interacting with the hardware of the ambient intelligence, and the surrounding environment, in a way simulating the emotional reaction/interaction between humans. Therefore, we refer to signals such as heart rate, face features, voice inflection, and other nonverbal indicators of the evolution of such indirect communication [2].

We will have a general discussion on the psychological and computational bases of this interaction, touching aspects coming both from the theory of mind and from theory of computation. As a result, we will discover that we cannot have an emotional interface that is separated from the rest of the processes involved in the adaptation of the ambient intelligence functionalities. Rather, the interface is a nested part of a cognitive system embedded in the ambient. Hence, we will discuss an enabling system introduced in the University of Milan for simulating emotional human-machine interactions and capturing this kind of nonverbal communication within a transparent communication pathway. We will present some experimental evidences addressed directly by a human operating on the part of the ambient, and will discuss ongoing steps to enable an artificial cognitive system (ACS) to efficiently replace him. In spite of some friendly and naive flavor, this ACS functionality is far from being realized since it requires the development of a new rigorous philosophy for conceiving and designing computational systems. On the one hand, we face input signals not completely understandable per se, as a practical counterpart of the theoretically assessed computational limits, such as biological and psychological signals, humoral attitudes of the subjects we want to serve, and complex reactions and interactions of the systems we are designing. On the other hand, we want precise and well-defined operational rules that we can manipulate in a totally understandable way; for instance, we want to comprehend the joint effect of a set of these rules, or the effect in the change of some free parameters. Without pretension of exhaustiveness, our approach is based on a series of theoretical tools, such as: cognitive psychology, algorithmic inference, sentry point theory, Boolean independent component analysis, hybrid learning algorithms, fuzzy relaxation, PAC-meditation, recently collected in a book by the authors [3].

12.2 Background

The capabilities of computers in their actual fashion appear to be close to the terminus as it concerns both hardware and software. On the one hand, the Moore's law for the computational power starts to be barred by the quantum phenomena posing problems that are yet to be solved. On the other hand, the utilities expected by computers, understood as executors of explicit commands, are almost saturated both in industrial and domestic environments. Rather wide rooms are open to new computer functionalities mainly related to the communications/interactions between computers and, over all, direct interaction between computers and human beings. With direct interaction it is understood that a nonsymbolic interaction that may paradigmatically denote the absence of keyboard and any its surrogates (such as speech-to-text devices). The main ingredients of this new kind of computer device are as follows:

- Hardware: conventional, powerful, and of reduced size;
- Peripherals: sensors of any kind (acoustic, biomedical, environmental, etc.);
- Logic: (possibly symbolic) rules directly inferred or at least adapted on the basis of examples of the expected functionality;
- Tasks: solving operational problems in a locally optimal way as learned from past experience on similar or analogous situations.

These points delineate the typical framework of recent computing frameworks commonly denoted as disappearing ubiquitous pervasive wearable computing [4], which constitute the operational counterpart of the ambient intelligence.

12.2.1 The General Framework

We assume a full HMI exploitation when a human is aware that the machine may understand him and cooperates with him in a task [5]. In this perspective, new-generation interfaces have to behave like skillful autonomous negotiators between the needs of the user and the capability of the machine. Their job is to facilitate and enhance the request of the former by recognizing main variations in his emotional states [6] that they perceive by stressing nonverbal messages, such as heart beat or eye and face movements, joined with verbal ones. In particular, we think of an interface providing the following distinguishing functionalities:

- An actual two ways interactivity;
- A processing of physiological and nonverbal communication signals;
- A reliable analysis of user emotional state as keen information at the basis of the interaction; the project considers a multidimensional approach to emotions represented by an hypercube with four axes:
  - Arousal: stress/relaxing;
  - Novelty: attention/boredom;
  - Hedonic value: pleasantness/unpleasantness;
  - Coping: frustration/satisfaction;
- A suitable feedback of the perceived emotional features.

12.2.2 The Key Role of Emotions

New theories of mind try to gather the continuity between cognition and action and their tight boundaries with motivation and emotion: Their goal is the elaboration of unified models, where these functions are merged to explain human behavior. In this perspective, those theories suggest that mind is embodied and situated as it originates in the interaction with the environment [7, 8]. Humans are then considered as biological agents capable of acting on the environment to change it and of functionally adapting themselves in relation to their own needs.

Procedural actions always involve emotion and motivation: Agents organize action sequences to achieve goals relying on the emotional and cognitive evaluation of events. Motivation, cognition, and emotion are no longer considered as separated functions and all support executing actions effectively. These issues raise interesting questions when applied to the context of the interaction between biological and artificial agents. The increasing use of computers and machines that support the human user in any kind of task has nowadays transformed these machines into important and constitutive members of the physical and social environment with which people interact. For this reason, researchers from several disciplines have turned their attention to emotion and its role in the interaction between human and artificial agents, trying to understand and recreate emotion in the user interface [9–13]. Emotional models have been proposed as a critical component of more effective human computer interaction: Explicit attention to the emotional aspects aims at increasing the system performance in terms of usability and acceptance. In this sense, emotions play an important role in the design of interfaces: They should not be considered a simple option, providing pleasantness, but they represent crucial cues since they are involved in the selection, regulation, and motivation to action.

12.2.3 The Emotion Physical Milieu

In human-machine interfaces (HMI), electronic interfaces are going to become an indispensable complement of any tool used by humans. LCD microscreens, coupled with human mimicking voice messages, possibly embedded in principled avatars, are the way by which messages will be transferred from machine to humans. The opposite flow, from human to machine, is committed either to buttons or to a limited set of voice commands. As a matter of fact, this kind of interaction is both primordial and intimately repellent for the user, who tries to use it as little as possible, i.e. for very elementary tasks (when using a videorecorder most people just use the three buttons: play, stop, and rewind). On the other hand, the real deep benefits expected and designed to come from the electronic interfaces will be actual (and a large market and technologic development of these devices will occur) only if a real two-way communication can be stated between human and machine. Human communication is multimodal by definition. For example, the voice tells a specific piece of information related to the context. Nevertheless, the intonation, the facial mimic, and other signals transmit complementary, rather the main in many cases, information with the respect to the content itself. For these reason, multimodal interfaces could be a powerful tool, especially in AmI applications where a strong interaction is required with the subjects. Furthermore, the improvement of the dialog through the utilization of alternative media, providing redundant interaction methods, can provide accessibility to users with limitations in specific operational fields. The interaction between the user and the system (and the task that the user wants to complete) will be so complex that the user can no longer be obliged to adhere to strict formal communication protocols imposed by the machine. The system must be able to support some kind of intelligent declarative interaction. In such a context, the machine must be capable of understanding the intentions of the customer and reacting appropriately. In the past, the component technologies (e.g., speech recognition, gesture recognition, natural language processing, biological signals acquisition, and processing) have been developed in isolation. Although none of those component technologies comes even close to human performance, all have reached the state where practical applications in well-delimited domains are feasible. Still, it is fair to say that all component technologies need major breakthroughs to approach human performance. The multidetection of these signals can offer the completeness of the information and, by understanding the role of the emotion, we may build bioinspired and adaptive HMI to maximize the efficiency on the communication and of the system performances.

12.2.4 The Emotion Psychological Milieu

We intend a communication that is agreeable and convincing to the human, on the one hand, and understandable in many of its nuances by the machine, on the other. Only in these conditions will pervasive computing develop in a way that is not prevaricating humans, but rather is at his real service. We figure a full exploitation of this communication when a human is aware that the machine may understand him and cooperates with it in this task, adapting his communication habit to the machines capabilities, as soon as the machine adapts its software to favor human communication. Since many of the tools for establishing this framework will be sophisticated, so much will the machine be adaptive toward human's high capabilities, rather than the human's yielding to machine's low capabilities. We identify the sophistication of these tools with the capability of automatically understanding human emotional features and the intelligence of designing a machine communicating strategies as a function of these features. The most important gap in our present understanding of the process in human–human communication is the lack of a comprehensive theory of the ways in which information coming from the various senses (auditory, visual, tactile) is integrated with the enormous store of prior knowledge about the world in general and the interacting partners in particular. It is difficult to imagine how a coherent perception of intentions can be synthesized without such integration or fusion. Considered by itself, the input from individual channels may be highly ambiguous. Straightforward, linear combination of all inputs does not guarantee disambiguation: In some cases the customer simply did not provide all information that is necessary to understand the intention, in other cases the different channels may seem to carry contradictory information (as when somebody says something negative while nodding). For all that is presently known about the ways in which humans understand intentions, it is clear that understanding is an active process (involving both signal-driven and interpretation-driven processes). Active processes are part and parcel of all individual components, even if they have been largely disregarded or modeled very poorly, as in existing speech recognition and natural language processing. The latest models of communication [14–18] present a circular vision of communication, where the role played by the receiver is just as important as that of the transmitter. The notion of communicative intention has been studied in the last 30 years through different approaches and has now a central role within communication models, deeply changing the concept of the message itself. The message meaning is therefore defined by the relationship of mutual intention and awareness of the participants. Its production involves different behavioral units strategically assembled in a coordinated and componential way [19–24]. As a matter of fact, according to these theories, the communicative act is multimodal, that is, it involves the whole body and is made up of expressive behaviors belonging to different semiotic systems. In *The Media Equation*, Reeves and Nass [25] argue that human–machine interaction is inherently natural and social, so that the rules of human-human interaction apply to it, as people's social response to technology mimics their social response to other human beings. In many fundamental ways, people respond psychologically to interactive computers as if they were human actors and not tools [26]. User interface designers should recognize that spoken conversation with a computer will generate social and emotional reactions in users, including expectations of appropriate emotional responses in the computer itself [27]. Anyway, it would be a mistake to implement just a few human like features and consider a system to be improved simply because it is more human-like: some HCI efforts have also noted problems with incautiously adding some human-like features, which caused more disturbing impressions [28].

12.2.5 No Emotion without Cognition

Over the last decade, user interface design has focused primarily on system capability and usability. The coming years see greatly increased attention to the subjective user experience, including the emotional impact of computer use and a reasearch effort to develop computers that have the skills involved in emotional intelligence [11, 12, 26]. A key part of this reasearch is focused on giving computers the ability to recognize human emotions and to respond intelligently to them. Researchers are working on the development of an adaptive system that observes users' emotional states, acquires awareness, and interprets the most probable users' current emotional states. The system is designed to observe physiological components associated with emotions. These components are to be identified and collected from observing the user via the three main sensory systems of machine perception: visual, kinesthetic, and auditory and via natural language processing. This recognition enables the interface to change adaptively some aspects of its interaction with the user. From past projects, we experienced that a real exploitation of these emotional signals may occur only through an emotional attuning between an artificial cognitive system (ACS) and human during its joint actions. Through attuning, the system is able both to induce, acquire and process emotional signals and to adapt its actions accordingly, with the ultimate goal of improving the effectiveness of the services it supplies the human. The attuning we look for cannot be attained at a simple level of an interface discovering the emotional state of the user per se, in view of exploiting this information on any cognitive system. Rather, the attuning is a process looping between human and ACS; hence, it cannot exploit the related nonverbal signals separately, rather they constitute relevant feedbacks in this process. In this sense, the architecture evolves and achieves completion, as shown in Figure 12.3. Here, we have a dynamic system running a three-corners loop between human, ACS, and goal that must bring to the optimisation of the objective function, a function that must explicitly depend on the goal symbolic description. For instance, with reference to Figure 12.1, the goal of the quizzes submitted to the human was to have fun. The answers to the quizzes, duly evaluated according to the goal cost function, constitute obvious feedback to both human and ACS, such as they are the both verbal and nonverbal human messages for the ACS, while the sole explicit answers are the feedback to the goal. More subtly the ACS feedback to the human (and to the goal as well) constitutes an anticipatory mechanism that regulates the sequence and cadence with which the quizzes are supplied to him in order to solicit special emotional reactions.

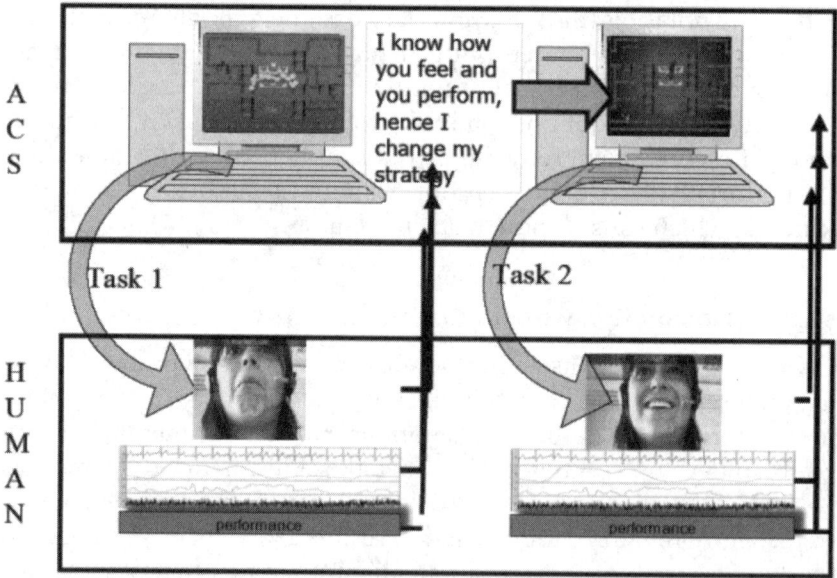

Figure 12.1 Human joins ACS in a conversation: Emotion is constantly changing during an effective tuning. The subject shows different emotional expressions according to the cognitive appraisal of the stimulus and performance.

12.3 Materials and Methods

An enabling system has been set up at the Psychology Communication Lab of the Catholic University of Milan—in cooperation with State University and Polytechnic of Milan, and ST Microelectronics—for implementing some game-playing sessions, where the computer provides a simulated intelligent feedback to the user in order to decode his emotional state and to adapt the tasks accordingly. Our initial goal was to assess what emotional information a computer can accurately obtain from the user and to describe where the latter may be set in the emotion axes space represented in Figure 12.2 through the four axes : novelty (attention/boredom); hedonic value (pleasantness/unpleasantness); coping (satisfaction/frustration); arousal (tension-relaxation).

In the workbench shown in Figure 12.3, subjects were asked to use the computer where an avatar [29] guided them through different kinds of tasks programmed to vary the subject's attentive level and to elicit specific emotional reactions (e.g., repetitive task boredom; negative/positive bonus unpleasant/pleasant surprise; impossible resolution frustration). See Figure 12.4.

12.4 The Data

The computer is equipped with the following:

1. Two high-resolution Web cameras: one Web cameras was placed in front of the subject to record facial movements, gaze direction, and posture changes. A second camera was placed behind the subject so that it was possible for the external experimenter to follow the subject's action on the screen.
2. Physiological recordings were taken using the BIOPAC System.
3. A high quality microphone to record vocal reports.

All recording devices are synchronized. See Figure 12.5.

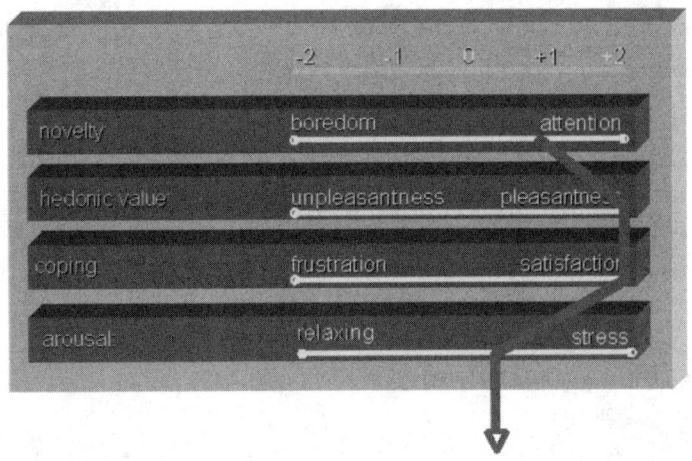

Figure 12.2 The four dimensional axes.

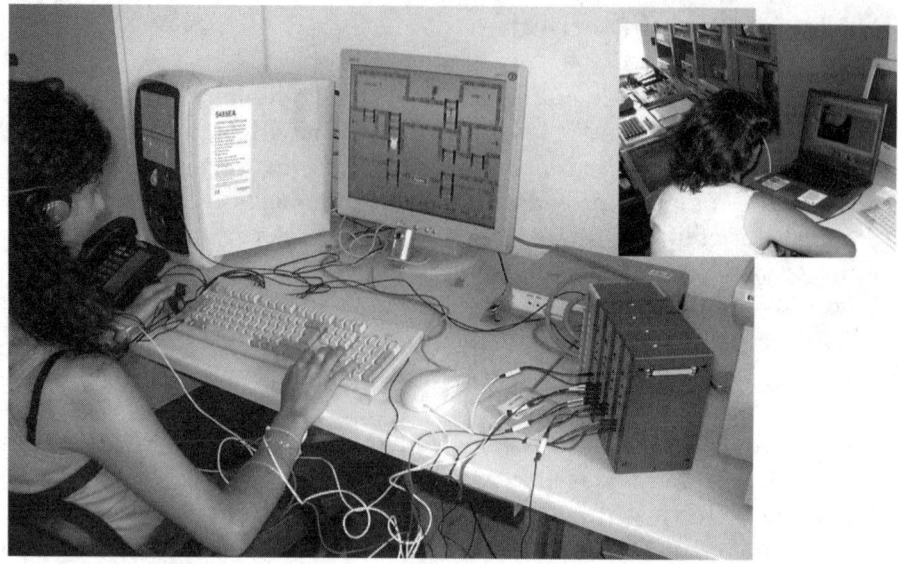

Figure 12.3 The workbench employed to capture the subject emotional state.

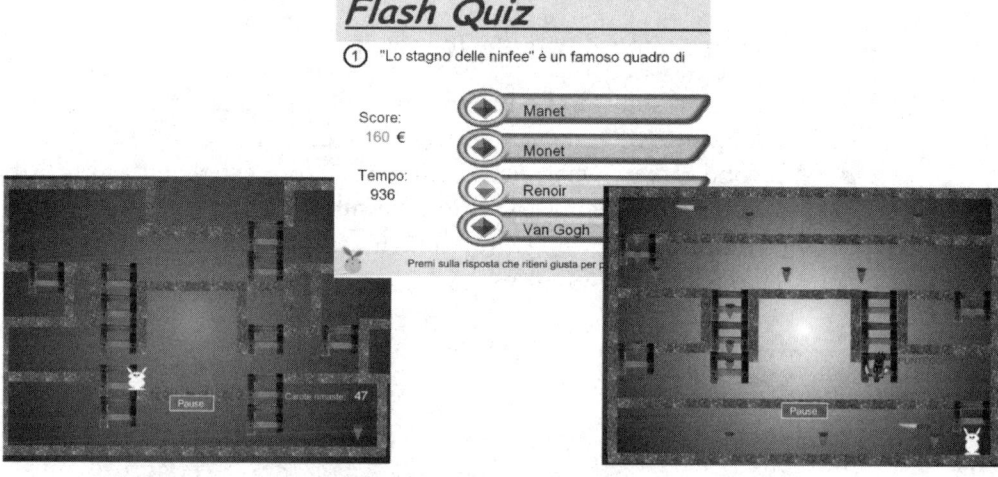

Figure 12.4 Some tasks supplied to the subject.

1. Physiological Signals
   - ElectroCardioGram (ECG)
   - Galvanic Skin Response (GSR)
   - Respiratory Effort (RSP)
   - Skin Temperature (SKT)

2. Non Verbal Signals
   - Posture
   - Face Expression
   - Gaze Movements

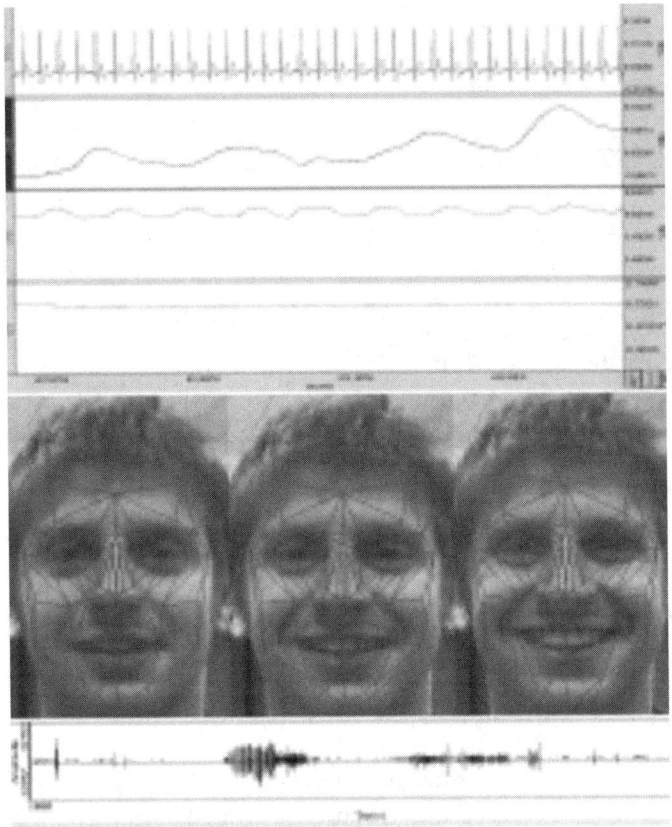

Figure 12.5 A sample of employed biosignals.

- Suprasegmental Features

3. Verbal Signals
   - Suprasegmental vocal cues

12.5 The General Architecture

We frame our cognitive system as a two-way pipe transmitting interactions between subject and machine within two nested loops. An outer one conveys the appropriate reaction of the machine in terms of adapted teaching strategy, the inner one exploits special functionalities of the general-purpose task of interfacing human with machine. Inner loop is partitioned into four blocks, as shown in Figure 12.6:

Block 1
Data Acquisition. We acquire physiological signals, such as galvanic skin response (GSR), electrocardiogram (ECG), skin temperature (SKT) and respiratory effort, that provide information that is strictly related to changes in the activity of the nervous system. We will also record voice and video streaming in order to complete a perceptual description of the user during his interaction with the machine. These

signals are acquired with absolutely noninvasive techniques maintaining synchronicity with the running task.

Block 2
From signals to features: data fusion and relevant features extraction. We will fuse data and extract joint nonlinear features using either standard or proprietary software. A key action will be the selection of relevant features for extracting information about the user-emotional state. We split this task in two hierarchical subtasks:

1. Arousal detection. We look at physiological correlates—occurring from electrodermal activity, heart and breath rate—denoting an autonomous nervous system (ANS) arousal and its alterations. ANS arousal modifications are processed mainly to highlight relevant and significant changes in the user physiologic condition in order to provide a Boolean signal that triggers the system, indicating that something has happened in his behavior.
2. Detection of relevant indicators of specific emotional states. Once arousal is up, we move to Block 3 for a finer analysis of above features coupled with more explicit ones deriving from the facial action coding system [30] [31], componential vocal patterns, gaze direction, and posture (Observer Noldus and Theme analysis software, frame by frame, 25 fps).

Both types of detection are committed to cognitive systems, such as neural networks or hybrid connectionist symbolic systems, that rely on short/long term statistics on physiological features and entropic incremental methods for identifying relevant features and their connections with the indicators (see Block 2.2). In principle, we do not expect to get totally explicit understandable rules for deciding arousal or, even more, emotions. Rather, the nonsymbolic memory of the cognitive systems—for instance, the weight connection of the neural network—synthesizes, with a more or less accurate definiteness, maps of arousal/nonarousal and emotional states in the feature space. In this way we expect to cope with nonlinearity and nonseparability of the maps.

Block 3
From features to symbols: Semantic encoder. In order to raise communicating strategies on the part of the machine, the interface must supply it with definite indicators acting as semantic symbols to be manipulated by the machine. The semantic is thought to refer to the mentioned emotional dimensions: stress/relaxing, attention/boredom, pleasantness/unpleasantness, coping.

Block 4
From symbols to signals. This block closes the loop bringing back from machine to the user specific stimuli aimed at enhancing the information content of signals flowing in the two directions between the actors. Also in this block we plan the intervention of the cognitive system to refine and tune the stimuli through functions directly learned from the experts' knowledge.

12.6 A Simulated Experiment

Combining human-machine interaction with a psychological view of emotion as a multiple component process [6, 32–34], complex relationships exist among emotion, motivation, and cognition in human-computer interaction, including the following:

1. Considering emotion as a process and its cognitive components [6, 35], we expect that:
    - Tasks involving different cognitive appraisals will excite active responses on the environment that have an expressive component.
    - These responses consist of actions rather than reactions and are aimed at adapting to the environment and changing it (e.g., the subject puffs to signal the need to change the task).
    - During the task the subject will exhibit dynamic and flexible responses based on a continuous evaluative monitoring of stimuli rather than fixed expressive patterns corresponding to an emotional label; each task will elicit congruent emotional expressive responses (e.g., boredom for the repetitive task).
2. A second research problem concerns the emotional interaction between the user and the machine. According to *The Media Equation* [25], we suppose that during the interaction with the computer subjects will exhibit communicative signals (verbal and nonverbal) to express their emotional state;
3. Finally, the following question is relevant: If the subject is aware of interacting with an understanding agent, is he encouraged to make use of emotional communicative signals? According to Fridlund [36], we suppose that in the experimental condition (when the computer tells the subject that it is able to

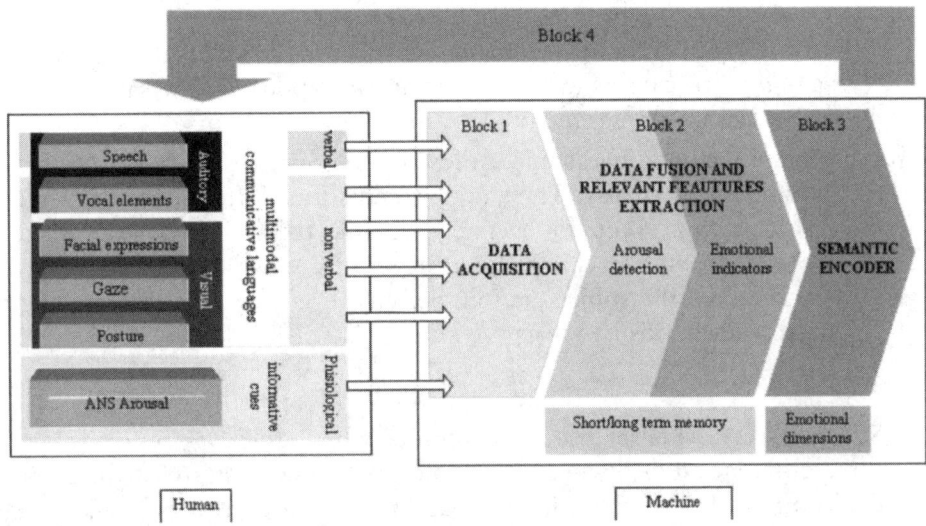

Figure 12.6 The logical architecture.

understand his/her emotional states) subjects will exhibit significantly more communicative signals than subjects in the control condition.

Addressing these points is a fundamental step for instantiating an efficient ambient intelligence system. We set up an express experiment on our workbench for getting preliminary answers, having the machine reaction simulated by a human operator. The experiment ran as follows.

Participants
A total of 30 university students (20 to 23 years old) were recruited from two different kinds of faculties (humanistic vs. scientific).

Stimuli Construction
Three different kinds of computer games were designed to modify the subjects' attention levels and to elicit specific emotional reactions. Systematic manipulation of appraisal dimensions was used through the selection of types of game events that were assumed to produce specific appraisals. Specifically, game events were supposed to support four emotional evaluation checks: novelty (a change in the situation that captures the subjects attention); hedonic value (intrinsic pleasantness or unpleasantness of the stimulus); goal conduciveness (events that help or hinder the subject to reach a goal); coping, (increasing levels of difficulty of tasks that change the estimated ability to deal with them). All games were previously tested on 10 subjects to assess their efficacy.

- Quiz game: 15 questions were presented to the subject, who has to select the right answer among four alternatives. The subject wins money for every correct answer and loses money when answering wrongly. Questions are divided into two series: A very easy one is followed by a very difficult one that makes the subject lose almost all the prize won. Selected events: correct/wrong answer; series of questions.
- Boring game: the subject moves a rabbit on the screen and has to collect a great number of carrots (50). Carrots always appear in the same positions (repetitive task). A message appears after every 10 carrots collected.
- Enemy game: The subject moves a rabbit that has to collect carrots while avoiding an enemy. The game presents four different levels of difficulty. The subject wins points for every carrot collected and every level successfully completed. In each level positive or negative bonuses appear randomly, independently from the subject action. Selected events: losing life, passing to the next game level, positive/negative bonus.

Procedure
Subjects were asked to use the computer where an avatar guided them across the three different kinds of computer games. All sessions started with 2 minutes of free exploration of a Web site for the baseline measure of physiological signals. Total duration was about 20 minutes. They were divided into two different groups, according to the kind of information received by the avatar. In particular, this experimental research will make use of the Wizard of Oz approach [11, 37], that is, it

employs an initially simulated prototype of emotional intelligent interface: The subjects were told they were interacting with an emotional intelligent computer, though in fact they were not. In the experimental condition, the subjects were exposed to a simulated emotional-intelligent computer, where the avatar provided a simulated intelligent feedback to the user to decode his emotional state and to adapt the tasks accordingly. For example, the avatar used sentences such as: <You seem to be bored, so here is a new game>, <You are in difficulties: I will repeat the instructions for you>. The simulated emotional-intelligent computer appeared to be automatic to the subjects, but it was actually controlled by an out-of-sight experimenter. In the control condition the avatar guided the subjects in the same activities but did not simulate to decode emotion. All subjects were alone in the room with the computer. At the end of the computer session, subjects were asked to answer a questionnaire and the coping inventory for stressful situations [38, 39]. In the questionnaire subjects were asked: to assess their own abilities at using a computer; to judge on a 7 point rating scale the computer games according to emotional dimensions (surprising/ boring; pleasant/unpleasant; frustrating/enjoying); to judge the efficacy of the interaction with the computer. In particular they were asked to judge to which extent the computer had been able to understand their emotional states. In this way it was possible to test the efficacy of the simulated prototype, as 13 to 15 subjects in the experimental condition said they believed in the PCs ability to understand.

Data: Different kinds of synchronized data were recorded into a proprietary database:

1. Physiological signals: ECG; respiration; galvanic skin response; skin temperature;
2. Nonverbal signals: Posture; gaze direction; facial movements;
3. Nonverbal vocal signals: suprasegmental cues;
4. Verbal signals speech.

It is a richer signal's repertoire with reference to the database used by Frank and Ekman [40] to study deception where a group of 20 subjects was recorded. Moreover, the expressions exhibited by our subjects are spontaneous while the 1,917 image sequences collected from 182 subjects in Kanade and collegue's database are not [41]. Our data will be available to the scientific community as a public database in a short time, after all the privacy procedures are adhered to.

Analysis: We developed a behavioral coding system (BCS) to code the subjects interaction. 30 behavioral units subdivided in four categories were mapped as in Table 12.1. Some units were identified by using specific action units (AU) defined by the facial action coding system [42]. Other units for posture, body movements, and vocal behavior were taken from grids used in previous studies on nonverbal behavior [43]. All video tapes were coded frame by frame (with a 25 fps rate) using the Theme 4.0 procedure of the Observer 5.0 Noldus software [44] for a preliminary inventory of recurrent patterns.

At a second level of analysis, we framed these patterns in a multidimensional emotional appraisal semantic space (MEAS) to detect users emotional states [45]. We used four dimensional axes (see Figure 12.7) scored on a 5 point rating scale from −2 to +2 by the interjudge agreement of a group of 3 judges.

Table 12.1 The Four Categories of the Behavioral Coding System.

Facial movements	Upper and lower face units were selected. For each unit intensity was rated (Low, Medium, High).
Gaze direction	To look at the screen, at the keyboard, around, etc.
Posture	Posture Behavioral units of moving near to /far from the screen were considered.
Vocal behavior	It was recorded when the subject speaks or uses other kind of vocalizations (grumbling, no-words, etc.)

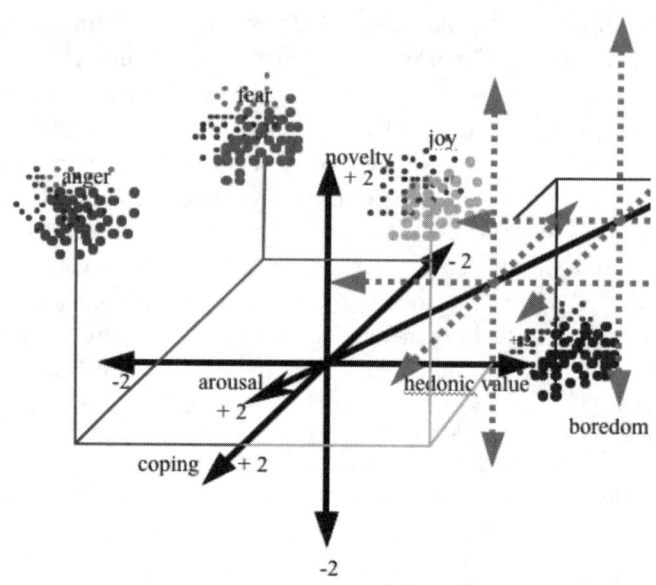

Figure 12.7 Multidimensional emotional appraisal semantic space axes.

Preliminary results from the analysis, 22 minutes and 27 seconds of one subject, for a total of 26,948 frames show that the subject exhibits communicative nonverbal behavior during the interaction with the computer. In particular, facial behavioral units (with a mean number of occurrences among the kinds of features $\_ = 47.5$ and an average total duration $\_ = 33, 85$ sec) are used more frequently than vocal ones ($\_ = 10.75; \_ = 10.54$ sec) and within the facial movements there is a higher exhibition of lower face units ($\_ = 51.6, \_ = 41.42$ sec). Nonverbal behavioral units are exhibited with different frequencies and durations, hence they seem to have different functions and relevance. The analysis of the most recurrent configurations of the emotional expressions highlights the presence of patterns of different levels of complexity that involve facial movements, posture, etc. (Figure 12.8).

Behavioral units are linked to each other in more complex pattern and expressive configurations. Hence, the subject exhibits dynamical facial responses based on the continuous monitoring of the task and performance rather than fixed patterns corresponding to an emotional label. Second, patterns denote a correlation between computer events and behavioral units. Figure 12.9 captures two episodes of emotional attuning between human and machine in terms of time position of the emotional behavioral response in relation to the antecedent action of the computer and

12.6 A Simulated Experiment

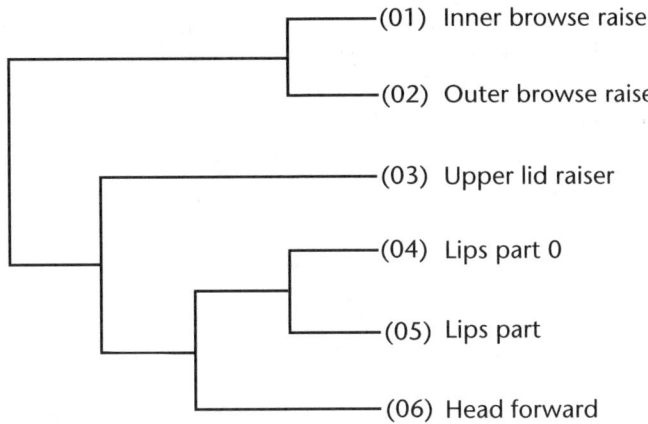

Figure 12.8 Hierarchical clustering of the most frequent behavioral patterns.

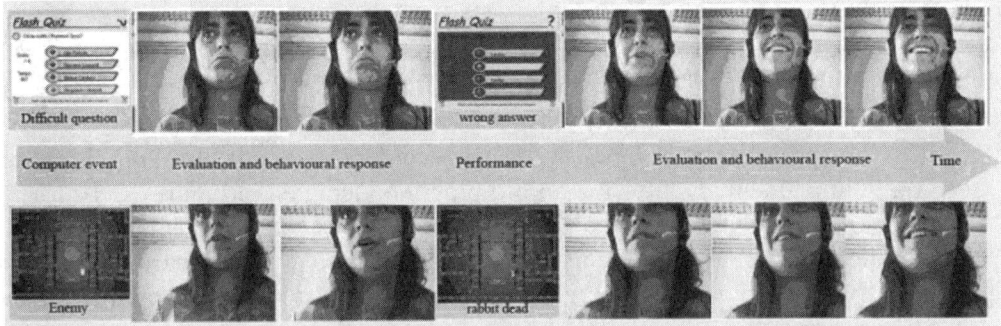

Figure 12.9 Frame by frame sequences and time structure.

the performance. It shows an example of two frame by frame sequences (6.89 sec; 5.32 sec), where the subject is dealing with the quiz game (a) and the enemy game (b). Each sequence starts with the computer event (difficult question; enemy); it is followed by its evaluation and a frustration behavioral response, which precedes the subjects' performance (wrong answer; rabbit dead). Finally, the emotional response to the mistake (smile). In this sense it has been possible to focus on sequences of events characterized by a specific timing and organization (antecedent, evaluation, performance, emotional response). In particular, the emotional response appears to have tight links with both the evaluation of the antecedent and the performance in the task. These links suggest in turn the emotional response role in the motivation to act. Accordingly, Table 12.2 shows the answers of subjects about the interaction they experienced with the computer. In the experimental condition subjects assign higher scores to the ability of the computer to understand emotion (a), to adapt the task according to them (b), and to answer to reactions (d), while in the control condition they assigned high scores to the ability to give information (c).

In conclusion, these preliminary results identify former significant features that can be extracted from the users' expressive behavior and udimentary traits of the dynamic changes of the users' emotions during the computer interaction. The traits

Table 12.2 Averages and standard deviations of the scores rated by subjects about the computer ability to understand.

Interaction Kind	Experimental		Control	
	Mean	Std	Mean	Std
a	3,93	1,62	2,35	1,49
b	3,87	1,64	2,41	1,42
c	3,73	1,79	4,29	1,53
d	3,53	1,59	2,76	1,68

may prime a formal emotional semantics that could be used by a machine to discern users' emotional states.

12.7 Current and Future Work

As mentioned in Section 12.1, work is in progress. Block 1 in Figure 12.6 is completed, while the labeling of the collected data in order to have suitable training sets is at hand.

The passages from signals to features and from features to symbols are only partially implemented. We have an old pipeline, depicted in Figure 12.10 for the complete processing of physiological signals, that has been experimented within other applications [46, 47] and must be updated and integrated in various directions. It is a hybrid multilayer perceptron, where first layers compute mainly subsymbolic functions such as conventional neural networks producing propositional variables from the signal features, while the latter are arrays of Boolean gates grouping these symbols in formulas of increasing complexity. Starting from the mentioned features, we give the neural network the task of producing symbols from them with the key

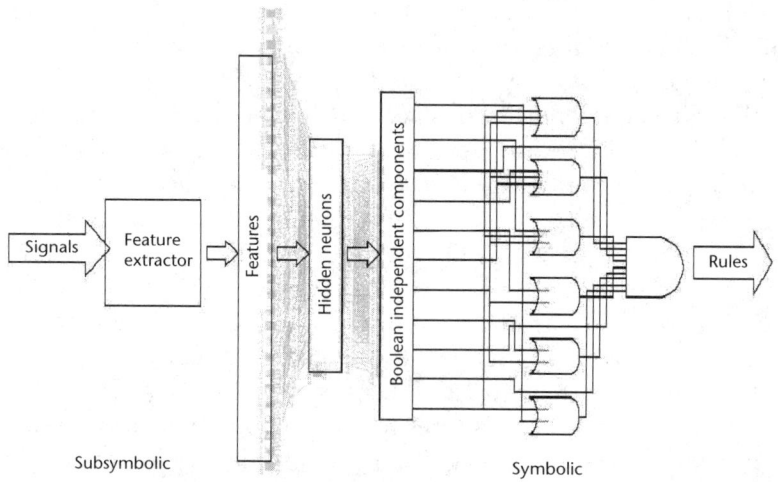

Figure 12.10 The hybrid subsymbolic-symbolic architecture.

commitment of getting a syntactic mirroring of this data. We look for a vector of Boolean variables whose assignments reflect the relevant features of the original (possibly continuous) data, where assignments may coincide when they code data patterns with the same values of the features of interest to us. In greater detail, we split the mirroring in two parts: a true mirroring of the data pattern and a projection of a compressed representation of the pattern (obtained as a side effect of the first part) into the space of Boolean assignments. Given a set of examples in the Boolean hypercube, we compute a pair of rough sets [48], respectively, and including the goal formula. Namely, we compute a minimal DNF (union of monotone monomials), D, including all positive examples with reference to a fixed emotional state such that no other DNF with the same property is included in D. Analogously, we compute a maximal CNF (intersection of monotone clauses), C, excluding all negative points such that no other CNF with the same property is included in C. Then we simplify these formulas in the aim of gaining them a better understandability at the expense of broadening their contour. Namely, we endow the formulas with membership functions spanning the gap between them. An optimal simplification (balancing length and fuzzyness of the formulas) is obtained via a simulated annealing procedure [49].

The model must be enriched and integrated in the following aspects:

- Extraction of meaningful features. We may expect that the neural network can both extract by itself independent unconventional features in a blind way and decide how to use and interpret the classical ones that are typically used in offline applications for exploiting their contributions in short-term analysis. Figure 12.11 shows some features extracted from the ECG signal and respiration signal using real-time algorithms.
- Addition of a vast set of features extracted from face expression analysis, such as the facial action coding system (FACS). Their robust automatic extraction is still an open problem when the face is neither fixed nor exactly in front of a Webcam. A new method has been assessed (based on the past works of Kanade and colleagues [50]) for retrieving the frontal posture of the face in case of small movements. The main idea consists for recovering the full motion of the head using the 3D head model Candide [51], and from that, to obtain the face in a normalized frontal position (Figure 12.12). Consequently, a set of features, related to the FACS, the gaze direction, and the head movements, can be extracted more easily.
- Identification of fuzzy rules from macrofeatures (such as FACS) to the single emotional states. A neuro-fuzzy system has been designed and tested on common benchmarks. In Figure 12.13 we report the network layout, where we distinguish an input layer, a rule layer, and a single output, plus a DEFuzzyfier block. Connection A_{ij} from input attribute i to rule node j expresses the membership of the input feature to this attribute—for instance gaze direction—involved in them "if..then.." rule R_j cast in the node. Connections B_k express the membership function of the whole input signal (through the input to rule connections) to the output attribute of R_k. Both As and Bs are triangular functions of their inputs and are merged through a $>$-Norm [52], the former within the single rule nodes and through a $>$-Conorm, the latter to supply the

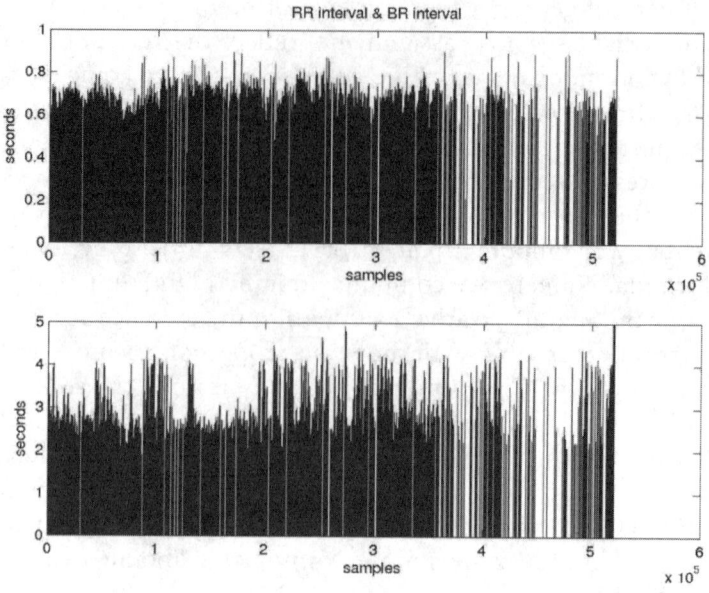

Figure 12.11 Synchronous heart and respiration intervals.

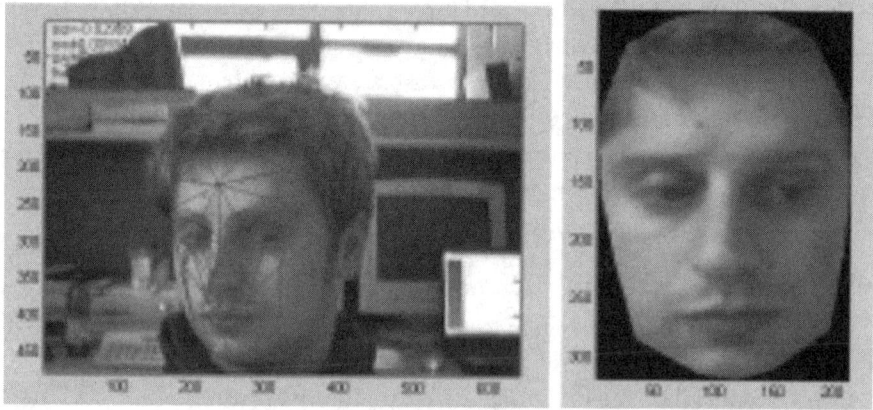

Figure 12.12 A face tracked and recovered in frontal and normalized position.

overall output membership function—for instance, to the boredom attribute. The true classification of the input features comes from the DEF block, which synthesizes the latter membership function through a centroid rule [52].

The layout will drastically change in respect to the one experimented within ORESTEIA, since we must introduce the dynamics of the attuning process running between human and ACS, having the goal of the joint action as a third corner of the loop. Then he have three local interventions to perform teh following tasks:

1. Embed symbolic nodes computing the rules. They will represent subroutines of the whole recursive function computed by the system that needs inputs

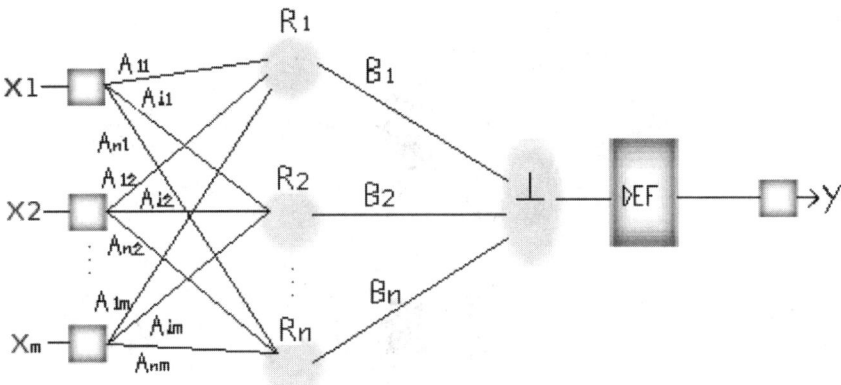

Figure 12.13 The neuro-fuzzy system.

supplied by other nodes of the network and free parameter assessment through the training of the whole system.

2. Dimensioning of the rest of the neural network in order to supply prepositional variables to old and new rules emerging from the network itself, and spread connections between these subroutines in order to convolve them into a unique function.

3. Prepare the Boolean gates into layers of "an" and "or" operators to be suitable aggregates for the learning algorithm to realize the Boolean rules in the object of this project. The whole architecture has the general property of connectionist systems of constituting the wiring translation by the rules it computes.

The general layout has the figure of an extended multilayer perceptron, as in Figure 12.10, where a vertical separator divides left subsymbolic layers, devoted to mapping from signals (rather, their features after elementary preprocessing) to symbols (in terms of propositional variables), from right symbolic layers mapping from symbols to decisions, collected in terms of Boolean variables in output of the whole network. A horizontal separator divides the agnostic part of the structure from the knowledge-based one. The former consists of purely symbolic neurons (implementing sigmoidal functions) on the left of the vertical separator and Boolean gates (hosting "and", "or", and "negation" operators) on the right. The knowledge-based part is a grid of symbolic computer units (CU) each variously connected to both symbolic and nonsymbolic CU in order to get input and forward ouput. Input may be either fixed functions of the signals or of the output of other symbolic CU, or functions nonsymbolically emerging from the training of the network. As said before, the knowledge-based part is plastic, being molded by adaptation rules of its parameters. The final effect of this composition is a symbolic function (actually the goal rule) with numerous free parameters and symbols occurring in a mostly nonsymbolic way from the input signals. (See Figure 12.14.)

Actually, the architecture is rich, since the MLP is embedded into a three-pole layout, as in Figure 12.15, where each pole acts as an autonomous agent—namely

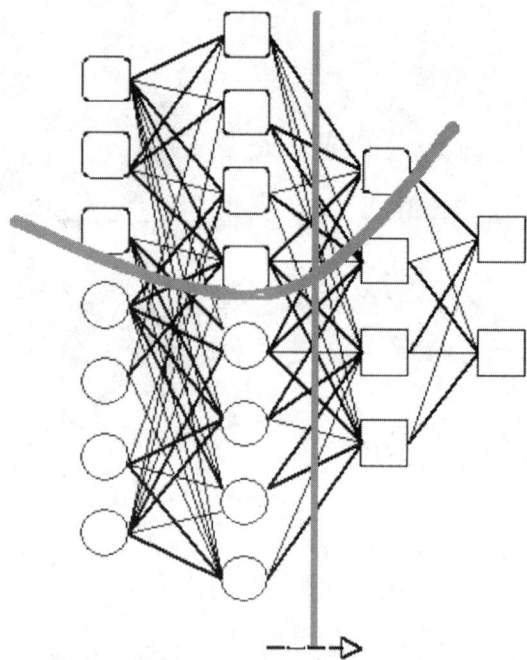

Figure 12.14 The multilayer structure. Boxes: Boolean gates; rounded boxes: symbolic functions; circles: sigmoidal functions. Vertical separator divides subsymbolic from symbolic layers; horizontal separators agnostic from the nonagnostic part.

Figure 12.15 The three-pole dynamic loop.

ACS, human, and goal—that receives solicitations from the others and reacts with a punishment/reinforcement on the actions of the sending agent. The suitability of this architecture is tightly dependent on the efficiency of its training. We logically segment the whole training process into three subtasks. The former concerns the mapping of input signals into Boolean vectors representing assignments to the propositional variables that will be gathered by the goal rules. This will be done by adapting to our dynamic architecture Boolean independent component analysis methods developed for static mappings elsewhere [53].

With this procedure we just force the network to produce Boolean variables, with the feature of having low mutual information for each other. Hence, the meaning of these variables is not given a priori; rather, it will emerge from the statistics of their use in filling the ACS goal. A second task concerns the training of a hybrid neural network equipped with both symbolic and nonsymbolic computing units. Also this task has been developed elsewhere [46] and is in general a spread matter of investigations in the literature. We will focus on two kinds of adjustments of the symbolic nodes: either free parameter identification or membership function adjustment of fuzzy sets involved in the units. Finally, a third subtask concerns the training of the Boolean rules. This is another favorite field of some consortium partners. A complete method has already been developed, called PAC meditation [54], that produces a pair of bordering formulas determining a gap inside the final rule, discovered with granular computing methods, where the gap is considered a fuzzy contour of the goal rule.

12.8 Conclusions

We described some steps of the overall trip for realizing an ambient intelligence capable of attuning with humans within the physical environment where it is situated. We set up an enabling system for eliciting and exploiting particular emotional states and analyzed, both theoretically and experimentally, the main features of the dynamical system realizing this cooperation between human and machine. We pointed out that attuning is a process and that automatically labeling with emotional states the nonverbal signals produced by human is a too ambitious goal. Rather, we look for a cognitive system capable of managing the signals produced by humans within his emotional states, having the goal of efficiently performing a joint action with him. We provided some evidence of this process, but admit that many steps has to be done yet to really build this autonomous cognitive system. Critical issues to be solved are, among the others:

- Assigning the human communicative expressions an intentional meaning (theory of mind) starting from the observation and analysis of the associated signals, recurrent patterns; evaluating a specific expression pattern (e.g., smile) in relation to the task underway, its performance, and the joint-action goal; defining a shell of synthesized knowledge and rules based on these issues.
- Applying attentive-anticipatory models for regulating and planning the relationships between: the subunits of a task and the goals and metagoals of the ACS; and the nonverbal message, the performance of the task, and the selection of the proper action unit.
- Designing the ACS logic in order to: incorporate the knowledge and rules provided in the previous point into the ACS ontology and update it dynamically; and improve its autonomy and efficiency, in terms of the quality of interaction with humans, by exploiting emotional attuning features.

- Implementing an architecture suited to: inferring rules from sensory data (coupled with verbal messages); hosting feedback that closes the loop between ACS, human and goal; memorizing existing rules based on past experience; modifying itself during its operational life; synthesizing new features and semantics based on the statistics of overall system dynamics; and implementing attentive-anticipatory signal processing within the emotional scenarios that emerge.

References

[1] Riva, G., et al., *Presence 2010: The Emergence of Ambient Intelligence*, IOS PRESS, Amsterdam, 2003.

[2] Coen, M. H., "National Conference on Artificial Intelligence—aaai98," *Proceedings of the Fifteenth National Conference on Artificial Intelligence*, 1998.

[3] Gaito, S., B. Apolloni, and D. Malchiodi, "Algorithmic Inference in Machine Learning," *Advanced Knowledge International*, Magill, Adelaide, 2003.

[4] Mattern, F., and M. Naghshineh, "Pervasive Computing," *First International Conference, Pervasive 2002*, Springer, Lecture Notes in Computer Science 2414, Zurich, Switzerland, 2002.

[5] Greene, J.O., *Message Production: Advances in Communication Theory*, Erlbaum, Mahwah N.J, 1997.

[6] Scherer, K. R., "Emotions as Episodes of Ssubsystems Ssynchronization Driven by non-linear Appraisal Processes," *Emotion, Development, and Self-organization: Dynamic Systems Approaches to Emotional Development*, Cambridge University Press, New York, NY, US, 2000.

[7] Barsalou, L.W., "Perceptual Symbol Systems," *Behavioral and Brain Sciences*, 22:577–609, 1999. Emotional Interfaces with Ambient Intelligence 23

[8] Johnson, M., and G. Lakoff, *Philosophy in the Flesh: The Embodied Mind and its Challenge to Western Thought,*. Basic Books, New York, 1999.

[9] Picard, R. W., *Affective Computing,*. MIT Press, Cambridge, 1997.

[10] Picard, R. W., E. Vyzas, and J. Healey, "Toward Machine Emotional Intelligence: Analysis of Affective Physiological State," *IEEE Transactions Pattern Analysis and Machine Intelligence*, Vol. 23, No. 10, 2001.

[11] Lisetti, C. L., "Personality, Affect and Emotion Taxonomy for Socially Intelligent Agents," *Proceedings of the 15th International Florida Artificial Intelligence Research Society Conference (FLAIRS'02)*, 2002.

[12] Lisetti, C. L., et al., "Intelligent Affective Interfaces: A Patient-Modelling Assessment for Tele-Home Health Care," *International Journal of Human-Computer Studies*, 59:245–255, 2003.

[13] Norman, D., *Emotional Design: Why We Love (or Hate) Everyday Things*, Basic Books, New York, 2004.

[14] Bratman, M.E., "What is Intention?," *Intention in Communication*, MIT Press, Cambridge, 1990.

[15] OKeefe, B.J., and B.L. Lambert, "Managing the Flow of Ideas: A Local Management Approach to Message Design," *Communication Yearbook 18.*, Thousand Oaks, CA: Sage, 1995.

[16] Ciceri, R., and L. Anolli, La Voce Delle Emozioni. Verso Una Semiosi Della Comunicazione Vocale Non-Verbale Delle Emozioni. Franco Angeli, Milano, 1995.

[17] Waldron, V.R., "Towards a Theory of Interactive Conversational Planning," in J.O. Greene, 1997.

[18] Searle, J.R., *Mind, Language and Society*, Basic Books, New York, NY, 1998.

[19] Sanders, R.E., "The Production of Symbolic Objects as Components of Larger Wholes," *Message Production: Advances in Communication Theory*, Mahwah, N.J., Erlbaum, 1997.

[20] Greene, J.O., "An Action Assembly Perspective of Verbal and Non Verbal Message Production: A Dancer's Message Unveiled," *The Cognitive Bases of Interpersonal Communication*, Lawrence Erlbaum Associates, Hillsdale, N.J., 1995.

[21] Santi, S., et al., "Oralit et Gestualit," *Communication Multimodale, Interaction*, Paris, 1998.

[22] McCullough, K.E., J. Cassel, and D. McNeill, "Speech-Gesture Mismatches: Evidence for one Underlying Representation of Linguistic and Non-Linguistic Information," *Pragmatics and Cognition*, Vol. 7, pp. 1–33, 1999.

[23] Sonesson, G., "Iconicity in the Ecology of Semiosis," *A Fundamental Problem in Semiotics*, NSU Press, Aarhus, 2000.

[24] Ciceri, R., Comunicare il pensiero. Omega Edizioni, Torino, 2001.

[25] Reeves, B., and C. Nass., *The Media Equation: How People Treat Computers, Television and New Media Like Real People and Places,*. CSLI Publications, Centre for the Study of Language and Information, Cambridge University Press, 1998.

[26] Picard, R.W., and J. Klein, "Computers That Recognise and Respond to User Emotion: Theoretical and Practical Implications," *Interacting with Computers*, Vol. 14, No. 2, pp. 141–169, 2002.

[27] Hayes-Roth, B., et al., "Panel on Affect and Emotion in the User Interface," *Proceedings of the International Conference on Intelligent User Interfaces (IUI'98)*, New-York, pp. 97–124, 1998.

[28] Sproull, L., et al., "When the Interface is a Face," *Human-Computer Interaction*, Vol. 11, pp. 97–124, 1996.

[29] Baldi, CSLU toolKit Website http://cslu.cse.ogi.edu/toolkit/ .

[30] Friesen, W., and P. Ekman, *Facial Action Coding System: A Technique for the Measurement of Facial Movement*, JConsulting Psychologists Press, Palo Alto, 1978.

[31] Friesen, W., P. Ekman, "Facial Expression of Emotion," *American Psychologist*, pp. 384–392, 1993.

[32] Frijda, N. H., *The Emotions*, Cambridge University Press, New York, 1986.

[33] Frijda, N. H., "The Self and Emotion," *Development through Self-Organization. Studies in Emotion and Social Interaction*, Cambridge University Press, NY, pp. 39–63, 2001.

[34] Roseman, I. J., "Appraisal Determinants of Discrete Emotions," *Cognition and Emotion*, Vol. 5, pp. 161–200, 1991.

[35] Lazarus. R. S., *Emotion and Adaptation*, Oxford University Press, New York, 1991.

[36] Fridlund, A., "Sociality of Solitary Smiling: Potentiation by an Implicit Audience 1991," *Emotions in Social Psychology: Essential Readings*, Vol. 14, Psychology Press, NY, 2001.

[37] Reilly, R., B. Kort, and R.W Picard, "An Affective Model of Interplay Between Emotions and Learning: Reengineering Educational Pedagogy-Building a Learning Companion," 2001.

[38] Endler, N. S., and J.D.A. Parker, "Multi Dimensional Assessment of Coping: Concepts, Issues and Measurement," *European Conference on Personality*, Ariccia. 1990.

[39] Pedrabissi, L., and M. Santinello, "Coping Inventory for Stressful Situations: Revision of Validity and Psychometric Properties," Vol. 4, No. 18. Ricerche di Psicologia, 1994.

[40] Frank, M., and P. Ekman, "The Ability to Detect Deceit Generalizes Across Different Types of High-Stake Lies," *Journal of Personality and Social Psychology, Proceedings of Conference on Automatic Face and Gesture Recognition*, pp. 1429–1439, 1997.

[41] Kanade, T., J. F. Cohn, and Y. Tian, "Comprehensive Database for Facial Expression Analysis," *Proceedings of Conference on Automatic Face and Gesture Recognition*, pp. 46–53, Mar. 2000.

[42] Hager, V., P. Ekman, and W. Friesen "Facial ACTION Coding Systems," The Manual, Research Nexus Division of Network Information Research Corporation, Salt Lake City, 2002.

[43] H.Wallbott, "Bodily Expression of Emotion," *European Journal of Social Psychology*, 28:8979–8996, 1998.

[44] Burfield, I., et al., *Theme: Powerful Tool for Detection and Analysis of Hidden Patterns in Behavior*, Reference Manual, v. 5.0., PatternVision, Ltd. and Noldus Information Technology, 2004.

[45] Colombo, P., R. Ciceri, and S Balzarotti, "Analysis of Human Physiological Responses and Multimodal Emotional Signals to an Interactive Computer," *Proceedings of AISB 2005 Annual Conference*, Hatfield, April 5–12, 2005.

[46] Apolloni, B., et al., "A General Framework for Learning Rules from Data," *IEEE Trans. on Neural Network*, 15, 2004.

[47] Apolloni, B., et al., "Detecting Driving Awareness," Lecture Notes in Computer Science, 3202:528–535, 2004.

[48] Pedrycz, W., "Granular Computing in Data Mining," *Data Mining & Computational Intelligence*, Physica-Verlag, *Studies in Fuzziness and Soft Computing*, Vol. 68. Springer-Verlag, 2001.

[49] Apolloni, B., et al., "Learning Rule Representations from Boolean Data," Lecture Notes in Computer Science, 2714:875–882, 2003.

[50] Kanade, T., J. Xiao, and J. F. Cohn. "Robust Full-Motion Recovery of Head by Dynamic Templates and Re-Registration Techniques," *Proceeding of the Fifth IEEE international Conference on Automatic Face and Gesture Recognition*, May,2002.

[51] Ahlberg, J., "An Active Model for Facial Feature Tracking," *EURASIP Journal on Applied Signal Processing*, Vol. 6, No. 6, pp. 566–571, 2002.

[52] Kruse, R., D. Nauck, and F. Klawonn, *Foundation of Neuro-Fuzzy Systems*, Chichester, New York, 1997.

[53] Apolloni, B., A. Brega, and D. Malchiodi, *BICA: A Boolean Independent Component Analysis Algorithm*, HIS'05, Rio de Janeiro, December 2005.

[54] Apolloni, B., F. Baraghini, and G. Palmas, "PAC Meditation on Boolean Formulas," *Abstraction, Reformulation and Approximation*, pp. 274–281, Berlin, 2002. Springer.

CHAPTER 13
A Sense of Context in Ubiquitous Computing

M.J. O'Grady, G.M.P. O'Hare, N. Hristova, R. Tynan

13.1 Introduction

Predicting how computer technology will evolve is a precarious task. Yet the evolutionary process continues. But an evolutionary leap may be only recognized as such in hindsight. At present, a significant transformation in people's concepts concerning the nature of computing and how they might interact with computational technology is ongoing. In short, a world saturated with technology such that people may access computational resources at any time and in any place is envisaged. But quite how this will materialize is speculative. A number of terms have been used to articulate this vision: *ubiquitous computing, pervasive computing* and *ambient intelligence* among others. While differing in particulars, their overall objectives are practically identical—namely, the seamless and transparent integration of computing into the very fabric of everyday life, as well as facilitating intuitive interaction with it. For the purposes of this chapter, the term *ubiquitous computing* will be used and understood as encapsulating these concepts.

One notion inextricably entwined with the ubiquitous computing paradigm is that of context. In effect, this is an attempt to gain a more complete picture of the end user and the prevailing situation at the time of interaction. In this way, services can be adapted according to circumstances, thereby, it is hoped, leading to a more satisfactory end-user experience. While context is an intuitive concept to many people, its meaningful interpretation, at least computationally, remains intriguingly difficult.

In this chapter, various interpretations of ubiquitous computing are initially presented. This is followed by a discussion on context. An overview of some of the pertinent issues that a prospective designer might consider is then presented. Finally, two examples of context-aware applications are discussed in detail.

13.2 Ubiquitous Computing: A Paradigm for the 21st Century

Ubiquitous computing was originally formulated at Xerox PARC in 1988 by the late computer scientist Mark Weiser [1]. Weiser observed that the modern computer interface was derived from the metaphor of the desktop. But why stop there? Why not make the desk itself a computer? This led to a vision of a world where computers would be embedded in everyday objects, could seamlessly communicate with each other and, in this way, provide more valuable services in combination than they could in isolation. Despite its importance, the network infrastructure itself should be transparent to the user. A suitable analogy is that of a signpost. People do not log in to signposts. They read the information at a glance, follow or ignore the sign as they desire, and continue their journey. The goal of ubiquitous computing is simply this: to make computing both as common and as intuitive to use as a signpost.

Even though the ubiquitous computing concept was, and remains to this day, impossible to implement in its entirety, it has had a profound effect on developments in computer science. It was soon realized that for ubiquitous computing systems to work efficiently and transparently, some knowledge of the user was essential. In short, the user's *context* needed to be considered. This has given rise to the development of the context-aware computing paradigm, an idea that has become increasingly important over the last few years. Context will be discussed further in Section 13.2.3. However, it is important to remember that a number of other computer usage paradigms also seek to address some of the issues that gave rise to Weiser's vision. One category of user is of particular interest: the mobile user. The mobile computing user community has grown significantly over the last decade and may well become the dominant computer usage paradigm in coming years. Catering to such users introduces further difficulties and complexities into the software engineering process. To delve further into these issues, two paradigms are now examined—namely, Mobile Computing and Wearable Computing.

13.2.1 Mobile Computing

Mobile computing may be regarded as synonymous with portable computing and, increasingly, handheld computing. The advent of wireless communication networks has been one of the defining events in the history of mobile computing. Though traditionally associated with laptops, current devices include PDAs as well as so-called smartphones that integrate traditional voice telephony with services that would normally be available to desktop users. Almost a decade ago, when researchers were attempting to define mobile computing, four constraints were identified that characterized it [2]. These included the resource differential when contrasted with static workstations, the hazardous nature of mobility, the limited connectively options, and finite energy sources. Though significant progress has been made, these issues still exist to some degree.

However, it is this pivotal issue of mobility that offers intriguing possibilities, as well as significant challenges, both to the research community and commercial companies. In the former case, principles and heuristics that had been developed over decades did not necessarily apply in the mobile case. The traditional computer usage scenarios, whether they involved mainframes or desktops, were inherently static and were well understood. Indeed, the first laptop users, while introducing additional issues, did not really challenge the status quo. Only with the widespread deployment of wireless data networks would this process commence. In the case of commercial companies, the ongoing deregulation of telecommunications networks has offered significant opportunities for service differentiation and attracting new customers. How best to design, implement, and deliver services to an inherently mobile end-user customer base remains an open question. In Section 13.2.4, a brief overview of some of the pertinent issues involved will be presented.

13.2.2 Wearable Computing

Wearable computing was conceived as a means of restoring the technological balance between people and their environment [3]. Mobile computing was perceived as failing users in that they had to stop what they were doing and make an explicitly conscious effort to use the device. In the case of ubiquitous computing as articulated by Weiser, an extensive and sophisticated technological infrastructure is required, and not all environments will ever be suitably equipped. Second, the ubiquitous (no pun intended!) issue of privacy arises due to the extraordinary amount of control the network operator wields. On the other hand, a wearable computing system offers users greater control over what information enters and leaves their personal information space. Augmented reality (AR) is an intuitive mechanism for augmenting the interface, thus leaving the user free to concentrate on the primary task at hand. Research in wearable computing currently embraces a number of diverse areas, including sensors, smart clothing, and so on.

13.3 The Question of Context

> Context: the interrelated conditions in which something exists or occurs.
> —*Merriam-Webster Dictionary*

Context is a curious term. People understand it intuitively yet have difficulty explaining it. Its formal definitions are terse, vague, and imprecise. Since studies in context occur in multiple disciplines, for example, philosophy, artificial intelligence (AI), and linguistics, it enjoys (or suffers?) multiple definitions. Conferences and workshops have focused on context, but an agreed definition has yet to emerge. For its part, the International Standards Organisation (ISO) has also attempted to formalize the notion of context [4]. But the generality and vagueness of current definitions have led some to query the practical use of the concept [5, 6]. Some have suggested that technological issues have driven the debate, and that other issues —for example, sociological, have been ignored. Dourish [7] is one exponent of this view and has suggested that the philosophical roots of context must be revisited before a clear definition can emerge. As part of his investigations, he proposes an alternative view, which he has termed embodied interaction. While acknowledging the importance of this issue, most researchers have, nonetheless, moved ahead by defining their own interpretation of the term and developing systems to explore and test their hypotheses.

13.3.1 Some Definitions of Context

Unhappy with the generality of some definitions of context, a number of researchers in the ubiquitous computing area have proposed their own definitions. One of the best known is that of Schilit et al in the mid 1990s who claim that the essential components of context are: where you are, who you are with and what resources are nearby [8]. A later definition along these lines is that of Schmidt who divides context into two categories: human factors and physical environment [9]. Human factors being further divided into information on users, their social environments, and their current tasks. Likewise, physical environment is subdivided into user location, available infrastructure, and physical conditions. Other researchers have proposed further definitions which follow a similar train of thought though stressing other features [10, 11]. However, a problem that became apparent with these definitions was that they were too specific—a reaction to the generality of the classic definitions of context, perhaps.

Unhappy with the prevailing tendency of defining context through specifics or synonyms, Dey [12] proceeds to define context as follows.

> Context is any information that can be used to characterize the situation of an entity. An entity is a person, place, or object that is considered relevant to the interaction between a user and an application, including the user and applications themselves.

Such a definition is a significant improvement in that it does not dwell on particulars while not being as terse as classic dictionary definitions. A problem with this definition, according to Greenberg, is that the inherent dynamism in the context construct is ignored [5] and what appear to be similar contextual situations at first sight, may actually differ substantially, thus requiring radically different courses of action. The net result is that in all but the most straightforward situations, a software designer may find it difficult to identify all possible contextual states, as well as specifying appropriate actions for particular states. Indeed, a casual browse of the literature demonstrates that practically all context-aware systems implemented to date are deployed in environments that are well defined and understood. Despite these inherent difficulties in defining and interpreting context, its use in numerous aspects of computing remains under active investigation. In the next section, some of the salient aspects of context that might be considered when deploying a generic mobile computing application or service are discussed.

13.4 Reflections on Context in Mobile Computing

Pragmatic use of context is a viable strategy when developing applications and services for mobile users. As an illustration of this, four elements of a mobile user's potential context that can be easily incorporated into a design are now considered.

13.4.1 Spatial Context

In essence, this involves the position and orientation of the user. Date and time might also be considered under this heading. In an outdoor environment, a GPS sensor would usually be adequate for acquiring spatial context. However, GPS does not give an instantaneous reading for the user heading; rather, it gives an average heading calculated from the user's previous positions. An electronic compass must be used for instantaneous readings. The accuracy of the desired position readings needs to be considered. In the case of GPS, this is 20m on average. However, ongoing deployment of satellite-based augmentation systems (SBASs), for example EGNOS and WAAS, may increase the available accuracy to about five meters. The current generation of GPS receivers, including PCMCIA and compact flash (CF) models include this facility.

Surprisingly, the situation indoors is more complicated. Satellite-based position technologies do not operate satisfactorily indoors. The problem is also exacerbated in that the position of the indoor user generally needs to be known with a greater precision. While a number of techniques are described in the literature [13], they are essentially proprietary solutions that require the availability of additional electronic infrastructure. While there is as yet no consensus about the best way forward, one attractive solution could involve the deployment of pseudolites [14], which, after essential calibration for a particular building, could mimic the GPS signal. One might also consider a scenario where a network of RFID tags is used.

Despite the intense debate over the nature of context, it is interesting to note that the availability of a user's spatial context has given rise to another computer usage paradigm—location-aware computing [15]. As well as academia, a number of commercial companies have launched various location-aware services. Some of these companies are closely associated with network operators since the E-911 directive in the United States, and a corresponding initiative—E-112 in the European Union, obligate network operators to provide location assistance in the case of emergency calls [16].

13.4.2 User Profile

A detailed model of a user's profile is invaluable when adapting a service for the user. At its simplest, this might just involve the user's language. But age may be considered in that certain context might not be appropriate for minors. Further aspects that could be gainfully considered might include the user's professional or cultural-interest profile. Depending on the nature of the service, of course, significant opportunities may exist for customizing and filtering content. A classic example is that of an electronic tourist guide in which multimedia presentations are customized according to a user's profile before being made available to the tourist [17].

13.4.3 Device Profile

The nature and characteristics of the user's device needs to be taken into account in the delivery of any service to mobile users. Given the plethora of devices in the market, this is becoming an increasingly urgent problem facing system designers and service providers alike. In the case of a service with a significant multimedia component, the problem has become more acute in that devices can differ radically in their capacity to support video and even basic audio. This issue is well documented in the literature and a number of solutions have been proposed [18, 19]. A primary motivation for this research concerned the desire to extend the reach of the WWW to mobile users.

13.4.4 Environment

An understanding of the user's environment offers significant opportunities for contextualising services. But interpretation of the term *environment* is subjective. However, three aspects may be considered: physical, electronic, and social. Aspects of the physical environment might include, for example, nearby shops, tourist attractions, public transport, and so on. Electronic infrastructure can likewise take many forms but for the most part would include the bandwidth of the local wireless telecommunication network. In the case of multimedia-related services, this is very important. A standard 2.5G network, for example, GPRS, might support 30 KB/s. A 3G network would increase this significantly, but in reality, 400 KB/s might a realistic expectation. In urban areas, one is, of course, more likely to encounter a WLAN hotspot, thus offering a 1 MB/s or more connection. One might also consider the availability of a satellite positioning system, for example, GPS, as being part of the prevailing electronic infrastructure. In the future, no doubt, it can be realistically expected that wireless sensor networks will be deployed and that these may offer further opportunities for gleaning more aspects of a user's context. In the case of social environment, this involves ascertaining what the user's social activities are. For example, if the user is talking to somebody, they may not appreciate receiving a voice call.

To illustrate some of the previous points, two examples of ubiquitous computing applications that harvest and process various aspects of a user's context are described. In the first case, a wireless advertising system, Ad-me, demonstrates how advertising can be selectively targeted at mobile users according to their emotional context. In the second example, ICE, an application more in line with the traditional view of ubiquitous computing, is outlined. This involves the use of wireless sensors embedded in the environment and how one aspect of the environment adapts according to the preferences of its occupants.

13.5 Wireless Advertising

Wireless advertising refers to advertising and marketing activities that deliver advertisements to mobile devices such as mobile phones, PDAs, and so on. A number of surveys conducted by leading companies have already shown that mobile devices offer an excellent medium for personalized advertising. At present, vendors in wireless advertising focus only on user location, and broadcast advertising SMS messages of common interest to users within a specific geographical region. While attempts have been made to focus on what the user is looking for, none of them, however, takes into account the current user situation and whether the user is potentially receptive to an advertisement. Furthermore, while some provide support for explicit user feedback (click-throughs and call-throughs), implicit user emotional feedback is ignored.

Measuring the effectiveness of advertising is one of the most unsatisfactory areas of market research. One dimension of the problem is whether the advertising stimulates low-involvement or high-involvement processing. Some consider that it is more important to measure consumers' emotional responses instead. Their reasoning is that emotional engagement is a prerequisite for behavioral change and that cognitive processing of information is secondary to the underlying emotional and behavioral effects [20]. Thus, measuring the emotional response pre-cognitively has become increasingly important [21].

13.5.1 Emotions in Context

A significant research challenge is the accurate recognition of user emotion. One major limitation of computer technology is that it cannot interpret the user's emotional state in the intuitive manner a human does. The area of ongoing research in computing that relates to, arises from, and influences emotion, is usually referred to as affective computing [22]. Normal approaches for measuring emotion usually consist of one or a combination of: audio, visual and physiological indicators. Three approaches exist for measuring the user affective state [23].

1. Self-reports require the user to fill in questionnaires that can only assess the conscious emotions; though much of the affective experience is, of course, nonconscious.
2. Behavioral methods utilize facial expressions, voice modulation, gestures, posture, and so forth. Facial recognition systems attempt to map pupil dilation, position of the eyebrows, and edges of the mouth into the emotion space. Such features may be extracted from a video stream, for example. Though this method can achieve very high emotion recognition, people consider recording devices such as video cameras obtrusive. Moreover, audio and video approaches can be computationally prohibitive for lightweight devices.
3. Physiological methods attempt to map physiological signals, such as heart rate, respiration and skin conduciveness, to a particular affective state [24]. The method correlates physiological variables together with the affective dimensions of both valence (pleasure-displeasure) and arousal (high-low arousal). Problems particular to the mapping of physiological signals include the necessity for a variety of sensors as well as requiring knowledge about the user's context. The advantages of this method are that the acquisition of data occurs in parallel with the task at hand, and this data is captured even when the user is not explicitly aware of their emotion. Traditionally, physiological signals have been used for giving emotional feedback in health care as an alternative method to drug psychological therapies. This process is frequently termed *biofeedback*. Physiological signals have also been used both for measuring the emotional immersion or presence in virtual reality (VR) environments [25] and for training people's emotional intelligence. Such systems attempt to influence user emotions and to change undesirable emotions into positive ones by showing empathy and encouragement.

13.5.2 Affective Computing Systems

A variety of affective systems that utilize the physiological method already exist for mobile lightweight devices. One notable example is SenSay (Sensing & Saying) [26]. This prototype uses a box, usually attached to a belt, and is fitted with light, motion, and microphone sensors. Another prototype is the affective DJ, developed at the MIT Media Lab, which utilizes the affective wearable system [27] and is hosted on a PalmPilot. This application aids in music selection by incorporating explicitly expressed user preferences, as well as the user's current mood, which is acquired by measuring the user's skin conductance. Distributed wireless system for stress monitoring is a personal health monitor based on a wireless body area network (BAN) of intelligent sensors [28]. This system uses a Polar chest belt as a heart-rate physiological sensor, a wireless heart rate monitor as a personal server, and an iPAQ PDA as the mobile gateway for transmitting the sensor data wirelessly onto a PC. Another project [29] has developed a method for unsupervised and dynamic identification of physiological and activity context in wearable computing using a Tablet PC.

13.5.3 Introducing Ad-me

Advertising for the mobile e-commerce user (Ad-me) [30, 31] is a context-sensitive advertising system for mobile handheld devices. The system provides context-sensitive tourist services, delivering advertisements to the users only when they need them, where they need them and how they need them. The motivation for Ad-me is that taking the user emotional state into consideration at any given time will subsequently increase the effectiveness of advertising. The system thus adapts its behavior to the user's emotion and only delivers advertisements when users are most disposed to receive them. Furthermore, the user emotional response is interpreted as an implicit feedback mechanism for further refinement of future behavior. Physiological methods are used for measuring the user's affective state, though such methods are not yet as widely accepted as self-reports in published advertising studies. Traditionally, the preferred method for measuring consumer affective response to advertisements was to use self-reports or behavioral responses such as click-through rates. Recently, however, the use of skin conductance in monitoring e-commerce activities has been described [32]. In summary, Ad-me matches selective aspects of the user's context—namely, location, time, historical interaction, expressed user preferences, and the properties of advertised objects—to target certain users with selective advertisements.

13.5.3.1 Architecture

Ad-me may be regarded as loosely conforming to a client-server architecture. More specifically, it is a multi-agent system (MAS) encompassing the PDA and a fixed network server (Figure 13.1).

Ad-me utilizes off-the-shelf, wireless, wearable sensor devices which operate in real time and offer support for a number of sensors that act as good indicators of the user valence and arousal. The blood volume pulse (BVP) sensor was considered as an appropriate indicator for valence, while the skin conductance (SC) sensor was considered appropriate for arousal. Most commercial wearable biometric sensor systems regrettably do not offer real-time feedback monitoring, but rather store the data for offline analysis. In such systems the sensor data is usually recorded, encrypted, and subsequently sent to a data center via a wireless telecommunication network. The data is then decrypted and posted in a database for the generation of

13.5 Wireless Advertising

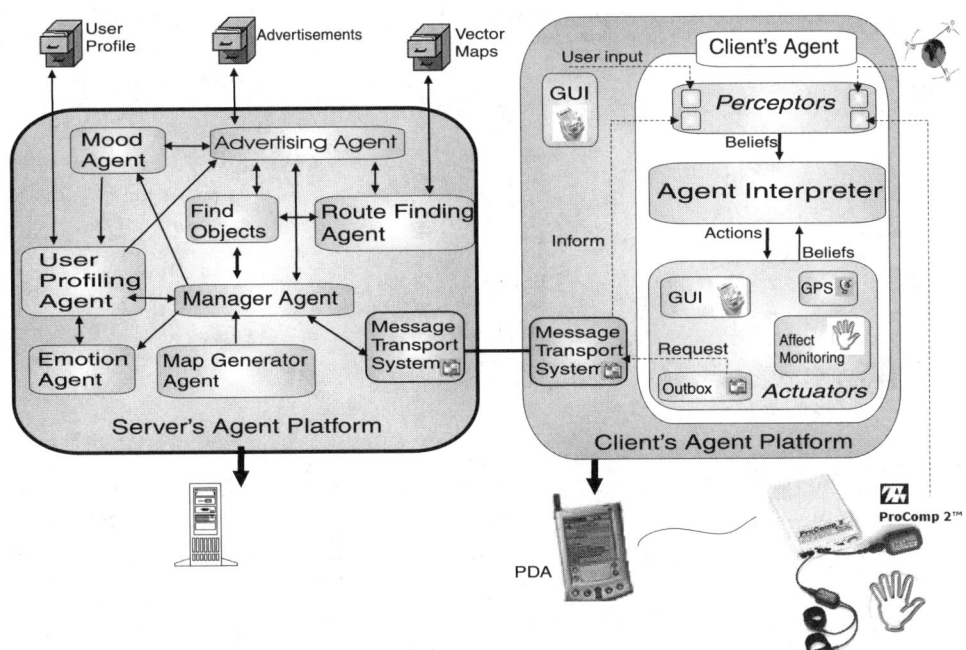

Figure 13.1 Ad-me architecture.

offline summary reports. Two candidate wearable sensors were considered for use within Ad-me. BodyMedia's SenseWear Pro Armband [33] is a wireless sensor that only supports off-line data analysis and does not support BVP measurement. The ProComp2 encoder device, manufactured by Thought Technology Ltd. [34], in contrast, supports input from any two sensors simultaneously. Though not wireless, it supports continuous, real-time monitoring, and is extremely portable. Unfortunately, the software supplied is not PDA comparable; this necessitated the development of a custom interface for the sensor-to-PDA connection. A ProComp+ device exists offering an eight-channel version of the ProComp2. The latter was selected since it is cheaper, smaller, and lighter.

Context within Ad-me comprises a rich set of features, including user profile, user location and user emotion. We will focus here on the latter of these since location and profiling are described extensively elsewhere in the literature. Location within Ad-me uses standard GPS localization techniques.

An emotion agent, residing on the user's device, receives data from both the skin conductance (SC) and the blood volume pulse (BVP) sensors, via the ProComp2 device. The agent then maps the sensor data, using the widely adopted Lang's methodology [35], to infer the user emotion by placing the data into the valence-arousal space. Suppose that the user valence is negative at some time and that, simultaneously, the user arousal is high. Then Ad-me infers that user emotion is angry (and inappropriate for advertisement). Alternatively, if the user's valence is positive (e.g., the user is relaxed, joyful, or excited), the emotion agent informs the server that the user's mood is appropriate for the delivery of an advertisement (see Figure 13.2). An Advertising Agent residing on the server correlates the emotional data together with the other aspects of the user's context to determine when it is appropriate to place

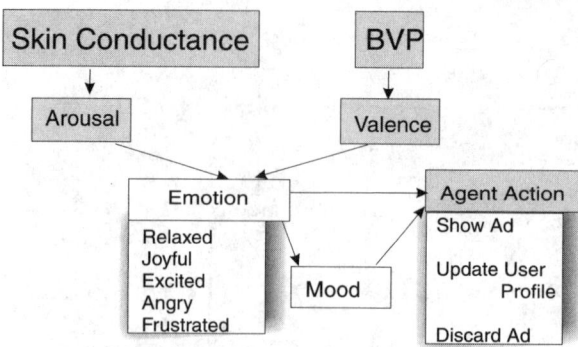

Figure 13.2 Ad-me decision diagram.

an advertisement. On presenting an advertisement, the Emotion Agent stores a one-minute record of the user emotion and sends it to the user profiling agent (on the server) which proceeds to update the user's profile. In the case where the emotion agent inferred that the user has got angry while viewing an advertisement, Ad-me immediately discards the advertisement. A typical scenario is as follows: Ad-me believes that the user's mood is relaxed, lunchtime is approaching and that the user likes cantonese food. An advertisement is then displayed that points out to the user's attention the name and location of a nearby restaurant that serves cantonese food.

13.6 Ambient Sensors: Foundations of a Smart Environment

Recent technological advances in wireless networking, IC fabrication, and sensor technology have led to the emergence of millimeter scale devices that collectively form a wireless sensor network (WSN). The cost of production for a single node has been reduced to less than one U.S. dollar, paving the way for large-scale networks consisting of thousands, if not millions, of nodes. Large-scale deployments of these networks have been used in many diverse fields, including wildlife habitat monitoring [36], traffic monitoring [37], and lighting control [38]. Clearly, WSNs represent a significant step toward realizing the ubiquitous computing vision. Essentially, a WSN is comprised of a set of sensor nodes and a base station to facilitate bidirectional communication between the network and, for example, a desktop computer (Figure 13.3). Due to an internal timer firing in the nodes or in response to a command issued by the base station, the node can sample its various sensory modalities and transmit these back to the base station either directly or through various intermediate nodes (termed multi-hopping).

The small, inexpensive, and unobtrusive nature of these devices is the key factor that allows them to be truly pervasive. Fusing this with their sensory capabilities means that contextual information on a much larger scale can be obtained. The abundance of this information can be somewhat of a burden, however. It can take a considerable time to process, which may lead to decisions being made based on stale data. In the next section, we introduce a context-sensitive system for an intelligent shopping mall. The temperature of the mall is maintained at a level that is tailored for the occupants of the mall. In reasoning about whether to increase the tempera-

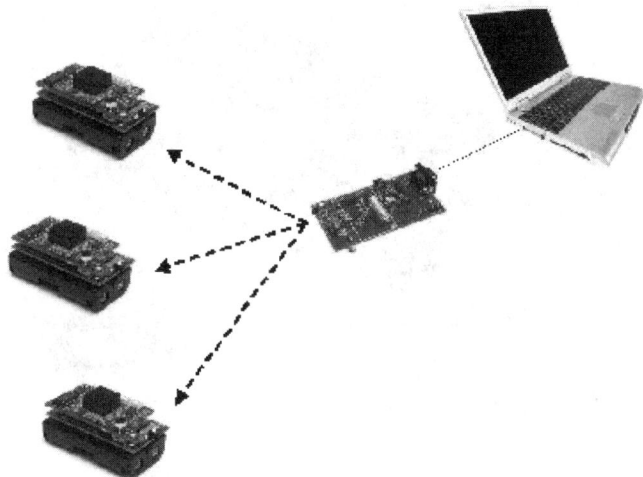

Figure 13.3 Example of a WSN topology. Dashed arrows indicate wireless links; solid line indicates wired connection between base station and base station computer.

ture in a given area, the system utilize the historical context of the users in order to predict how best to act. Users who have previously proved insensitive to the changes in temperature of the mall will have less impact on the decision taken to increase or decrease the internal temperature of the mall.

13.6.1 Introducing the Intelligent Climitization Environment (ICE)

The intelligent climitization environment (ICE) augments an existing environment with networked sensors so as to receive real time telemetry regarding its current state and the state of the users within it. This acquired data can in turn be used to control various actuators to enhance people's experience and comfort within this environment by adapting it to suit their preferences. Initial investigations into these so-called intelligent environments [39] led to the implementation of a personalized climate control system for a shopping mall, that monitored customers' temperature fluctuations using WSNs and adjusted the temperature accordingly to meet shoppers' requirements. Consider the following simple scenario: The average temperatures for the customers in a section of a mall, are above their preferred temperatures, so the air conditioning unit switches on to cool down the entire area. The system uses the historical context of both the users and environment to selectively remove users whose temperature proves unresponsive to changes in the environmental temperature. Such unresponsiveness could be due to illness or excessive clothing. This fact could be relayed to the user via their mobile phone or PDA.

13.6.2 Architecture

ICE requires a set of statically positioned WSN nodes to provide environmental context, such as Mica2 Motes [40] (Figure 13.4), for sampling the air temperature around the mall. This incurs minimal installation cost due to the wireless capability. There are a number of potential candidates for sensing the customers' temperature

Figure 13.4 Mica2 and Mica2Dot.

and location: RFID and the WSN nodes themselves. Using an RFID-based approach, the spatial context of the user can be calculated using multiple scanners. Alternatively, the signal strength of the node on the user at the various statically located nodes can be used to determine location. An alternative to this uses an acoustic sounder for ranging and subsequent triangulation. It operates by a node sounding its sounder and sending a packet to the target node it is trying to locate. The receiving node can then measure the time between receiving the message and hearing the sounder to calculate the distance. Other approaches make use of simple sonar hardware for localization [41, 42]. However, such acoustic methods could prove undesirable for shoppers due to the increased noise levels in the mall. As mentioned previously in Section 13.4, GPS would not prove very useful in this application, since it would be used in an indoor environment. In the case of temperature, each solution can sense the body temperature of the user quite simply through a standard temperature sensor. The user profile can be stored on each user's sensor node as a simple temperature value which the user can set.

Considering the remaining element of a mobile user's context from Section 13.4, the device profile in this application is not necessary. In the previous example, this was used to tailor various advertisements for a given user and device. No such device-specific behavior is required; ICE aligns itself more closely with an ambient computing paradigm where there is no direct interaction with the user, the environment adapts itself to the user based primarily on their presence in a given locality. While a body temperature sensor could quite feasibly be integrated into the undergarments of an individual, it may be unreasonable to assume that the outer garments in contact with the outside air are also augmented in such a fashion. To solve this, an interpolation mechanism to interpolate the air temperature at the user's current location using the statically located nodes that surround the user was designed. Interpolation is a mathematical technique for estimating the value of a function between known values of that function. Many interpolation models exist for different domains [43, 44]. These can be of varying complexity and accuracy. In this case, the Weighted Average Algorithm was used:

$$T(x,y) = \frac{\sum_{i=1}^{N} d_i T_i}{\sum_{i=1}^{N} d_i} \qquad (13.1)$$

where $T(x, y)$ is the interpolated temperature at the point (x, y), d_i is the distance of sensor i to the point (x, y), T_i is the sensor's corresponding temperature value, and N is the number of sensors used for the interpolation. This equation implies that a sensor that is far away from the point of interpolation will have less impact on the interpolated temperature.

So by placing customers' temperature fluctuations in the context of the interpolated temperature at their locations, we are better positioned to decide whether or not to increase/decrease the temperature of the area. A customer who has a temperature that is high for a prolonged period of time, while the temperature surrounding them is coo,l may be suffering an illness and so decreasing the temperature using their data would be unwise. In fact, such customers could be ruled out of the calculations altogether. These factors can also be incorporated into the decision-making process when one or more users are above their desired temperature and others are below. Highly sensitive customers can be weighted higher, meaning that the system would attempt to accommodate their needs at the expense of the less sensitive ones.

To test our system, a scale model of a shopping mall in Park City, PA was developed (Figure 13.5). This mall was in the shape of a cross, so an individual wing of the mall was initially focused on. To simulate users, WSN nodes were encased in an insulating material with varying properties to simulate actual users' responses to temperature. To cool down the model, small tubes were fitted to the sides of the model and a PC cooling fan was used to pump cold air into the mall. To heat up the mall model, the same fan system was used, only this time a small heater was activated in front of the air intake for the fan. The WSN node used was the Mica2 Mote interfaced through a base station to a desktop computer where the decision-making process was undertaken. At present, the system is incapable of catering to the migration of customers around the mall; however, the sensors could have been endowed

Figure 13.5 Arial photo of Park City Mall.

with some mobility—for example, RoboMote [45]. Instead, the nodes representing customers were manually moved when investigating how the system responds and reconfigures the temperature.

13.7 Conclusions

Context is a rich and sometimes nebulous term. It is very sensitive to a given situation and richness of sensory modalities. However, the richer the sense of context the greater the relevance and pertinence of the content or service delivered to the mobile user. It is, therefore, a key enabler in the delivery of user acceptance of ubiquitous systems.

Though there remain difficulties with the formal nature of context, researchers are continuing to tease out the issues involved. It is a valid observation that most of the applications described in the research literature are targeted at very specific niches. Two such scenarios have been described—in this chapter, namely, Ad-me and ICE—each representing divergent user contexts. Context can be observed to be a far from trivial concept, albeit somewhat curious. In the long term, an agreed definition of context is essential if context-aware computing is to become a paradigm in its own right. Indeed, it may well be that the widespread deployment of wireless sensor networks is an essential precursor to this. Such networks will increase the quantity and quality of contextual information available to system designers, providing a rich context landscape within which more personalization and contextualization can be provided in a cost-effective manner. In the meantime, context will continue to be harvested and integrated into applications in various application domains. In this way, the ubiquitous computing vision will become an incrementally more important component of everyday life.

References

[1] Weiser, M., "The Computer of the 21st Century," *Scientific American*, Vol. 265, No. 3, pp. 94–100, 1991.

[2] Satyanarayanan, M., "Pervasive Computing: Vision and Challenges," *IEEE Personal Communications*, Vol. 8, No. 4, pp. 10–17, 2001.

[3] Mann, S., "Wearable Computing as Means of Personal Empowerment," In *Proc. 1st International Conference on Wearable Computing (ICWC'98)*, Fairdax, Va, USA, 1998.

[4] ISO 9241. International Standards Organization (ISO), http:// www.iso.org.

[5] Greenberg, S., "Context as a Dynamic Construct," *Human-Computer Interaction*, Vol. 16, pp. 257–268, 2001.

[6] Erickson, T., "Some Problems with the Notion of Context-Aware Computing," *Communications of the ACM*, Vol. 45, pp.102–104, 2002.

[7] Dourish, P., "What we talk about when we talk about context," *Personal & Ubiquitous Computing*, Vol. 8, pp. 19–30, 2004.

[8] Schilit, B., N. Adams, and R. Want, "Context-Aware Computing Applications," *Proc. of the Workshop on Mobile Computing Systems and Applications*, Santa Cruz, CA, USA. 1994.

[9] Schmidt, A., M. Beigl, and H-W. Gellersen, "There is More to Context than Location," *Computers and Graphics*, Vol. 23, No. 6, pp. 893–901, 1999.

[10] Pascoe, J., "Adding Generic Contextual Capabilities to Wearable Computers," *Proceedings of 2nd International Symposium on Wearable Computers*, Pittsburgh, Pennsylvania, USA, pp. 92–99, 1998.

[11] Chen, G. and D. Kotz, "A Survey of Context-Aware Mobile Computing Research," *Dartmouth Computer Science Technical Report TR2000-381*, 2000.

[12] Dey, A. K., "Understanding and Using Context," *Personal and Ubiquitous Computing*, Vol. 5, No. 1, pp. 4–7, 2001.

[13] Hightower, J., and G. Borriello, "Location Systems for Ubiquitous Computing," *IEEE Computer*, Vol. 34, No. 8, pp. 57–66, 2001.

[14] Kee, C., et al., "Centimeter-Accuracy Indoor Navigation Using GPS-Like Pseudolites," *GPS World*, November, 2001.

[15] Patterson, C. A., R. R. Muntz and C. M. Pancake, "Challenges in Location-Aware Computing," *IEEE Pervasive Computing*, Vol. 2, No. 2, April–June 2003.

[16] Zhao, Y., "Standardization of Mobile Phone Positioning for 3G Systems," *IEEE Communications*, Vol. 40, No. 7, 2002.

[17] O'Hare, G. M. P., and M.J. O'Grady, "Gulliver's Genie: A Multi-Agent System for Ubiquitous and Intelligent Content Delivery," *Computer Communications*, Vol. 26, No. 11, pp. 1177–1187, 2003.

[18] Lemlouma, T., and N. Layaida, "Context-aware Adaptation for Mobile Devices," *Proc. of the International Conference on Mobile Data Management*, Berkeley, California, USA, 2004.

[19] Smith, J. R., R. Mohan and C-S. Li, "Scalable Multimedia Delivery for Pervasive Computing," *ACM Multimedia '99*, Vol. 10, pp. 131–140, Orlando, FL, USA, 1999.

[20] Hall, B. F., "Is Cognitive Processing the Right Dimension?," *Admap*, No. 435, pp. 39–41, 2003.

[21] Xie, F. T., et al., "Emotional Appeal and Incentive Offering in Banner Advertisements," *Journal of Interactive Advertising*, Vol. 4, 2004.

[22] Picard, R. W., "Affective Computing," *MIT Press*, 1997.

[23] Zimmermann, P., et al., "Affective Computing—A Rationale for Measuring Mood with Mouse and Keyboard," *Intl. Journal of Occupational Safety and Ergonomics*, No. 9, pp. 539–551, 2003.

[24] Prendinger, H., et al., "Character-Based Interfaces Adapting to Users' Autonomic Nervous System Activity." *Proc. Joint Agent Workshop (JAWS-03)*, Awaji, Japan, pp. 375–380, 2003.

[25] Nasoz, F., et al., "Emotion Recognition from Physiological Signals for Presence Technologies," *International Journal of Cognition, Technology and Work—Special Issue on Presence*, Vol. 6, No. 1, 2003.

[26] Siewiorek, D., et al., "Sensay: A Context-Aware Mobile Phone." *Proc. of the 7th IEEE International. Symposium on Wearable Computers*, 2003.

[27] Picard, R. W., and J. Healey, "Affective Wearables," *Personal Technologies*, Vol 1, pp. 231–240, 1997.

[28] Jovanov, E., et al., "Stress Monitoring Using a Distributed Wireless Intelligent Sensor System," *IEEE Engineering in Medicine and Biology Magazine*, Vol. 22, No. 3, pp. 49–55, 2003.

[29] Krause, A., et al., "Unsupervised, Dynamic Identification of Physiological and Activity Context in Wearable Computing." *Proc. 7th IEEE International. Symposium on Wearable Computers*, 2003.

[30] Hristova, N., and G.M.P. O'Hare, "Ad-me: Wireless Advertising Adapted to the User Location, Device and Emotions," *Proc. 37th Hawaii International Conference. on System Sciences*, 2004.

[31] Hristova, N., and G.M.P. O'Hare, "Ad-me: A Context Sensitive Advertising System," *Proc. of 3rd International Conference on Information Integration and Web-based Applications & Services*, 2001.

[32] Goulev, P., and E. Mamdani, "Investigation of Human Affect During Electronic Commerce Activities," *In: IEE Eurowearable'03*, 2003.

[33] BodyMedia, Inc. www.bodymedia.com.

[34] Thought Technology Ltd. http://www.thoughttechnology.com.

[35] Lang, P. J., "The Emotion Probe: Studies of Motivation and Attention," *American Psychologist*, Vol. 50, No. 5, pp. 372–385, 1995.

[36] Mainwaring, A., et al., Wireless Sensor Networks for Habitat Monitoring, *International Workshop on Wireless Sensor Networks and Applications*, 2002.

[37] Coleri, S., S. Y. Cheung, and P. Varaiya, "Sensor Networks for Monitoring Traffic," *Proc. 42nd Allerton Conference on Communication, Control and Computing*, Illinois, 2004.

[38] Sandhu, J., and A. Agogino, "Wireless Sensor Networks for Commercial Lighting Control: Decision Making with Multi-agent Systems," *AAAI Workshop on Sensor Networks*, 2004.

[39] Droege, P., "Intelligent Environments," Elsevier Science Publishing Co. 1997

[40] Hill. J., "System Architecture for Wireless Sensor Networks," PhD thesis, UC Berkeley, 2003.

[41] Whitehouse. K., "The Design of Calamari: An Ad-Hoc Localization Systems for Sensor Networks," Masters Thesis, University of California at Berkeley, 2002.

[42] Priyantha, N. B., et al., "The Cricket Compass for Context-Aware Mobile Applications," *Proceedings of 7th ACM MOBICOM*, Rome, 2001.

[43] Shepard, D., "A Two-Dimensional Interpolation Function for Irregularly-Spaced Data," *Proc. 23rd ACM National Conference*, 1968.

[44] Meijering, E., "A Chronology of Interpolation: From Ancient Astronomy to Modern Signal and Image Processing," *Proceedings of the IEEE*, Vol. 90, No. 3, pp. 319–342, March 2002.

[45] Sibley, G. T., et al., "Robomote: A Tiny Mobile Robot Platform for Large-Scale Sensor Networks," *Proceedings of the IEEE International Conference on Robotics and Automation (ICRA2002)*, 2002.

CHAPTER 14
Ad Hoc On-Demand Fuzzy Routing for Wireless Mobile Ad Hoc Networks

Shivanajay Marwaha, Dipti Srinivasan, Chen Khong Tham, Athanasios Vasilakos

14.1 Introduction

Intelligent environments are opening new scenarios where people interact with electronic devices embedded in environments that are sensitive and responsive to the presence of people. The term ambient intelligence (or, AmI) reflects this tendency, gathering best results from three key technologies: ubiquitous computing, ubiquitous communication, and intelligent user friendly interfaces. A mobile ad hoc network (MANET) [1] is one such technology enabling ubiquitous communication consisting of many wireless mobile devices (probably in an AmI environment), that are able to communicate with each other over wireless links without the aid of any infrastructure or centralized administration. The emergence of MANET has created opportunities for new forms of mobile collaboration involving interaction between people who are colocated and organized in unforeseeable ways such as conferences, seminars, campus and office networks, battlefields, law enforcement, and postdisaster relief scenarios. MANETs have also started proliferating with multimedia and real-time services, making versatile routing protocols very critical. Networked multimedia applications such as voice and video are becoming increasingly popular and are making a good case for the deployment of quality-of-service-based network protocols that require multiple routing constraints [2]. Simple routing algorithms that determine routes based on a single metric like minimum number of hops, remaining battery power or link stability between intermediate nodes are no longer sufficient [3]. One of the key issues in MANETs is how to determine appropriate paths that fulfill the multiple objectives of the transported traffic. Multiple performance requirements, load balancing and scalability have thus become increasingly significant factors in routing for MANET. However, most MANET routing protocols such as the ones proposed in [4–10] for MANETs, are not multiobjective and multimetric, thus compromising network performance on the overlooked objectives and suffering from incomplete and inaccurate information of the route. Some of the important MANET routing objectives are as follows:

1. Reducing end-to-end delay by selecting the shortest route (minimum number of intermediate hops) or shortest path with least congestion. Some

on-demand routing protocols [4] select the first route discovery that reaches the destination as the route discovery that reaches the destination first must have come through the path with the least number of intermediate hops and was least congested compared with other routes.
2. Increasing battery lifetime by avoiding routes in which the intermediate nodes have very little remaining battery power and by balancing the packet-forwarding responsibility among all the nodes in the MANET [7, 9].
3. Increasing packet delivery by selecting routes that are long lived and stable. Protocols such as associativity based routing (ABR) [8] and signal strength analysis (SSA) [10] use the link stability between intermediate hops as a route selection metric to select routes in MANETs in order to reduce packet loss due to frequent route failures.

While single objective routing in MANETs is relatively simple, routing problems subject to multiple constraints are computationally difficult to solve and can only be dealt with using heuristics and approximate solutions [2]. Multiobjective routing requires the exact state of the network to be known to the nodes performing the route computation. In practice, however, network state is not known for certain due to many reasons as described in this chapter. This chapter thus utilizes fuzzy logic [11] for solving the multiobjective routing problem in MANETs due to its remarkably simple way to draw definite conclusions from imprecise information.

14.2 Problem Statement

14.2.1 Limitations of Single Metric Single Objective Routing Schemes

Most current routing protocols in MANETs try to achieve a single routing objective, such as reducing end-to-end delay, increasing packet delivery fraction, or increasing the lifetime of battery-powered mobile nodes using a single route selection metric such as the number of intermediate hops or remaining battery power of the intermediate nodes. Since the various routing objectives are closely interlinked, a gain in one can only be achieved at the expense of the others. Hence, a routing protocol that tries to achieve only one routing objective, can severely compromise network performance on the remaining overlooked objectives. Therefore, efficient routing in MANETs requires selecting routes that meet multiple objectives.

A routing objective cannot be fully achieved if a single metric, such as the number of hops or remaining battery power of intermediate nodes, is used for route selection, since it cannot provide complete link state information of a route. Therefore, it is important for the route selection algorithm to utilize multiple routing metrics to describe the link state information more accurately. For example, packet delivery fraction depends on selection of routes that are stable, least congested, and in which the intermediate nodes have high remaining battery power. It would also depend on the number of hops in the route, because as the hop count increases so would the possibility of route failures due to the increased number of intermediate hops. Similarly, end-to-end delay will depend on the number of intermediate hops and traffic congestion. It would also indirectly depend on the stability of routes and battery power of the nodes in the route, which are the two most probable reasons for route failures causing rerouting and could increase the time spent by packets in the send buffer of the source node as they wait for new routes to be discovered. The battery lifetime of nodes depends on the current values of remaining battery power at the time of route selection and the amount of data traffic waiting to be transmitted. Last, overhead, another important objective, depends not only on the route length but also on the stability of the routes, which, if they fail, would cause an increased number of rerouting operations, thus resulting in increased overhead.

Consider the network shown in Figure 14.1. Using the shortest path metric will always lead to the selection of route 1-2-3 even if node 2 is congested or its remaining battery power is low or even if the links 1-2 or 2-3 are unstable. MBCR will select route 1-2-3 if the remaining battery power of node 2 is not much less compared with the remaining battery power of nodes 4 and 5, but it is blind to the congestion and link stability situation in routes 1-2-3 and 1-4-5-3. AOFR, on the other hand, will make the routing decision based on all three parameters, link stability, traffic congestion, and remaining battery power to select the best routing path.

14.2.2 Complexity in Multiobjective Routing in MANETs

Multiple routing objectives can be met together only if multiple routing metrics that give detailed information on the state of the links are considered, such as traffic congestion, route stability, remaining battery power at intermediate nodes, and the number of intermediate hops. However, rules used to describe the best-cost path for one objective may not be the same for the other objectives. For example, delay uses an additive rule while throughput uses a convex or bottleneck rule.

The multiobjective route selection problem in MANET, which relies on multiple parameters, is similar to the multiple-input multiple-output (MIMO) system shown in Figure 14.2. The inputs to the system, i.e. the metrics, are the parameters that describe the state of a route, such as the remaining battery power (B) of the intermediate nodes, the traffic congestion at intermediate nodes (Q), the number of hops in a route (N) and the received signal strength (SS) which is a measure of signal quality, while the outputs are the various routing objectives such as packet delivery ratio (PDR), end-to-end delay (EED), battery lifetime (B.Life) and control overhead (OH).

Developing an accurate input-output relationship between the objectives and the route selection metrics is quite difficult. Conventional model-based approaches often fail to truly capture the intricacies and interplay between the various inputs

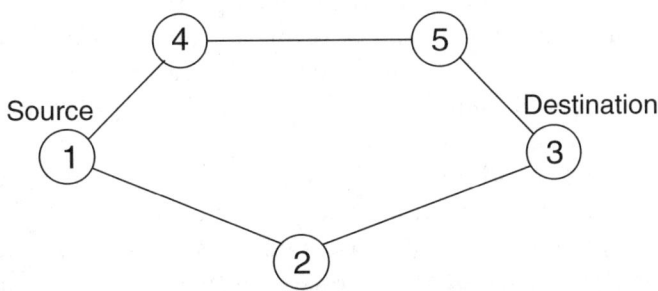

Figure 14.1 Example of a five-node network.

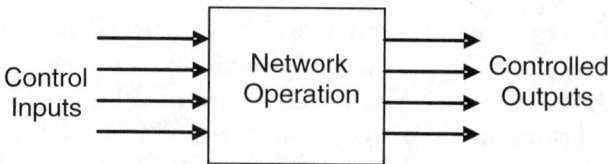

Figure 14.2 A multiple-input, multiple-output (MIMO) system.

and outputs of such complex systems. It may not be feasible to execute a very complex mathematical model, even if it is developed on mobile devices with limited computational and battery power. Such a complicated model would slow down the route discovery process, increase computational complexity, and require more battery power, which could limit the scalability of such a technique. It would also involve making optimistic assumptions about the mobility and traffic patterns, thus making it difficult to adapt in highly dynamic networks such as MANETs.

14.2.3 Applicability of Fuzzy Logic for Multiobjective Routing in MANETs

Fuzzy systems are suitable for approximate reasoning in systems where it is difficult, if not impossible, to derive an accurate mathematical model. Many design and control problems in communication systems are well suited for analysis using fuzzy logic, due to their characteristics of having multiple performance criteria, some of which are often conflicting, where the gain in one performance metric can only be achieved at the expense of the other. This has led to the application of Fuzzy logic in many complex problems in communications networks such as distributed medium access control, queuing, policing and routing [12–14].

Heuristics and approximation algorithms are often used for Multi-constrained routing, which is a computationally difficult problem [2]. The problem of multi-objective routing becomes even more complicated since exact network state information is not always available for the node that is performing the path computation function. There are also many uncertainties under which MANET routing decisions are made including the following:

1. Unknown node mobility leading to unpredictable link changes.
2. Latency caused due to the delay in route requests and routine table updates reaching the destination or source nodes, respectively, thus making the routing information to the destination for making routing decisions obsolete.
3. State aggregation. In most routing protocols the state of a group of nodes is summarized (aggregated) before reaching the destination node both in table-driven as well as in on-demand routing protocols. Hence, the exact state information of the network is not available to the node performing the path computation function.
4. In on-demand routing [4] there is also incomplete routing information available at the destination node, since the multiple copies of the same route discovery received at the intermediate node are discarded after the first route discovery has been forwarded by it (which had the same source ID and route request sequence number), due to the possibility that the subsequently arriv-

ing route discoveries may cause routing loops and also to reduce the routing overhead. However, some of these route discoveries arriving late at an intermediate node may have reached that node via a different route, which could better achieve the various routing objectives together compared to the first route discovery, which came via the shortest path that was least congested without considering the remaining battery power of the intermediate nodes.

The difficulty of characterizing a multiobjective route selection problem with many uncertainties for highly dynamic multihop ad hoc networks led us to explore alternative solutions based on artificial intelligence techniques, specifically, in the field of fuzzy systems due to their simplicity in implementation and proven performance [11].

The complex interactions among the various routing objectives and metrics in MANETs described earlier can be modeled using fuzzy rules. Fuzzy logic permits easy modeling of the complex relationship between the objective function and the metrics. Last, some objectives such as route stability and packet delivery ratio are not exactly crisp; hence, they can be more efficiently modeled using fuzzy logic (see Figure 14.3.)

14.3 Brief Background of Fuzzy Logic

Fuzzy sets are a generalization of conventional set theory that was introduced as a mathematical way to represent vagueness of parameters [11]. The basic idea in fuzzy logic is that statements are not just true or false, but partial truth is also accepted. In the same way, in fuzzy set theory, partial membership of a set, called a fuzzy set, is possible. Fuzzy sets are critical to the granulation of information and are characterized by fuzzy membership functions. Each fuzzy set is an example of an information granule with well-defined semantics. The size of the information granule is clearly linked with the character of the membership function. For instance, by changing (increasing or decreasing) the spread of a Gaussian membership function, we can model a size of the granule and make it more general (abstract) or specific

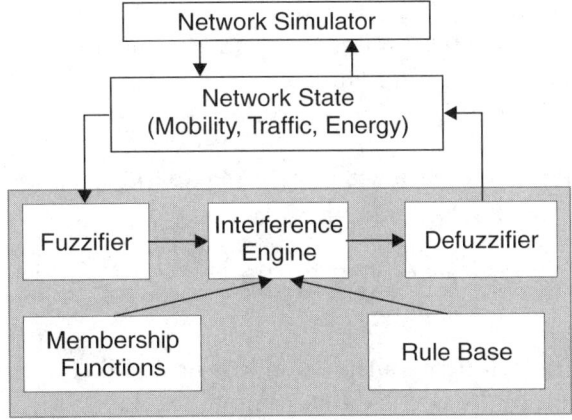

Figure 14.3 Fuzzy logic system used in MANETs.

(detailed). The granularity of information helps to develop a model that is optimized with regard to its generality, development effort, and computational capabilities.

The concept of fuzzy sets makes it possible to use fuzzy inference, in which the knowledge of an expert, who is skilled in a field of application and can make informed decisions, is expressed as a set of IF–THEN rules, leading to algorithms describing what action should be taken based on currently observed information.

Fuzzy logic operation is typically divided into the following steps.

1. Fuzzification is a procedure for membership mappings from the measured values of each input to a set of linguistic values for that input.
2. Inference is a procedure that produces a new fuzzy set from the result of the fuzzification, i.e., the linguistic values using a set of rules.
3. Defuzzification is the procedure that produces a crisp output from the result of the inference.

14.4 Cost Function for MANET Multiobjective Routing

To select the routes in MANETs a figure of merit or a cost function is required to select the route that best meets all the objectives. A route is selected that has the least-cost metric. This section provides the relationship between the combined cost metric and the various routing metrics to simultaneously meet the objectives of minimizing the delay, maximizing packet delivery and (iii) increasing the lifetime of battery powered nodes.

14.4.1 Objective 1 (O_1): Minimizing End-to-End Delay

$$T_{total} = \min\left\{\sum_{i=1}^{N+1}(T_{Q_i} + T_{W_i} + T_{C_i} + T_{P_i})\right\} \quad (14.1)$$

where T_{Q_i} is the queuing delay at node i, T_{W_i} is the wireless transmission time from $i-1$ to the ith node, T_{C_i} is the waiting time to access the wireless channel at node i, T_{P_i} is the processing time at node i, N is the number of intermediate nodes in a route, and T_{total} is the total end-to-end delay.

14.4.2 Objective 2 (O_2): Maximizing Probability of Successful Packet Delivery

$$P_{packetdelivery} = \max\left\{\prod_{i=1}^{N+1}(1 - P_{W_i} - P_{Q_i})\right\} \quad (14.2)$$

where P_{W_i} is the wireless transmission loss probability at link before node i, P_{Q_i} is the queue congestion loss probability at node i and $P_{packetdelivery}$ is the probability of successful packet delivery at the destination.

14.4.3 Objective 3 (O_3): Minimizing Total Battery Cost of the Route

$$E_{BatteryCost} = \min\left\{\sum_{i=1}^{N+1}\left(\frac{1}{B_i^t}\right)\right\} \quad (14.3)$$

where B_i^t is the remaining battery capacity of a node n_i at time t and is the remaining battery cost of using a route.

For an n objective decision-making problem, the decision is given by:

$$D = O_1 \wedge O_2 \wedge \cdots \wedge O_n \quad (14.4)$$

which satisfies all the objectives. Since there are three different objectives, the decision variable which is the cost C to be minimized, can be expressed as:

$$C \propto \frac{O_1 \times O_3}{O_2} \quad (14.5)$$

O_2 has an inverse relationship with cost, since cost has to be minimized, and O_2 is maximized (higher O_2 should result is less cost). Combining (14.1), (14.2), (14.3) and (14.5) we get:

$$C \propto \frac{\left\{\sum_{i=1}^{N+1}(T_{Q_i} + T_{W_i} + T_{C_i} + T_{P_i})\right\} \times \left\{\sum_{i=1}^{N+1}\left(\frac{1}{B_i^t}\right)\right\}}{\left\{\prod_{i=1}^{N+1}(1 - P_{W_i} - P_{Q_i})\right\}} \quad (14.6)$$

or,

$$C = K_1 \frac{\left\{\sum_{i=1}^{N+1}(T_{Q_i} + T_{W_i} + T_{C_i} + T_{P_i})\right\} \times \left\{\sum_{i=1}^{N+1}\left(\frac{1}{B_i^t}\right)\right\}}{\left\{\prod_{i=1}^{N+1}(1 - P_{W_i} - P_{Q_i})\right\}} \quad (14.7)$$

where K_1 is the constant of proportionality. Assuming that the wireless channel access time at a host is solely dependent on data traffic conditions in that particular wireless range, and packet processing time is constant at a node (assuming packets being of equal size), we have:

$$C = K_1 \frac{\left\{\sum_{i=1}^{N+1}(T_{Q_i} + T_{W_i} + K_2)\right\} \times \left\{\sum_{i=1}^{N+1}\left(\frac{1}{B_i^t}\right)\right\}}{\left\{\prod_{i=1}^{N+1}(1 - P_{W_i} - P_{Q_i})\right\}} \quad (14.8)$$

where K_2 is a constant. Since queuing delay is directly proportional to the length of the queue at node i at time t, we have:

$$T_{Q_i} \propto Q_i^t \tag{14.9}$$

or,
$$T_{Q_i} = K_P Q_i^t \tag{14.10}$$

where K_p is the average processing time per packet at the host node for the queued packets. From equations (14.8) and (14.10) we get:

$$C = K_1 \frac{\left\{\sum_{i=1}^{N+1}\left(K_P Q_i^t + T_{W_i} + K_2\right)\right\} \times \left\{\sum_{i=1}^{N+1}\left(\frac{1}{B_i^t}\right)\right\}}{\left\{\prod_{i=1}^{N+1}\left(1 - P_{W_i} - P_{Q_i}\right)\right\}} \tag{14.11}$$

Distance between two wireless hosts can be calculated from the free space wireless propagation model [15], which is:

$$P_r(d) = \frac{P_t G_t G_r \lambda^2}{(4\pi)^2 d^2 L} \tag{14.12}$$

where P_t is the transmitted power, which can be assumed to be constant, or, if it is a variable value, it can be specified in the packet header for calculating the transmitter–receiver (T-R) distance; $P_r(d)$ is the received power, which is a function of T-R separation, G_t is the transmitter antenna gain; G_r is the receiver antenna gain, d is the T-R separation distance in meters, L is a system loss factor not related to propagation ($L \geq 1$); and λ is the wavelength in meters. Considering Pt, Gt, Gr, L, λ to be known parameters, we have:

$$P_r(d) \propto \frac{1}{d^2} \quad \text{or} \quad SS \propto \frac{1}{d^2} \tag{14.13}$$

If v is the speed of radio propagation, then wireless transmission time can be expressed as:

$$T_{W_i} = \frac{d}{v} \quad \text{or} \quad T_{W_i} \propto d \tag{14.14}$$

Combining (14.13) and (14.14) we get:

$$T_{W_i} \propto \frac{1}{\sqrt{SS}} \quad \text{or} \quad T_{W_i} = \frac{K_w}{\sqrt{SS}} \tag{14.15}$$

where SS denotes received signal strength and is the proportionality constant. Combining (14.11) and (14.15) we get:

$$C = K_1 \frac{\left\{\sum_{i=1}^{N+1}\left(K_P Q_i^t + \frac{K_w}{\sqrt{(SS)}} + K_2\right)\right\} \times \left\{\sum_{i=1}^{N+1}\left(\frac{1}{B_i^t}\right)\right\}}{\left\{\prod_{i=1}^{N+1}\left(1 - P_{W_i} - P_{Q_i}\right)\right\}} \tag{14.16}$$

14.4 Cost Function for MANET Multiobjective Routing

Wireless transmission loss depends on signal to noise ratio (SNR). If the received signal strength is low, the probability of wireless loss increases. Hence, probability of wireless loss can be expressed as (for a given SNR value):

$$P_{W_i} \propto \frac{1}{SS} \quad or \quad P_{W_i} = \frac{K_{wloss}}{SS} \qquad (14.17)$$

where, K_{wloss} is the proportionality constant in equation (14.17). Combining (14.16) and (14.17) we get:

$$C = K_1 \frac{\left\{\sum_{i=1}^{N+1}\left(K_P Q_i^t + \frac{K_w}{\sqrt{(SS)}} + K_2\right)\right\} \times \left\{\sum_{i=1}^{N+1}\left(\frac{1}{B_i^t}\right)\right\}}{\left\{\prod_{i=1}^{N+1}\left(1 - \frac{K_{wloss}}{SS} - P_{Q_i}\right)\right\}} \qquad (14.18)$$

Congestion loss depends on the buffer occupancy at time t. If the buffer is full at time t, the incoming packets would be dropped (assuming FIFO buffer management scheme). Therefore, the queue congestion loss probability, which is proportional to buffer occupancy at time t and at node i, is given by:

$$P_{Q_i} \propto Q_i^t \quad or \quad P_{Q_i} = K_Q Q_i^t \qquad (14.19)$$

where K_Q is the proportionality constant in equation (14.19). Combining equations (14.18) and (14.19):

$$C = K_1 \frac{\left\{\sum_{i=1}^{N+1}\left(K_P Q_i^t + \frac{K_w}{\sqrt{(SS)}} + K_2\right)\right\} \times \left\{\sum_{i=1}^{N+1}\left(\frac{1}{B_i^t}\right)\right\}}{\left\{\prod_{i=1}^{N+1}\left(1 - \frac{K_{wloss}}{SS} - K_Q Q_i^t\right)\right\}} \qquad (14.20)$$

From (14.20), it has been established that the combined cost function for simultaneously achieving the three objectives is a complex function f of N, SS, Q, and B:

$$C = f(N, SS, Q_i^t, B_i^t) \qquad (14.21)$$

where N is the number of intermediate nodes. In particular, the combined route cost C_i can be calculated at each intermediate node i using a fuzzy cost function F:

$$C_i = F_i(B_i^t, Q_i, SS_i) + N_i \qquad (14.22)$$

In the proposed AOFR scheme, fuzzy logic is used to determine and express the cost function. The value N_i is added to the fuzzy cost derived from the other three metrics at each intermediate node. For a delay-sensitive application, which requires fewer intermediate hops, e.g., a real-time voice or video application, a linearly or exponentially increasing link cost N_i with respect to number of nodes traversed by the route discovery process (as shown in Figure 14.4) so far, can be added to the

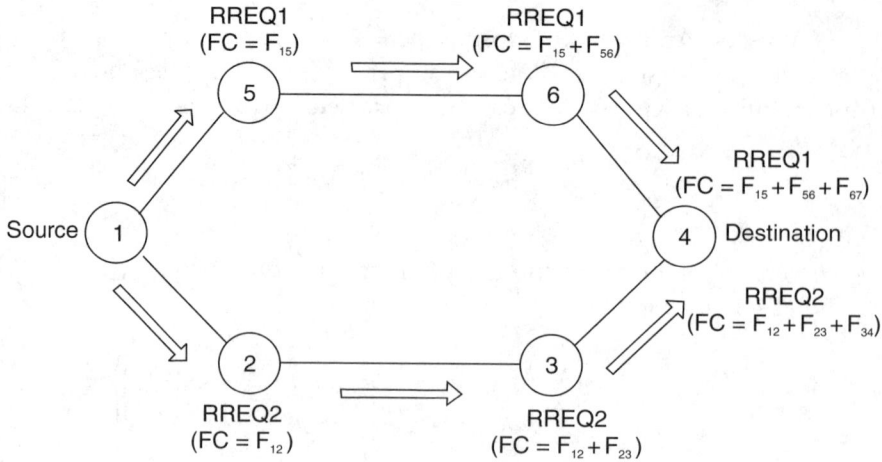

Figure 14.4 Route discovery phase (FC = Fuzzy cost, Fij = Fuzzy cost of link between nodes i and j).

defuzzified value at each intermediate node. Finally, the total fuzzy cost C_n of nth route with N intermediate nodes is the summation of all individual link costs:

$$C_n = \sum_{i=1}^{N+1} C_i \qquad (14.23)$$

14.5 Ad Hoc On-Demand Fuzzy Routing (AOFR)

As mentioned in the previous section, the ad hoc on-demand fuzzy routing (AOFR) scheme proposed in this chapter is designed to achieve various objectives while selecting a route. The objectives that are considered for route selection are to minimize end-to-end delay; maximize packet delivery; and maximize the lifetime of the batteries.

Several metrics have been chosen to meet these objectives and to produce a single cost metric (C) for selecting routes. The various routing metrics chosen are: remaining battery capacity (B) at an intermediate node in a route; buffer length (Q) at an intermediate node; link stability (SS) between two nodes in a route; and the number of intermediate hops in a route (N). The relationship between the cost function C and the other metrics is given by equation (14.22).

Using remaining battery capacity as one of the route selection metrics enables routes with intermediate nodes that have higher battery power to be selected. This helps in balancing the packet-forwarding task in the network in such a way that a few nodes do not get depleted of their batteries early compared with other nodes. Buffer length as another metric helps in selecting routes that are not congested, thus decreasing congestion loss and end-to-end delay (due to queuing delay). Last, link stability metric allows the route selection process to select routes that are comparatively more stable and long lived, which would ensure a lower packet loss rate occurring from the movement of intermediate nodes and result in fewer route failures. Figure 14.3 shows the block diagram of the AOFR routing scheme proposed in this chapter. As on-demand routing protocols have shown to outperform the table-driven schemes (Broch et al., 1998) in MANETs for highly mobile networks, AOFR is implemented on top of an on-demand route discovery mechanism. It consists of a route discovery and a route reply phase. AOFR can also be easily extended to work with proactive routing schemes. Link stability can be found using either a positioning system or by measuring the signal strength. This chapter uses the signal strength of received packets to determine the link stability.

14.5.1 AOFR Route Discovery Phase

When a source needs to start a communication with a node, it sends a route discovery (RREQ) packet, which collects the sum of fuzzy costs for all the individual links along the path. When an intermediate node receives an RREQ packet, it calculates its cost of participating in the route and adds that cost to the cost calculated at the previous node using the same procedure and stored in the RREQ packet. The intermediate nodes make use of fuzzy logic to calculate the cost, which is dependent on multiple metrics mentioned earlier. Figure 14.4 shows the route discovery phase to discover the appropriate route from node 1 to node 4. It is important to emphasize that the AOFR only determines the best forwarding path that simultaneously meets multiple objectives. It does not reserve any resources on that path.

14.5.2 AOFR Route Reply Phase

When the destination node receives the first RREQ packet, it starts a timer for the route selection time window. In this period different RREQ packets reach the destination. After the time window expires, the destination selects the route R with least cost. A route reply (RREP) packet is sent from the destination back to the source along the opposite direction of the selected route, as shown in Figure 14.5.

14.5.3 Fuzzy Cost Calculation in AOFR

A fuzzy logic system is used to calculate the combined cost metric at each node as the route discovery progresses. The parameters of the membership functions and fuzzy control rules are set via fuzzy set manipulation (linguistically stated but math-

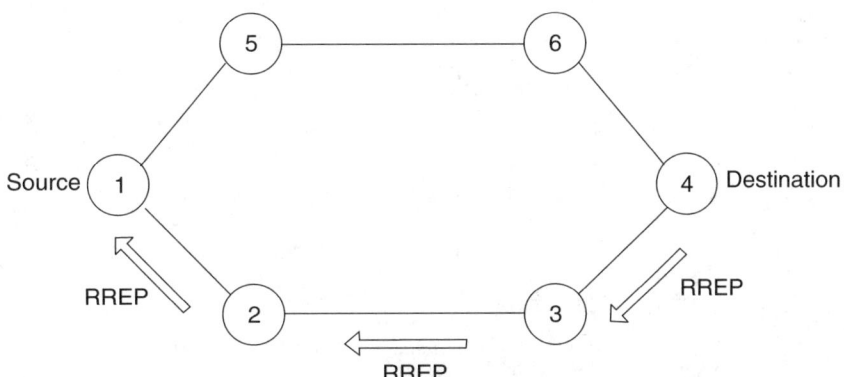

Figure 14.5 Route reply (RREP) packets being sent from the destination to the source on the path with minimum fuzzy cost.

ematically treated). The inputs to the fuzzy controller are: buffer occupancy, remaining battery power, and signal stability. The cost is updated as the route discovery progresses. Finally, when the RREQ packet reaches the destination, the destination selects the route with the minimum fuzzy cost. The steps involved in the calculation of fuzzy cost are shown in Figure 14.6 and are elaborated as follows:

14.5.3.1 Step 1: Fuzzification of Inputs and Outputs

The underlying reasoning scheme dwells on information granules, which are fuzzy sets that are regarded as generic building blocks. Subsequently, all transformations and relationships between fuzzy sets are used to carry out inferences.

Each input (routing metric) is first fuzzified using fuzzy set manipulation, as shown in Figure 14.6. The three inputs fuzzified are received signal strength, remaining battery power, and buffer length, as shown in Figure 14.6. Fuzzy membership functions allow each condition of each rule to be satisfied to some degree, so that the output is a weighted combination of these values. The membership function can be

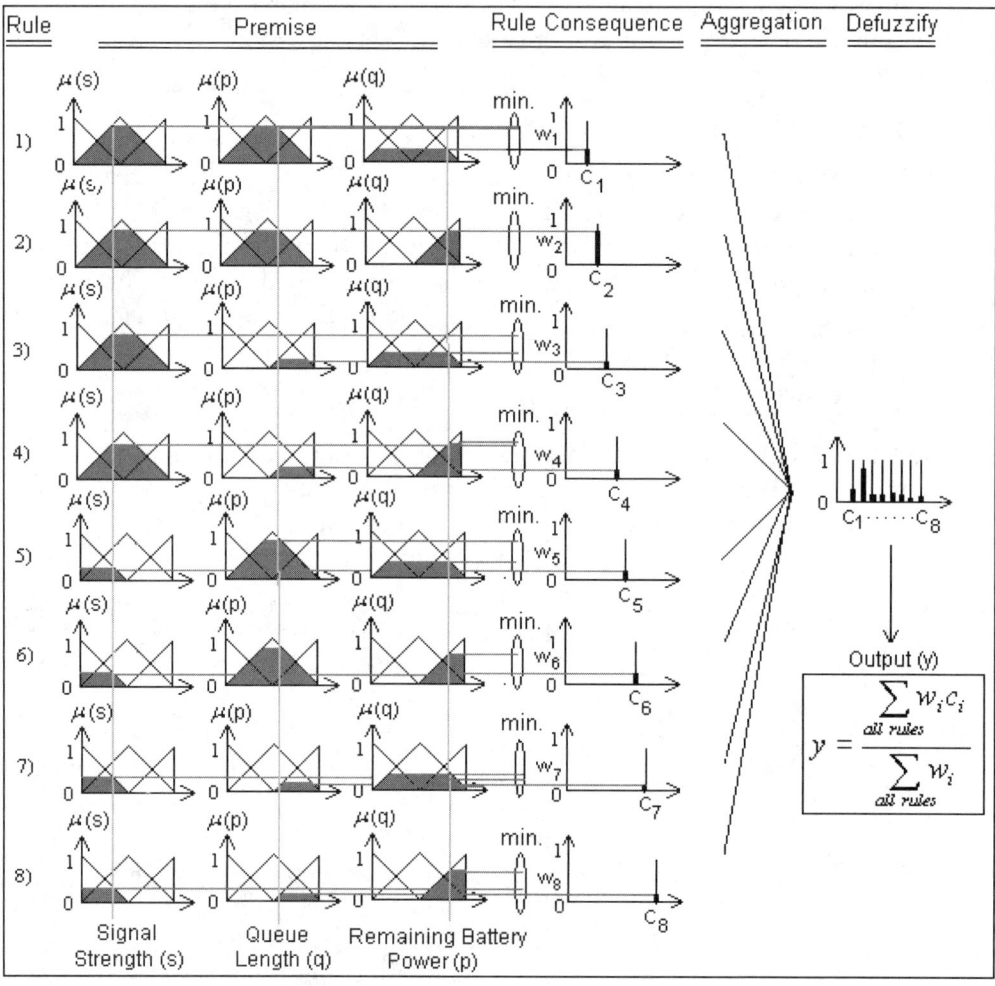

Figure 14.6 Fuzzy cost calculation procedure.

defined in a subjective manner using linguistic variables (such as high, medium, and low). Subjective judgment, intuition, and expert knowledge are commonly used in constructing membership functions. Triangular-shape membership functions have been extensively used for real-time operation since they provide simple formulas and computational efficiency [13]. In this chapter the universe of discourse (range on x-axis) of the inputs is divided into three fuzzy sets of desired shapes, triangular in the present case as shown in Figure 14.6. In Figure 14.6, $\mu(s)$, $\mu(p)$ and $\mu(q)$ denote the membership functions for received signal strength, queue length and remaining battery power.

14.5.3.2 Step 2: Inference Engine and Knowledge Base

The knowledge base (KB) is a set of rules (developed using expert knowledge) required to achieve the objective. The rules connecting the inputs and the output (rule consequence) are designed based on a thorough understanding of the system. Fuzzy rules have an IF-THEN structure. The inputs are combined using the AND operator. An example of a rule that describes the input-output mapping is as follows:

IF {*signal strength* is *low* (A_i)} AND {*battery power* is *high* (B_i)} AND {*buffer occupancy* is *medium* (D_i)} THEN {the cost is C_i}

with A_i, B_i, and so on, being fuzzy sets defined in the corresponding input spaces.

The fuzzy set parameters and rules developed in AOFR are initially set by expert knowledge and heuristics, including the following:

- Higher cost for extreme values of signal strength. The interpretation is that as the received signal strength decreases, link stability worsens. On the other hand, when the signal strength is too high, it would mean that the nodes (transmitter and receiver) are too close to each other. This would lead to a higher number of intermediate hops, thus resulting in a higher end-to-end delay.
- The cost of using a route should be higher as the buffer length (occupancy) at the intermediate node increases. This is because increased buffer length would adversely affect the first two objectives, which are to achieve reduced delay and increased packet delivery. An increase in buffer length means more waiting time for data packets to be processed and routed as well as increased probability of buffer or congestion loss (if the buffer is full, the data packets are simply dropped). So an increase in cost with increase in buffer length would reduce the probability of a congested node joining any new routes and vice versa.
- The cost of using a route increases as the remaining battery power decreases. This is in order to achieve the third objective, which is to increase the lifetime of battery-powered nodes. Furthermore, using nodes with lower battery power for routing packets would result in frequent route failures due to the expiration of batteries of intermediate nodes in the route. This would have a negative impact on both delay and packet delivery ratio, since the packets

would have to wait for a new route to be discovered and the packets already on their way would get dropped by intermediate nodes due to an expiration of the battery of a downstream node in the route.

The parameters of the fuzzy membership functions and fuzzy rules are refined through simulations using an evolutionary algorithm (EA) to obtain the optimal rule structure (at which the fitness value of EA is maximum), which is shown in Tables 14.1, 14.2 and 14.3.

In these three tables, B is the remaining battery power, Q is the queue length, SS is the received signal quality (strength), L means low, M means medium, and H means high.

Since there are three inputs (signal strength, queue length, and remaining battery power) to the fuzzy logic system, and the universe of discourse (range) is divided intro three fuzzy sets (low, medium and high), there are a total of 27 () fuzzy rules. Since there are three inputs and every input can only belong to two fuzzy sets for any value of that input (low and medium or medium and high), a total of eight () rules are fired for any given values of the three inputs. The shaded portion of the fuzzy membership functions in Figure 14.6 indicates the degree to which each rule is satisfied.

14.5.3.3 Step 3: Defuzzification

The standard inference mechanism computes the output of the fuzzy system on the basis of rules provided in advance and a given new input datum that describes a current state of the system. Outputs are mapped into several fuzzy regions of desired

Table 14.1 Fuzzy Rules Table for Low SS

Q	B		
	L	M	H
L	0.66	0.5	0.33
M	0.83	0.66	0.5
H	1	0.83	0.66

Table 14.2 Fuzzy Rules Table for Medium SS

Q	B		
	L	M	H
L	0.3072	0.2076	0.1538
M	0.3846	0.3072	0.2076
H	0.46152	0.3846	0.3072

Table 14.3 Fuzzy Rules Table for High SS

Q	B		
	L	M	H
L	0.4358	0.32691	0.21794
M	0.54485	0.4358	0.32691
H	0.6538	0.54485	0.4358

shapes (for a Mamdani-type system) or several singletons (for a Sugeno-type system).

This chapter utilizes the Sugeno-type fuzzy regulator [11], in which fuzzy sets are involved only in rule premises. The advantages offered by Sugeno-type fuzzy logic compared with other types are: it is computationally efficient, it works well with optimization and adaptive techniques, it has guaranteed continuity of the output surface, and it is well suited to mathematical analysis.

Rule consequences in a Sugeno-type fuzzy system are crisp functions of the output variables. Since there are 27 rules, as mentioned in the previous section, 27 Sugeno-type fuzzy output singletons have been developed to represent the output membership functions, since every rule corresponds to one output singleton. The values of the 27 Sugeno-type fuzzy singletons are shown in Tables 14.1, 14.2 and 14.3 corresponding to the 27 fuzzy rules taking values between 0 and 1. The final fuzzy cost value (y) at a node is calculated by combining all the points of intersection on the Sugeno singletons using center of gravity method. Here, the output y of the inference scheme is shown in (14.24) where is the degree of activation of the condition part of the rule for the given input (signal strength, queue length and remaining battery power), which is also the degree to which a rule holds good and is found using the fuzzy AND operator.

$$y = \frac{\sum_{all\ rules} w_i c_i}{\sum_{all\ rules} w_i} \qquad (14.24)$$

These computations are straightforward: we determine the extent to which each sub-condition of the rule is invoked (computed as $A_i(x_1)$, $B_i(x_2)$ … etc.) and the degrees are combined AND-wise, using a product or min operation. The terms denote the numerical representation of the fuzzy set in the conclusion part of the rule; this can be the modal value of the respective membership function.

14.6 Simulation Parameters

Extensive simulations were carried out to compare the performance of the AOFR scheme proposed in this chapter with the conventional shortest path (SP), maximum battery cost routing (MBCR) [9] and cost adaptive mechanism (CAM) [16] route selection schemes.

SP selects the route with the minimum number of intermediate hops utilizing on-demand route discovery similar to [4]. CAM [16] calculates a cost for each route using a weighted summation of individual metrics:

$$\phi_K = \alpha K_L + \beta K_A + (1 - \alpha - \beta) K_P \qquad (14.25)$$

The metrics considered in CAM are transmitting power (K_P), wireless activity (K_A) and load (K_L). CAM[16] does not provide any method to calculate the weights for the various cost coefficients in equation (14.25). It arbitrarily sets these weights using trial and error. CAM also makes use of the transmit power, whereas AOFR makes use of the remaining battery cost metric to describe the lifetime of each host to select routes. Although the transmit power metric can reduce the total power consumption of the overall network, it does not reflect directly the lifetime of each host. If the minimum total transmission power routes are via a specific host, the battery of this host will be exhausted quickly, and this host will fall due to battery exhaustion. Therefore, the remaining battery capacity of each host is a more accurate metric to describe the lifetime of each host [9]. AOFR also does not require accurate location information for calculation of transmission power. The implementation of CAM in this chapter uses the same routing metrics as AOFR. The routing metrics in CAM have been kept similar to AOFR so as to compare the multiobjective routing algorithm of AOFR rather than comparing the metrics. The implementation of MBCR is exactly similar to the actual MBCR algorithm proposed in [9], which minimizes the summation of inverse of the remaining battery capacity for all nodes in the routing path:

$$E_{BatteryCost} = \min\left\{\sum_{i=1}^{N+1}\left(\frac{1}{B_i^t}\right)\right\} \qquad (14.26)$$

where, B_i^t is the remaining battery capacity of a node at time t, and N is the number of intermediate nodes.

Network simulator (NS-2) [17] is used to simulate these protocols. NS-2 is a discrete event simulator. The latest version of NS-2 (ns-2.26), which can model and simulate multihop wireless ad hoc networks, was used for the simulations. The physical layer for the simulation uses two-ray ground reflection as the radio propagation model. The link layer is implemented using the IEEE 802.11 distributed coordination function (DCF) as the media access control protocol (MAC). It uses RTS/CTS/Data/ACK pattern for unicast packets and data for broadcast packets. Carrier sense multiple access with collision avoidance (CSMA/CA) is used to transmit these packets.

Figure 14.7 shows a view of the simulated MANET generated by NS-2's network animator. All protocols simulated maintain a send buffer of 64 data packets, containing the data packets waiting for a route. Packets sent by the routing layer are queued at the interface queue until the MAC layer can transmit them; this has a maximum size of 50 data packets. The interface queue gives priority to routing packets. The transmission range for each of the mobile nodes is set to 250m and the channel capacity is 2 Mbps. Simulations were run for 900 simulated seconds. The routing table used for CAM, AOFR, and MBCR are similar having a cost field associated with every routing entry. SP, on the other, hand has the number of hops field instead of the cost field in its routing table. The remaining fields in the routing table entries are similar for all the four protocols simulated. Every route entry in the routing table has a destination, node address, the next hop to route the packets, the sequence number of the destination and the time to live for that route. The simulation environment used in this chapter, closely matches the specifications of previous performance comparisons using the same simulator (Broch et al., 1998). All protocols detect link failures using feedback from MAC layer. A signal is sent to the routing layer when the MAC layer fails to deliver a unicast packet to the next hop. No

14.6 Simulation Parameters

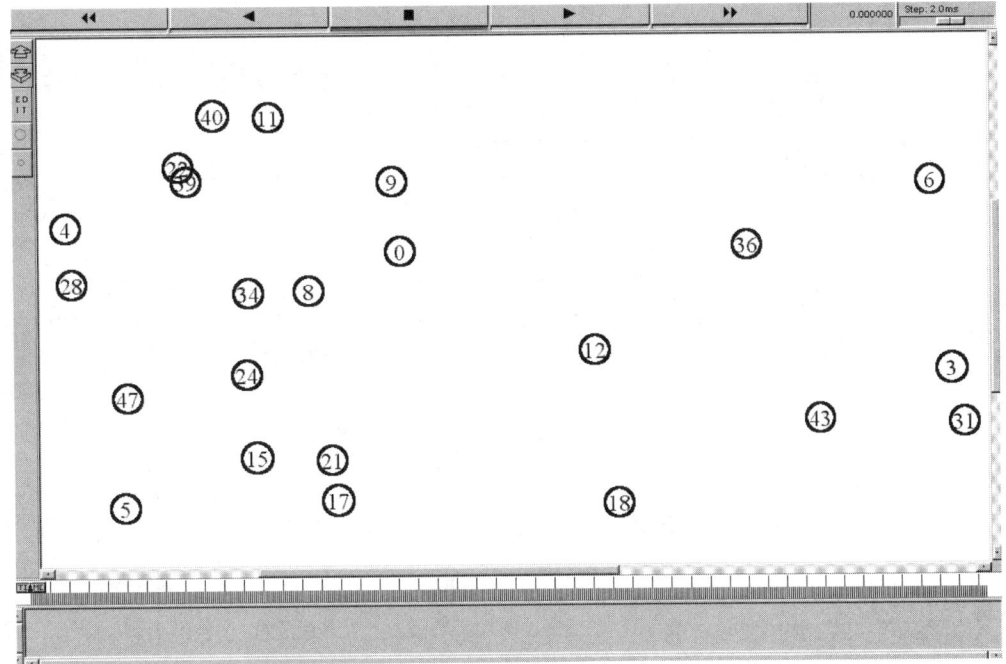

Figure 14.7 Network Animator view of the random node movement in the MANET.

additional network layer mechanism, such as "hello" messages is used. The traffic and mobility models used are similar to previous reported results using this simulator [18].

14.6.1 Mobility Model

A network of 50 mobile nodes migrating within an area of 1500m x 300m with a speed of 0 to 20m/s was simulated. A rectangular space was chosen in order to force the use of longer routes between nodes than there would be in a square space with the same amount of nodes [18]. The mobility model uses the random waypoint model in the rectangular field. The simulations were run multiple times for 7 different pause times: 0, 30, 60, 120, 300, 600, and 900 seconds. After pausing for pause time seconds the mobile nodes again select a new destination and proceed at a speed distributed uniformly between 0 and the maximum speed.

14.6.2 Traffic Model

The simulations carried out consisted of 20 continuous bit rate (CBR) sources. The source-destination pairs are spread randomly over the network. CBR traffic sources were chosen, since the aim was to test the routing protocols. Source nodes and destination nodes were chosen at random with uniform probabilities. The sending rate used was 4 packets per second with a packet size of 64 bytes. Each data point in the comparison results represents an average of multiple runs with identical traffic models but with different movement scenarios. The same movement and traffic scenarios were used for all the three protocols simulated.

14.6.3 Energy Model

The energy model in a node has an initial value that is the level of energy the node has at the beginning of the simulation. It also has a given energy usage for every packet it transmits and receives. At the beginning of simulation, energy is set to an initial value, which is then decremented for every transmission and reception of packets at the node. When the energy level at the node goes down to zero, no more packets can be received or transmitted by the node. The numbers of mobile devices that are simulated are divided into two categories. The first category of nodes is the one with higher battery capacity, such as, for example, laptop users, and so the initial value of their battery energy is much higher compared with the second. The second category of nodes is assumed to have less initial battery capacity, such as, for example, palm devices.

14.7 Performance Evaluation

AOFR's performance is compared with MBCR [9], CAM [16], and SP routing schemes, and various important performance measures for mobile wireless ad hoc networks are evaluated.

14.7.1 End-to-End Delay

This includes all possible delays caused by buffering and wireless transmission during route discovery, queuing at the interface queue, retransmission delays at the MAC, propagation, and transfer times. The average end-to-end delay for AOFR is least compared with that of SP, CAM, and MBCR (Figure 14.8). SP has the highest packet delay because of high congestion, as seen in Figure 14.9. The packet delay in CAM is more than that of AOFR because of higher route failures and overhead as well as higher congestion in CAM. As the mobility decreases (with increasing pause time) the delay also decreases due to less number, of route failures at low mobility. Route failures have an impact on the delay, because route failures require rerouting

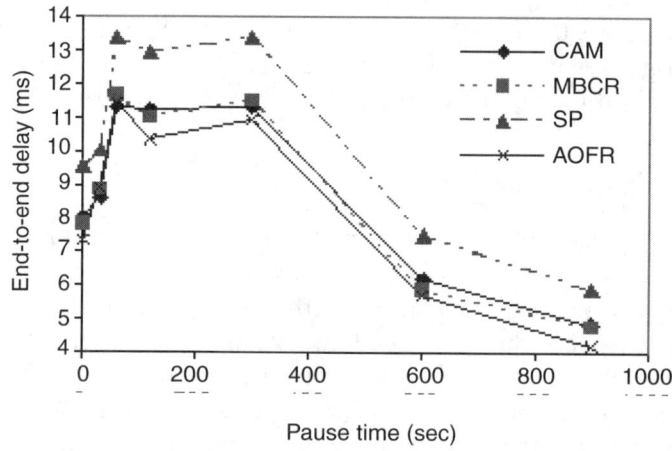

Figure 14.8 Comparison of end-to-end delay.

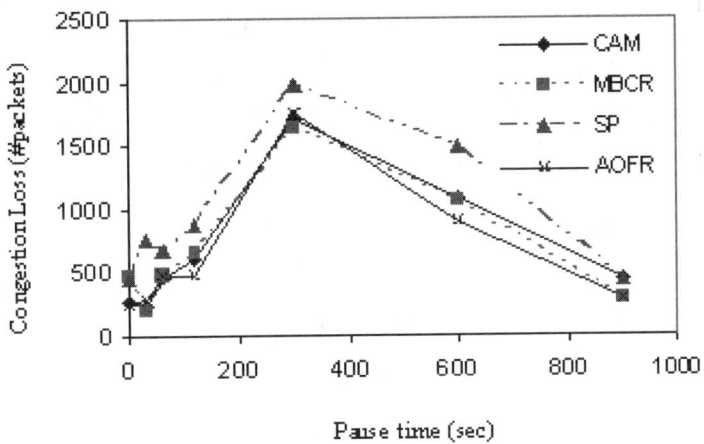

Figure 14.9 Comparison of congestion loss.

and storing of packets in the send buffer. It also leads to an increase in overhead and hence contention for the shared wireless channel.

14.7.2 Congestion Loss

Congestion loss is the number of packets dropped at mobile terminals due to congestion in the buffers. As seen in Figure 14.9, the congestion loss for AOFR is least compared with that of shortest path, CAM and MBCR due to its inherent feature of calculating cost using buffer length of intermediate nodes, hence reducing congestion spots in the network (Figure 14.9). The highest congestion loss is for SP since the shorter routes get overused leading to congestion at bottleneck links. CAM considers congestion as one of the route selection criteria, hence its congestion loss is not very high. There is an almost 33 % gain in congestion loss with AOFR compared to SP.

14.7.3 Packet Delivery Fraction

Packet delivery fraction is the ratio between the number of packets originated by the application layer CBR sources and the number of packets received by the CBR sink at the final destination. Packet delivery ratio is highest for AOFR (Figure 14.10), which is due to its ability to select stable and least congested routes (Figure 14.9).

14.7.4 Expiration Sequence

This is the sequence in which nodes in the network become devoid of energy (exhausted) with respect to the time. As can be seen in the results (Figure 14.11) the expiration sequence for SP, AOFR, and MBCR is quite similar for the set of simulation conditions selected in this chapter. It is also seen, that CAM leads to battery-powered nodes becoming drained out very quickly.

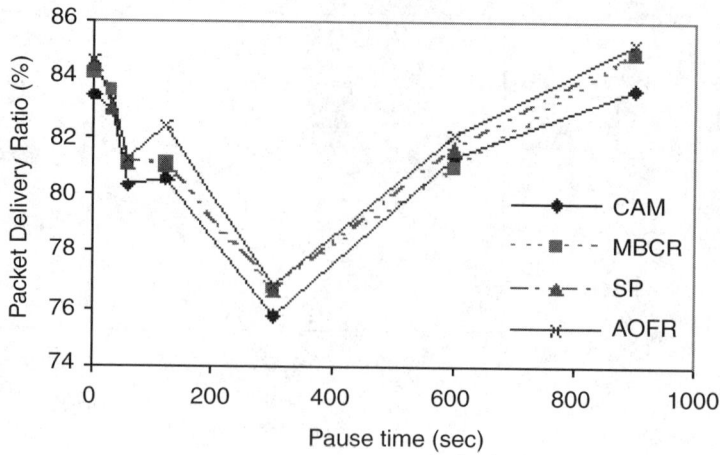

Figure 14.10 Packet delivery ratio comparison.

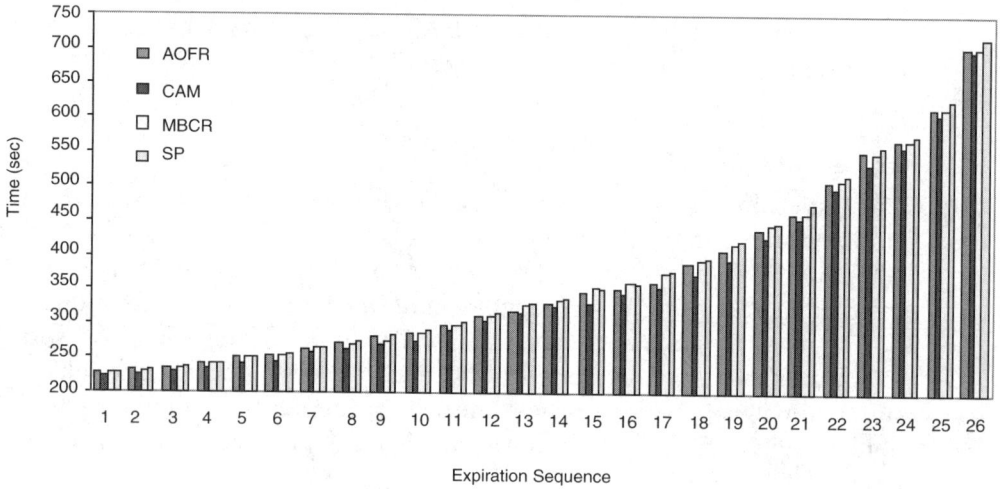

Figure 14.11 Expiration time versus expiration sequence of a mobile node's batteries.

14.7.5 Normalized Routing Load

This is the number of routing packets transmitted per data packet delivered at the destination. Each hop-wise transmission of a routing packet is counted as one transmission. For packets sent over multiple hops, each transmission of the packet (every hop) counts as one transmission. Normalized routing load is the number of routing packets transmitted per data packet received at the destination.

Routing packets include route request, route replies, and route errors. The routing load is highest in CAM (Figure 14.12). SP and AOFR have the lowest overheads. CAM has the highest number of route failures, hence its overhead is very high.

14.7 Performance Evaluation

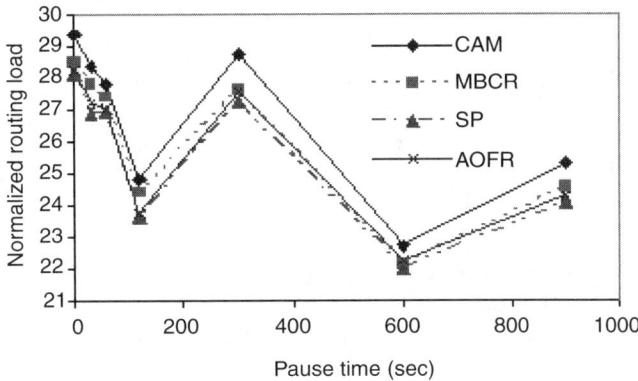

Figure 14.12 Comparison of routing load.

14.7.6 Route Stability

Route stability is a very important performance parameter for a routing protocol. Route stability can be measured in terms of number of route failures. Figure 14.13 shows that the number of route failures for SP and AOFR are least compared with MBCR and CAM. Route stability is worse for CAM. SP has least number of intermediate hops compared to the other protocols; hence, the probability of route failures is less, since the probability of fewer nodes going away from the route is less compared to routes with many intermediate hops.

14.7.7 Packets-per-Joule

This is the number of data packets that are successfully delivered per joule of energy spent. It is also regarded as a measure of the energy efficiency of the routing protocol. As can be seen from Figure 14.14, the highest packet-per-joule ratio is for SP, since the routes selected by it are shortest and hence the energy spent is least. Almost 5 % gain is achieved with AOFR compared with CAM. CAM has the lowest

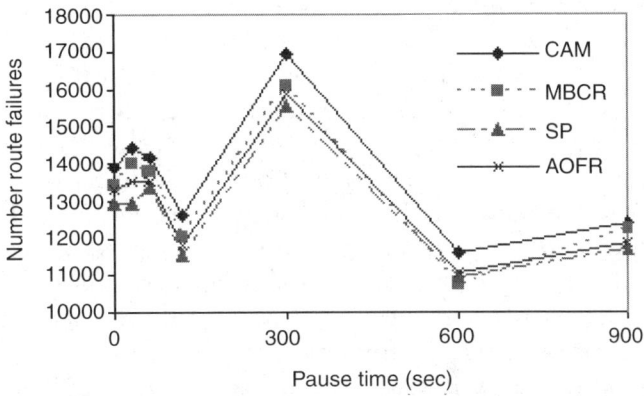

Figure 14.13 Route stability comparison.

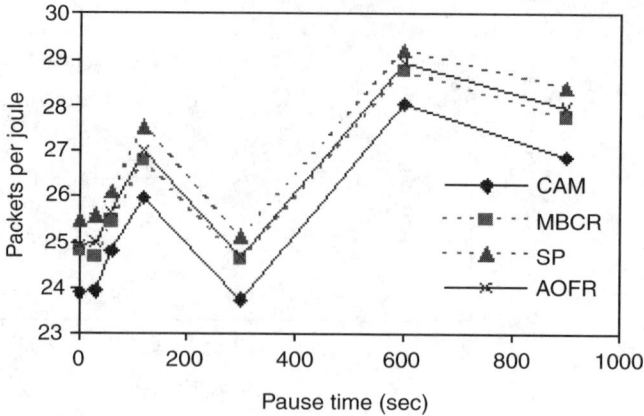

Figure 14.14 Packets-per-joule comparison.

packets-per-joule energy efficiency due to high route failures (Figure 14.13) and low packet delivery ratio (Figure 14.10).

14.8 Discussion

From the simulation results it is seen that the fuzzy system of AOFR effectively captures the interplay of the various metrics and objectives for the purpose of multiobjective route selection in MANETs. As a direct result of multiobjective route selection, AOFR does not have to compromise network performance on any objective, as is evident in the comparison results.

It is seen from the simulation results that AOFR can perform better for most of the objectives. This is because most of the single objective schemes use single criteria, which means that they only use one metric for route selection; this may not lead to the best solution. For example, shortest path routing scheme only utilizes the number of intermediate hops between a source and destination to discover shortest routes, and MBCR selects routes depending upon remaining battery lifetime to prolong battery lifetime of the nodes in the network. It may seem that since the various objectives can be conflicting, the performance should degrade for a multiobjective routing scheme compared to a single objective scheme. However, schemes such as SP and MBCR, which do not consider multiple routing metrics, cannot perform very well compared with AOFR, since the various routing objectives depend on multiple metrics, and AOFR utilizes these multiple metrics by combining them using fuzzy logic for route selection. Even if a routing scheme that considered all the routing metrics were to be developed to meet one single objective, it would compromise the network performance on the overlooked objectives, since the various objectives are interlinked and a gain in one can only be achieved at the expense of the others. Therefore, a multiple-objective routing scheme, which can consider multiple routing metrics, seems to be the best approach for route selection in MANETs and can achieve the best overall performance.

The proposed AOFR routing scheme is robust and highly efficient. The robustness of AOFR is demonstrated in its capability to perform well under low as well as high mobility rates compared with SP, CAM, and MBCR. The routing load metric evaluates the efficiency of the routing protocol. AOFR has a very low routing load compared to the other route selection schemes. Low routing load affects both packet delivery fraction and delay, since it causes less network congestion and multiple-access interference. Routing overhead is also an important metric for comparing these protocols, since it also measures the scalability of a protocol, the degree to which it will function in congested or low-bandwidth environments, and its efficiency in terms of consuming node battery power

AOFR has a low routing overhead, which helps to enhance its scalability. Furthermore, AOFR's route discovery mechanism is based on on-demand routing protocols, which have proven performance [4] with respect to scalability in bandwidth constraint wireless networks as compared with proactive schemes [1].

Packet delivery ratio and end-to-end delay are the two most important metrics required for best-effort traffic. These performance metrics also contribute to the efficiency of the routing protocol. Packet delivery ratio and end-to-end delay are also not completely independent. For example, lower packet delivery fraction means that the delay metric is evaluated with a lesser number of samples, as is the case in MBCR. The end-to-end delay in AOFR is less than that of SP and MBCR, since it considers both route length and congestion at intermediate nodes while selecting routes, whereas SP and MBCR do not consider these two parameters together.

Congestion loss is also less in AOFR due to its capability to balance traffic congestion as it considers queue length to calculate the combined cost metric used for selecting routes. Traffic balancing increases the system capacity and enhances resource utilization. It also leads to reduction in wireless contention in a localized region.

Last, the stability of the routing protocol depends on its ability to find routes, that are long lived and stable and do not require frequent rerouting. Route stability has an impact on the efficiency of the routing protocol, since it can seriously affect routing load as well as the packet delivery and end-to-end delay. The more the number of route failures, the more would be the amount of packet loss and buffer delay experienced by data packets waiting to be sent while rerouting is performed. The importance of efficient multiple metric and multiple-objective routing in MANETs is clearly evident from the comparison results presented in this chapter.

14.9 Conclusions

This chapter proposes a computationally intelligent algorithm named AOFR for multiobjective routing in the presence of link-state inaccuracies for MANETs. Heuristics and approximation algorithms have been used before [2, 3] to deal with multiconstrained path selection, which is considered a very difficult problem. Unlike conventional MANET routing schemes, AOFR does not impose strict preference for some routing metrics over others, which helps in improving network performance. Multiplemetrics are combined together using a fuzzy rule base system in AOFR to generate a single cost value that is used for route selection. The route with the minimum fuzzy cost is then selected that best meets the multiple objectives simultaneously.

The performance of AOFR has been evaluated through several simulations and compared with that of MBCR, SP, and CAM. It has been found that the use of fuzzy set theory to represent various uncertainties associated with multiobjective routing problems for ad hoc networks is very promising. Simulation results show that the proposed AOFR multiobjective route selection scheme provides less end-to-end delay and enhances packet delivery ratio without increasing routing overhead. It is also seen from the simulation results that AOFR reduces congestion probability.

There are several potentially fruitful areas for future work. The focus of one direction would be on extending the AOFR framework to include more metrics and objectives. Another possible area of future work would be provisioning to achieve quality of service by assigning preferential weights to various objectives, along with an admission control scheme for bandwidth estimation to provide feedback to the application about the network status. We also plan to expand the proposed scheme to networks with multiple service classes.

References

[1] Perkins, C.E., 2001, *Ad Hoc Networking*, Addison–Wesley, United States of America.

[2] Korkmaz, T., Krunz, M., 2003, "Bandwidth-Delay Constrained Path Selection Under Inaccurate State Information," *IEEE/ACM Trans. on Networking*, Vol. 11, Issue 3, pp. 384–398.

[3] Younis, O., Fahmy, S., 2003, "Constraint-Based Routing in the Internet: Basic Principles and Recent Research," *IEEE Communications Society (COMSOC) Surveys & Tutorial*, Vol. 5, No.1.

[4] Perkins, C.E., Royer, E.M., Das, S.R., 1999, "Ad Hoc On-Demand Distance Vector (AODV) Routing," *Proceedings of IEEE Workshop on Mobile Computing Systems and Applications*, pp. 90–100.

[5] Johnson, D. B., Maltz, D. A., 1996, "Dynamic Source Routing in Ad Hoc Wireless Networks," *Mobile Computing*, Kluwer Academic Publishers, pp. 153–181.

[6] Marwaha, S., Tham C.K., Srinivasan, D., 2002, "Mobile Agents based Routing Protocol for Mobile Ad Hoc Networks," *Proceedings of IEEE Globecom International Conference*, Taiwan, Vol. 1, pp. 163–167.

[7] Singh, S., Woo, M., Raghavendra, C.S., 1998, "Power-Aware Routing in Mobile Ad Hoc Networks," *Proceedings of IEEE/ACM MOBICOM International Conference*, pp. 181–190.

[8] Toh, C.K., 1997, "Associativity Based Routing for Ad Hoc Mobile Networks," *Wireless Personal Communications Journal*, Vol. 4, No. 2, 103–139.

[9] Toh, C.K., 2001 "Maximum Battery Life Routing to Ssupport Ubiquitous Mobile Computing in Wireless Ad Hoc Networks," *IEEE Communications Magazine*, Vol. 39, Issue 6, 138–147.

[10] Dube, R., Rais, C., Wang, K.Y., Tripathi, S., 1997, "Signal Stability based Adaptive Routing (SSA) for Ad-Hoc Mobile Networks," *IEEE Personal Communications*, Vol. 4, Issue 1, 36–45.

[11] Ross, T.J., 1995, *Fuzzy Logic with Engineering Applications*, McGraw Hill, USA.

[12] Pedrycz, W., Vasilakos, A., 2000, *Computational Intelligence in Telecommunications Networks*, CRC Press, Florida, USA.

[13] Ghosh, S., Razouqi, Q., Schumacher H.J., Celmins, A., 1998, "A Survey of Recent Advances in Fuzzy Logic in Telecommunications Networks and New Challenges," *IEEE Transactions on Fuzzy Systems*, Vol. 6, Issue 3, 443–447.

[14] Aboelela, E., Douligeris, C., 1999, "Fuzzy Generalized Network Approach for Solving an Optimization Model for Routing in B-ISDN," *Journal of Telecommunication Systems*, pp. 237–263.

[15] Rappaport, T.S., 1996, *Wireless Communications: Principles and Practice*, Prentice Hall Publication, USA.

[16] Roux, N., Pegon, J.S., Subbarao, M.W., 2000, "Cost Adaptive Mechanism to Provide Network Diversity for MANET Reactive Routing Protocols," *Proceedings of IEEE MILCOM International Conference*, Vol.1, pp.287–291.

[17] Fall, K., Varadhan, K., 2000, "The ns Manual," *The VINT Project*, UC Berkeley, LBL, USC/ISI and XEROX PARC. http://www.isi.edu/nsnam/ns/ns-documentation.html

[18] Broch, J., Maltz, D., Johnson, D., Hu, Y.C., Jetcheva, J., 1998, "A Performance Comparison of Multi-Hop Wireless Ad Hoc Network Routing Protocols," *Proceedings of the Fourth Annual International Conference on Mobile Computing and Networking*, pp. 85–97.

CHAPTER 15
Authentication and Security Protocols for Ubiquitous Wireless Networks

Wenye Wang, Janise Y. McNair, Jiang (Linda) Xie

15.1 Introduction

Wireless networks have evolved into a heterogeneous collection of network infrastructures providing a wide variety of options for user access, such as WiFi and cellular systems, which have facilitated ubiquitous wireless services, anytime, anywhere, and anything. Therefore, ubiquitous wireless networks are part of ambient intelligence. The successful deployment of wireless local area networks (WLANs) for high-speed data transmission and cellular systems for wide coverage and global roaming has emerged to be a complementary platform for wireless data communications. In order to fully support intersystem roaming for maximizing system revenue as well as resource utilization, authentication needs be enforced to fully explore the potentials of portable devices, in addition to mobility support in mobile environments. In the past decade, WLANs have demonstrated an exceptional success because of their high data rates and flexible deployment. Mobile users can easily access the Internet through a laptop or a portable digital assistant (PDA) with embedded or removable 802.11a/b/g cards, thus experiencing data communications over the IP backbones. However, WLAN systems are not capable of providing mobility and roaming support due to local authorization and registration. This limit is complemented perfectly by established cellular networks on which many users have depended for ubiquitous access. On the other hand, even with the advances in general packet radio service (GPRS) as 2.5G systems, as well as universal mobile telecommunication systems (UMTS) specified by third generation partnership project (3GPP) and cdma2000 specified by 3GPP2, low transmission rate and expensive data service have made 3G cellular networks not preferable for mobile data applications compared with WLAN systems [1].

However, enabling ubiquitous wireless communications introduces many challenges for network management, since a network cannot authorize an unknown user without having the identification and history of mobile users, which may create a security breach in misusing system resources. In particular, due to the open medium of wireless and the mobility of roaming terminals, along with the use of IP backbone networks, sensitive data information and user identity, even a wireless network itself, are vulnerable to attacks. Although security issues can occur in both wired and wireless networks, risks in wireless domain are greater than those in wired networks [2].

The increasing concerns about security and universal access demand efficient authentication and security protocols for mobile environments. There are many security concerns in ubiquitous, wireless networks, such as user identity privacy, data integrity, and confidentiality as described in [1]. In this chapter, we focus on the issue of authentication and security protocols for wireless networks. Authentication is inherently a security technique, which is designed to protect networks against acceptance of a fraudulent transmission by establishing the validity of a transmission, a message, or an originator. During the authentication process, a user must provide verifiable credentials. When a mobile node requests service from a network other than its home network from which it subscribes services, it must provide sufficient individual information for authorization and register its locations to the home network for subsequent service. This process of authentication and registration plays a very important role in protecting the confidentiality and integrity of wireless networks through denying an unauthorized transmission and preventing intrusions [3–6].

Therefore, authentication has a great impact on network security and mobility management in wireless data networks. First, authentication is aimed to ensure that network resources are used by authorized users and to prevent resources from any unauthorized use/damage. Second, authentication involves negotiation of credentials for secure communications. While the major purpose of authentication in wireless networks is to authorize networking access, it also has significant influence on the quality of ongoing service, because authentication may introduce signaling overhead. Moreover, the delay caused by authentication may increase packet loss and even decrease system throughput.

Authentication as a procedure to authorize mobile users involves two major issues: the design of authentication architecture and authentication protocols or procedures. The former is concerned with what entities are needed for authentication and how to distribute them in different networks, while the latter is concerned with signaling messages used for transferring requirements and credentials. Therefore, we introduce a background of authentication and design concerns for wireless networks in Section 15.2. Then we present an overview of authentication architecture for the integration of 3G and WLAN systems in Section 15.3. The details of security architecture for individual 3G and WLAN systems are not covered in this chapter, because we intend to highlight the ambient or ubiquitous perspective, that is, heterogeneous mobile environments. The authentication procedures that are usually included in security protocols are described in Section 15.5 to show the interaction of security protocols at different layers with respect to data streams, delay, and throughput. Finally, we conclude the chapter and discuss future challenges in Section 15.6.

15.2 System Architecture and Design Issues

Mobility support in wireless networks has been researched extensively [7]. For 3G systems such as UMTS, mobile services are secured by a subscriber identity module

(SIM). A SIM contains subscriber-related information, including its identity, authentication key, and encryption key. The subscriber identity and authentication key are used for authorization of mobile access [8]. In UMTS, authentication is achieved by sharing a secret key between the SIM on a mobile device and an authentication center (AuC) [9]. This procedure follows a challenge/response protocol combined with a sequence-number-based protocol for network authentication derived from ISO/IEC 9798-4 [10]. In UMTS, authentication is typically performed through registration, a process for validating user records in a centralized database home location register (HLR); AuC, call origination, a process of originating an outgoing call; and call termination, a process of receiving an incoming call. Therefore, security architecture and authentication functions are embedded in 3G systems. In other words, there are no separate protocols or architectures for authentication in 3G systems, which is one of the features of centralized cellular systems.

For distributed or localized wireless systems, Mobile IP is proposed by IETF as a network layer protocol to facilitate macromobility in WLAN systems, which is not only recommended for WLAN systems but, more important, is considered as a networking protocol by 3GPP2 for 3G cellular systems [11]. In Mobile IP, each mobile node is associated with a home agent (HA) which behaves as a router for the mobile node and keeps all the information of this mobile node. If a mobile node moves out the coverage of a home agent, then it needs to contact a local agent, a so-called foreign agent (FA). However, the basic Mobile IP does not provide security protection on the communication links from an FA to an HA and from an FA to a mobile node (MN). Thus, Mobile IP with authentication, authorization, and accounting (AAA) extensions is developed to provide secure communications on these two links [12, 13].

Current research efforts on authentication and mobility management are focused on authentication architecture and signaling interoperation to reduce handoff latency [14–17]. Note that authentication in wireless networks brings about new challenges and design considerations, that go far beyond conventional security solutions for wired networks and mobility management for wireless networks. These design considerations, which are the driving force for developing new solutions rather than using existing mobility support technologies, can be summarized as follows.

First, in distributed or heterogeneous wireless networks, there may be many mobile users roaming among network domains with different technical specifications, signaling formats, identity authorization credentials, network protocols, and so on. The coverage of each autonomous network varies from tens of meters in WLANs to tens of kilometers in 3G systems, depending on the design and architecture of each network. Inside an autonomous network, there is an authentication server (AS), which is a centralized server for authenticating users within the coverage of the network, as shown in Figure 15.1, where an access router (AR) can be either a radio network subsystem (RNS) in UMTS or an access point (AP) in WLAN. Each mobile node, the device used by mobile users, has a permanent authentication association with the AS in its home network from which a mobile user subscribes service. Authentication servers trust each other based on security associations (SAs), which are a one-way relationship between a sender and a receiver for security services defined in IP security (IPSec). Authentication architec-

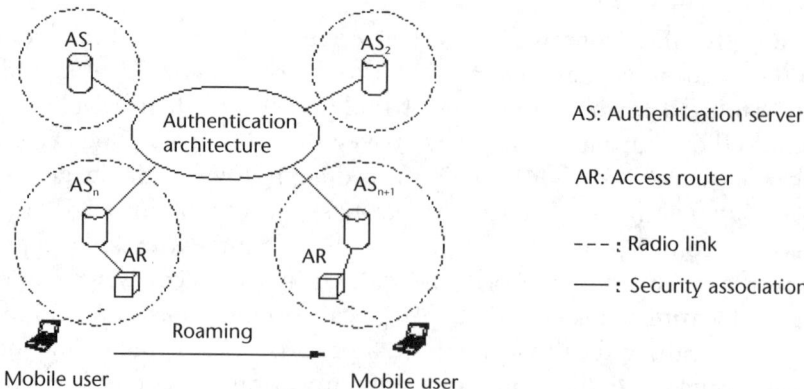

Figure 15.1 System architecture of authentication in wireless networks.

ture, which interconnects authentication servers, has a great impact on performance of authentication mechanisms and protocols for heterogeneous environments, because user credentials will be transmitted through the authentication architecture; thus, it is an important design issue in 3G/WLAN integration.

Second, ambient intelligence requires every device be involved with pervasive computing with sufficient security functions. In order to protect information secrecy, data integrity, and resource availability for users, security architecture and protocols are needed. Information secrecy means to prevent improper disclosure of information; data integrity is concerned about improper modification of data; and availability is to prevent denial-of-service attacks that may inject useless packets into the network on purpose to block legitimate network traffic. During an authentication process, especially for interdomain roaming, a mobile node negotiates cryptography algorithms with an authentication server and obtains keys for subsequent data transmission. In addition, authentication can mitigate the attack of denial of service. Complicated authentication protocols can use security associations for each connection segment and enable data encryption throughout the entire session of data delivery.

Finally, since authentication is a necessary procedure before actual transmissions over radio channels, it unavoidably introduces overhead into the network. An authentication process includes negotiation of encryption/decryption algorithms, encryption/decryption of messages, transmission of messages, and credential verification. Therefore, the overhead includes signaling, verification, and transmission cost used to exchange credentials between mobile clients, home networks, and authentication servers. Also, encryption/decryption algorithms require strong processing and computation capabilities, which must be considered for battery-operated portable devices.

15.3 Authentication Architecture for Interworking 3G/WLAN

Mobile IP is developed for terminal mobility over the Internet, so it enables roaming in WLAN, i. e., mobile terminals can move from one subnet to another without communication interruption [18]. Mobile IP is of particular important because it is the basis for the 3G/WLAN integration [1, 15]. Therefore, we focus on the authentication architectures that are related to Mobile IP networks since roaming capability is one of the most important features in ambient intelligence applications. In this chapter, we introduce three architectures proposed recently.

15.3.1 Mobile IP with AAA Extensions

In basic Mobile IP architecture, authentication extension (AE) is defined for registration messages, consisting of a security parameter index (SPI) and an authenticator calculated by using a keyed hash function. It is designed to provide entity authentication, which protects home agent (HA) and mobile nodes (MNs) against replay attacks, either by timestamps or by nonces (i.e., random numbers). However, AE does not provide data protection between a foreign agent (FA) and MNs. The protocol assumes that security associations (SAs) between FAs and HAs already exist. This assumption requires a huge effort to manage a large networking environment, and it may not be effective for scaling up networks. To strengthen relay protection, Mobile IPv4 challenge/response extensions (MICRE) have been developed. This protocol provides replay protection for all messages exchanged with the Mobile IP protocol by defining two new types of message extensions: challenge extension for FA advertisement messages and mobile challenge response extension for registration messages [19]. When an MN wants to authenticate itself, it must send an authentication request message with the challenge value received from the FA advertisement. By checking the challenge value the FA can avoid a malicious relay attack from an MN.

The verification of a challenge value depends on the security association between the MN and its HA, while the security association between the FA and the MN may or may not exist. A secure scalable authentication (SSA) is aimed to provide Mobile IP with a strong, scalable authentication mechanism based on public-key cryptography. When an MN moves close to an FA, it receives an advertisement with authentication extension and certificate extension broadcast by the FA. The MN then extracts and validates the certificate with a public key issued by a certificate authority. After the verification, the MN uses the public key of an FA to verify the digital signature in the FA authentication extension created by the FA's private key. Then, the MN will obtain the secret key of the FA; thus, the communication between the MN and the FA will be protected.

From this description, we can see that the authentication process may incur a significant delay due to transmitting credentials, which is critical to MNs that are far away from their home networks. Often, it is necessary for access routers to keep track of pending requests while a local authority contacts an appropriate remote authority. With these requirements, RFC 2977 further provides authentication architectures for Mobile IP networks with AAA extensions [17]. In the proposed basic architecture shown in Figure 15.2(a), each local AAA server (AAAL) should share security association with a home AAA server (AAAH) of a roaming MN in the current domain, so that the AAAL can securely transmit MN's credentials. This configuration, however, may cause a quadratic growth in the number of trust relationships, as the number of AAA authorities (AAAL and AAAH) increases. This problem has been identified by the IETF roaming working group [17, 20]. A possi-

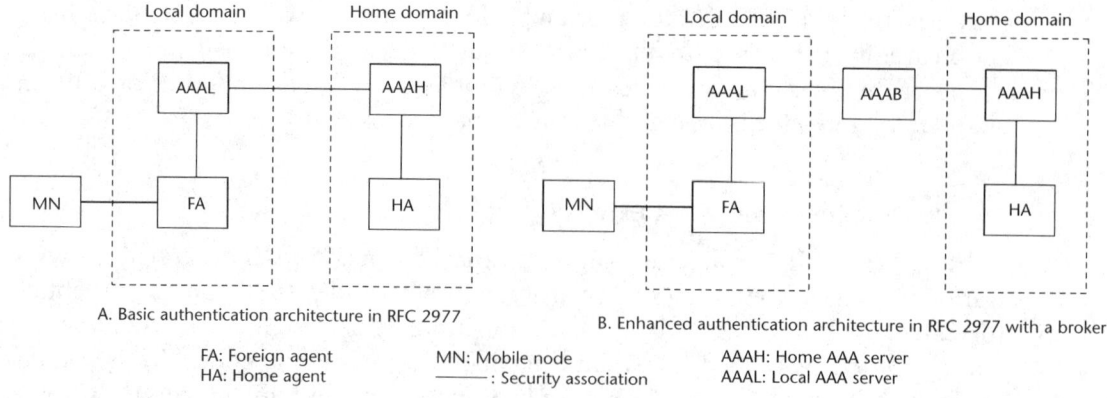

Figure 15.2 Authentication, authorization, and accounting in mobile IP networks.

ble solution to this problem is to use brokers to avoid overwhelming trust relationships between every pair of administrative domains. One example with multiple layers of brokers is shown in Figure 15.2(b). In this model, integrity or privacy of information between the home and serving domains may be achieved by either hop-by-hop security associations or end-to-end security associations established with the help of the broker infrastructure. A broker may play the role of a proxy between two administrative domains, which have security associations with the broker, and be able to relay AAA messages back and forth.

In addition, the Diameter protocol, published by the IETF, is as a practical solution for AAA in Mobile IP networks [3]. A Diameter server is defined as an authority center, which is able to authenticate, authorize, and collect accounting information for Mobile IPv4 service rendered to a mobile node. The Diameter base protocol is intended to provide an AAA framework for applications, such as network access or IP mobility, and work in both local AAA and roaming situations. Nowadays, Diameter is being deployed as a more flexible successor to the widely-deployed RADIUS protocol for authentication, authorization, and accounting. Security is enhanced between AR and either HA or MNs during AAA and the registration process. With these advantages, the Diameter protocol is recommended to be the authentication standard in Mobile IP networks.

15.3.2 Authentication Servers and Proxy

For interworking of 3G/WLAN, the main issue is how to identify a valid user and how to validate a user's credentials, which are maintained through authentication servers. In a 3G authentication server is proposed as a new functional component in 3G systems, which behaves as a gateway between WLANs and 3G systems to support interworking. The AAA server will terminates all AAA signaling from WLANs and route to other components in 3G systems. In this case, an AAA servers is also called AAA Proxy. The counterpart in WLANs is WLAN AAA proxy, which routes AAA messages to 3G servers. WLANs are identified based on network address identifier (NAI) included in an access request sent by mobile nodes. A similar idea that uses a security gateway (SGW) instead of authentication proxies to integrate IP mobility and security management together is shown in Figure 15.3. IPSec tunnel mode is enabled between the SGW and its MNs by which the SGW sets up an SA for each MN in its network. The MN maintains a single SA between itself and the SGW in its home network, whether the MN is in its home subnet or it roams to a foreign subnet. The HA is responsible for only Mobile IP registration and relaying packets to an MN's care of address (CoA). The MN is protected by the IPSec tunnel between the SGW and an MN. While the MN is roaming, there is no window of clear data transmission over wireless links, and there is no need to reestablish an IPSec tunnel between the SGW and an MN. Therefore, this scheme provides a secure communication segment between a roaming MN and its HA without requiring the participation of foreign networks.

15.3.3 AAA and Inter-Domain Roaming

The most important objective of Mobile IP with AAA extension is to support mobility for interdomain roaming. When a mobile object moves out of the coverage of its home network, the network address, which was assigned previously, such as an active IP session, is useless. However, a permanent IP address is the key to Mobile IP, so that regardless of an MN's current location, the packets can always be delivered to its permanent IP address. Therefore, interdomain roaming becomes a problem for Mobile IP users. To efficiently solve these problems, a common architecture for handling inter-system terminal mobility has been developed by us with the Mobile IP authentication architecture, as shown in Figure 15.4 [16]. In this architecture, mobility support is integrated with AAA functions through carefully designed

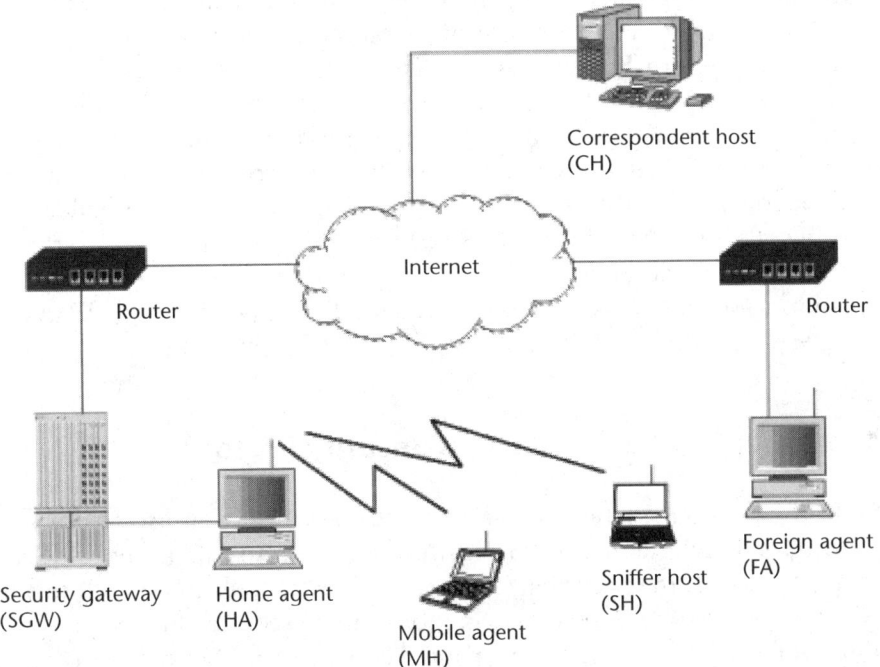

Figure 15.3 Integration of IP mobility and security mechanism.

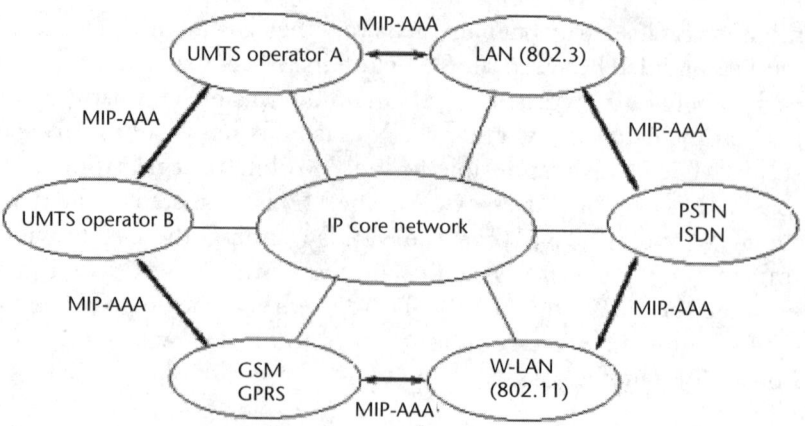

Figure 15.4 Mobility support in heterogeneous networks.

signaling messages. In other words, before an FA confirms the registration of a visiting node, it contacts a foreign AAA server with an access request message. Therefore, the AAA functions are completed along with the registration. Since no separate signaling is needed for authentication and registration, the number of packets exchanged is reduced.

In [15], two architectures for 802.11 and 3G integration are proposed: tightly-coupled and loosely-coupled interworking. In a tightly coupled architecture, the 802.11 network would emulate functions in 3G; that is the 802.11 hides all details for the 3G networks and appears as either a packet control function (PCF) in cdma2000, or as a serving GPRS support node (SGSN) to the core network. As a result, two domains would share the same authentication server for billing and accounting. To avoid the use of a common authentication mechanism, based on UMTS SIM (or USIM) for 3G or removable user identity module (R-UIM) cards for authentication on WLANs, a loosely coupled architecture can be used in which two domains can use their individual authentication solutions. The interworking of authentication requires that a new 802.11 network gateway support Mobile IP functionalities and new AAA servers in 3G networks. Consequently, 3G networks can collect records of users in 802.11 networks through AAA servers in two different network domains. This would not be a serious problem for interworking cdma2000 and WLANs, because cdma2000 supports Mobile IP and AAA, whereas it requires additional specifications for UMTS networks.

15.4 Authentication in Wireless Security Protocols

In the previous section, we introduced authentication architectures for integrated wireless networks. For each architecture, it is also important to design corresponding protocols that deliever messages and credentials in an architecture. In this section, we introduce several commonly used security protocols. Both UMTS and Mobile IP are designed to provide wireless services in mobile environments, especially for intra- and intersystem roaming. Although these technologies are going to be mature, their commercial deployment is still in the preliminary stage. WLAN technology and WLAN industry for nomadic roaming, which originated in the mid 1980s, are experiencing a tremendous growth due to the increased bandwidth made possible by the IEEE 802.11 standard. Considering that the Internet access is dominant in WLAN systems, many wireless security protocols are designed to protect link-level data without mobility support.

15.4.1 Wired Equivalent Privacy 802.11 LANs

Wired equivalent privacy (WEP) protocol is a built-in security feature of 802.11standards such as 802.11a/b/g. WEP protocol is designed to protect link-level data during wireless transmission, i.e., only the wireless portion of connections, between clients and access points; but, it does not provide end-to-end security. In particular, the WEP algorithm prevents unauthorized access to a wireless network by relying on a secret key shared between mobile stations and an access point. The WEP secret key encrypts packets before they are transmitted. Also, the WEP uses integrity check to prevent packets from being modified in transit. Most installations use a single WEP key between the mobile stations and access points, even though multi-WEP key techniques can enhance security.

15.4.2 Extensible Authentication Protocol and its Variants

Extensible authentication protocol (EAP) is the mechanism that is used between a client and an authenticator. Standard 802.1x can use EAP protocol as a transport mechanism for exchanging messages. The 802.1x standard specifies encapsulation methods for transmitting EAP messages so they can be carried over different media type. EAP is a framework for providing centralized authentication and dynamic key distribution, and it is a general protocol supporting multiple authentication methods, such as token cards, Kerberos certificates, public-key authentication, etc. EAP enables wireless client adapters that may support different authentication types to communicate with different back-end servers, such as remote authentication dial-in user service (RADIUS). When used with 802.1x, it provides end-to-end authentication when a wireless client that associates with an AP cannot gain access to the network until the user performs a network logon. In wireless communications using EAP, a user requests connection to a WLAN through an AP, which then requests the identity of the user and transmits that identity to an authentication server, such as RADIUS. The server asks the AP for proof of identity, which the AP obtains from the user and then sends back to the server to complete the authentication. EAP performs mutual authentication, so each side is required to prove its identity to the other, using its certificate and private key. When both client and server authenticate each other, resulting in stronger security, EAP protects against man-in-the-middle and sniffing attacks.

15.4.3 802.1x Authentication Protocol

Standard 802.1x provides network login capabilities between PCs and edge networking entities such as an access point. It offers an architectural framework for implementing various authentication schemes. Standard 802.1x itself does not provide encryption, and it is not an alternative to WEP, 3DES, AES, or any other cipher. Therefore, 802.1x is focused only on authentication and key management, and it can be used in combination with any ciphers. Also, 802.1x is not a single authentication method; rather, it utilizes EAP as its authentication framework. This means that 802.1x-enabled switches and access points can support a wide variety of authentication methods, including certificate-based authentication, smart cards, token cards, one-time passwords, and so on. Standard 802.1x supports open standards for authentication, authorization, and accounting much like AAA in Mobile IP (including RADIUS, so it works with the existing infrastructure for managing remote and mobile users. It can be combined with an authentication protocol, such as EAP-TLS, LEAP, or EAP-TTLS,

Standard 802.1x provides port-based access control and mutual authentication between clients and access points via an authentication server. It's authentication protocol consists of three entities:

1. Authenticator: An access point that has 802.1x authentication enabled. This includes LAN switch ports and wireless access points.
2. Authentication server: A server that performs authentication, allowing or denying access to the network based on username or password. The 802.1x standard specifies that the RADIUS is the required Authentication Server.
3. Supplicant (client): An access device requesting LAN service.

Once 802.1x authentication is enabled (both in the client and authenticator), a successful authentication must be completed before any traffic is allowed to transit the network from the client, including critical traffic, such as DHCP requests, regardless of whether a link is established between the client and authenticator (switch port). The 802.1x client will transmit appropriate EAP messages to the authenticator (switch port). The switch port with 802.1x authentication enabled is set to an uncontrolled state, accepting only EAP messages (all other traffic will be discarded). Upon receipt of the client's EAP message, the switch forwards the request to the authentication (RADIUS) server without changing its contents. Although the EAP contents have not been changed, the encapsulation must be translated from the originating EAP message to a RADIUS request; therefore, the only supported RADIUS servers are ones that support EAP. In response to the RADIUS message, the authentication server will grant or deny access to the network. A RADIUS response will then be transmitted back to the switch, which will determine whether the port remains in an uncontrolled state (access denied) or changes to a controlled state (access granted). Security flaws in 802.1x architecture are concerned with the absence of mutual authentication and may cause a man-in-the-middle attack. Another issue is session hijacking in which the original user is not able to proceed, but the attacker acts as an original user and starts using the session created by original user.

15.4.4 WiFi Protected Access and 802.11i

WiFi Protected Access (WPA) is a standards-based, interoperable security specification, which significantly increases the level of data protection and access control for existing and future wireless LAN systems. WPA is a subset of the 802.11i draft standard and will maintain forward compatibility. It will replace WEP as standard Wi-Fi security. The task group's short-term solution is to improve WiFi protected access (WPA) to address the problems of WEP. The group also defines the temporal key integrity protocol (TKIP) to address the problems without requiring hardware changes—that is, requiring only changes to firmware and software drivers.

To strengthen user authentication, WPA implements 802.1x and EAP together by using a central authentication server, such as RADIUS, to authenticate each user on the network before they join it, and also employs mutual authentication so that the wireless user does not accidentally join a rogue network that might steal its network credentials.

The IEEE 802.11i draft standard defines additional capabilities required for secure implementation of IEEE 802.1X on 802.11 networks. These include a requirement for using an EAP method in supporting mutual authentication, key management, and dictionary attack resistance. In addition, 802.11i defines the hierarchy for use with the TKIP and AES ciphers and a four-way key management handshake used to ensure that the station is authenticated to the AP and a back-end authentication server, if present. As a result, to provide adequate security, it is important that IEEE 802.1x are implemented in 802.11 networks for security enhancements.

15.4.5 Virtual Private Network

A virtual private network (VPN) creates a private network over a publicly accessible medium, such as a WLAN. It provides private communication between users, remote offices, and businesses. A VPN works by maintaining privacy through security procedures, including encryption, keying, authentication, and tunneling protocols. These protocols verify users and servers, encrypt data at the sending end, and decrypt it at the receiving end. VPN creates a tunnel that cannot be entered by data or users who are not properly encrypted or authenticated. VPNs use security protocols such as IPSec, layer two tunneling protocol (L2TP), and Point-to-point tunneling protocol (PPTP). VPN uses encryption protocols such as triple DES (3DES) and AES, which are much stronger than those used in WEP. VPN is a strong solution, enabling authorized wireless users to connect securely from virtually anywhere, and it is transparent to applications.

15.5 Comparison Study of Wireless Security Protocols

There are many research efforts on security protocols from a functional perspective; however, there is a lack of quantitative results demonstrating the impact of security protocols on system performance that can be affected dramatically by applying security policies in combination with mobility. Therefore, we conducted an experimental study on a wireless IP test bed and analyzed the interaction of security protocols at different layers with respect to data streams, delay, and throughput [21].

15.5.1 Security Policies

Security policies are designed to demonstrate potential security services provided by each security protocol. Each protocol uses various authentication and encryption mechanisms to provide security. Therefore, by configuring different security mechanisms for each protocol, a variety of security policies are implemented in the test bed. Besides individual policies, hybrid security policies are also configured, involving multiple security protocols at different network layers. In Table 15.1, a total of 12 policies—namely, PN-1 to PN-12 are listed. We consider that no security case as a special policy, PN-1, which is used in evaluating authentication overhead.

We evaluate security policies in different mobile scenarios by considering the current location of the mobile node (MN) in the network. Therefore, we investigate both "no roaming" (NR) and "with roaming" (WR) scenarios. "With roaming" (WR) refers to the situation when one of the mobile nodes is visiting a foreign network, whereas "no roaming" (NR) means when all MNs stay in their home network. Moreover, those mobility scenarios take into account the presence of correspondent nodes (CN) also. In our test bed, we have considered correspondent nodes as both wireless and wired. Table 15.2 shows all the scenarios considered.

Table 15.3 shows authentication time (seconds) for IPSec and 802.1x security policies. Since WEP does not involve exchange of control messages, so there is no authentication time involved in it. Moreover, authentication time for IPSec and 802.1x also involves Mobile IP authentication time. We observe that when an MN is not roaming, IPSec authentication takes longer than 802.116. However, when an MN roams, the 802.1x authentication time is longer. This is due to the fact that when an MN roams, MN re-authenticates with an FA using the 802.1x mechanism, whereas this is not the case with IPSec protocol, because the IPSec tunnel is already established between the MN and the HA. It is also observed that 802.1x with IPSec policies causes longer authentication delay than 802.1x without IPSec policies. Furthermore, Table 15.3 shows that 802.1x-EAP-TLS authentication time is longer than 802.1x-EAP-MD5 because 802.1x-EAP-TLS uses digital certificates for mutual authentication, which involves exchange of several control packets.

Table 15.1 Security Policies

Policy Number	Description of Security Policies
PN-1	No Security
PN-2	WEP-128 bit key
PN-3	IPSec-3DES-SHA
PN-4	IPSec-3DES-SHA-WEP-128
PN-5	8021x-EAP-MD5
PN-6	8021x-EAP-TLS
PN-7	8021X-EAP-MD5-WEP-128
PN-8	8021X-EAP-ELS-WEP-128
PN-9	8021X-EAP-MD5-WEP-128-IPSec-3DES-MD5
PN-10	8021X-EAP-TLS-WEP-128-IPSec-3DES-MD5
PN-11	8021X-EAP-MD5-WEP-1280IPSec-3DES-SHA
PN-12	8021X-EAP-TLS-WEP-1280IPSec-3DES-SHA

Table 15.2 Mobility Scenarios

Number	Scenario	Roaming or No-Roaming
M1	Mobile to mobile node in the same domain	No Roaming (NR)
M2	Mobile node to home agent	
M3	Mobile node to corresponding node (fixed) in the same tomain (register to HA)	
M4	Mobile node to mobile node in different subnets	
M5	Mobile node to corresondeing node (fixed) in the same domain	
M6	Mobile node to mobile node in different domains	With Roaming (WR)
M7	Mobile node to corresponding node (fixed) in different domain (register to FA)	
M8	Mobile node and corresponding node (fixed) in different domains	
M9	Mobile to mobile node in the same Domain	

Table 15.3 Authentication Time Measurements for Different Security Policies

Policy	IPsec (seconds)	802.1x-EAP(MD5) with IPsec (seconds)	802.1x-EAP(MD5) without Ipsec (seconds)	802.1x-EAP(TLS) without Ipsec (seconds)	802.1x-EAP(TLS) with IPsec (seconds)
Non-Roaming	1.41	0.43	1.72	1.82	3.12
Roaming	2.84	2.18	3.47	4.97	6.28

Since WEP does not involve exchange of control messages, there is no authentication time involved with it. Since Mobile IP is used for enabling mobility in the test bed, authentication time AT for IPsec and 802.1x involves Mobile IP authentication time as well. In Figures 15.5 and 15.6, we demonstrate throughput variations for TCP and UDP data streams for some security policies in all mobility scenarios. Here we have presented only one security policy for each security protocol. We observe that overall IPSec security policies cause greater decrease in throughput than WEP and 802.1x security policies. This is because IPSec uses the 3DES encryption algorithm which is computationally slower than the encryption algorithm used in WEP and 802.116. But IPSec provides stronger security services, which compensates for the higher encryption overhead.

15.6 Conclusions

In this chapter, we introduced major authentication and security protocols for ubiquitous wireless communications in integrated 3G and WLAN systems. Authentication is inherently a security technique, which is designed to protect networks against acceptance of a fraudulent transmission by establishing the validity of a transmission, a message, or an originator. Therefore, it is an important issue in ambient intelligence that involves pervasive computing of portable devices. In this chapter, we first described the new challenges and design considerations of authentication

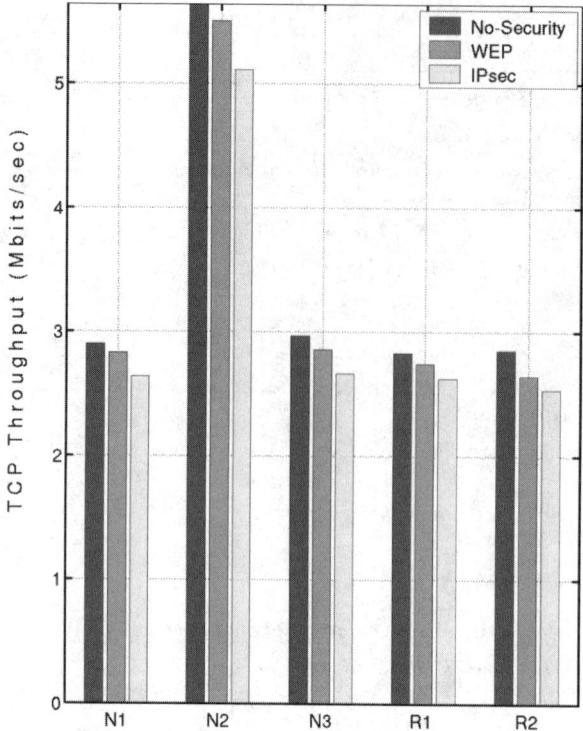

Figure 15.5 TCP Throughput.

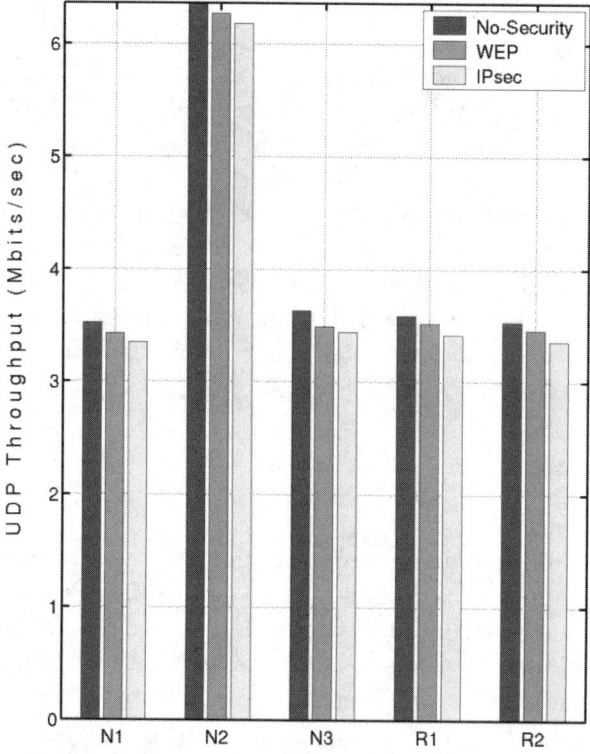

Figure 15.6 UDP Throughput.

protocols in wireless networks. We then presented different authentication architectures and security protocols in UMTS, WLANs, Mobile IP, and the integrated 3G and WLAN systems. We also showed experimental results of a wireless LAN test bed with cross-layer security solutions to demonstrate the impact of security protocols on authentication time and system throughput.

References

[1] Køien, G., and T. Haslestad, "Security Aspects of 3G-WLAN Interworking," *IEEE Communications Magazine*, Vol. 41, pp. 82–88, November 2003.

[2] Karygiannis, T. and L. Owens, "Wireless Network Security 802.11, Bluetooth and Handheld Devices," *NIST Special Publications 800-48*, November 2002.

[3] Calhoun, P., et al., "Diameter Base Protocol," draft-ietf-aaadiameter-17.txt, December 2002.

[4] Jacobs, S., "Security and Authentication in Mobile IP," *IEEE International Symposium on Personal, Indoor and Mobile Radio Communications (PIMRC)*, Vol. 3, pp. 1103–1108, September 1999.

[5] Perkins, C., and P. Calhoun, "Mobile IPv4 Challenge/Response Extensions," *IETF RFC 3012*, November 2000.

[6] Xu, M., and S. Upadhyaya, "Secure Communication in PCS," *IEEE Vehicular Technology Conference*, Vol. 3, pp. 2193–2197, 2001.

[7] Akyildiz, I., et al., "Mobility Management in Next-Generation Wireless Systems," *Proceedings of the IEEE*, Vol. 87, pp. 1347–1384, August 1999.

[8] Lin, Y-B, M-F Chen, and H. C-H Rao, "Potential Fraudulent Usage in Mobile Telecommunications Networks," *IEEE Trans. Mobile Computing*, Vol. 1, No.2, pp. 123–131, 2002.

[9] 3GPP. "3rd Generation Partnership Project; Technical Specification Group Services and Systems Aspects; 3G Security; Security Architecture," *Technical Specification 3G TS 33.102 V3.7.0 (2000-12)*, 2000.

[10] ISO/IEC. "Information Technology Security Techniques—Entity Authentication—Part 4: Mechanisms Using a Cryptographic Check Function," *Technical Report ISO/IEC 9798-4*, ISO/IEC, 1999.

[11] Patel, G., and S. Dennett, "The 3GPP and 3GPP2 Movements toward an All-IP Mobile Network," *IEEE Wireless Communications*, Vol. 7, pp. 62–64, August 2000.

[12] Johnson, D. B., C. Perkins, and J. Arkko, "Mobility Support in IPv6," *IETF Internet Draft*, draft-ietfmobileipv6-17.txt, May 2003.

[13] Calhoun, P., J. Loughney, G. Zorn, and J. Arkko, "Diameter Base Protocol (Request for Comments 3588)," http://www.ietf.org/rfc/rfc3588.txt, September 2003.

[14] Misra, A., S. Das, A. McAuley, and S. Das, "Autoconfiguration, Registration, and Mobility Management for Pervasive Computing," IEEE Personal Communications, Vol. 8, pp. 24–31, August 2001.

[15] Buddhikot, M., et al., "Integration of 802.11 and Third-Generation Wireless Data Networks," *IEEE INFOCOM'03*, April 2003.

[16] Cappiello, M.. A. Floris, and L. Veltri, "Mobility amongst Heterogeneous Networks with AAA Support," *IEEE ICC 2002*, Vol. 4, pp. 2064–2069, 2002.

[17] Glass, S., et al., "Mobile IP Authentication, Authorization and Accounting Requirements," RFC2977, October 2000.

[18] Perkins, C., "IP Mobility Support for IPv4," Request for Comments (RFC) 3220, January 2002.

[19] Perkins, C., and P. Calhoun, "Mobile IP Challenge/Response Extensions," draft-ietf-mobileip-challenge-09.txt, February 2000.

[20] Aboba, B., and J. Vollbrecht, "Proxy Chaining and Policy Implementation in Roaming," RFC2607, June 1999.

[21] Agarwal, A. K., and W. Wang, "Measuring Performance Impact of Security Protocols in Wireless Local Area Networks," *2nd IEEE International Conference on Broadband Networks—Broadband Wireless Networking Symposium*, Boston, USA, October, 2005.

CHAPTER 16
Learning in the AmI: from Web-Based Education to Ubiquitous Learning Experiences

Charalampos Karagiannidis, Athanasios Vasilakos

This chapter discusses learning in the ambient intelligence (AmI) environment. The chapter begins with a short review of the current paradigm in learning technologies. It then discusses mobile learning, as the current emerging paradigm for learning. The chapter concludes with some theoretical and technical issues for the exploitation of AmI toward the delivery of personalized, ubiquitous learning experiences, which are seamlessly integrated within our everyday activities.

16.1 Introduction

The emergence of the knowledge society poses new requirements for education and training: the knowledge-based economy requires a flexible, very well trained workforce; and the citizens of the information society need to be continuously (re)trained in order to remain competitive within this workforce and to fully exploit the knowledge society for their personal development.

The rapid evolution of learning technologies—exploiting the respective developments in information and communication technologies (ICT)—create numerous new opportunities for meeting these requirements: Web-based learning environments (learning management systems, learning content management systems, etc.) deliver life-long education and training applications and services to anyone, anytime, anyplace. However, most of these applications realize a learning context that is rather traditional in nature: It is based on the notion of one (or many) tutors, who help learners acquire a specific body of knowledge (through learning material, learning activities, etc.), which can be measured through specific assessment methods.

Recent developments in ICT facilitate a departure from this model, since learning can be embedded into our everyday environment (objects, devices, activities, etc.). The exploitation of this potential can bring about a new era for learning: just-in-time, just-enough, on-demand personalized learning experiences, seamlessly integrated within our everyday activities.

This chapter reviews some theoretical and technical issues toward the exploitation of the ambient intelligence (AmI) for the delivery of learning experiences that go beyond the current state of the art.

16.2 The Present Paradigm in Learning Technologies

Learning technologies[1] are attracting increasing interest worldwide, since they can meet the requirements of the knowledge society and knowledge-based economy for high-quality, lifelong learning. Over the past decades, a number of applications and services have been made available out of these efforts, that reflect the current paradigm of learning: from computer-aided instruction, to Web-based learning.

Over the last decade, a major transformation has taken place, mainly due to the wide adoption of the Internet. The main drivers for this transformation can be summarized as follows: [2]

- Demand: rapid obsolescence of knowledge and training; need for just-in-time training delivery; search for cost-effective ways to meet learning needs of a globally distributed workforce; skills gap and demographic changes, which drive the need for new learning models; demand for flexible access to lifelong learning; and so on;
- Supply: Internet access is becoming standard at work and at home; advances in digital technologies enable the creation of interactive, media-rich content; increasing bandwidth and better delivery platforms make e-learning more attractive; a growing selection of high-quality e-learning products and services are made available; emerging technology standards facilitate compatibility and reusability of e-learning products; etc.

As a result, the interest in learning technologies in the past few years has turned mainly into: the development of learning material, activities and software, of high-quality, exploiting multimedia, interactive, immersive, technologies, and the delivery and management of learning material, activities, and software.

The current paradigm[3] of learning technologies can be summarized in , which depicts the idea of learning content management systems (LCMSs): reusable learning objects (any digital resource that can be reused to support learning [1]), learning activities; software, are created by instructional designers and published within a repository, together with their description (which is based on a common format, e.g., through learning technologies specifications and standards); learners can search this repository, and retrieve and access learning objects according to their profile (which can also be described through a common format, through the same specifications and standards). LCMSs offer a number of advantages, including re-usability of learning material (facilitating the development of economies of scale); personalized access to learning material; and just-in-time, on-demand, and just-enough learning.

1. We use this term to refer to applications of ICT for learning.
2. Source: SRI Consulting and WR Hambrecht + Co.
3. We use this term to refer to mainstream technologies, rather than the state-of-the-art.

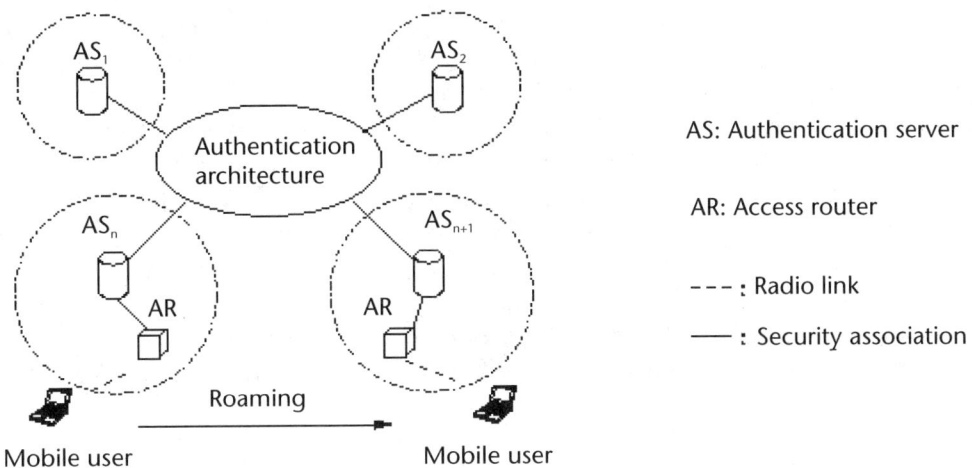

Figure 16.1 Learning content management systems (*From*: [1]).

These developments facilitate the departure from a number of constraints, relating to time, place, and so on: learning (mainly material) is available to anyone, anytime, anyplace. However, the current state of the art still realizes a learning context that is rather traditional in nature: Learners access a common repository to acquire a body of knowledge, which can be measured through specific measures. Moreover, a number of constraints are not yet overcome: learners need to be in specific places, accessing the Internet through a desktop or laptop machine.

The next section describes mobile learning, as an emerging paradigm for overcoming one additional constraint: Learners can learn anywhere.

16.3 The Emerging Paradigm: Mobile Learning

Mobile learning is usually defined as learning that takes place via wireless devices, such as mobile phones, personal digital assistants, laptop computers, and so on. That is, in most definitions encountered in the literature, it is only the employment of specific types of technology that differentiates mobile learning from other forms of learning. However, when considering the learner's point of view, it can be argued that the main difference is that mobile learning can take place everywhere: students can learn while in outdoor activities, engineers can update their knowledge while in the field, and so forth. In this context, a definition of mobile learning should be widened to include any sort of learning that happens when the learner is not at a fixed, predetermined location, as well as learning that happens when the learner takes advantage of the opportunities offered by mobile technologies [3].

There are a number of reasons that make mobile technologies and devices attractive for learning, including palmtops, which are relatively inexpensive, compared with full-sized desktop or laptop computers; they offer the possibility of ubiquitous computing; they facilitate access to information and promote the development of information literacy; and they offer the possibility of collaborative learning and independent learning [3].

Table 16.1 summarizes the current use of mobile technologies and devices for learning.

During the past few years (since mobile devices and technologies have become mainstream) a number of prototype learning applications have been deployed and tested in different contexts (extensive reviews are included in [3–6]). Table 16.2 is mainly based on [5], which categorizes these efforts according to their underlying learning model.

Table 16.2 demonstrates that learning and teaching with mobile technologies is beginning to make a breakthrough from small-scale pilots to institution-wide implementations. A number of key issues need to be taken into account in the development of these implementations, which can be summarized as follows [5]:

- Context: gathering and utilizing contextual information may clash with the learner's wish for anonymity and privacy;
- Mobility: the ability to link to activities in the outside world also provides students with the capability to escape the classroom and engage in activities that do not correspond with either the teacher's agenda or the curriculum;
- Learning over time: effective tools are needed for the recording, organization and retrieval of (mobile) learning experiences;
- Informality: students may abandon their use of certain technologies if they perceive their social networks to be under attack;
- Ownership: students want to own and control their personal technology, but this presents a challenge when they bring it in to the classroom.

Finally, related literature includes a number of guidelines that can help developers to address these issues, along with more practical concerns such as cost, usability, and technical and institutional support, including the following [3]:

Table 16.1 Levels of Objectives: Mobile Computers in Education (*From:* [4]).

Level 1	Level 2	Level 3	Level 4
Productivity	Flexible physical access	Capturing and integrating data	Communication and collaboration
Sample applications			
Calendars	Local database	Network database	Real-time chat
Schedule	Interactive prompting	Data collection	Annotations
Contact	Just-in-time instruction	Data synthesis	Data sharing
Grading		Mobile library	Wireless email
Content-intensive			Communication-intensive
Users: individual			Users: group
Mostly asynchronous			Mostly synchronous
Information storage			Knowledge construction
Hardware-centered			Network-centered
Isolation			Interconnection

16.3 The Emerging Paradigm: Mobile Learning

Table 16.2 Review of Mobile Technologies for Learning (*From:* [5])

Theme	Key Theorists	Activities	Example Systems
Behaviorist learning	Skinner, Pavlov	Drill and feedback Classroom response systems	Skills Arena: a mathematics video game where drills in addition and subtraction are presented as a game with advanced scoring and recordkeeping, character creation and variable difficulty level
			BBC Bitesize: provides revision materials via mobile phones; it has been running since 2003, and has proved to be very popular (over 650,000 GCSE students, as well as a number of curious adult learners)
			m-phones for language learning: SMS is used as part of an English language course, where students receive frequent vocabulary messages (which also act as reminders to revise)
			Classroom response systems: e.g. Classtalk which engages students in communication with the classroom, for articulating and presenting their ideas
Constructivist learning	Piaget, Bruner, Papert	Participatory simulations	The virus game: each student wears a badge-tag which shows whether they are "infected"; students can watch the "spread of the disease", through their communication
			Savannah: students play the role of lions roaming in the wild in an area 100m x 50m; each student carries a PDA that gives them a "window" into the gameworld, displaying content and actions that were appropriate to their current location and what was going on in the rest of the game
			Environmental detectives: a scenario was built around a spill of a toxin; students develop a suitable remediation plan, assisted by their PDAs which allow them virtual activities based on their virtual location
Situated learning	Lave, Brown	Problem and case-based learning Context awareness	Ambient wood: integrates physical and digital interaction; digital information is coupled with novel arrangements of electronically-embedded physical objects; a series of activities are designed around the topic of habitats, focusing on the plants and animals in the different habitats of woodland and the relationships between them
			Natural science learning: a butterfly-watching system; a database of different butterfly species is used with a content-based image retrieval system; students visit a butterfly farm, take photographs of the butterflies and query the database for possible matches
			Tate modern: allows visitors to view video and still images, listen to expert commentary and reflect on their experience by answering questions or mixing a collection of sound clips to create their own soundtrack for an artwork; the wireless network is location-sensitive, which means that users do not have to search out the information
			MOBIlearn: context-awareness is being explored, not just as a way to deliver appropriate content, but to enable appropriate actions and activities, including interactions with other learners in the same or similar contexts
Collaborative learning	Vygotsky	Mobile computer-supported collaborative learning (MCSCL)	MCSCL: activities are distributed through the teacher's hand-held device; the teacher downloads the activity from the project website, and then transmits the activity to the students; the students are automatically assigned to collaborative groups; upon completion of the activity, the teacher's Pocket PC collects the students work, which can then be downloaded to the school's PC for analysis
Informal and lifelong learning	Eraut	Supporting intentional and accidental learning episodes	m-learn: mobile technology to teach basic literacy and numeracy skills; custom content was created, for example an urban soap opera about two characters moving into a flat for the first time to help with language and provide advice about how to set up a home
			Breast cancer care: delivers personalized education of breast cancer patients; the users can query specific subject knowledge bases through a content specialist, to gain the information they need
Learning and teaching support	n/a	Personal organization Support for administrative duties (e.g. attendance)	Student learning organizer: an integrated suite of software tools enabling students to create, delete and view timetable events and deadlines, as well as download course material packages
			Support for teachers and administrators: managing teachers' workloads and supporting teaching and learning
			SMS supports computing students at risk: develops, delivers and evaluates blending learning opportunities that exploit SMS, WAP and VLE technologies; students use SMS text messaging, receive noticeboard information such as room changes, appointments, feedback and exam tips via SMS

- Investigate a cost model for infrastructure, technology, and services.
- Study the requirements of all those involved in the use of the technology (learners, teachers, content creators) to ensure it is usable and acceptable.
- Assess that the technology is suited to the learning task and examine advantages and disadvantages of each technology before making a decision on which one to use.
- Assign the necessary roles for initiating and thereafter supporting mobile learning.
- Develop procedures and strategies for the management of equipment when it is provided by the institution.
- Provide training and (ongoing) technical support to the teachers to enable them to use mobile technologies to enhance current and enable new instructional activities.
- Consider the use of mobile technologies for student administration tasks.
- Consider the use of mobile technologies to support collaborative and group learning.
- Discover and adopt suitable applications that match the needs of your specific classroom and map directly to your curriculum needs.
- Ensure security and privacy for the end users.

16.4 The Future Paradigm: Learning in the AmI

The concept of ambient intelligence (AmI) provides a wide-ranging vision on how the knowledge society will develop: People are surrounded by intelligent intuitive interfaces which are embedded in all kinds of everyday objects; AmI is capable of recognizing and responding to the presence of different individuals; and, most important, the AmI works in a seamless, unobtrusive, and often invisible way. That is, the emphasis of AmI is on user friendliness, more efficient services support, user empowerment, and support for human interactions [7].

Perhaps the most comprehensive description of what learning might be like in AmI is included in the Information Society Technologies Advisory Group (EC ISTAG) report [7]. The scenario is centered around the concept of the ambient for social learning (ASL), which can empower learners and facilitate learning in a number of different ways. From a technical point of view, the ASL integrates in a smooth and seamless way a large variety of technologies, which are described in different chapters of this book (e.g., [8]). From a learning/learner point of view, the main innovations of the ASL can be summarized as follows:

- *Learner knowledge:* the ASL can recognize and identify the profile of each learner, i.e., his/her background, interests, goals, preferences, etc, as well as his/her everyday agenda; the ASL utilizes this knowledge so as to facilitate learning, empower individual learners, form learning communities, plan individual and collaborative learning experiences, and so on.
- *Domain knowledge:* the ASL can have access to a variety of knowledge sources, and can negotiate with other ambients in order to gain knowledge for

different domains; the ASL utilizes this knowledge to act as an expert who can guide the learner (or group of learners) during the learning experience.
- *Tutoring knowledge:* the ASL is capable of adopting a variety of flexible (individual or collaborative) tutoring strategies, which can be continuously adapted to the profile of each individual learner, the dynamics of learner communities, and so on.
- *Common sense knowledge:* The ASL demonstrates common sense knowledge, including knowledge about the discourse of communication, so as to get involved in natural dialogs with the learners; it can collaborate and negotiate with other ambients, so as to get additional knowledge, e.g. concerning other learners, learning domains.

Based on these resources, the ASL is capable of supporting innovative learning experiences:

- The ASL can create learning experiences that are adapted to each individual learner (profile, agenda, etc.); it can negotiate with the learner its degree of participation in the learning experience: it can answer learner's questions, make suggestions, take initiatives, and so on; in fact, the ASL can share experience and synchronize its mental state with the individual learner; moreover, the ASL can support learning communities, both at the semantic level (based on the profile of each individual learner, the learning domain, etc.), as well as at the management level; it can schedule meetings, set the agenda, monitor the meetings, and provide advice at any stage.
- These services are available to learners through intuitive interfaces, which are embedded within everyday objects and are seamlessly integrated in everyday activities.

The next section discusses some theoretical and technical issues that can contribute toward the realization of AmI for learning, both from a technical, and from a user and commercial point of view.

16.5 Enabling Technologies, Models, and Standards

The realization of AmI requires a number of technological innovations, which span from network technologies to advanced intelligent user interfaces. For example, the EC ISTAG has identified the following main technologies, which are required for the realization of AmI: recognition (tracing and identification) of individuals, groups, and objects; interactive aids for negotiating targets and challenges (goal synchronization); natural language and speech interfaces and dialog modeling; projection facilities for light and sound fields; tangible/tactile and sensorial interfacing; content design facilities, simulation and visualization aids; and knowledge management tools to build community memory.

In the following text, we briefly describe some areas that are particularly related to learning; other issues are described in other sections of this book.

16.5.1 Personalization Technologies and Computational Intelligence

Personalization technologies are attracting increasing interest, since they can provide high-quality applications and services (for information retrieval, e-commerce, performance support systems, etc) that can be adapted to the profile of each individual user. Especially in learning environments, personalization is in line with the basic premise of all learning theories: There is no average learner; learners are diversified according to a variety of dimensions, therefore learning should be adapted to their individual profile.

Research and development in personalized learning started more than three decades ago and has produced a number of systems, mainly in the areas of intelligent tutoring systems, intelligent pedagogical agents adaptive educational hypermedia, and so forth. Personalized learning systems are based on the utilization of knowledge concerning the learner (learner model), the learning domain (domain model), the tutoring strategies (tutoring model) and the user interaction (interface model), for dynamically adapting the learning experience to the learner requirements, which are modified continuously [9].

Despite these intensive research efforts, personalization is not yet a mainstream technology; most existing systems are still prototypes, and have not reached commercial exploitation. This is due to a number of reasons, including the following:

- *Models and authoring:* the theoretical basis for personalized learning has not yet matured at the level of providing specific guidelines for building personalized learning environments; a direct result of this fact is that it is very difficult for "authors" (tutors, instructional design experts, etc.) to design material, and activities, for such personalized learning environments, even with the existence of powerful authoring tools [10].
- *Evaluation, reusability, and standards:* the lack of evaluation data, as well as the difficulty in its generalization, when available, and the resulting difficulty in the reuse of successful design practices constitutes, one of the main barriers for adaptive applications and services to become mainstream technology; also, there are currently no standards that can promote the re-use of personalization methods and techniques, and the interoperability of tools [11].

Our objective is to study how computational intelligence (CI) [12] can complement, augment, and expand human wisdom. In particular, we emphasize the intertwining linkages between CI and AmI. Several key features that come hand in hand with the paradigm of CI are worth emphasizing:

- Flexibility of expressing granularity of information acquired through processes of learning; this high versatility directly opens up various opportunities of seamless operation at the most suitable level of abstraction; this feature becomes of paramount relevance when realizing communication mechanisms among different participants of the learning.
- Adaptive properties of information granules whose size (granularity) could be made adjustable; various mechanisms of learning are anticipated to help deliver the required flexibility of the learning environment.

- Structural reconfiguration of the mechanisms of learning, selective focus on various learning resources, establishing new mechanisms of collaboration; all of these could be completed within the realm of intelligent agents, whose topology and functionality could be structurally optimized within the framework of evolutionary computing or other biologically inspired mechanisms.
- Transparency of developed topologies, calibration mechanisms, and topological refinements that is fully supported by the ideas of logic-driven modeling—one of the cornerstones of CI; the inherent logic character of the underlying mechanisms of collaborative learning, selective perception, and stepwise refinement of learning interactions are of the outstanding features in the context of mobile learning.

There are numerous opportunities along the line of mobile learning. In what follows, we elaborate on several interesting and the most promising pursuits:

- Virtual participation and participants: Mobile learning can be realized between various participants as well as virtual participants (software agents); the objective is to identify and investigate various forms of interaction, quantify their intensity and looking into detailed realization and further assessment of their ensuing efficiency.
- AmI and mobile learning: It becomes evident that with the omnipresence of diverse sources of knowledge, their accessibility needs to be highly selective to avoid a very likely possibility of overload, leading to inefficient and reduced learning; this implies that CI can play a role of a sensitive, intelligent, and learner-aware filtering layer; the AmI environment endowed with the operational framework of CI is to be investigated in this setting.
- Building effective mechanisms of learner feedback (relevance feedback) and optimization of a level of interaction along with its dynamics: Learning requires feedback; the quest is to form efficient ways of forming relevance feedback by using that the learner provides as feedback—this triggers some dynamic changes to the learning process; here, the role of CI is to quantify the relevant feedback, form fuzzy associations between outcomes of learning and key parameters of the learning environment, and finally arrive at the change strategy,
- Emotion–enhanced learning processes: Emotions (e.g., expressed by a level of interest) are critical to the overall effectiveness of learning; here, our objective is to embark on this challenging issue, investigating how emotions could be captured, quantified, and converted into the most suitable mechanism of high-quality and specialized relevant feedback.
- Multiuser distributed and collaborative processes: The objective of this pursuit is to concentrate on the multiuser and highly collaborative learning environment and develop its detailed models.
- quantitative assessment of mechanisms of mobile and interactive learning: We are concerned with a comprehensive, in-depth, and numeric quantification of the effectiveness of mobile learning; in this context, CI is being regarded as a vehicle to build models of learning and its quality.

16.5.2 Learning Technologies Standards

During the past few years, a number of powerful authoring tools have been made available in the market for the development of learning resources, i.e., learning material, learning activities, assessment questions, etc. However, the development of high-quality learning material is still a challenging task, requiring a variety of expertise and intensive investment. As a result, if these resources cannot be easily reused across different learning applications and services, then the learning technology industry cannot reach economies of scale, and cannot therefore deliver applications and services that are cost-effective.

In this context, a number of international efforts have been initiated during the past decade for the development of standards that facilitate reusability in learning technologies. The major committees responsible for the development of these standards are: IMS Global Learning Consortium (http://www.imsglobal.org); IEEE LTSC—Learning Technologies Standards Committee (http://ltsc.ieee.org); CEN/ISSS LTW—European Standardization Committee, Information Societies Standardization Committee, Learning Technologies Workshop (index.htm); and ADLNet—Advanced Learning Technologies (http://www.adlnet.org) [13].

A number of standards (or specifications) are being developed as the outcome of these efforts, which can facilitate the description in a common format and thus the reuse of the following:[4]

- Learning content characteristics, at the level of atomic learning objects (learning resource metadata specification), assessment questions and tests (question and test interoperability specification), sets of learning objects together with their structure (content packaging specification), and sequencing (simple sequencing specification);
- Learner characteristics, including the learner profile (learner information package specification), the learner portfolio (ePortfolio specification), as well as the aspects of the learning context that can create accessibility problems to the learner (AccessForAll meta-data specification);
- The design of a learning environment, including the roles, activities, and resources comprising the learning environment (learning design specification);
- Other information which can be reused in learning environments, such as learner competencies or objectives (reusable definition of competency or educational objective specification), metadata sets (resource list interoperability specification), and vocabularies (vocabulary definition exchange);
- Information concerning the interoperation of the most common repository functions (digital repositories specification), the structures that can be used to exchange data between different systems (enterprise specification), and the exchange and management of information that describes people, groups, and memberships within the context of learning (enterprise services specification).

The development of these standards, as well as their integration with similar standardization efforts in other technological fields related to AmI technologies

---

4. The description includes the specifications which are developed by the IMS Global Learning Consortium, Inc.

(e.g., semantic Web technologies, mobile technologies), can provide the necessary technical infrastructure for the delivery of learning experiences in the AmI.

16.5.3 Learning Theories and Models

Educational research has produced a number of learning and instructional theories and models over the past decades, that can guide the development of effective learning environments with a sound pedagogical basis. These models reflect the current paradigm of learning theories, which has been modified during the past few years toward more social, collaborative and situated settings.

These theories are utilized in the current paradigm of learning technologies: As described in the previous section, current mobile learning applications are based on behaviorist learning, constructivist learning, situated learning, and collaborative learning, models. The learning experiences in the AmI will also need to be based on similar theories, so as to be pedagogically sound.

It should be noted, however, that learning theories and instructional models are usually rather descriptive in nature, in the sense that they offer guidelines as to which methods to use to best attain a given educational goal; that is, they are not usually prescriptive, in the sense of spelling out in great detail exactly what must be done and allowing no variation: prescription only applies to deterministic or positivistic theories, which are almost nonexistent in the social sciences [14]. In this context, a lot of work needs to be carried out, so as to make learning theories and models more usable for learning systems development.

Moreover, it could be argued that some of the existing theories and models may not fit within the learning experiences supported in AmI: existing theories are based on the notion of a tutor, who helps learners attain a specific body of knowledge, through learning material, activities, assessment questions; or, as Falk and Dierking argue, most of what is known about learning is based on studies from either classrooms or psychology laboratories, and so may be inappropriate as a basis for considering learning outside of these settings [15]. This model is not in line with the new opportunities offered in AmI, such as those outlines in the previous section. In this sense, new theories and models of learning may need to be developed, which can form the educational, as well as the social, basis for the delivery of learning experiences in AmI.

For example, most learning theories and instructional design models assume that learners enter a learning session, with a specific motivation and goal, knowledge gap, available time frame, and so on. Therefore, they can assume that learning will follow specific cycles, and they can make specific suggestions about these cycles (e.g., attract motivation, introduce concepts, elaborate concepts, provide examples, assess knowledge, etc.). However, learning in AmI may be quite different, since learners may enter learning experiences at any time, while involved in a variety of everyday activities and tasks, with everyday objects:

> Instead of looking at training as linear processions with a beginning, middle and end, we must now look at training as clusters of independent, standalone bits of knowledge. They are certainly related to each other and they may be viewed together, but they may also be viewed singly. Just as you can enter a Web site at any page and leave at any point, so too can training consumers. Lose the notion of a

class of eager learners trapped before you for a day. These new consumers of training can come in at nearly any point in the training, stay as long or as short as they wish, and leave when either they are bored or they have learned what they want. Bits of the training may be used in dozens of different trainings for different people. Training must be developed to allow, indeed to help, the learners get to exactly the point they wish, and then helping them learn and understand that exact piece of information, knowing that once they get what they want, they will leave: no evaluations, no thanks, no flowers [2].

16.6 Conclusions

The AmI environment signifies a departure from the current paradigm of desktop, direct manipulation office applications. The realization of AmI comprises a great technical challenge with a tremendous potential, which has initiated a large number of international R&D efforts [16, 17].

From a user perspective, a series of necessary characteristics are required to facilitate the eventual societal acceptance of AmI: AmI should facilitate human contact; it should be orientated towards community and cultural enhancement; it should help to build knowledge and skills for work, better quality of work, citizenship, and consumer choice; it should inspire trust and confidence; it should be consistent with long-term sustainability; it should be controllable by ordinary people [7].

Moreover, in order for AmI to be acceptable by citizens of the knowledge society, a number of ethical, social, and other issues need to be resolved. For example, AmI needs to continuously monitor users and draw conclusions about their profiles, so as to adapt the learning environment to their individual (or group) goals, interests, and preferences. The security and privacy of this information may raise a number of issues that can directly affect the acceptability of AmI (this issue is addressed in a number of chapters in this book).

This chapter has discussed some theoretical and technical issues that are involved in the realization of an AmI for learning, i.e., for the delivery of learning experiences which are seamlessly integrated within out everyday activities, through intuitive intelligent interfaces.

The main idea behind the description of this chapter is that learning technologies need to build on novel learning theories and models that take into account the unique characteristics of the AmI environment and interoperate with the variety of technologies that are required in order to realize AmI. This is a great theoretical and technical challenge, with a tremendous potential. (See Figure 16.2.)

As stressed, there are a significant number of possible interactions between the contributing technologies in the realm of CI. Bearing in mind the main objectives of granular computing and neural networks, we can envision a general layered type of the model in which any interaction with the external world (including users) is done through the granular interface (external layers), whereas the core computing part is implemented as a neural network or a neuro-fuzzy structure (in which case we may be emphasizing the logic facet of ongoing processing faculties). (See Figure 16.3.)

16.6 Conclusions

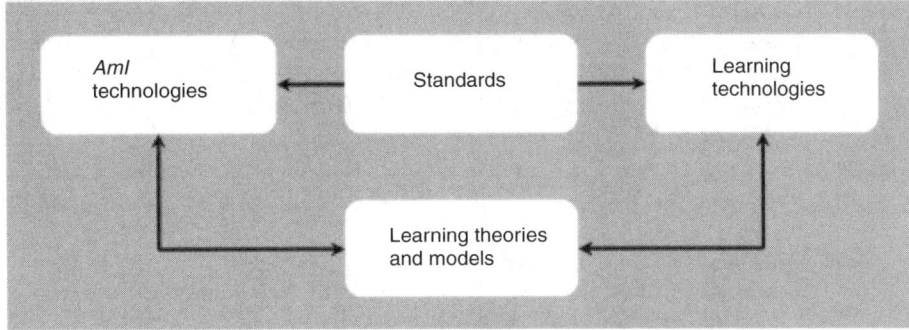

Figure 16.2 Some issues toward the realization of learning in AmI.

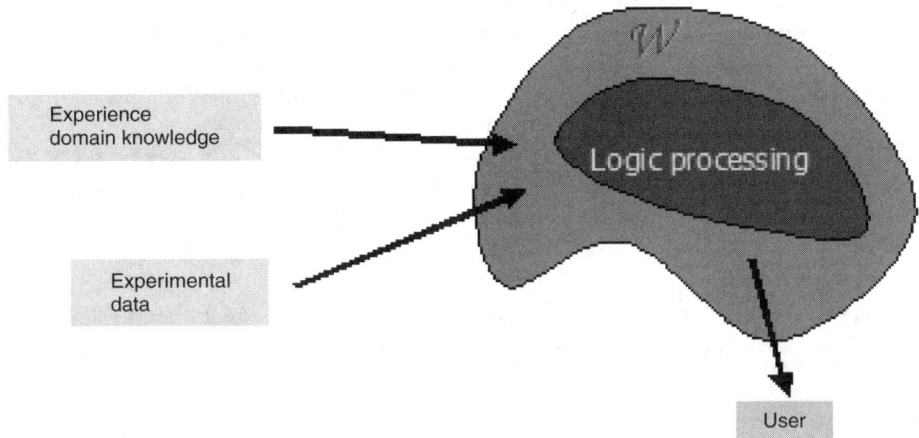

Figure 16.3 A layered style of CI constructs (*From:* [18].)

References

[1] Wiley D., "Connecting Learning Objects to Instructional Design Theory: A Definition, a Metaphor, and a Taxonomy", *The Instructional Use of Learning Objects*, 2001. Electronically available at http://reusability.org/read.

[2] Nichani M., "LCMS = LMS + CMS [RLOs]", *elearningpost*, 2001. Electronically available at 001022.asp.

[3] O'Malley C., et al.., "Guidelines for Learning, Teaching and Tutoring in a Mobile Environment," *MOBIlearn Project Deliverable D4.1*, 2003.

[4] Gay G., R. Rieger and T. Bennington, "Using Mobile Computing to Enhance Field Study", *International Conference on Computer-Supported Collaborative Learning (CSCL 2002)*, Boulder, CO, USA, January 7–11, 2002.

[5] Naismith L., et al., "Literature Review in Mobile Technologies and Learning," *NESTA FutureLab Report*, 2004.

[6] Savill-Smith C. and P. Kent, "The Use of Palmtop Computers for Learning: A Review of the Literature," *Learning and Skills Development Agency Research Report*, 2003.

[7] Ducatel K., et al., "Scenarios for Ambient Intelligence in 2010," *ISTAG Final Report*, February 2001.

[8] Okada, M., and A. Vasilakos, "Social Networking of Ubiquitous Experiences for Mobile Learning in an Ambient Intelligence Environment", *Ambient Intelligence*, Artech House, Norwood MA, 2006.

[9] Brusilovsky P., "Adaptive Hypermedia", *User Modeling and User-Adapted Interaction*, Vol. 11, No. 1, 2001.

[10] Sampson, D., C. Karagiannidis, and Y. Kinshuk, "Personalized Learning: Educational, Technological and Standardization Perspective", *Interactive Educational Multimedia*, Vol. 4, 2002.

[11] Brusilovsky, P., C. Karagiannidis, and D. Sampson, "Layered Evaluation of Adaptive Learning Systems", *International Journal of Continuing Engineering Education and Life-Long Learning*, Vol. 14, No. 4-5, 2004.

[12] Pedrycz, W., and A. Vasilakos (eds.), *Computational Intelligence in Telecommunications Networks*, CRC Press, 2001.

[13] The MASIE Center, "Making Sense of Learning Specifications and Standards: A Decision Maker's Guide to their Adoption," Industry Report, 2002. Electronically available at http://www.masie.com/standards/S3_Guide.pdf.

[14] Reigeluth C., "The Elaboration Theory: Guidance for Scope and Sequence Decisions", *Instructional Design Theories and Models*, Lawrence Erlbaum Associates, 1999.

[15] Falk, J., and L. Dierking, *Lessons Without Limit: How Free-Choice Learning is Transforming Education*, AltaMira Press, 2002.

[16] Markopoulos, P., et al., *Ambient Intelligence*, Springer Verlag LNCS 3295 (EUSAI 2004 Conference Proceedings, Eindhoven, The Netherlands, November 8–11, 2004).

[17] Riva G., et al., *Ambient Intelligence: The Evolution of Technology, Communication and Cognition towards the Future of Human-Computer Interaction*, IOS Press, 2005. Electronically available at communication.com.

[18] Vasilakos A., et al., "Computational Intelligence in Web-based Education: A Tutorial", *Journal of Interactive Learning Research*, Vol. 15, No. 4, 2004.

CHAPTER 17
Meetings and Meeting Support in Ambient Intelligence

Rutger Rienks, Anton Nijholt, and Dennis Reidsma

17.1 Introduction

Environments equipped with ambient intelligence technology provide social and intelligent support to their inhabitants. The majority of ambient intelligence research is on providing support to individuals living or working in these smart environments [1]. However, in home and office environments we also have people interacting with each other. Can the environment support this interaction as well?

Looking at smart environments from the point of view of supporting multiparty interaction adds some interesting research issues to the area of ambient intelligence research. First, in order to be able to provide support, the environment is asked to understand the interactions between its inhabitants and between inhabitants and the environment or smart and maybe mobile objects available in the environment. Although we see the development of theories of interaction and behavior, these theories are rather poor from a computational point of view and therefore they hardly contribute to the design of tools and environments that support activities of human inhabitants. Hence, the need for computational theories of behavior and interactions must be emphasized.

A second research issue that needs to be mentioned is the real-time monitoring of activities, the online access to information about activities taking place, and also the online remote participation in activities or influencing activities in smart environments. The third research issue concerns the off-line access to stored information about activities in smart environments. This latter issue may involve retrieval, summarization, and browsing.

Certainly, not all three research issues need to be considered for every type of smart environment. Sometimes we are only interested in providing real-time support to an individual entering an ambient intelligence environment. At other times we just want to monitor what is happening and having an alert when something unusual is going on. Sometimes we just want to know what has been going on while we were not present.

There is one important domain of application of ambient intelligence technology where all these research issues play an important role. This is the domain of meetings supported by smart environment technology.

In this domain it is useful to provide support during the meeting, to allow people who can not be present to view what is going on, to allow people to remotely participate, and to provide access to captured multimedia information about a previous meeting, both for people who were present and want to recall part of a meeting and for people who could not attend. Hence, on the edge of multimodal research and ambient intelligence lies the domain of meetings.

We cannot think of a world without meetings, although sometimes we wish we could do without them. In the best case all these meetings would be efficient, effective, manageable, and, afterward, accessible. Although this seems the perfect example of what can be called wishful thinking, these goals would certainly be in accord with any management strateg. The reality is that, among other things, meetings are expensive, have an unpredictable outcome, and are hard to manage. To improve the quality of meetings, they need to be assisted by a wide variety of technology.

It appears that the environment can play an important role in helping the meeting. Auxiliary devices such as microphones and data projectors have augmented traditional meetings making things more convenient for participants as well as more accessible for others. Indeed, during the last decade we have seen smart meeting rooms appear where information can, apart from being captured and presented, also be interpreted. Since support can now be designed to assist proactively and in any combination of modalities, it is at this point that the role of ambient intelligence comes in.

These supportive ambient intelligent technologies all have one thing in common: They depend on interpreting the incoming data from a multitude of sensors. A system could, for example. be triggered by the start of a presentation to switch off the lights, or aim to record the speakers of the meeting using close-up cameras. If, in this last case, a system is capable of recording only one camera stream, it is fully dependent on the determination of the speaker. For various other types of support, more complex input is required which cannot be obtained by direct observations. Such support requires a system to reason about the information, making use of contextual or historic knowledge or analyzing the input to discover the underlying ideas and motives, which is often referred to as the beliefs, desires and intentions (BDI) state of the users [2, 3]. Based on such knowledge a system could make more complicated decisions about appropriate responses to the situation at hand.

Figure 17.1 shows how we propose such a system could work for various applications. The examples in the general model are tailored to the meeting domain and show various levels of complexity. The environment and the people are captured using cameras, microphones, and other sensors. The resulting data is analyzed, and recognition technology is used to detect the directly observable events. This class of events contains things such as body postures, facial expressions, or hand movements. Subsequent subsystems analyze these observable events on progressively higher levels of interpretation [4]. The knowledge that the system now has about the world can be used to trigger a certain response.

This chapter starts with some general background information about meetings and the role prospective technology could play to assist meetings. We will discuss the role of technology along the virtuality continuum, along which meeting rooms become smart rooms, which in turn become virtual meeting rooms that allow immersive remote presence. As an example of the latter we will discuss our own vir-

17.2 What Are Meetings?

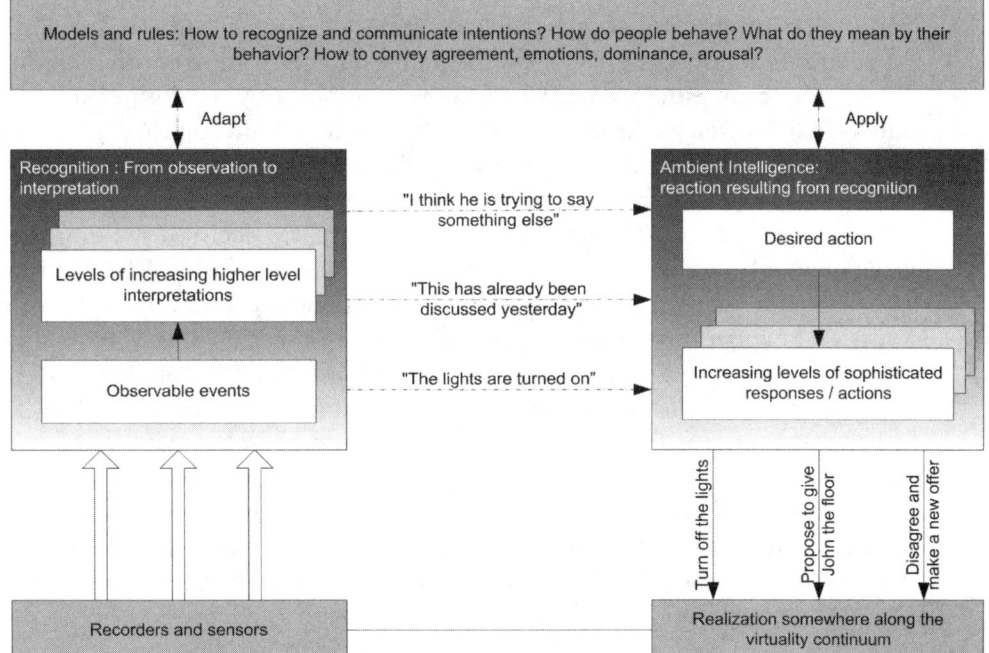

Figure 17.1 Ambient intelligence: recognition resulting in reaction throughout the virtuality continuum.

tual meeting room [5]. The chapter ends with an overview of current state of the art research projects focusing on meetings and their support. These projects are related and compared using the multi-modal research framework [4].

17.2 What Are Meetings?

This is a fundamental question that needs to be addressed before we can start talking about meetings. A very basic definition would be to say that a meeting is when people meet. But what does it mean to meet someone? When people say they have met someone, there must have been at least some sort of interaction. One cannot say "I met a table": it makes hardly any sense, since this group of passive entities is not able to exchange information or interact. This then is a prerequisite for a meeting. Op den Akker et al. [6] define a meeting to be when individual entities interact. This seems quite reasonable; however we need to be a little more precise.

We cannot say that we have met, let us say, an ATM, which *could* be defined as an individual entity. On the other hand, it seems less strange to say that we have met a robot. So we are able to meet not just humans but also nonhuman entities. Anthropomorphism appears to be a key concept here. If we are able to ascribe to an entity a certain set of human affordances, we can say that we are able to meet this entity. Therefore, we define a meeting as anthropomorphic entities that interact. This could be embodied agents or humans but also a mixture of these.

A second question is then: Why do these entities meet, what is the reason behind this? The answer, at least for humans, is due to the basic need to get together, or group. The ability to exchange ideas, make decisions, and work collaboratively facilitates functioning as a society. Exchange and generation of information in meetings leads to an enhanced level of knowledge, improving the performance of the individuals as well as the group [7]. Since these ideas and solutions generally emerge faster when more minds share the same thoughts, a group becomes more than the sum of its parts, which is beneficial for the interacting entities. Generally, a meeting is held in order to move group actions forward by decision making through the exchange of thoughts by both information presentation and collaboration.

Throughout this chapter we focus on interacting humans that communicate ideas and express their emotions and feelings. In these meetings, the humans could however be assisted by, or even completely represented or replaced by virtual humans. We will discuss this later. Essential here is the existence of copresence of other anthropomorphic entities. If there are none, there is no interaction and hence no meeting.

Antunes and Carrio [8] describe three main aspects that pertain to meetings. The *resources* of the meeting, the meeting *process* and the *roles* in the meeting. Their subdivision gives a good and structured overview of the main meeting components. We will use their terminology throughout the rest of this chapter.

17.2.1 Meeting Resources

The resources are split up into two distinct sets, logistics and group memory.

Meeting Logistics include the generic meeting facilities such as the location, the table setting and the physical appearance.

Group memory concerns the shared information resources that a group uses to accomplish work. Examples are agendas, meeting reports and support documents.

17.2.2 Meeting Process

The meeting process structures the set of activities that the participants must execute in order to achieve some common goal. Possible approaches to structuring a meeting are the the genre approach, the decomposition approach, and the individual intervention approach.

The genre approach focuses on the purposes and the communication patterns based on recurrent communicative actions; typical examples are a briefing, a progress report meeting, or a staff meeting.

The decomposition approach regards meetings as decomposable in multiple levels of detail with goals and sub goals, such as a (recursive) combination of divergent, convergent, and closure phases.

7The individual intervention approach structures the meeting process according to individual (process or content) interventions produced by the participants and the facilitator. Examples are: defining the agenda, opening and closing the meeting, and making a statement.

17.2.3 Meeting Roles

Meeting Roles are the roles assumed or played by attendees of the meeting, such as participant, secretary, or observer.

This arrangement is useful to categorize aspects pertaining to certain views of a meeting. But there is more one can say about meetings. Some meetings, for instance, may be nice or chaotic and others are possibly even effective.

17.2.4 Problems with Meetings

There are several things to say about meetings. When looking at meeting resources, for example, they are notably expensive. It is said that on average, business organizations spend around seven to fifteen percent of their budgets directly on meetings. With respect to meeting logistics, the question of where and when to hold a meeting is the first contributing factor. Large amounts of money are spent on transportation and hotel costs. Another contributing factor, apart from the fact that time itself has to be paid for, is that in the time of the meeting itself the attendees could have put their effort somewhere else. A second issue is mentioned by Gordon [9]. He found that up to fifty percent of the meeting's productivity is wasted. When having a meeting, it might happen that due to time constraints the meeting functions of being together to create ideas and decisions, and plans and actions are under high pressure. People sometimes leap into problem solving before it is clear what the actual problem is [10]. A U.K. meeting study mentions that four out of five professionals admit to daydreaming during meetings, while nearly a quarter dozes off [11]. With respect to the meeting process, for instance, group size is an issue. Depending on the meeting genre, the meeting group size is important in order to maintain order and make sure everybody is able to have his or her say. More aspects in this sense are, for example, that some people can dominate a meeting, whereas others do not contribute at all. As a final example of all these possible problems, when looking at meeting roles, confusion might arise about which roles there are during a meeting and which responsibilities are associated with them.

A lot of literature about meetings has been written trying to approach these problems, aiming at an increase of efficiency and effectiveness [12, 13] All of them mention more or less similar issues that have to be taken care of before, during, and after the meeting. Before the meeting, recommendations stress the importance of the meeting logistics and a thorough preparation of the meeting process, knowing the expected meeting genre. Several points to take care of are the predefinition of the meeting objectives, the fact that an overview is provided containing the expected effort of all participants, and the presence of an agenda including a global time frame. Techniques can be used in order to maximize output. (Typical techniques to make sure everyone is able to give his or her opinion are: brainstorming, silent writing down of ideas, discussion, etc.) During a meeting, the meeting process should be managed to maintain order in such a way that the sought after results such as striving for a timely and effective execution of the agenda are achieved. This is usually done by assigning meeting roles and their accompanying responsibilities. After the meeting it is important that the contents of the meeting are made available for the attendees as well as other interested persons. Minute taking and writing down notes are considered important in this respect to prevent the organizational group memory from disappearing. The next section will elaborate on how technology can play a part in assisting meetings.

17.3 Technology: Mediation and Support

Reasons that technology has found its way into meetings might originate from the fact that technology enables humans to approach the problems mentioned in Section 17.2.4. Technology facilitates people in saving money and time and creates opportunities that, without technology, humans on their own could have never created.

Meetings can nowadays be assisted by a huge variety of tools and technology, ranging from completely passive objects, such as a microphone, to completely autonomous actors, such as virtual meeting participants. There are several dimensions defining this spectrum. There is the dimension reactive/proactive, revealing information about whether or not a piece of technology is able to act by itself. A light switch, for instance, will typically undergo an external trigger before acting by turning on the light, whereas a radiator might regulate the temperature proactively. Other dimensions are the sensing ability (how and which information can be captured), the reasoning ability (the complexity level of the reasoning about information), and the acting ability (the flexibility to present information). Although somewhere in this space a boundary between object and actor can be defined, for our purpose there is no need to make this distinction and we will therefore stay away from this discussion. Typical labels given to technology that assists meetings are: meetingware [8], groupware [14], or meeting helper agents [15].

17.3.1 The Virtuality Continuum

The notion of having a meeting remotely without being physically present in the meeting, but having a representation (e.g., a video stream) there where the actual meeting takes place, introduces a new dimension, which can help us in structuring our view on meeting support: the virtuality continuum. The concept of the virtuality continuum was introduced by Milgram and Kishino [16]. In this continuum (see Figure 17.2), there is an increasing degree of computer-produced stimuli from left to right.

At the extreme left the environment is completely real, whereas at the extreme right there are completely (immersive) virtual environments, where all stimuli are computer generated. Recasting this in meeting context, we find at the left side face-to-face meetings with real humans in one location and real equipment. The more we move to the right, the more mediated meetings will emerge. At the far right we find the immersive virtual meetings where everything is virtual: humans are

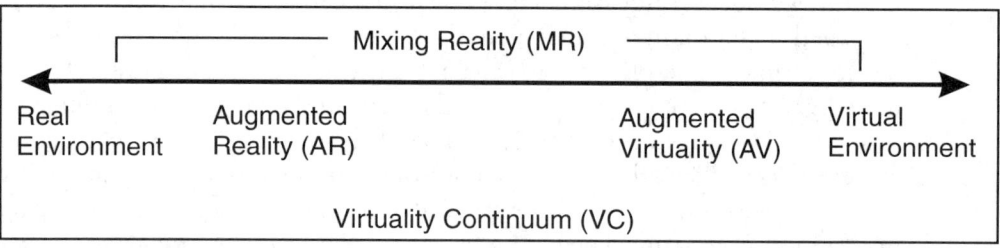

Figure 17.2 Virtuality continuum.

replaced by avatars, the location is a virtual environment, and all communication signals are technology mediated.

The responses from ambient intelligent systems can take place anywhere along the virtuality continuum. However, when a system needs to change the environment—that is that the action performed changes the way the world is arranged—this is harder to achieve in the physical world than in a virtual one.

We will organize our focus when describing the technological meeting support using the three previously mentioned meeting aspects (meeting resources, process, and roles) along the dimension of the virtuality continuum. But first we will elaborate on the role of the continuum in social interaction.

17.3.2 Meetings in the Virtuality Continuum

Greenhalgh and Benford [17] were among the first developers of a virtual reality teleconferencing system (MASSIVE), where participants can participate in a virtual meeting wearing a head-mounted display (HMD). They mention a number of disadvantages in comparison with real face-to-face meetings. Typical examples are limited peripheral awareness due to the HMD, lack of engagement, and participants not feeling fully present in the meeting.

It is clear that these drawbacks all relate to the state that technology had advanced. Lack of engagement was explained due to the uncertainty of whether the participant had been heard by the others due to inconsistent audio quality. Other technological issues that played a role were the level of detail (LOD) of the virtual environment and time delay issues such as delayed updates of the HMD viewpoint. The main problem here is that communication toward the right side of the continuum is still confronted with issues resulting from a leaner medium in a sense that not all social cues can yet be conveyed. As a consequence of this, the hierarchical structure is flattened, social cues are reduced, participants are depersonalized, and the overall volume of communication is less [18].

The sense of realism, or the feeling of reality, relates to the degree one is able to express oneself and perceive others. These issues are covered by the term *presence*, being a communicator's sense of awareness of the presence of an interaction partner [19]. Presence is important for the process by which man comes to know and think about other persons, their characteristics, qualities and their inner mental states. Lombard and Ditton [20] give a good overview of aspects related to the perception of presence. Presence by itself is determined by the richness of the media, and the abilities it affords to the user. Once we are able to increase presence, this will lead to enhanced person perception and increase possibilities to express oneself.

So, whenever one moves along the virtuality continuum, presence is an important issue to deal with. Increased mediation usually leads to a decreased feeling of presence. That is to say, this holds whenever technological mediation is noticed, when social cues such as turn signaling or the ability to express agreement are lost in the communication process. In some cases technological mediation might lead to an increased feeling of presence—for example when particular social cues are amplified [21].

The question of how to determine the appropriate set of cues for a specific task seems to be the main challenge to be resolved. Vertegaal [22], for instance, stresses the importance of proper representation of gaze in teleconferencing systems. On the other hand, one could imagine that leaving out disturbing cues could also possibly lead to increased sense of presence, since other cues get more attention.

Although the focus of this chapter is to elaborate on how technology might advance in order to resolve the mentioned drawbacks, users of technology have already come up with ideas to approach some of the problems. The idea of using smileys as a representation for an emotion is an example of how communication is adapted to the medium instead of the other way around. It is also possible to compensate for technological deficiencies by explicit verbalization of other cues, e.g., by giving a textual description of your emotions.

We will now discuss different types of meeting support along the virtuality continuum describing the space between passive and proactive technology by using the categories from Section 17.2: resources, process, and roles.

17.3.3 Technology and Meeting Resources

The development of many new kinds of peripheral devices has extended the meeting logistics in a way that they facilitate more possibilities. Starting from the earliest microphone, we nowadays might encounter smartboards, data projectors, cameras, and all other kinds of equipment facilitating the participants' needs.

The development of ambient intelligent devices allows proactive systems to provide useful services for the participant. A typical example is Easy Meeting [23], which among other things, can automatically adjust the light of a meeting room or display a PowerPoint presentation in reaction to sensor information. Another example is found in the work of Chen et al. [24]; they discuss a system where presentations are automatically scheduled for the data projector making use of the meeting agenda. Ubiquitous computing techniques enable communication between various devices. An example of a ubiquitous interface can be found in the iRoom of the interactive workspace project [25], where large display devices could connect to participants' portable devices. Oh et al. [26] describe several technological components that can work together to assist the meeting. During the meeting the light might be geared to the presentation and after the meeting a summarizer and a browser could help to retrieve the information generated in that meeting.

But apart from technology being out there there is also a trend toward personification of the intelligent environment, enabling a more natural interaction. The robotic assistant from Nuttin et al. [27], for instance, can be used for videoconferences, either remotely controlled by a user or controlling itself in autonomous operation.

Being remotely present is perhaps the greatest benefit from technology so far, resulting in a huge reduction of time and money spent on meetings. Technological developments supporting remote participation have gone through a series of rapid developments. The traditional audio-only teleconferencing was replaced by videoconferencing, which then again was soon augmented by additional advanced services such as instant messaging, file transfer, and application sharing.

The video streams used for these teleconferences can be automatically chosen by an automatic director [28] and edited before being presented at the other end using augmented reality techniques. The presented information can be augmented to highlight important parts in video to attract attention, or to hide or manipulate parts of a signal [29]. This technique also enables modification of the participants' representation. A person's face might, for instance, be represented by another face, or any other preferred representation [21]. At one extreme end of the virtuality continuum, it is also possible to meet remotely at a virtual location. These locations can vary from a text-based chat environment such as IRC[1], up to a complete 3D virtual meeting environment such as the Active Worlds [30]. The more modalities supported by the environment, the more possible ways of communication [31].

One can equip a virtual world with all the technology discussed so far. Humans in this world can be represented by avatars. Assistants in this virtual world can be completely generated virtual human characters, showing intelligence and emotions and knowing how to interact with human users [32, 33]. In the near future, these virtual humans could be completely autonomous pieces of software, indistinguishable from real human representations and able to communicate in several modalities. 3D virtual worlds are being extended with technology increasing the feeling of presence, such as head mounted displays, spatial audio mediation (see Rodenstein and Donath [34] for a 2D version and, Aoki et al. [35] for a 3D version) enabling sound from objects or people to get louder as you move closer and haptic interfaces that enable people to touch objects in virtual words [36].

The preservation of group memory is, as stressed before, a problem, due to the volatile nature of meetings. Once a meeting is over, the group memory or the knowledge from that meeting is dependent on the participants and perhaps on notes taken during that meeting. Notes typically contain nothing more than captured decisions, action points, and perhaps some issues that were left open. Apart from the fact that people might be interested in things not captured in the notes, it might take hours to find answers digging through piles of hard-copy notes. There is an information need here and the question is: Where and how can technology help us in satisfying this need?

The key issue is to provide access to representations of conveyed information from the meeting, as mentioned in [37]. The problem of how to do this can, according to Shum [38], be classified into what Rittel [39] called wicked problems, in contrast to tame problems which can be analyzed using existing methods. According to Rittel [39], the key to approaching wicked problems is to frame them using key questions and defining key priorities. Once meeting information can be accessed, it is shown that people will adapt their way of working based on what they have available in order to increase efficiency [40].

A very important question here is what information should be captured [38], or what do people want to remember from meetings? There are three categories of people showing interest in the content of a meeting: the actual participants, customers, and analyzers. Lisowska [41] gives an overview of typical queries for meeting retrieval systems, obtained through questionnaires, that in the future will be evaluated using a Wizard of Oz experiment. Similar research was conducted by Jaimes et al. [42]. The ultimate piece of technology in this sense would be able to answer all questions in a clear and comprehensible manner. Approaches to capturing informa-

---

1. See http://www.irc.org.

tion from meetings are the current topic of many research projects. We will elaborate on this in Section 17.4.

Since it might be hard for people to express their need to a system able to answer their queries, the interface is of utmost importance. Jaimes et al. [42] describe an implementation of a system that helps users to easily express cues people recall about a particular meeting. Tucker and Whittaker [43] provide an overview of systems grouped into four categories able to browse through meetings. The first three categories of meeting browsers can be grouped around classes immediately presenting itself: browsers focusing on audio (including both presentation and navigation), browsers focusing on video and the third class of browsers focusing on meeting artifacts, such as slides and documents pertaining to a particular meeting. However, a fourth category, and probably the most useful class of possible browsers, can be grouped around derived data forms. The Ferret browser [44] is a browser where several plugins can work together in order to answer a specific query. A related area of research is the automatic generation of multimodal summaries of a meeting [45]. In fact, a summary can be seen as an answer to a question, where the best summary is perhaps the one that answers the most frequently asked questions.

17.3.4 Supporting Meeting Processes

The meeting process deals with the meeting content, the meeting organization and the social aspects of the meeting. In an ordinary meeting the chairman has to manage the meeting process in order to e.g. maximize the output of the meeting, stick to the agenda or to improve the meeting atmosphere. DiMicco [46] describes a system called Second Messenger, which shows real-time text summaries of participants' contributions. After increasing the visibility of the group members speaking less frequently, it appeared that they started to speak more frequently than before, while the more dominant people started to speak less. One could imagine that in the long run systems such as these can make real-time inquiries during a meeting by questioning participants through an input panel or interface in order to create an inventory of aspects relevant for the meeting. Such a system might, for example, ask about the attitude toward the ongoing discussion (level of interest, level of detail, etc.) or about whether one finds it time to break. This information could be used for support functions, for example presenting it afterward in a summary to the chairman. For him, as authority, this information can be very helpful to regulate the meeting process. Altering the organization or the content of the meeting can be done, for instance, by manipulating the form of the discussion (converge, diverge, end by voting). Other ways to steer the meeting process are selective turn giving, or interrupting. For more elaborate information about leadership issues and virtual teams the reader is referred to [18, 47].

Ellis et al. [48] describe, in their article about agent-augmented meetings in the Neem project, how they envisage meeting process support. A collection of agents are continuously interpreting what is going on in the meeting: what is currently under discussion, which arguments are used, who contributes most (least), what the social atmosphere is, is the meeting still on schedule, and so on. Other agents use this information to make judgments: are there large gaps in the information presented or in the argument map, is there a lot of redundancy or contradiction in the arguments so far, is it time for a break, does everybody get an equal chance to contribute to the meeting, is it time to take a vote on the current topic, etc. Finally, these judgments are presented to the participants. The presented information here could be a proposal to hold a break, encouragement to people to give their opinion, possible gaps in the argument structure of the debate so far, or relevant information from the Web or from a previous meeting.

The mentioned systems so far are in a way all passive they do nothing but present their findings to (some of) the participants, who then have to decide how to deal with the presented information. One step further, proactive systems can be introduced that take action based upon the sensed information. A very primitive version of organizational support could in this case be a microphone system providing participants with fixed time slots to speak.

Most of this meeting process support builds upon the availability of extensive information about the topics and the argumentation structures. As an extension to what was mentioned before about the preservation of group memory, the argument chain in this example is to be monitored in real-time in order to inform a chairman about possible flaws. In order to actually realize this type of support new ways of analyzing discourse content are necessary. As Purver et al. [49] argue, this problem is much harder than conventional discourse analysis in human-machine dialogs, since the computer cannot steer the process by posing the right questions or try to understand something by initiating some clarification dialog.

17.3.5 Supporting Meeting Roles

Antunes and Carrio [8] define meeting roles as the organization resulting from institutionalized practices, negotiated and agreed before the meeting, and assumed during the meeting by meeting participants. As described by Biddle [50], roles allow well-understood division of work among a group, and mediate expectations of who will do what. There are several roles that can be identified in a meeting. In reality the roles of the participants may not always be formally and explicitly defined, participants may have more than one role, and there can also be implicit and informal roles. Technology can assist several roles by passively providing resources to the participants having that role. We already gave examples of how a chairman could be assisted in his task to monitor and control the meeting process. As another example we will now elaborate on the possible assistance of the secretary.

A secretary could use technology to aid him or her with the creation of notes or minutes varying from the early sketch pad [51] to the latest tablet PC [52]. Also, the mentioned recording technologies such as cameras and microphones can be used for playback and reinterpretation while creating or revising the notes. Once notes are available in a digital format, it is possible to make them available to a wide public using modern communication technology. Browsers in combination with information retrieval technology can be used here to help in fulfilling the information need from the users. An example system that can be used by a secretary is the SoniClear MeetingPro Assistant from Trio Systems[2] that is able to create digital recordings

---

2. See http://soniclear.com/ProductsMeetingPro.html.

and has automation for entering meeting notes. Advances in automatic speech recognition and summary creation define the biggest hurdles on the road to full automatic minute creation, or ideally, of the input generation required by the meeting information systems mentioned above in Section 17.3.3.

As more and more participant tasks such as secretary or chairman can be replaced or assisted by technology, in the future the support could even go as far as replacing a real human with a complete virtual replica. This may be especially useful in a case where there is no competent or unbiased human participant available for that role.

17.3.6 Learning How to Respond

As we have seen, ambient intelligence is all about responding to interpreted events. The responses, as well as the interpretation of the events, are based on world models and their interacting inhabitants. The models may either be hard coded in the system or formed adaptively.

These world models are often difficult to derive beforehand. Especially models that combine several input modalities and make complex inferences are hard to obtain. One approach to derive these models is to use annotated corpora. Corpora are large collections of data, such as video and audio recordings of meetings or other interaction situations, or text collections, such as several years of newspaper issues. Annotations are what some call metadata' on this data namely, descriptions of properties of the data and of the content of the data. In the case of meetings, a corpus would contain recordings of many meetings, possibly also including copies of documents and other artifacts used during the meeting, in an electronic data format. The annotations defined in such a corpus might include transcriptions of the speech, annotations of the dialog structures occurring in the data, labeling of visual and other events that happened during the meeting (gestures, other movements, slide changes, etc.), descriptions of the structure of the argumentation used during a decision-making meeting, labeling of the emotions of the participants during the meeting, and many other information sources. Through careful analysis of the corpus and its annotations, one can derive models of human behavior in meetings. Examples of higher-level models developed are models for addressee and dominance detection [53, 54].

All derived world models can be tested in a virtual world. Reidsma et al., [5] describe the usage of an annotated corpus and a virtual meeting room in this context. For this they introduce the virtual meeting room (VMR), developed in the context of the AMI project (see Figure 17.3) by the Human Media Interaction Group of the University of Twente. Technically seen, the VMR is simply a 3D environment with support for controlling humanoid avatars. However, as a meeting support tool one can view the VMR as a teleconference facility that allows the avatars to regenerate the behavior of remote participants or to embody a meeting assistant. Furthermore, it can also serve as a test-bed for the development of all kinds of ambient intelligent applications.

One nice aspect of the VMR is that it can be used to build models of human behavior that can be used for proper interpretation of obtained signals and for the generation of adequate actions and responses. In order to find out which signals are

17.4 Projects on Meetings

Figure 17.3 Real and virtual meeting room. Upper left: Real setting of recorded meeting. Upper right: VMR view seen through the eyes of a participant. Below left: VMR central view. Below right: VMR extended view with visualization of head orientation, body pose, speaker, and addressees.

important, the VMR enables experiments where generated stimuli can be controlled (see also [55] for the advantages of VR in experiments). Any combination of channels (generated or regenerated), such as head orientation, arm movements, or body poses, can be shown. This allows us to perform experiments to validate or elicit models of human behavior. Rienks et al. [56] present an example of a conducted experiment in the VMR. The acquisition of these types of models used by ambient intelligent systems is perhaps the greatest challenge for the upcoming years.

17.4 Projects on Meetings

Meetings, and the development of new technology to support meetings, have long been a subject of research. In this section, we review some projects from the past and the present on the topic of meetings, meeting support, and collaborative smart environments. One of the first projects that we discuss is the work carried out at the Fuji-Xerox Palo Alto Laboratory (FXPal). Already in the early 1990s they started their work on collaborative notetaking, camera control, and browsers for video content in lectures, a line of research that is still very much active [7, 57].

A few years later, smart meeting rooms began to appear at several institutions. These are being used to record large corpora of meeting data, in many cases still focusing on the audio. Work on meetings goes on at, for example, CMU, NIST, and ICSI [58–60].

Meeting research, as well as ambient intelligence, continues to receive more attention. Large-scale collaborations grow, resulting in more smart meeting rooms and international projects. At the moment there are several new, heavily funded, projects that have meetings as their main topic. AMI, CALO, CHIL, and IM2 all work on collecting and researching corpora of meetings, modeling meetings and developing technology to support meetings [61–64].

The rest of this section will discuss some highlights from those projects, centered on three groups of characteristics (sensors, layers of annotation, and tasks). The distinction is derived from the framework discussed in [4], which was explicitly developed to facilitate comparison of projects that focus on different parts of the general field of (corpus based) multimodal research. Note that we will not discuss each project in exhaustive detail along each dimension. We will simply describe the characteristics and how they can be seen within the projects, describing their role in specific projects in the most conspicuous cases.

17.4.1 Recordings and Sensors

The first group of characteristics deals with the types of sensors and recording devices that are used. In many projects the meetings are recorded in a smart room. The range of devices present in these rooms can vary a lot. Audio is in most cases recorded with lapel microphones, binaural manikins, or microphone array technology. A number of video cameras can capture the whole meeting room, parts of it from different viewpoints, individual participants, or even close-ups of their faces. Documents pertinent to the meeting can be captured with scanners during or after the meeting, presentations can be captured through the data projector or just copied from the computer. Other devices may involve smart whiteboards, which record every stroke, smart paper for taking notes and (collaborative) note-taking applications on handheld devices.

The more recent projects are all very much alike in this respect. They all facilitate audio recordings and most meeting rooms facilitate video recordings as well. AMI even has close-up cameras for each participant, to facilitate emotion research based on facial expressions. The IM2 project has a whole subproject dedicated to mining the data present in parallel documents [65]. At FXPAL much work has been done on collaborative note taking [66, 67]. For the projects that target real-time support as their main task (most notably CHIL and CALO), the requirements for real-time availability of the recorded/sensed data are, of course, especially stringent.

17.4.2 Annotations and Layers of Analysis

The second group of characteristics concerns the layers of annotation and analysis (observable events and higher levels of interpretation) that are considered by a project. Strictly speaking, this point could be covered quite adequately by simply listing all layers that are automatically recognized and/or manually annotated in a project, as in the generic version in Figure 17.1. This would, however, not present the complete picture. One could also determine why certain layers are included and in what way they are approached. If the main research topic of a certain part of a project is the development of machine-learning algorithms, a layer of information is just a collection of features to be used by the algorithms. The information present in the different layers could, however, also be approached from the other side, in an attempt to make sense of the meeting in structures that are really useful and informative for the users who will actually use the technologies developed in the project. Moran et al. [40] coined the term *salvaging* for this type of information structuring. The different resources captured during a meeting are analyzed, described on higher levels of interpretation, and restructured in an attempt to understand their content.

Issues that play a role here are, among others: a distinction between manual and automatic annotation, adapting old technology to new domains or the development of radically new technology and the level of interpretation of the layers, as well as the question as to which particular modalities are under consideration (also dependent on the nature of the recordings). The projects using the ICSI corpus (which contains audio recordings only), for example, stand out with an exceedingly detailed analysis of the speech audio data. Dialog analysis is a topic in most projects, but it is especially important in IM2 (shallow parsing, work on the MALTUS tag set), ICSI (MRDA, and the mapping meetings project) and AMI (dialog analysis, partly with a focus on argumentation structuring in discourse).

17.4.3 Applications and Tasks

The last group of issues centers on applications and tasks. One question that is raised here is what type of meetings is considered in a certain project. Some projects clearly delineate the type of meetings that they want to research, e.g., meetings with a lot of decision making (CALO), using artifact design meetings as a specific instance (M4, AMI), or lectures and presentations to a larger audience (CHIL). Choosing a specific type of meeting helps focus the development of scenarios that are used for collecting the corpus and the tasks that the system is supposed to perform in support of the meeting participants. The scenarios define what will be recorded as corpus: real meetings (e.g., ICSI), completely scripted (acted) meetings (M4), or something in between these extremes. As was already discussed in Section 17.3, the support tasks can take many forms, not all of which can be considered in all projects. Projects can work on a wide range of tasks, such as the search and retrieval of meeting content after the meeting (most projects), the process of sense-making from the unstructured recordings and other artifacts and derived data forms resulting from the meeting capture [40], the presentation of meeting content [44, 68], and the many types of real-time support discussed in Section 17.3 (CHIL, AMI, CALO). Evaluating scenarios and tasks assumes the existence of user requirements. User requirement collection is discussed for example in [41, 42]. In some projects the user requirements elicitation process has been made a research topic in itself [69]. One extra project worth mentioning in this section is the NEEM project. Ellis et al. [15] describe a very broad vision of meeting support, using embodied conversational agents (virtual humans) to communicate the support to the human participants of a meeting.

17.5 Conclusions

In this chapter we have given an overview of meetings and their everyday problems. We have shown that meetings all along the virtuality continuum can benefit from upcoming ambient intelligent technologies. In particular, meetings involve multiparty interaction, and modeling multiparty interaction is a theme that needs to be investigated in all kinds of ambient intelligence environments. An overview of the current state of the art of meeting applications has been given, divided into three main categories: the meeting resources, the meeting process, and the meeting roles. These applications that act upon interpreted sensor information need models in order to obtain a comprehensive understanding of meeting events. The development of these models is the main aim of the current international research projects dealing with meetings and multiparty interaction. As an example application where these models can be evaluated, we have presented the virtual meeting room. This VMR also gives an idea of what future meetings might be like, with participants being remotely present through video or even completely represented by virtual humans and browsers making the meeting information available afterward preventing group memory from dissolving. Ambient intelligent systems will analyze the information being conveyed and behave proactively in order to help realize making meetings as comfortable, efficient, and effective as possible.

Acknowledgments

This work was partly supported by the European Union 6th FWP IST Integrated Project AMI (Augmented Multiparty Interaction, FP6-506811, publication AMI-46).

References

[1] Vasilakos, A., and W. Pedrycz, *Ambient Intelligence, Wireless Networking, and Ubiquitous Computing,* Artech House, Norwood MA, 2006.

[2] Rao, A. S. and M. P. Georgeff, "Modeling Rational Agents Within a BDI-Architecture," *Proceedings of the 2nd International Conference on Principles of Knowledge Representation and Reasoning* (KR'91), pp. 473–484. Morgan Kaufmann: San Mateo, CA, USA, 1991.

[3] Jennings, N. R., "Specification and Implementation of a Belief-Desire Joint Intention Architecture for Collaborative Problem Solving," *Int. Journal of Intelligent and Co-operative Information Systems,* Vol. 2, No. 3: pp. 289–318, 1993.

[4] Reidsma, D., R. Rienks and N. Jovanovic, "Meeting Modeling in the Context of Multimodal Research," *Proceedings of the 1st Joint Workshop on Multimodal Interaction and Related Machine Learning Algorithms* (MLMI 2004), pp. 22–35. Springer-Verlag, Berlin, 2005.

[5] Reidsma, D.,et al., "Virtual Meeting Rooms: From Observation to Simulation," *Proc. of the Social Intelligence Design Workshop,* 2005, CD-Rom, Stanford University, 2005.

[6] Op den Akker, R., et al., "A Glossary on Meetings and Selected Meeting Scenarios (AMI d.1.1)," *Technical report, AMI Project,* January 2005.

[7] Moran, T.P.,et al., "Evolutionary Engagement in an Ongoing Collaborative Work Process: A Case Study," *Proceedings of the 1996 ACM Conference on Computer Supported Cooperative Work*, pp. 150–159, 1996.

[8] Antunes, P., and L. Carrio, "Modeling the Information Structures of Meetingware," *Proc. of Workshop de Sistemas de Informao Multimedia e Cooperativos (COOP-MEDIA'03)*, October 2003.

[9] Gordon, M., *How to Plan and Conduct a Successful Meeting*, Sterling Publishing Co., 1985.

[10] Doyle, M., and D. Straus, *How to make Meetings Work*, Berkeley Publishing group, 1976.

[11] MCI WorldCom, 1998. "A Study of Trends, Costs, and Attitudes Toward Business Travel and Teleconferencing, and Their Impact on Productivity," MCI WorldCom Conferencing White paper, 1998.

[12] Robert, H. M., *Roberts Rules of Order Revised*, Bartleby.com, 2000.

[13] Hocking, D., *Hockings Rules: The Essential Guide to Conduction Meetings*, Simon and Schuster, 1996.

[14] Johansen, R.,et al., *Leading Business Teams: How Teams Can Use Technology and Group Process Tools to Enhance Performance*, Addison Wesley, 1991.

[15] Ellis, C. S., and P. Barthelmess, "The Neem Dream," *Proceedings of the 2003 Conference on Diversity in Computing*, pp. 23–29. ACM Press, 2003.

[16] Milgram, P., and F. Kishino, "A Taxonomy of Mixed Reality Visual Displays," *IEICE Transactions on Information Systems*, E77-D(12), pp. 1321–1329, December 1994.

[17] Greenhalgh, C. M., and S.D. Benford "Virtual reality tele-conferencing: Implementation and experience," *Proc. Fourth European Conference on Computer Supported Cooperative Work* (ECSCW95), September 1995.

[18] Sudweeks, F., and S. Simoff, "Leading Conversations: Communication Behaviors of Emergent Leaders in Virtual Teams," *Proceedings of the Thirty-Eighth Annual Hawaii International Conference on System Sciences*, January 2005.

[19] Short, J. A., E. Williams and B. Christie, *The Social Psychology of Telecommunications*. John Wiley & Sons, 1976.

[20] Lombard, M., and T. Ditton, "At the Heart of it All: The Concept of Presence, *J. Computer-Mediated Communication*, Vol. 3, No. 2, pp. 1–43, 1997.

[21] Bailenson, J. N.,et al., "Transformed Social Interaction: Decoupling Representation from Behavior and Form in Collaborative Virtual Environments," *Presence*, Vol. 13, No. 4, pp. 428–441, August 2004.

[22] Vertegaal, R., "Look Who is Talking to Whom," PhD thesis, University of Twente, September 1998.

[23] Chen, H., and F. Perich, "Intelligent Agents Meet Semantic Web in a Smart Meeting Room," *Proceedings of the Third International Joint Conference on Autonomous Agents and Multi Agent Systems* (AAMAS 2004), pp. 854–861, July 2004.

[24] Chen, H., T. Finin, and A. Joshi, "A Context Broker for Building Smart Meeting Rooms," *Proceedings of the Knowledge Representation and Ontology for Autonomous Systems Symposium*, (AAAI Spring Symposium), pp. 53–61, March 2004.

[25] Johanson, B., A. Fox, and T. Winograd, "The Interactive Workspaces Project: Experiences with Ubiquitous Computing Rooms," *IEEE Pervasive Computing*, Vol. 1, No. 2, 67–74, 2002.

[26] Oh, A., R. Tuchinda, and L. Wu, "Meeting Manager: A Collaborative Tool in the Intelligent Room. *Proc. of the MIT Student Oxygen workshop*, 2001.

[27] Nuttin, M.,et al., "A Robotic Assistant for Ambient Intelligent Meeting Rooms," *Proceedings of the First European Symposium on Ambient Intelligence* (EUSAI 2003), pp. 304–317, November 2003.

[28] Sumec, S., "Multi Camera Automatic Video Editing," *Proceedings of the International Conference on Computer Vision and Graphics 2004* (ICCVG 2004), pp. 935–945, 2004.

[29] Barakonyi, I., W. Frieb, and D. Schmalstieg, "Augmented Reality Videoconferencing for Collaborative Work, *Proceedings of the 2nd Hungarian Conference on Computer Graphics and Geometry*, May 2003.

[30] Tatum, M., "Active Worlds," *SIGGRAPH Computer Graphics*, Vol. 34, No. 2, pp. 56–57, 2000.

[31] Fisher, S. S., et al., "Virtual environment display system," *Symposium on Interactive 3D Graphics*, pp. 77–87, October 1986.

[32] Rist, T., E. André, and S. Baldes, "A Flexible Platform for Building Applications with Life-Like Characters," *Proceedings of the 8th International Conference on Intelligent User Interfaces*, pp. 158–165. ACM Press, 2003.

[33] Traum, D., and J. Rickel, "Embodied Agents for Multi-Party Dialogue in Immersive Virtual Worlds," *Proc. of the 1st Int. Joint Conf. on Autonomous Agents and Multi-agent Systems* (AAMAS 2002), pp. 766–773, July 2002.

[34] Rodenstein, R., and J.S. Donath, "Talking in Circles: Designing a Spatially Grounded Audio Conferencing Environment," *Proceedings of the CHI 2000*, pp. 81–88. ACM Press, 2000.

[35] Aoki, P. M., et al., "The Mad Hatters Cocktail Party: A Social Mobile Audio Space Supporting Multiple Simultaneous Conversations," *Proc. ACM SIGCHI Conf. on Human Factors in Computing Systems*, pp. 425–432. ACM Press, April 2003.

[36] Mark, W., et al., "Adding Force Feedback to Graphics Systems: Issues and Solutions," *Computer Graphics*, Vol. 30 (Annual Conference Series), pp. 447–452, 1996.

[37] Palotta, V., et al., "Towards Meeting Information Systems," *Proceedings of the ICEIS 2004: 6th International Conference on Enterprise Information Systems*, Vol. 4, pp. 464–469, 2004.

[38] Shum, S., "Negotiating the Construction and Reconstruction of Organizational Memories," *Journal of Universal Computer Science*, Vol. 3, No. 8, pp. 899–928, 1997.

[39] Rittel, H., "Second Generation Design Methods," *Design Methods Group 5th Anniversary Report: DMG Occasional Paper*, Vol. 1, pp. 5–10, 1972.

[40] Moran, T., et al., "I'll Get That Off the Audio: a Case Study of Salvaging Multimedia Meeting Records," *Proceedings of the SIGCHI Conference on Human Factors in Computing Systems*, pp. 202–209. ACM Press, 1997.

[41] Lisowska, A., "Multimodal Interface Design for the Multimodal Meeting Domain: Preliminary Indications from a Query Analysis Study," *Technical report, ISSCO/TIM/ETI*, University of Geneva, Switzerland, November 2003. IM2.MDM Report 11.

[42] Jaimes, A., et al., "Memory Cues for Meeting Video Retrieval," CARPE'04, *Proceedings of the 1st ACM Workshop on Continuous Archival and Retrieval of Personal Experiences*, pp. 74–85. ACM Press, 2004.

[43] Tucker, S. and S. Whittaker, "Accessing Multimodal Meeting Data: Systems, Problems, and Possibilities," *Proceedings of the 1st Joint Workshop on Multimodal Interaction and Related Machine Learning Algorithms* (MLMI 2004), pp. 1–11. Springer-Verlag, Berlin, 2005.

[44] Wellner, P., M. Flynn, and M. Guillemot, "Browsing Recorded Meetings with Ferret" *Proceedings of the 1st Joint Workshop on Multimodal Interaction and Related Machine Learning Algorithms* (MLMI 2004), pp. 12–21. Springer-Verlag, Berlin, 2005.

[45] Erol, B., D. Lee, and J. Hull, "Multimodal summarization of meeting recordings," *Proceedings of the IEEE International Conference on Multimedia and Expo* (ICME 2003), July 2003.

[46] DiMicco, J. M., "Designing Interfaces That Influence Group Processes," *Proceedings of the Conference on Human Factors in Computer Systems* (CHI 2004), pp. 1041–1042, April 2004.

[47] Misiolek, N., and R. Heckman, "Patterns of Emergent Leadership in Virtual Teams" *Proceedings of the Thirty-Eighth Annual Hawaii International Conference on System Sciences*, January 2005.

[48] Ellis, C., J. Wainer, and P. Barthelmess, Agent Augmented Meetings, Volume 8 of Multi Agent Systems, Artificial Societies and Simulated Organizations. Chapter 2, 2003.

[49] Purver, M., J. Niekrasz, and S. Peters, "Ontology-Based Multi-Party Meeting Understanding," Position papers of CHI 2005 Workshop: CHI Virtuality 2005, July 2005.

[50] Biddle, B., *Role Theory: Expectations, Identities, and Behaviors,* Academic Press, 1979.

[51] Sutherland, I. E., "Sketchpad, a Man-Machine Graphical Communication System," PhD thesis, Massachusetts Institute of Technology, January 1963.

[52] Cochrane, T., "Do Tablets Dream of Electric Ink?" *Proceedings of the 16th Annual NACCQ,* pp. 239–243, July 2003.

[53] Jovanovic, N., R. Op den Akker, and A. Nijholt, "A Corpus for Studying Addressing Behavior in Multi-Party Dialogues," *Proc. of The sixth SIGdial conference on Discourse and Dialogue,* Lisbon, 2005.

[54] Rienks, R. J., and D. Heylen, "Dominance detection in meetings, using support vector machines," *Proceedings of the 2nd Joint Workshop on Multimodal Interaction and Related Machine Learning Algorithms* (MLMI 2005). Springer-Verlag, Berlin, 2005.

[55] Loomis, J., J.J. Blascovich, and A.C. Beall, "Immersive Virtual Environment Technology as a Basic Research Tool in Psychology," *Behavior Research Methods, Instruments and Computers,* Vol. 31, No. 4, 557–564, 1999.

[56] Rienks, R. J., R.W. Poppe, and M. Poel. "Speaker Prediction Based on Head Orientations," *Proceedings of the Fourteenth Annual Machine Learning Conference of Belgium and the Netherlands* (Benelearn 2005), pp. 73–79, Enschede, The Netherlands, 2005.

[57] Minneman, S. and S. Harrison, "Where Were We: Making and Using Near Synchronous, Pre-Narrative Video," *MULTIMEDIA '93: Proceedings of the first ACM international conference on Multimedia,* pp. 207–214, 1993.

[58] Schultz, T.,et al., "The ISL Meeting Room System," *Proceedings of the Workshop on Hands-Free Speech Communication,* 2001.

[59] Garofolo, J. S.,et al., "The NIST Meeting Room Pilot Corpus," *Proc. of the LREC2004,* 2004.

[60] Morgan, N.,et al., "Meetings About Meetings: Research at ICSI on Speech in Multiparty Conversations," *Proc of the ICASSP'03,* 2003.

[61] Nijholt, A., H.J.A. op den Akker, and D. Heylen, "Meetings and Meeting Modeling in Smart Surroundings," *Social Intelligence Design, Proceedings Third International Workshop,* Vol. 2, pp. 145–158. CTIT, 2004.

[62] Waibel, A., H. Steusloff, and R. Stiefelhagen, "Chil—Computers in the Human Interaction Loop," NIST ICASSP Meeting Recognition Workshop, Montreal, Canada, 2004.

[63] IM2 Website. http://www.im2.ch/.

[64] CALO Website. http://www.ai.sri.com/project/calo/.

[65] Lalanne, D.,et al, "Using Static Documents as Structured and Thematic Interfaces to Multimedia Meeting Archives," *Proceedings of the 1st Joint Workshop on Multimodal Interaction and Related Machine Learning Algorithms* (MLMI 2004), pp. 87–100. Springer-Verlag, Berlin, 2005.

[66] Foote, J., and D. Kimber, "Remote Interactive Graffiti," *MULTIMEDIA '04: Proceedings of the 12th annual ACM international conference on Multimedia,* pp. 762–763, 2004.

[67] Singh, G., L. Denoue, and A. Das, "Collaborative Note Taking," *WMTE '04: Proceedings of the 2nd IEEE International Workshop on Wireless and Mobile Technologies in Education* (WMTE'04), pp. 163–167, IEEE Computer Society, 2004.

[68] Girgensohn, A., J. Boreczky, and L. Wilcox. "Key Frame-Based User Interfaces for Digital Video," *Computer,* Vol. 34, No. 9, pp. 61–67, 2001.

[69] Tucker, S., S. Whittaker, and R. Laban, "Identifying User Requirements for Novel Interaction Capture," Symposium *Annotating and Measuring Meeting Behaviour* at Measuring Behaviour 2005, Wageningen, the Netherlands, 2005.

CHAPTER 18
Handling Uncertain Context Information in Pervasive Computing Environments

Mohamed Khedr

An infrastructure that supports context-aware applications must address two crucial objectives. First, it has to provide a means for effective representation of context, and second, it has to support techniques for managing contextual information. We will argue in this chapter that representing the uncertainties found in context information using ontologies , [2, 3] will grant context-aware applications the common understanding of vague context information and will provide the reusability and extensibility requirements. In addition, the decision to use ontologies for representing uncertain context simplified the second objective. Clarifying what we mean by managing context explains the reason behind this premise.

Context management, in this chapter, stands for the ability of the infrastructure to gather, manipulate, disseminate, and infer context and consequently make right decisions. These decisions are based on the level of certainty about the correctness of contextual information. This ensures that all entities in the environment know what they need to know and are able to influence others with what they know. Levels of certainty are acquired using inductive fuzzy logic reasoning, which when integrated with the context manipulation process, will enable the infrastructure to adapt its behavior automatically and manage context effectively in uncertain and vague environments.

With this clarification, it is easily deduced that the semantic representation of context facilitates the manipulation of contexts as entities share this common understanding of what contexts mean. However, semantic modeling of context using ontologies lacks the appropriate method to represent the level of certainty in captured context. It is also weak in providing adaptability and approximate reasoning about context information in order to manage the requirements of entities in the environments automatically.

This chapter discusses our approach to support levels of certainty in context-aware environment using inductive fuzzy reasoning over semantically modeled context. In the first section, we explain the reason for using fuzzy logic in a context-aware infrastructure. Next, we illustrate the process of fuzzifying ontology-based contexts with the developed fuzzy ontology. The last part of the chapter deals with the process of reasoning about context using the inductive fuzzy reasoning mechanism.

18.1 Extending Ontologies of Context with Fuzzy Logic

The concept of a fuzzy logic was introduced by Zadeh [4, 5] to deal with the representation of classes whose boundaries are ill defined or flexible. It does so by means of characteristic functions that take values from the interval [0,1]. A fuzzy set F in referential U is thus characterized by a membership function (MF) $\mu_F: U \rightarrow [0, 1]$. The value $\mu_F(u)$ represents the grade of membership of u in F. Specifically, $\mu_F(u) = 1$ reflects full membership of u in F, while $\mu_F(u) = 0$ expresses absolute non-membership in F.

Classical sets can be viewed as special case of fuzzy sets, where only full membership and absolute non-membership are allowed (they are called crisp sets, or Boolean sets).

Based on this definition of fuzzy logic, extending ontology-based context information will have the following general model:

$$F = \{(x, \mu_F(x)) | x \in X\}$$

where

- F is a fuzzy set in X.
- X is the universe of discourse of the context variable (constructs in the ontology).
- x is the allowed values for the contextvariable, i.e., its ontology range.
- $\mu_F(x)$ is the grade of membership function associated with the context variable.

This model applies to any type of context such as location, time, activity, users and device preferences. For example, suppose that by using an actor and location ontology, I can define the following context: *locatedIn*(khedr, E234). This crisp semantic representation states khedr is located in E234 and with a default certainty value equal to one. If the system is uncertain about the location of the actor, it can represent this uncertainty using the following fuzzy-based information.

locatedIn(khedr, B502) = F = 0.1/E238 + 0.4/E236 + 0.8/E324 + 0.4E232

where

- F is a discrete fuzzy set.
- X is universe of discourse of the context variable locatedIn.
- x allowed values = {E238, E236, E234, E232}.
- $\mu_F(x) = (0.1, 0.4, 0.8, 0.4)$.

This fuzzy-based, semantically modeled, context information means that khedr can be located in $E238$ with certainty value equal to 0.1, he can be located in $E236$ with certainty value 0.4, or he can be located in $E234$ with certainty value equal to 0.8, or he can be located in $E232$ with certainty value equal to 0.4.

Noticeable observations are that the system is no longer as certain as it was in the crisp case, and its certainty that khedr is located in *E236* is the same as that of *E232*. This enriches any context-aware infrastructure with the ability to reason about uncertain context since it can now take alternative decisions for actions that rely on the context of the user, and can cope with the uncertainty in this context information.

Three steps are required for representing uncertain contexts. The first step is extending the ontologies, which represent crisp contextual information, with the adequate properties to incorporate fuzzy logic. The second step is designing an ontology for representing the vocabularies used in fuzzy logic and approximate reasoning. The third step is developing the fuzzy inference mechanism needed to reason about context.

The first step is carried out by defining built-in classes of similarity in every context ontology and adding the property *hasSimilarity*, which has the *Similar* class defined in the fuzzy ontology as its range and the property *hasMF*, i.e., has membership function, which relates context classes to the fuzzy ontology.

For the second step, we modeled common fuzzy terms and vocabularies in the form of OWL ontology[1] which allows other ontologies to import and to subclass concepts defined in this fuzzy ontology.

18.2 The Fuzzy Ontology

The fuzzy ontology defines the common concepts and terms used in fuzzy logic theory. It also defines appropriate constructs for representing fuzzy rules. These rules can be used by entities in context-aware environments to set conditions or trigger actions. For example, an administrator can use the following fuzzy rule to set the condition required for granting services to users in a meeting.

> **If** *UserLocation is SmallerOrEqual to MeetingRoomA_Proximity with Gaussian MF that has mean=10, variance=5* **Then** *ServiceGrant is PrinterGrant with a crisp MF equal true* ....(Rule_1)

The fuzzy ontology, shown in Figure 18.1, has four key classes:

18.2.1 The Membership Function Class

The *FuzzyMembershipFn* class is used to transform any crisp context to a fuzzy term that takes its graded values from some characteristic function. This function could have continuous or discrete values. The *Trapezoidal*, *Triangular*, *Gaussian*, *Sigmoid*, and *UserDefined* classes are types of the continuous functions defined in the ontology. They are also arranged in the form of an *is a* relationship, for example, *Triangular* is a *Trapezoidal*, which gives the inference engine (IE) the ability to reason with alternatives in case the reasoning results are inadequate. For example, in Rule_1, the inference engine can use the *Sigmoid* function instead of the *Gaussian* in

---

1. http://www.schemaweb.info/schema/SchemaDetails.aspx?id=159.

order to have a wider range in the input location values and thus can extend this service granting rule to more users.

The discrete function is a special case of the membership functions where the input values are finite and specific. It is used with context information that is disconnected in nature such as number of copies in the printer preferences, or rooms in a certain floor as was illustrated in example 18.1.

The *UserDefined* membership function provides the extensibility requirement to this ontology. A developer uses this class if his context information cannot be fuzzified with one of the defined membership functions. In this case, the developer provides the inference engine with a set of $(x, \mu_F(x))$ describing context information. Consequently, the inference engine applies the Lagrange interpolation method

- FuzzyRule (hasAntecedent*, hasConsequent*, hasID*, hasInferenceType*)
    - Complement ()
    - Teonorm ()
        - Max ()
        - Sum ()
    - Tnorm ()
        - Min ()
        - Product ()
- Consequent (hasOperationType*)
- Antecedent (hasOperationType*)
- ConflictResolution ()
    - instance MostFrequent
    - instance Priority
- FuzzyInference (hasResolutionMethod*)
    - Inactivefuzzy (hasIFT_URL*)
        - Node (isaLeaf*)
            - Child (isaLeaf)
            - parent (lessNodes*, isaLeaf[0])
        - Tree (hasHeuristicType*)
            - Branch ()
    - MinMax ()
        - Conditional ()
            - Unqualified ()
            - Qualified (hasHedge*)
- FuzzyMembershipFn (hasAssociatedFuzzyForm*, operatedUpon*)
    - Continuous (hasEndPoint*, hasStartPoint*)
        - Gaussian (hasVariance*, hasmean*)
            - Sigmoid ()
        - UserDefined ()
        - Trapezoidal (hasFirstpnt*, hasSendpnt*)
            - Triangular (hasCenter*)
    - Discrete (hasInputValue, hasMembershipValue)
- FuzzyNumber (hasValue*)
- FuzzyOperation ()
    - Declassification (hasDeclassificationType*)
    - Fuzzification (hasFuzzificationType*)
    - Hedges (hasHedgeFormula*, hasHedgeName*)
    - Operation ()
        - instance Ana
        - instance BiggerOrEqual
        - instance Equal
        - instance Or
        - instance Smaller
- FuzzySet (AggregationOf*, alphacutSupport*, hasCardinality*, hasMembershipFn*)
- Similar (degreeOfContain*, similarhr*, similarTo*)

Figure 18.1 Class-property hierarchy diagram of the fuzzy ontology.

on this set of values to provide a continuous, user-defined, fuzzy membership function.

18.2.2 The Fuzzy Rule Class

The *FuzzyRule* class defines rules in the form of *If Antecedents (operand used) with MFs then consequents with MFs*, where *Antecedent* and *consequent* have the form of (variable, operand, value). Variables and values have *owl:Thing* as their range, which means that they can be linked to any external instance defined in the context ontologies.

For example, in the event-notification application, which we will discuss later in this chapter, an inference engine can use the following rule to decide the level of alert to send to the users. This alert is based on the users' location and the time of the activity.

The rule states: If (UserLocation, Equal, Far) and (meetingA, at, meetingtime) then (Alert, Equals, High). The rule also illustrates how ontologies are combined together. This is shown in the temporal entity *meetingA*, which we defined using the time ontology, and is fuzzified with the fuzzy variable TimeVariable.

18.2.3 The Similar Class

The *Similar* class is the link used by other ontologies to describe the similarity between concepts. The Similar class has three properties for defining fuzzy similarity:

1. similarTo property: Used to related the original concept with another concept. For example, a location instance could have the following triple (E234, similarTo, PERLAB), which states that E234 is similar to PERLAB, or an actor instance could have the triple (Khedr, similarTo, mkhedr), which means that Khedr is similar to mkhedr.
2. similarIn property: Used to state what these concepts are similar to. For example, (E234, similarIn, Room) which means that E234 is similar to PERLAB and they are similar in their type, being a room. Also, (khedr, similarIn, PhDStudent) means that khedr and mkhedr are similar in their role, being a PhD student.
3. degreeOfCertain property: Used to describe how accurate the similarity is between the concepts. For example, (E234, degreeOfCertain, 0.8) means that E234 is similar to PERLAB with 80% degree of certainty.

18.2.4 The Fuzzy Inference Class

Many methods were developed for reasoning using fuzzy logic [6] and each has its approach of deducing knowledge from existing facts. However, no single inference method is adequate for all situations. Hence, we semantically represented these methods in the fuzzy ontology in order for the inference engine to execute the appropriate method based on context. In particular, the fuzzy ontology defines the approximate and inductive reasoning mechanisms, since they are the two most common types of inference used in fuzzy logic. However, other developers can benefit from the extensibility feature in our ontology to define their own inference process. The approximate reasoning class addresses the conditional-unqualified and conditional-qualified methods.

In the first case, rules are expressed in the form p: *If χ is A, then y is B*, where χ, y are the context variables and A, B are the fuzzy sets that represent the universe of discourse (X, Y) of these contexts respectively. This type of inference can be viewed in the form of fuzzy relation $R(\chi, y) = p[A(x), B(y)] =$ and $\wp(a,b) = \min(1, 1 - a + b)$ where R is a fuzzy set on the cartesian product $X \cdot Y$ that is calculated over each $x \in X$ and $y \in Y$ using the fuzzy implication operation $\wp$.

Thus, for a rule such as Rule_1, we can say that $A =$ SmallerOrEqual to MeetingRoomA_Proximity while $B =$ PrinterGrant. If we assume that A has the MF of $\{0.1/x_1 + 0.8/x_2 + 1/x_3\}$ and B has MF $\{0.5/y_1 + 1/y_2\}$ then $\wp(a,b)$ will be equal to $\{1/x_1,y_1 + 1/x_1,y_2 + 0.7/x_2,y_1 + 1/x_2,y_2 + 0.5/x_3,y_1 + 1/x_3,y_2\}$. This means that if a user has the location x_2, then the context-aware infrastructure can grant him access to printer y_1 with a degree of certainty, $T(p)$, equal to 0.7 or to printer y_2 with $T(p) = 1$. After using the approximate reasoning, the inference engine can successfully grant the user access to printer y_2 since it has a higher membership value.

The conditional-qualified method has the form of p: *If χ is A, then y is B is S* and applies the same implication process as in the conditional-unqualified method except for the last step where the output is passed through a hedging filter, S, in order to increase or decrease the degree of certainty of the outcome.

An inference engine uses the approximate reasoning when the number of context variables is relatively small. This is because the number of rules increases exponentially with the number of variables used in this type of fuzzy inference method.

In situations where the number of context variables is large, the inference engine will use the inductive reasoning. This is because the inductive reasoning can detect patterns in the input values and remove irrelevant contexts, which will have the result of reducing the number of generated rules and enhance system dynamism.

In both cases, the inference engine must perform the following steps:

1. Define linguistic terms and generate membership functions.
2. Build the fuzzy inductive inference tree (case of inductive reasoning only).
3. Generate rules from fuzzified context information.
4. Perform the inference process.

Figure 18.2 illustrates these steps and depicts the process of binding fuzzy logic with the context information described using ontologies.

18.3 The Fuzzy Inference of Context

Fuzzy inference works on input values that have the form of $\sum_{i=1}^{N} \mu(x_i) / x_i | x \in X$ in case of discrete input or the $\int_X \mu_F(x) / x$ form of in case the input is continuous. This

18.3 The Fuzzy Inference of Context

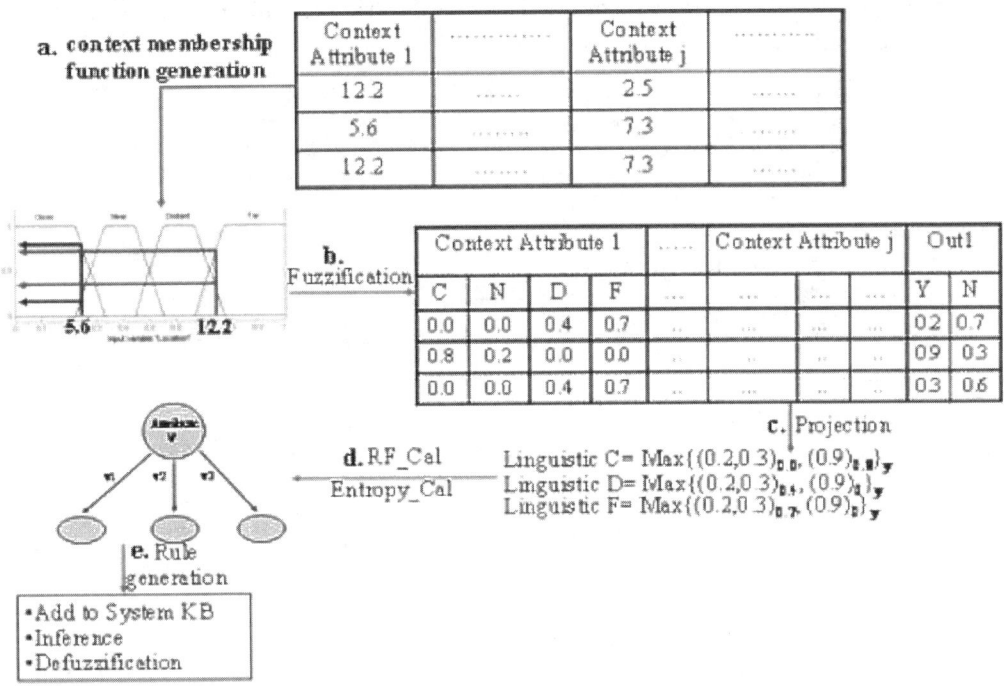

Figure 18.2 Steps used in the fuzzy inference of context information.

means that an inference engine needs first to loosen the semantically modeled context before starting the inference.

Context fuzzification (steps a and b in Figure 18.2) means transforming the ontology-based context instances into a set of linguistic terms. Each linguistic term has a certain membership function that is used in the inference instead of the original context instance as in the case of standard logic.

A relation between a device and its owner is an example that illustrates the fuzzification of context. A crisp relation will have the following ontological representation.

The captured context indicates that PERLAB-10 is a device of type Desktop, it is located in E234, its vendor is IBM and it belongs to user Khedr.

```
<Desktop rdf:ID="PERLAB-10">
  <hasLocation rdf:resource="&Loc;E324"/>
  <belongsTo rdf:resource="&Confoaf;Khedr"/>
  <vendor>IBM</vendor>
</Desktop>
```

This information is acceptable if user Khedr is actually the only person who is using this desktop all the time. However, he might be using it only during the morning hours, which will make this fact inaccurate and might cause unacceptable consequences. When fuzzy logic is used in this situation to represent the uncertainty in the relation between the desktop and the user, the semantic representation will include the property hasMF that will link the crisp context ontology to the fuzzy ontology;

then linguistic terms are defined which describe the membership functions used in the relation.

```
<DeskTop rdf:ID="PERLAB-10">
  <hasLocation rdf:resorce="&Loc;E234"/>
  <belongsTo rdf:resource="&Confoaf;Khedr"/>
  <vendor>IBM</vendor>
  <hasMF>  ….

<Fuzz:hasFuzzyTerm >
<Fuzz:FuzzyTerm rdf:ID="Morning">
  <Fuzz:hasMembershipFn>
    <Fuzz:trapezoidal>
      <Fuzz:hasFirstpnt>9.0</Fuzz:hasFirstpnt>
      <Fuzz:hasEndPoint>12.0</Fuzz:hasEndPoint>
      <Fuzz:hasScndpnt>11.0</Fuzz:hasScndpnt>
      <Fuzz:hasStartPoint>8.0</Fuzz:hasStartPoint>
    </Fuzz:trapezoidal>
  </Fuzz:hasMembershipFn>
</Fuzz:FuzzyTerm>

<Fuzz:hasFuzzyTerm>
  <Fuzz:FuzzyTerm rdf:ID="Afternoon">
    <Fuzz:hasMembershipFn>
      <Fuzz:triangular>
        <Fuzz:hasCenter>13.0</Fuzz:hasCenter>
        <Fuzz:hasEndPoint>15.0</Fuzz:hasEndPoint>
        <Fuzz:hasStartPoint>11.0</Fuzz:hasStartPoint>
      </Fuzz:triangular>
    </Fuzz:hasMembershipFn>
  </Fuzz:FuzzyTerm>
</Fuzz:hasFuzzyTerm>

<Fuzz:hasFuzzyTerm>
    <Fuzz:FuzzyTerm rdf:ID="Night">
      <Fuzz:hasMembershipFn>
        <Fuzz:Crisp rdf:ID="Zero"/>
      </Fuzz:hasMembershipFn>
    </Fuzz:FuzzyTerm>
  </Fuzz:hasFuzzyTerm>
```

In this example, we defined three linguistic terms that describe the temporal relation between the desktop and the user. The first term is *Morning* which defines the degree of certainty that *Khedr* is using *PERLAB-10* during the morning hours from 8:00 a.m. to 12:00 p.m. The MF is represented as a trapezoidal function indicating that between 9:00 and 11:00, the context-aware infrastructure is certain that *PERLAB-10* belongs to user *Khedr*. The second term is *Afternoon* which is described as a triangular function between the hours 11:00 a.m. to 3:00 p.m. and the last term is *Night* which is described as a crisp membership function with value equal to zero. This means that *PERLAB-10* does not belong to *Khedr* during the night hours.

With the aforementioned example that clarifies our intention of using fuzzy and fuzzy ontology to represent uncertainty in the context information, the next sections describe the steps performed by an inference engine in a context-aware environment for reasoning about uncertain context information.

18.3.1 Defining Linguistic Terms and Generating Membership Functions

Similar to the notions of variables and ranges of variables that exist in traditional mathematical theory, fuzzy set theory has the notion of linguistic terms that represent the universe of discourse of a linguistic variable.

Zadeh [4, 5] defined linguistic variables as a quintuple in the form of (ω, T(ω), X, G, M);
where

- ω is the name of the linguistic variable.
- T(ω) is the term set or set of fuzzy reference sets of ω.
- X is the universe of discourse on which all member sets of T(ω) are defined.
- G is the grammar comprising a set of syntactic rules to generate the terms in T(ω).
- M is the semantic rule for associating with each value of ω its meaning.

Applying the quintuple to the *PERLAB-10 belongsTo Khedr in the Morning hours* example, ω will represent *PERLAB-10 belongsTo Khedr*, T(ω) will represent {Morning, Afternoon, Night}, X is the time range from [8:00,20:00], G is the same as T(ω) and M will be the membership functions {trapezoidal, triangular and crisp}describing each term in the T(ω).

The critical part in the quintuple is to decide the term set, T(ω), and the membership functions, M, in such a way that will provide good modeling to the uncertainty in context and with low overhead in computation.

The inference engine is responsible for performing this task, since users and application developers are assumed not to be expert in fuzzy logic and domain mapping techniques. Application developers will only specify the number of linguistic terms to be used by the inference engine to map the context information to them.

For example, a developer may state that the location context inside a meeting room is specified by two linguistic terms near and far with respect to sensor positions in the room. These two linguistic terms, T(ω), need to be mapped to the standard membership functions defined in the fuzzy ontology and stored in the infrastructure repository. Therfore, the two linguistic terms can be represented by Gaussian functions that have mean and variance determined by the range of the location and the overlap region.

Pseudocode 18.1 illustrates the sequence used for fuzzifying context and eliciting membership functions representing this fuzzification of context.

To generate the required membership functions, the IE first checks the context ontology to determine the range and domain of the context construct to be fuzzified. Then the IE uses a simple algorithm to generate the membership functions, which requires least computational overhead.

The algorithm starts by dividing the context range into 2n + 1 equal slots, where n is the number of linguistic terms required (in the previous location example, n = 2). It then assigns three slots to each linguistic term and each set of adjacent terms share one slot in common that will represent the overlapping region in the fuzzy sets. The three-slots concept is chosen because the trapezoidal function, which is considered the general membership function from which other member-

Inputs: T(ω), C where:
  T(ω) is the set of linguistic terms representing a context instance.
  C is the context instance
Procedure:
  Context Preprocessing
  1 Fetch the Range associated with the context C.
  Range Clustering
  1 Divide the range into 2n + 1 equal slots. n is the number of linguistic terms required
  2 Assigns three slots to each linguistic term.
  3 Each adjacent terms share one slot in common that will represent the overlapping region in the fuzzy sets
  Membership function Approximation
  1 Sample each of the 2n + 1 regions that represent the range of the context C.
  2 Randomly choose samples from each of these regions.
  3 Project these samples to the user for setting expected membership values for them.
  Membership function Elicitation
  1 Validate the received data
  2 If Consistent
    a. Apply the interpolation function on the received values.
    b. Decide the corresponding standard membership function using the similarity function
  3 Else
    a. Add/Remove slots from regions that include inconsistencies → non-uniform division
    b. Apply the interpolation function on the received values.
    c. Decide the corresponding standard membership function using the similarity function
Output: A set of Membership functions M representing the linguistic terms in the T(ω)

Pseudocode 18.1 Generating membership functions for context information.

ship functions can be derived, needs four points to be defined. A Triangular, Gaussian, or S-function can easily be assigned to these regions by executing a transformation function that maps the generated trapezoidal function to any of these functions.

After successfully dividing the context range into its 2n + 1 regions, the IE randomly selects samples from each region and requests from the application developer, or user, to specify the expected membership value for each of these samples. This step is required in order for the inference engine to deduce correctly the best membership functions that resemble user expectation. This step also allows the inference engine to decide if the tentative membership functions should remain equally divided as initially proposed, or if they should be recalculated and as a result become non-uniformly spaced. After this decision, the inference engine applies the Lagrange interpolation mechanism on this sampled data followed by Euclidian measure to decide the standard membership functions that best resemble the user-defined data. If the context range is not equally divided, the context membership generation starts with the same procedure as stated above, but assigns a different number of slots to each linguistic term according to the required ratio. This ratio is calculated from the sampled data taken in the **Membership function Approximation** step shown in Pseudocode 18.2.

Figure 18.3 demonstrates the output membership function generated from the described algorithm. The left figure represents the context information, location,

18.3 The Fuzzy Inference of Context

Inputs: (D, A, C), where:
- D is the training set of instances
- A is the context attributes used to build the inductive fuzzy tree.
- C is the output context/action attribute

Output: an inductive fuzzy tree S representing a sequential decision process for classifying instances (Predicting the values of the class attribute C); each node of S is labeled with a non-class attribute of A

Informal Inductive Bias: minimize the average height of the tree

Procedure:
  If the set of remaining non-class attributes is empty or if all of the instances in D are in the same class
    return an empty tree
  else {
  Compute the class entropy of each attribute over the dataset D
  let a^* be an attribute with minimum class entropy
  create a root node for a tree S; label it with a^*
  for each linguistic term b of attribute a^* {
    let $S(a^* = b)$ be the tree computed recursively by FID on input $(D|a^* = b, A\text{-}a^*, C)$, where:
    $D|a^* = b$ contains all instances of D for which a^* has the value b
    $A\text{-}a^*$ consists of all attributes of A except a^*
    attach $S(a^* = b)$ to the root of S as a subtree
  }
  return the resulting decision tree S

Pseudocode 18.2 Building the fuzzy induction tree of context information.

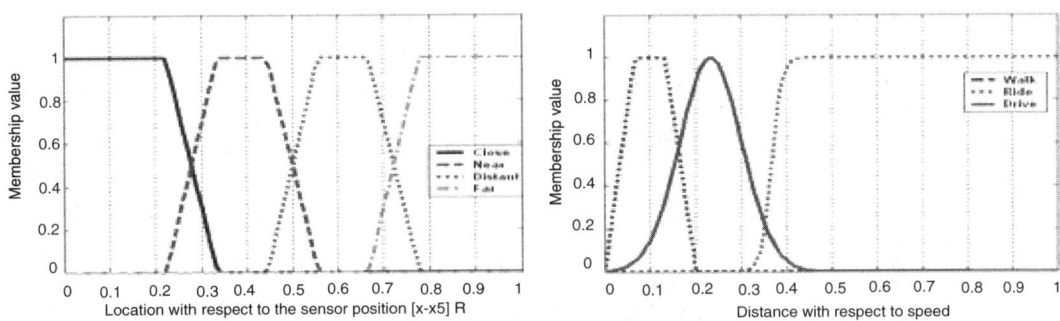

Figure 18.3 Elicitation of fuzzy membership functions, a) Uniform membership function representation of location. b) Nonuniform membership function representation of distance

using four linguistic terms with equal regions. The right figure shows the context information, distance, using three unequal membership functions according to a specified ratio between them, 1:2:4.

18.3.2 Building the Inductive Fuzzy Tree

After deciding the membership functions for the context information captured and derived by the inference engine, the inference engine builds the inductive fuzzy tree, shown as steps c and d in Figure 18.2.

In general, building an inductive fuzzy tree requires a set of training examples, a heuristic method for choosing the best attribute to split at any node of the tree, and an inference mechanism [7]. Training examples in our infrastructure are acquired either from experts in the application domain or from sensors and historical actions of users. These examples are stored in the infrastructure repository, which is under the control of the inference engine and can be manipulated by application developers.

The IE uses a heuristic method based on entropy measurement. This method has two purposes. First, it reduces the training example space through which the computation is performed, thus reducing complexity and increasing the speed of convergence in calculations. Second, because repeated values are averaged, it reduces the effect of a dominant linguistic term in the calculation.

The inference engine performs the steps in Pseudocode 18.2 to build the inductive fuzzy tree. The root of the tree contains all the training examples. It represents the whole description space since no restrictions are imposed yet. Each node is recursively split by partitioning its training examples. Training examples at the node to be split are distributed on its child nodes according to the degree with which they match the linguistic terms at the parent node. A node becomes a leaf if its associated examples come from a unique class, or when all context attributes are used along the path from the root to this node.

Specifically, the inference engine performs the following calculations in order to build the inductive fuzzy tree.

- For every linguistic term T_i of context attribute A, calculate its relative frequency, p_{ij}, $p_{ij} = M(Tnorm(T_i, C_j)) / M(C_j)$ with respect to each linguistic term, C_j, of the output class C. M is the cardinality measure for all projected examples.

- Calculate the total relative frequency of every linguistic term T_i of context attribute A $P_i = \sum_{j=1}^{J} p_{ij}$ where J is total number of projected examples.

- Calculate the weighted entropy $E_A = \sum w_i \cdot P_i$ of each context attribute A over its total linguistic terms N. The weight factor is equal to $w_i = \dfrac{p_i}{\sum_{l=1..N} p_l}$

- Choose the context attribute with the minimum entropy as the winning attribute.

- Classify the examples along each linguistic branch of that context attribute.

- Calculate the stopping factor, beta; if beta is bigger than a certain threshold, then this node is a leaf; else repeat for the new node.

For each branch in the tress, repeat the process until all training examples are classified.

We illustrate the fuzzy induction process through a sport recommendation system that uses the small training examples shown in Table 18.1. The training examples represent the fuzzified contextual information needed to induce the type of sport that can be recommended to the users. The types of contextual information

18.3 The Fuzzy Inference of Context

Table 18.1 Fuzzified Context Attributes and the Output Decision Values

Look Sunny	Look cloudy	Look rainy	Temp Hot	Temp mild	Temp cool	Hum Hum	Hum Normal	Wind Wind	Wind No	Volley	Swim	Weigh Lift
0	0	1	0	0	1	1	0	0.8	0.2	0	0	1
0	0.1	0.9	0.7	0.3	0	0.5	0.5	0.5	0.5	0	0	1
0	0.3	0.7	0	0	1	0	1	0.1	0.9	0	0	1
0	0.7	0.3	0	0.3	0.7	0.7	0.3	0.4	0.6	0.2	0	0.8
0	0.7	0.3	0.8	0.2	0	0.1	0.9	0.2	0.8	0.7	0.3	0.1
0	0.9	0.1	0	0.9	0.1	0.1	0.9	0.7	0.3	0	0	1
0	1	0	0	0.2	0.8	0.2	0.8	0	1	0.7	0	0.3
0.2	0.6	0.1	0.3	0.7	0	0.3	0.7	0.3	0.7	0.6	0.4	0
0.2	0.6	0.2	0	1	0	0.3	0.7	0.3	0.7	0.7	0.2	0.1
0.3	0.8	0	0.6	0.4	0	0	1	0	1	0.9	0.3	0
0.5	0.6	0	0	0.3	0.7	0	1	0.9	0.1	0	0.1	0.9
0.5	0.7	0	1	0	0	0.6	0.4	0.7	0.3	0.2	0	0.7
0.8	0.3	0	1	0	0	0.8	0.2	0.4	0.6	0.4	0.8	0
0.8	0.3	0	1	0	0	1	0	0.2	0.8	0.3	0.7	0
0.9	0.1	0	0.2	0.8	0	0.1	0.9	1	0	0	0	1
1	0	0	0.8	0.5	0	0	1	0	1	0.8	0.5	0

used in this application are temperature, outlook, wind, and humidity, while the output is to recommend for the users playing one of the following sports: volleyball, swimming, or weightlifting, according to these contextual information.

Figure 18.4 shows an excerpt of the semantic modeling of the sport activity. The figure also shows how Temperature, one of the contexts used in this activity, was modeled using the property necessaryFeature and the class ActivityFeature.

Each of these context terms is fuzzified to their linguistic terms using the process discussed in Section 18.3.1. For example, temperature is fuzzified into three linguistic terms {Hot, Mild, Cool} with membership functions {Gaussian, Trapezoidal, Sigmoid} respectively.

Every row in Table 18.1 represents a situation where different values were taken for the four contexts and the type of sport that is adequate with these context values. After the inference engine performs the entropy calculation on these examples, as shown in Table 18.2, it decides that the outlook attribute is the most appropriate context to be used as the root of the inductive fuzzy tree, since it has the minimum entropy. The IE continues the process of splitting context attributes and categorizing training examples until all the training examples are distributed on the nodes of the generated tree.

Figure 18.5 shows the inductive fuzzy tree formed using the training examples in Table 18.1.

Several observations can be recorded from the generated tree. First, outlook was the most important context for this type of application, followed by wind and then temperature contexts. Second, although many factors may influence the output decision, the IE found that with rainy conditions the pattern is always toward indoor sports. Third, humidity was an important context in case of sunny outlook, while it was not in case of a cloudy outlook, since the impact of humidity vanishes with cooling conditions. Finally, the inference engine from this tree can generate the rules that are used in the inference process.

```xml
<Contivity:Activity rdf:ID="Sports">
    <rdfs:comment>illustrates a sample of inductive
        fuzzy</rdfs:comment>
    <Contivity:describesedBy>
      <ActivityProfile rdf:ID="SportProfile">
        <necessaryFeatures>
          <ActivityFeatures rdf:ID="Tempereture_H">
            <featureValue>
              <Fuzz:Gaussian rdf:ID="GaussianHot">
                <Fuzz:hasmean>27.0</Fuzz:hasmean>
                <Fuzz:hasVariance>5.0</Fuzz:hasVariance>
              </Fuzz:Gaussian>
            </featureValue>
            <featureName>Hot</featureName>
          </ActivityFeatures>
        </necessaryFeatures>
        <withActivityType rdf:resource="# Volley"/>
        <withActivityType rdf:resource="#WeightLifting"/>
        <withActivityType rdf:resource="#Swimming"/>
        <necessaryFeatures>
          <ActivityFeatures rdf:ID="Temperature_C">
            <featureValue rdf:resource="#Sigm_C"/>
            <featureName>Cool</featureName>
          </ActivityFeatures>
        </necessaryFeatures>
        <necessaryFeatures>
          <ActivityFeatures rdf:ID="Tempereture_M">
            <featureName>Medium</featureName>
            <featureValue>
              <Fuzz:trapezoidal rdf:ID="Trap_M">
                <Fuzz:hasEndPoint>25.0</Fuzz:hasEndPoint>
                <Fuzz:hasFirstpnt>18.0</Fuzz:hasFirstpnt>
                <Fuzz:hasScndpnt>22.0<</Fuzz:hasScndpnt>
                <Fuzz:hasStartPoint>15.0</Fuzz:hasStartPoint>
              </Fuzz:trapezoidal>
            </featureValue>
          </ActivityFeatures>
        </necessaryFeatures>
        <Contivity:describes rdf:resource="#Sports"/>
      </ActivityProfile>
    </Contivity:describesedBy>
</Contivity:Activity>
<Fuzz:Sigmoid rdf:ID="Sigm_C">
  <Fuzz:hasStartPoint>10.0</Fuzz:hasStartPoint>
  <Fuzz:hasEndPoint>15.0</Fuzz:hasEndPoint>
</Fuzz:Sigmoid>
```

Figure 18.4 The fuzzified semantic representation of context.

18.3.3 Rule Generation and Inference Process

Each path along the generated tree from root to leaf is converted into a rule, with the antecedent part representing the context attributes of each succeeding branch, and the consequent part representing the class at the leaf with highest membership value. The membership value of the every output at each leaf is taken to be equal to the maximum membership value of all instances reaching that leaf and having that class as their output.

For instance, the tree in Figure 18.5 will generate the following rules:

18.3 The Fuzzy Inference of Context

Table 18.2 Excerpt of the Process of Building the Induction Tree.

	Look Sunny	Look cloudy	Look Rainy	Volley	Swim	W.L	M(T1, U1,S)	M(T1, U2,S)	M(T1, U3,S)	M(T2, U1,S)	M(T2, U2,S)	M(T2, U3,S)	M(T3, U1,S)	M(T3, U2,S)	M(T3, U3,S)
	0	0	1	0	0	1	0	0	0	0	0	0	0	0	1
	0	0.1	0.9	0	0	1	0	0	0	0	0	0.1	0	0	0.9
	0	0.3	0.7	0	0	1	0	0	0	0	0	0.3	0	0	0.7
	0	0.7	0.3	0.7	0.3	0.8	0	0	0	0.7	0.3	0.7	0.3	0.3	0.3
	0	0.9	0.1	0	0	1	0	0	0	0	0	0.9	0	0	0.1
	0	1	0	0.7	0	0.3	0	0	0	0.7	0	0.3	0	0	0
	0.2	0.6	0.1	0.6	0.4	0	0.2	0.2	0	0.6	0.4	0	0.1	0.1	0
	0.2	0.6	0.2	0.7	0.2	0.1	0.2	0.2	0.1	0.6	0.2	0.2	0.2	0.2	0.1
	0.3	0.8	0	0.9	0.3	0	0.3	0.3	0	0.8	0.3	0	0	0	0
	0.5	0.6	0	0	0.1	0.9	0	0	0	0	0.1	0.6	0	0	0
	0.5	0.7	0	0.2	0	0.7	0.2	0	0.5	0.2	0	0.7	0	0	0
	0.8	0.3	0	0.4	0.8	0	0.4	0.8	0	0.3	0.3	0	0	0	0
	0.9	0.1	0	0	0	1	0	0	0.9	0	0	0.1	0	0	0
	1	0	0	0.8	0.5	0	0.8	0.5	0	0.3	0	0	0	0	0
M(T,S)	4.4	6.7	3.3				2.1	2	1.5	3.9	1.6	3.9	0.6	0.6	3.1
Pij							0.477	0.454	0.34	0.582	0.238	0.582	0.181	0.181	0.93
Entropy(i)	1.548	1.401	0.99												
Weight	0.305	0.465	0.229												
Total entropy of this attribute			1.35												

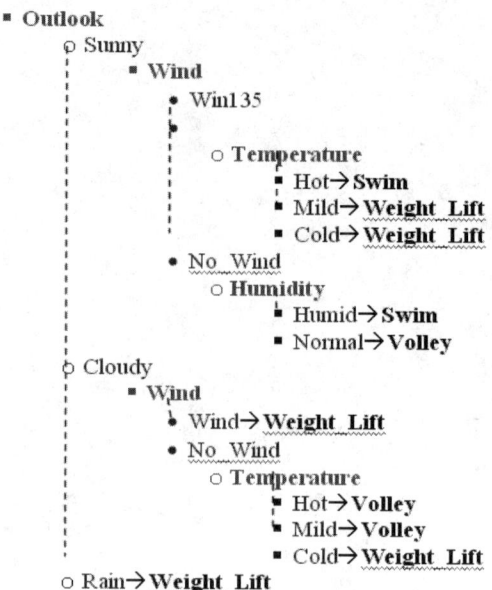

Figure 18.5 The inductive fuzzy tree.

RULE 0: IF (Look is Look_Sunny) And (Wind is Wind_Wind) And (Temp is Temp_Hot) THEN (OutPut is Sport_Swim) with MF value [0.211, 0.421, 0.368]

RULE 1: IF (Look is Look_Sunny) And (Wind is Wind_Wind) And (Temp is Temp_Mild) THEN (OutPut is Sport_Lift) with MF value [0.0, 0.0, 1.0]

RULE 2: IF (Look is Look_Sunny) And (Wind is Wind_Wind) And (Temp is Temp_Cool) THEN (OutPut is Sport_Lift) with MF value [0.0, 0.0, 1.0]

RULE 3: IF (Look is Look_Sunny) And (Wind is Wind_No) And (Hum is Hum_Hum) THEN (OutPut is Sport_Swim) with MF value [0.375, 0.625, 0.0]

RULE 4: IF (Look is Look_Sunny) And (Wind is Wind_No) And (Hum is Hum_Norm) THEN (OutPut is Sport_Volley) with MF value [0.615, 0.385, 0.0]

RULE 5: IF (Look is Look_Cloudy) And (Wind is Wind_Wind) THEN (OutPut is Sport_Lift) with MF value [0.0, 0.0, 1.0]

RULE 6: IF (Look is Look_Cloudy) And (Wind is Wind_No) And (Temp is Temp_Hot) THEN (OutPut is Sport_Volley) with MF value [0.75, 0.083, 0.167]

RULE 7: IF (Look is Look_Cloudy) And (Wind is Wind_No) And (Temp is Temp_Mild) THEN (OutPut is Sport_Volley) with MF value [0.846, 0.154, 0.0]

RULE 8: IF (Look is Look_Cloudy) And (Wind is Wind_No) And (Temp is Temp_Cool) THEN (OutPut is Sport_Lift) with MF value [0.467, 0.0, 0.533]

RULE 9: IF (Look is Look_Rainy) THEN (OutPut is Sport_Lift) with MF value [0.0, 0.0, 1.0]

18.3 The Fuzzy Inference of Context

The process of inferring new context instances is carried out from the generated rules using the compositional rule of inference method [6]. First, context instances are fuzzified by comparing their captured values with their linguistic terms. Then, for each rule, the minimum value between all the fuzzified contexts in its antecedent part is selected and used as the membership value of the rule's consequent part. The outcome of these steps will be one of the following:

1. All rules have the same output consequent. In this case, the IE takes this consequent as the result of the fuzzy inference, and its membership value equals the minimum membership value of all the rules output
2. Rules have different output consequents and different membership values. In this case, the IE takes the maximum value of all similar consequents and then selects the output consequent with the highest membership value

For example, using the generated tree shown in Figure 18.5, let a new fuzzified context instance for recommending a sport to users based on outlook, temperature, wind, and humidity be equal to {Outlook (0.6, 0.4, 0.1), Temperature (0.8, 0.2, 0.0), Humidity (0.4, 0.6), Wind (0.2, 0.9)}. Then the output of the generated rules will be equal to

Rule 1: min(0.6,0.2,0.8) = 0.2 → Swim with MF of 0.2.

Rule 2: min(0.6,0.2,0.2) = 0.2 → Lift with MF of 0.2.

Rule 3: min(0.6,0.2,0.0) = 0.0 → Lift with MF of 0.0.

Rule 4: min(0.6,0.9,0.4) = 0.4 → Swim with MF of 0.4.

Rule 5: min(0.6,0.9,0.6) = 0.6 → Volley with MF of 0.6.

Rule 6: min(0.4,0.2) = 0.2 → Lift with MF of 0.2

Rule 7: min(0.4,0.9,0.8) = 0.4 → Volley with MF of 0.4

Rule 8: min(0.4,0.9,0.2) = 0.2 → Volley with MF of 0.2.

Rule 9: min(0.4,0.9,0.0) = 0.0 → Lift with MF of 0.0.

Rule 10: min(0.0) = 0.0 → Lift with MF of 0.0.

Since there is more than one consequent that has the same output class, the IE will take the maximum value of each group of classes that have the same consequent, then the maximum of the resultant is as follows:

$\max(\max(0.2,0.4)_{swim}, \max(0.6,0.4,0.2)_{Volley}, \max(0.0,0.0,0.2,0.0,0.2)_{Lift}) = \max(0.4S_{wim}, 0.6_{Volley}, 0.2_{Lift}) = 0.6$ → the inference output of that context instance has an outcome equal to volleyball with membership value equal to 0.6.

Thus, a recommendation for users who would like to do sports is to play volleyball followed by swimming as an alternative according to the context values in this new instance.

18.4 Prototype Implementation: The Event-Notification Service

A typical type of application that was implemented to demonstrate the uncertainty concept in context-aware environments deals with event monitoring and notification.

In this application, the IE reasons about user's context and notifies the user about his scheduled events. The reasoning process is based on fuzzy logic, because the two contexts required in the reasoning process are vague in nature (location, time). Moreover, the process of notification is not crisp but has a range of transition in its outcome.

18.4.1 Fuzzy Inference in the Event-Notification Service

From study, location as a context can be represented as a position and a direction. Position was mapped to {near, within, far} linguistic terms while direction was mapped to {left and right}. This produced five linguistic terms, which are shown in . In addition, users specify time in standard format, e.g., notify me at 12:00 p.m. However, if the user is far from the location where this event will take place, he might be delighted to be informed before 12:00 p.m. so he can catch the event on time. Thus, we fuzzified the time values to five linguistic terms, {too early, early, about to start, late, very late}.

Finally, the action taken by the notification service is itself uncertain. The user may specify to show text on his PDA as a kind of alert but he might be already late for his scheduled event and thus playing a tone would be more applicable to get his attention. Thus, we modeled the alert process to the following linguistic terms {no notification, low notification, medium notification, and high notification}. The IE uses the compositional rule of inference method to combine these inputs that represent the contextual information needed in this service and generates the rule surface diagram shown in Figure 18.6. It uses this rule surface to decide the best action to take for alerting the subscribed users about their events according to their current locations and the time of these events.

18.4.2 Semantic Inference in the Event-Notification Service

For the fuzzy inference to take place, the IE has first to reason semantically over these contexts to deduce the appropriate contexts that will be used in the fuzzy inference.

To illustrate this, a user in the deployed conferencing environment may issue an event-notification request that requires the context-aware infrastructure to notify him about the PERLAB_Meeting event that will take place in the PERLAB_LAB.

The context-aware infrastructure will use the fuzzy inference if it has the fact that the user is in PERLAB_LAB and that the PERLAB_Meeting will take place in

18.4 Prototype Implementation: The Event-Notification Service 397

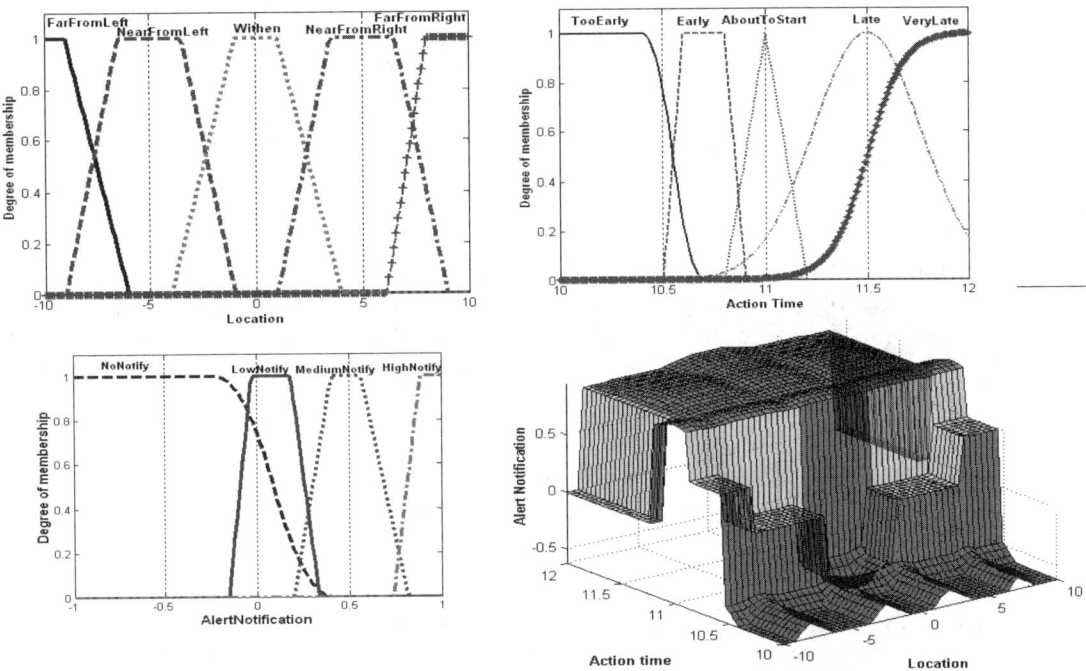

Figure 18.6 Membership functions and rule surface of the event-notification service.

that location. The problem is that the context-aware infrastructure's knowledge about this event and its location differs from what the user issued in his request. The following is a snapshot from the knowledgebase repository about the event, user location, and event location.

```
<!--IE knows that this is the location where PERLAB_Meeting is
    taking place-->
<Loc:Complex rdf:about="&Loc;E234">>
  <Loc:hasLocationRole>
    <Loc:Educational rdf:about="&Loc;PERLAB_Meeting">
      <rdf:type rdf:resource="&Loc;Complex"/>
    </Loc:Educational>
  </Loc:hasLocationRole>
  <Loc:hasLocationDataType>
    <Loc:CartesianType rdf:about="&Loc;SiteCartesian">
      <Loc:hasLocData>
        <Loc:LocationData rdf:about="&Loc;SiteLocData">
          <Loc:hasHorizontal>1250</Loc:hasHorizontal>
          <Loc:hasVertical>2000</Loc:hasVertical>
        </Loc:LocationData>
      </Loc:hasLocData>
    </Loc:CartesianType>
  </Loc:hasLocationDataType>
</Loc:Complex>
<<!--the user knows that this is the location where PERLAB_
    Meeting is taking place-->
<Loc:Indoor rdf:about="&Loc;PERLAB_LAB">
  <Loc:hasLocationDataType rdf:resource="&Loc;SiteCartesian"/>
</Loc:Indoor>
  <!--IE knows that this is the location where the user is
      currently in-->
```

```
<Loc:Office rdf:about="&Loc;Office@2ndFloor">
  <Loc:hasLocationDataType>
    <Loc:VectorType rdf:about="&Loc;CLbyVectorType">
      <Loc:hasAlternativeTrans rdf:resource="
        &Loc;SiteCartesian"/>
    </Loc:VectorType>
  </Loc:hasLocationDataType>
</Loc:Office>
```

As shown from the repository snapshot, the PERLAB_Meeting will take place in E234 and not in PERLAB_LAB. It also shows that the location of the user is Office@2ndFloor and not PERLAB_LAB. Thus, the context-aware infrastructure is required to infer and relate these locations and events semantically before it can apply the fuzzy inference process. As a result, the IE has to prove the following:

- PERLAB_LAB is the same as E234.
- PERLAB_Meeting will take place in PERLAB_LAB.
- User location is the same as E234.

This is shown in the following reasoning snapshot:

***begin derivation of [Loc:PERLAB_LAB, owl:sameAs, Loc:E234]

Rule RULE_LOC_SameAS_LocData

    concluded (Loc:PERLAB_LAB owl:sameAs Loc:E234) <-

Fact (Loc:PERLAB_LAB rdf:type Loc:PlaceInstance)

Fact (Loc:E234 rdf:type Loc:PlaceInstance)

Fact (Loc:PERLAB_LAB Loc:hasLocationDataType Loc:SiteCartesian)

Fact (Loc:E234 Loc:hasLocationDataType Loc:SiteCartesian)

Fact (Loc:SiteCartesian Loc:hasLocData Loc:SiteLocData)

Fact (Loc:SiteCartesian Loc:hasLocData Loc:SiteLocData)

***end derivation of [Loc:PERLAB_LAB, owl:sameAs, Loc:E234]

***begin derivation of [Loc:PERLAB_LAB, Loc:hasLocationRole, Loc:PERLAB_Meeting]

Rule RULE_LOC_SameAS_LocRole

    concluded (Loc:PERLAB_LAB Loc:hasLocationRole Loc:PERLAB_Meeting) <-

Fact (Loc:E234 rdf:type Loc:PlaceInstance)

Fact (Loc:PERLAB_LAB rdf:type Loc:PlaceInstance)

Rule RULE_LOC_SameAS_DataType

    concluded (Loc:E234 owl:sameAs Loc:PERLAB_LAB) <-

Fact (Loc:E234 rdf:type Loc:PlaceInstance)

Fact (Loc:PERLAB_LAB rdf:type Loc:PlaceInstance)

Fact (Loc:E234 Loc:hasLocationDataType Loc:SiteCartesian)

Fact (Loc:PERLAB_LAB Loc:hasLocationDataType Loc:SiteCartesian)

Fact (Loc:E234 Loc:hasLocationRole Loc:PERLAB_Meeting)

***end derivation of [Loc:PERLAB_LAB, Loc:hasLocationRole, Loc:PERLAB_Meeting]

***begin derivation of [Loc:Office@2ndFloor, owl:sameAs, Loc:PERLAB_LAB]

Rule RULE_LOC_SameAs_Alternative

concluded (Loc:Office@2ndFloor owl:sameAs Loc:PERLAB_LAB) <-

Fact (Loc:PERLAB_LAB Loc:hasLocationDataType Loc:SiteCartesian)

Fact (Loc:Office@2ndFloor Loc:hasLocationDataType Loc:CLbyVectorType)

Fact (Loc:CLbyVectorType Loc:hasAlternativeTrans Loc:SiteCartesian)

****end derivation of [Loc:Office@2ndFloor, owl:sameAs, Loc:PERLAB_LAB]

From these chains of semantic reasoning about location and event contexts, the inference engine can successfully proceed and reason about the uncertainty in these contexts using the fuzzy inference.

18.5 Conclusions

The semantic modeling of context using ontologies enhances the development of context-aware applications, since it allows for the common understanding of context and facilitates reusability and interoperability. However, it also lacks the essential mechanisms to deal with uncertainty found in these types of contexts and falls short when reasoning about context information in the presence of uncertainty.

In this chapter, we presented a fuzzy reasoning approach that enables the infrastructure to leverage the semantically modeled contexts to include methods for representing, also using ontologies, the uncertainty in these types of context information.

The proposed fuzzy reasoning is an inductive reasoning approach, which has the ability to discover the patterns in the context information used in the applications and automatically generates fuzzy rules that are used in the inference process.

We presented a detailed analysis of the inductive fuzzy reasoning mechanism. We described the process of fuzzifing context, elicitating membership functions, and performing the fuzzy inference mechanism. We also illustrated this inference approach using a simple example to emphasize the novelty and effectiveness of this approach to deal with uncertainty in context-aware applications.

References

[1] Perich, F., et al., On data management in pervasive computing environments," *Knowledge and Data Engineering, IEEE Transactions on*, Vol. 16, No. 5, May 2004, pp.621–633.

[2] OWL Web Language Ontology Reference.

[3] Masuoka, R.; et al., "Ontology-enabled pervasive computing applications," *Intelligent Systems, IEEE*, Vol. 18, No. 5, Sep–Oct 2003, pp. 68–72.

[4] Zadeh, L.A., "Fuzzy Algorithms," *Information and Control*, Vol. 12, 1968, pp. 94–102.

[5] Zadeh, L.A., "Fuzzy Logic," *Computer*, Vol. 21, No. 4, April 1988, pp. 83–93.

[6] Klir, G. J., et al, *Fuzzy Sets, Uncertainty and Information*, Prentice-Hall, NJ, 1988.

[7] Janikow, C., "Fuzzy Decision Trees: Issues and Methods," *IEEE Transactions on Systems, Man, and Cybernetics*, Vol. 28, Issue 1, pp.1–14, 1998.

CHAPTER 19
Anomaly Detection in Web Documents Using Computationally Intelligent Methods of Fuzzy-Based Clustering

Menahem Friedman, Abraham Kandel, Mark Last, Moti Schneider, Omer Zaafrany

19.1 Introduction

Intelligent environments, such as WWW, are leading to unprecedented scenarios where people (e.g., Web users) interact with electronic devices embedded in environments that are sensitive and responsive to the presence of people. The term ambient intelligence [1] (or, AmI) reflects this tendency, gathering best results from three key technologies: ubiquitous computing, ubiquitous communication, and intelligent user friendly Interfaces. The emphasis of AmI is on greater user-friendliness, more efficient services support, user empowerment, and support for human interactions. An AmI environment is capable of recognizing and responding to the presence of different individuals, working in a seamless, unobtrusive, and often invisible way. An important function of the AmI environment is security, i.e., using AmI technologies to protect people from internal and external threats ranging from natural disasters to cyberattacks by malicious hackers.

In recent years, the Internet has become an efficient communication infrastructure that can be seamlessly used at a workplace, a private home, or an Internet cafe. Its connection speed, accessibility, and user-friendliness, built upon behind-the-scenes data mining and information retrieval techniques, provide it with more and more characteristics of the ambient intelligence environment. Unfortunately, the Web is also increasingly used by illegal organizations to safely communicate with their affiliates, coordinate action plans, spread propaganda, raise funds, and introduce possible new supporters to their networks. Facing this reality, governments and intelligence agencies around the globe are calling for major efforts in developing new methods, technologies, and tools that will enable them to identify these illegal activities in time. An example of such tool is a terrorist detection system (TDS) [2], which is aimed at detecting suspicious users on the Internet, by the content of the information they access. The use of content monitoring and analysis of Web pages downloaded by a group of users from the Internet, enables us to infer their typical areas of interest and identify those users who access potentially illegal information on the Internet. However, tracing and tracking down terrorists is only a single application of the methods and procedures presented in this chapter. Our major purpose is to enhance an ambient intelligent environment on local networks by detecting anomalies in Web documents accessed by an arbitrary user connected to the network.

Prior to the detection stage, one must first define the anomaly by presenting one set of Web documents, all of which are normal and a second set of anomalous documents. A successful procedure for detection must be able to test an arbitrary incoming document and decide whether it belongs to the first set of normal documents or to the second set of anomalous documents. In order to construct such procedure, one must naturally apply cluster theory.

Cluster analysis, or clustering, is defined as the unsupervised process of grouping a set of data objects into clusters [3]. Intuitively, a good clustering method should produce clusters with high intra-class similarity and low inter-class similarity. However, evaluating the quality of a clustering result is a highly subjective task, since it depends on both the similarity measure used for the clustering and its implementation [4]. Cluster analysis can be applied either as an exploratory tool (to discover previously unknown pattern in data), or as an input to a decision-making process. In the latter case, it is reasonable to measure the clustering quality as the quality of the best decisions that can be made on the basis of clustering results. In this work the ultimate goal is to detect anomalous objects with maximum accuracy and simultaneously recognize each incoming normal object and assign it to one of the existing normal clusters.

The main steps of a typical clustering activity include object representation (optionally including feature extraction and/or selection), definition of an object similarity measure appropriate to the data domain, clustering or grouping objects by a given clustering algorithm, and assessment of output [3]. All classical popular clustering methods (such as k-means and ISODATA) are limited by the assumption that every object is represented as a fixed-size vector of measurements, or a point in a multidimensional space. The fixed number of vector components is usually called the dimensionality m of the feature space. Due to the fixed-size assumption, any collection of n vectors can be represented by an $n \times m$ data matrix.

In the information retrieval (IR) domain, clustering of a document collection can solve various problems, such as determining the set of documents relevant to the user query [5], automatically generating taxonomy of document topics [6], Web activity monitoring [2], and so on. Document clustering is usually based on the vector-space model [4], where each document is represented by a fixed-size vector of key terms' weights. Consequently, traditional clustering methods can be easily applied to matrix-based representations of document collections. The number of components in a document vector equals the total number of index terms in the corpus, where a zero weight is associated with each term that does not appear in a particular document [4]. Since a document generally contains only a small subset of system terms, the term-document matrix is typically very sparse, with almost all the entries being zero [7]. This generates a tremendous computational overload on a clustering algorithm, which considers every term in its similarity calculations. In order to overcome this shortcoming, inverted files where only the terms, that actually appear in a given document, i.e., possess nonzero weights are considered, were applied to the vector-space approach. Yet, the clustering methods, using the vector-space model, assume a relatively small prefixed number of clusters. This assumption, though advantageous from a computing cost point of view, usually decreases the clustering performance in the anomaly detection phase, when classifying new incoming documents within a set of clusters.

We propose a new approach for clustering Web documents that are represented by vectors of variable size, that does not restrict the final number of clusters. A formal presentation of the problem is given in Section 19.2. Several algorithms, both fuzzy based and crisp, are described in Sections 19.3 and 19.4. Section 19.5 presents the results of initial experiments performed in an ambient intelligent environment that performs Web traffic monitoring using the fuzzy-based global clustering (FGC) method (presented in Section 19.4), where each document is represented by its most important key phrases. Section 19.6 concludes this work and examines possible applications of the new methodology.

19.2 The Problem

The purpose of this study is to design a system that, given a large set of documents (e.g., Web documents), related to various topics from all walks of life and defined as normal, would cluster the documents and use the generated centroids (cluster centers) to detect an arbitrary new incoming anomalous document that is not similar to any of the normal clusters.

We consider a set S of n vectors with variable lengths, with maximum length m. Each vector represents a whole or a part of a Web document by k components called key phrases, where $1 \leq k \leq m$. Each key phrase is a single word or a meaningful expression consisting of a very small number of words—usually no more than three or four. The total number of distinct key phrases in all the vectors is n_0 and for an arbitrary vector $x = (t1, t2,\ldots,tk)T$ we assign to each key phrase t_i a real-valued score $ù_i$, usually referred to as importance weight (or simply weight) which can be calculated by a frequency-based indexing model [5] or by some key phrase extraction algorithm [8]. There are two main approaches for defining ω_i: (1) a quantity that measures the local (intra-vector) importance of a key phrase ti within a vector x (e.g., the raw term frequency tf [4]). Thus, if a similar document is twice as long, this quantity is expected to be approximately doubled. Therefore, in this case, prior to looking for similarity between vectors, the vectors should be normalized. (2) a quantity that measures the global (inter-vector) importance of t_i in x (e.g., the normalized term frequency f [4]). Here no normalization process is needed and for two arbitrary vectors to be similar, they need to share a large number of key phrases with close scores.

Our purpose is to divide the set S of n vectors into several clusters, as many as necessary. The final number of clusters is determined not by the user, but rather by the algorithm, its tuning parameters that control the process, the similarity measure and the nature of S. The only requirement is that each cluster includes vectors that are similar to each other (intersimilarity) and that vectors that belong to separate clusters will indeed be radically different (intradissimilarity). Removing the restriction on the final number of clusters increases the computing costs on one hand but at the same time improves the method's performance in classifying new incoming vectors. Interpretation of the weight ω_i is closely related to the choice of the clustering method to be used in a specific situation. We present here four clustering algorithms, where the first three are based on the local weights and the cosine similarity measure [7]. These algorithms, two of which apply fuzzy-based clustering [9], are described in the next section. The fourth algorithm, based on global importance weights, is also fuzzy based. This algorithm is presented in detail in Section 1.4 and its implementation for detecting anomalous Web pages is illustrated in section 19.5.

19.3 Cosine-Based Algorithms

We may refer to our clustering methods as incremental clustering algorithms, meaning that incoming vectors are processed one at a time for clustering [3]. Only the cluster centers, usually referred to as centroids and basically calculated as the averages of the vectors in each cluster, are stored in main memory. By using a cosine-based method for clustering, we accept the concept that similarity between two vectors is measured by the $\cos(\theta)$, the inner product of their normalized forms, as follows: maximum similarity is attained when the vectors are parallel, i.e., identical, and $\cos\theta = 1$ while similarity is minimal when the vectors are orthogonal and $\cos\theta = 0$. Prior to measuring the similarity, we define a centroid c as the normalized sum of all the vectors already in the cluster C:

$$c = \sum y_i / \left|\sum y_i\right|, y_i \in c \qquad (19.1)$$

The components of each vector participating in the computational process are the importance weights associated with the respective key phrases.

Then, to get the similarity between an incoming vector $x = (x_1, ..., x_m)$ and a centroid $c = (c_1, ... ,c_m)$, we normalize x and calculate the inner product

$$c \cdot x = \sum_k (c_k \cdot x_k) \qquad (19.2)$$

considering only key phrases that appear in both vectors. Finally, we assign the vector to the cluster that produces the maximum inner product, i.e., maximum similarity, provided it is above a prefixed similarity threshold. If all inner products are below this threshold, the incoming vector starts a new cluster. The inner product, which equals the cosine of the angle between the two vectors, is a value between 0 and 1. Thus, instead of considering the similarity s between arbitrary vectors x and y, defined as

$$s = (x \cdot y) / (|x| \cdot |y|) \qquad (19.3)$$

we usually apply the distance function

$$d = 1 - s \qquad (19.4)$$

which is also between 0 and 1 and assigns distance 0 to maximum similarity (i.e., similarity 1) and distance 1 to minimum similarity. The three cosine-based algorithms for clustering are as follows.

19.3.1 Crisp Cosine Clustering (CCC)

The similarity between a centroid c and a normalized vector x is defined by (19.2). The generated clusters depend on the similarity threshold s_0 and on the order of the vectors. A major disadvantage of this algorithm is that vectors such as (100, 300, 200,...) and (1, 3, 2,...) are considered similar in spite of the fact that they present two documents that enormously differ in size (at least according to the first three key phrases) and therefore may relate to two totally different subjects (clusters).

19.3.2 Fuzzy-Based Cosine Clustering (FCC)

In order to avoid the shortcoming of the CCC, mentioned previously, we assign to each incoming vector x a grade of membership, which is a number $\chi(x, c)$ between 0 and 1, depending on the ratio $|x|/|c|$. The similarity is then defined, for the originally non-normalized x and c, as

$$s = \chi(x,c)(c \cdot x) / (|x| \cdot |c|) \tag{19.5}$$

If $|x|$ is small with respect to $|c|$ the grade of membership is small and the similarity decreases. If $|x|/|c|$ is above some prefixed threshold, we usually take $\chi(x, c) = 1$. A possible choice for $\chi(x, c)$ is, for example:

$$\chi(x,c) = \begin{cases} 4\alpha(1-\alpha), \alpha \leq 1/2 \\ 1, \alpha > 1/2 \end{cases} \tag{19.6}$$

where $\alpha = |x|/|c|$ is the relative size of x with respect to c. Note that α may be larger than 1, in which case it is only reasonable to take $\chi(x, c) = 1$.

Thus, the fuzzy-based cosine similarity function considers also the relative size of the incoming vector such that a relatively very small vector could not be considered similar to a given sizable centroid consisting of the same key phrases, but with much larger weights.

19.3.3 Local Fuzzy-Based Cosine Clustering (LFCC)

This algorithm is a modified FCC. The single grade of membership $\chi_k(x, c)$ is replaced with m grades of membership $\chi(x, c)$, $1 \leq k \leq m$ which are quantities between 0 and 1 such that each $\chi_k(x, c)$ is determined only by $\alpha_k = |x_k|/|c_k|$. Then, while calculating the inner product, the term $(c_k \cdot x_k)$ is multiplied by $\chi_k(x, c)$ and therefore usually decreased. Each grade of membership is defined by a relation similar to the one given Eq. (19.6).

The following examples illustrate how to generate a centroid and calculate its distance from an incoming vector. Each vector is represented by an arbitrary number of pairs where each pair contains a key phrase and its associated importance weight.

19.3.3.1 Example 1 (CCC).

Consider a cluster C consisting of the following 100 vectors: 20 identical vectors each equals to $\{(t_1,6), (t_2,4), (t_3,15), (t_4,10)\}$, 30 identical vectors each equals to $\{(t_2,10), (t_3,20), (t_4,15), (t_5,20)\}$ and 50 that are equal to $\{(t_1,20), (t_4,20), (t_5,10), (t_6,25)\}$. The total sum of the vectors in the cluster is:

$$\{(t_1,1120), (t_2,380), (t_3,900), (t_4,1650), (t_5,1100), (t_6,1250)\} \tag{19.7}$$

producing the normalized centroid

$$c = \{(t_1,0.404), (t_2,0.137), (t_3,0.324), (t_4,0.594), (t_5,0.396), (t_6, 0.450)\} \tag{19.8}$$

Let $x = \{(t_2,30), (t_3,40), (t_7,20)\}$ be an incoming vector. Its normalized form is $x' = \{(t_2,0.557), (t_3,0.743), (t_7,0.371)\}$ and the similarity between x and c is:

$$s = (x' \cdot c) = 0.137 \cdot 0.557 + 0.324 \cdot 0.743 = 0.317 \tag{19.9}$$

producing a distance $d = 0.683$. Whether this similarity is sufficient for assigning x to C, depends on the prefixed threshold, which is application dependent and chosen by the user to maintain acceptable performance of the clustering method. The performance quality is determined by the values obtained for true positives (TP) and false positives (FP) in the validation phase.

The clustering procedure is as follows: Consider the existing clusters $C_1, C_2, ..., C_k$ with centroids $c_1, c_2, ..., c_k$ and an incoming vector x normalized to $x' = x/|x|$. Let s_0 denote a similarity threshold for determining the membership in a cluster and let $s' = (x', cj)$ where:

$$s' = \min(x' \cdot c_k), 1 \leq i \leq k \tag{19.10}$$

Then, if $s' \geq s_0$ we assign x to C_j, otherwise, we start a new cluster C_{k+1}, of which x is the first member.

As previously stated, the CCC algorithm, as well as the other two cosine-based algorithms should be considered whenever the most dominant feature for similarity between vectors, is the closeness of their inner product (of normalized representations) to 1. Whether this is the case depends overwhelmingly on the particular application and certainly on the accumulated experience of the user.

19.3.3.2 Example 2 (FCC)

Consider the cluster of section 19.3.3.1. The non-normalized centroid is the total sum divided by 100:

$$c' = \{(t_1,11.2), (t_2,3.8), (t_3,9), (t_4,16.5), (t_5,11), (t_6,12.5)\} \tag{19.11}$$

with $|c'| = 27.76$. An incoming vector $x = \{(t_2,5), (t_3,6), (t_7,4)\}$ satisfies $|x| = 8.77$ and if (19.5) is applied, we obtain $\alpha = 0.32$ and $\chi(x,c) = 0.87$. The inner product of x and c' is:

$$(c' \cdot x) = 3.8 \cdot 5 + 9 \cdot 6 = 73 \qquad (19.12)$$

and the similarity between them is:

$$s = \chi(x,c')(c' \cdot x)/(|x| \cdot |c'|) = 0.87 \cdot 73/(27.76 \cdot 8.77) = 0.26 \qquad (19.13)$$

The size of the incoming vector $y = \{(t_2,10), (t_3,2), (t_7,10)\}$ is $|y| = 14.28$ and since $\alpha = 0.51$ we get $\chi(y,c) = 1$.

The clustering principle given by (19.10) applies for all the three algorithms.

In the next example we assign a different grade of membership to each single key phrase.

19.3.3.3 Example 3 (LFCC)

Consider the cluster presented in Section 19.3.3.1 and 19.3.3.2 and the incoming vector $x = \{(t_2,5), (t_3,1), (t_7,4)\}$. Here we have $\alpha_2 = |x_2|/|c_2| = 1.32$ hence $\chi_2(x,c) = 1$ and $\alpha_3 = |x_3|/|c_3| = 0.11$ which provides $\chi_3(x,c) = 0.39$. The similarity between x and the centroid is therefore:

$$s = (1 \cdot 5 \cdot 3.8 + 0.39 \cdot 1 \cdot 9)/(6.48 \cdot 27.76) = 0.125 \qquad (19.14)$$

We next introduce a new fuzzy clustering method, which is based on global interpretation of the importance weights and on a nonstandard observation of the average concept. Also, in this fuzzy clustering approach, we measure distance rather than similarity, between an incoming vector and a centroid.

19.4 Fuzzy-based Global Clustering

19.4.1 A New Distance Measure

The phrase fuzzy based clustering is usually applied when a given object (vector) is classified in several clusters with various grades of membership. Here we apply this terminology to define the contribution of a single key phrase in an incoming vector to the total similarity between this vector and an arbitrary existing centroid.

Each key phrase t in an incoming vector x, whose possible classification in a given cluster C is examined, may have already appeared several times in C and thus possesses a distribution function. At each stage we maintain two numbers for each key phrase in C: its average importance weight in the cluster and the maximum value of all the importance weights of this key phrase within C. This information is applied to assign t a grade of membership $\chi(t)$, $0 \leq \chi(t) \leq 1$ within x, with respect to membership of x in C. The distance between x and C is defined as the fraction:

$$d(x,C) = \frac{\sum \left[\omega(t) \cdot |\chi_x(t) - \chi_C(t)|\right]^r}{\sum \left[\omega(t) \cdot \min(\chi_x(t), \chi_C(t))\right]^r} \qquad (19.15)$$

where t is an arbitrary key phrase in either x or C. The number $\omega(t)$ is the average weight of t if this key phrase already appeared in existing vectors within C. Otherwise, $\omega(t)$ is defined as the weight of t in x. The quantities $\chi x(t)$ and $\chi C(t)$ are the grades of membership of t within x and within C respectively. Their calculation is illustrated in Section 19.4.1.1. The parameter r is a positive number empirically chosen in to improve the algorithm's performance.

Let w denote the average importance weight of an arbitrary key phrase in C. Then, we increase the denominator by $[\omega \chi x(t)]r$ which measures the similarity between x and C with regard to t. Simultaneously the numerator is increased by $[\omega(1 - \chi x(t))]r$ which measures the dissimilarity between the two. This procedure is referred to as fuzzy based global clustering (FGC). It is illustrated in detail through Section 19.4.1.1, followed by a representation of its general scheme.

19.4.1.1 Example 4

Consider the previous example shown in Section 19.3.3.1. For FGC the importance weights average is calculated differently than for the cosine methods (19.1) and is not the standard algebraic average. It is defined as the total sum of the weights of the particular key phrase divided by the number of vectors in the cluster where the key phrase has actually appeared (referred to as inter-document frequency). Thus, the average $\omega_{av}(t)$ of the importance weights of an arbitrary key phrase within a cluster C is:

$$\omega_{av}(t) = \frac{1}{k} \sum_{x \in C, t \in x} \omega_x(t) \qquad (19.16)$$

where k is the number of vectors in C that include the key phrase t.

The logic behind (19.15) might be simply explained by the following example: Appearance of a key phrase only in 50 out of the 100 vectors that are already in the cluster, indicates that the relative frequency of the particular key phrase within the cluster is about 50%. Therefore, the real expected value for this key phrase, should it appear in a new member of the cluster, is the current sum of its weights divided by 50. However, if any of the cosine methods are applied, then ideally the vectors in a cluster will be all proportional to each other and each key phrase is expected to appear in most of the cluster members.

Thus, when using FGC, the cluster center is defined as the vector:

$$\{(t_1, 1120/70), (t_2, 380/50), (t_3, 900/50), (t_4, 1650/100), (t_5, 1100/80), (t_6, 1250/50)\} \qquad (19.17)$$

i.e.,

$$c = \{(t_1, 16), (t_2, 7.6), (t_3, 18), (t_4, 16.5), (t_5, 13.75), (t_6, 25)\} \qquad (19.18)$$

Consider now a procedure where for each importance weight of a key phrase in an incoming vector that is larger than its average in the cluster (i.e., in the centroid), we assign a grade of membership 1, while for a weight smaller than the average we assign a linearly decreasing value. Then, for the incoming vector $x = \{(t_2,8),(t_3,9),(t_7,30)\}$, we have $\chi x(t_2) = 1$, $\chi x(t_3) = 0.5$ and for $r = 1$ we obtain:

$$d(x,C) = \frac{16 \cdot 1 + 7.6 \cdot 0 + 18 \cdot 0.5 + 16.5 \cdot 1 + 13.75 \cdot 1 + 25 \cdot 1 + 30 \cdot 1}{16 \cdot 0 + 7.6 \cdot 1 + 18 \cdot 0.5 + 16.5 \cdot 0 + 13.75 \cdot 0 + 25 \cdot 0 + 30 \cdot 0} = 6.64 \quad (19.19)$$

19.4.2 The General Scheme

Consider a cluster C. Let t_i, $1 \leq i \leq k$ denote all the key phrases in the cluster and let $\omega_{i,av}$ and $\omega_{i,max}$ denote the averages and maxima of their importance weights respectively. Let x denote an incoming vector to be classified and let denote a key phrase in x which appeared already in C, say $t = t_j$, that has an associated importance weight $\omega(t)$ in x. We assign to t a grade of membership (in x) as follows:

$$\chi(t) = \begin{cases} \left(\omega(t)/\omega_{j,av}\right)^p, \omega(t) < \omega_{j,av} \\ 1 - \left(\omega(t) - \omega_{j,av}\right)(1-q)/\left(\omega_{j,max} - \omega_{j,av}\right), \omega(t) > \omega_{j,av}, \omega_{j,max} > \omega_{j,av} \\ \left(\omega_{j,av}/\omega(t)\right)^p, \omega(t) > \omega_{j,av}, \omega_{j,max} = \omega_{j,av} \end{cases} \quad (19.20)$$

where $p > 0$, $0 < q = 1$ are parameters to be determined. If $\chi(t) < 0$ (19.20) we fix $\chi(t) = 0$. The grade of membership of t in C is obviously taken as 1. The motivation for (19.20) is to assign grade of membership 1 to $\omega = \omega_{i,av}$ and monotonic decreasing functions on either side with tuning parameters p (left) and q (right).

We could have certainly applied various memberships rather than the one given by (19.20), but felt it was unnecessary at this preliminary stage of the research, especially when just by varying p and q, one obtains an enormously large family of applicable membership functions.

Each key phrase in C that does not appear in x obtains a grade of membership 0 in x but 1 in C. Each key phrase in x that is not yet in C, has a grade of membership 1 in x and 0 in C. Using this concept, we may observe the centroid of C and the vector x as possessing the same key phrases, some of them with grade of membership 0. Let t denote an arbitrary key phrase in either C or x and let $\chi x(t)$ and $\chi C(t)$ denote its grades of membership in x and C respectively. Define its weight as:

$$w(t) = \begin{cases} \omega_{j,av}, t = t_j \text{ for some } j \\ \omega(t), t \neq t_j \text{ for all } j \end{cases} \quad (19.21)$$

and substitute it instead of $\omega(t)$ in (19.15) to get the distance between x and the cluster C, where r is another tuning parameter to be determined by the user. An r larger than 1 increases the reward for a key phrase with high $\chi_x(t)$. The final classification of x is determined by comparing the minimum distance of x from all the current centroids with a prefixed threshold d_0. If this minimum does not exceed d_0, the new vector is assigned to an existing cluster. If it does, the vector forms a new cluster, or, if all normal clusters are exhausted, the new vector is declared anomalous.

The parameters p,q,r are empirically determined by experimenting with the given data. A generally safe though not optimal choice, is $p = q = r = 1$. The choice $q = 0$ is excluded, since it assigns zero membership to a key phrase that has a weight equal to $\omega_{i,max}$. An alternative choice such as $q = 1/2$ is more realistic. However, the optimal choice for p, q and r is application dependent.

After assigning a vector x to a cluster C, the cluster's statistics are updated. This includes the averages and the maxima of all the key phrases that are members in x, now assigned to C. The next example consists of an incoming vector to be classified in one of two existing clusters, using the FGC method.

19.4.2.1 Example 5

Let C_1 contain the vectors:

$$y_1 = \{(t_1,20),(t_2,10),(t_4,10)\}$$
$$y_2 = \{(t_1,10),(t_2,25),(t_3,5)\}$$
$$y_3 = \{(t_2,15),(t_3,30),(t_5,16)\}$$

and let C2 contain the vectors

$$u_1 = \{(t_1,24),(t_2,15),(t_3,10)\}$$
$$u_2 = \{(t_1,16),(t_2,20),(t_4,25)\}$$
$$u_3 = \{(t_2,20),(t_3,35),(t_4,5)\}$$
$$u_4 = \{(t_3,10),(t_4,12),(t_5,20)\}$$

(19.22)

Consider an incoming vector:

$$x = \{(t_1,15),(t_2,8),(t_5,10),(t_6,20)\} \quad (19.23)$$

The cluster centers of C1 and C2 are:

$$c_1 = \{(t_1,15),(t_2,16.7),(t_3,17.5),(t_4,10),(t_5,16)\}$$
$$c_2 = \{(t_1,20),(t_2,18.3),(t_3,18.3),(t_4,14),(t_5,20)\}$$

(19.24)

respectively. Assume the parameters $p = r = 1$, $q = 0.5$ and $d_0 = 2.5$. The grades of membership of the key phrases in x with respect to C_1, calculated by (19.20) are $\chi_x(t_1) = 1, \chi_x(t_2) = 0.48, \chi_x(t_5) = 0.625, \chi_x(t_6) = 1$. Applying (19.5-19.10) we obtain:

$$d(x,C_1) = \frac{15 \cdot 0 + 16.7 \cdot 0.52 + 17.5 \cdot 1 + 10 \cdot 1 + 16 \cdot 0.375 + 20 \cdot 1}{15 \cdot 1 + 16.7 \cdot 0.48 + 17.5 \cdot 0 + 10 \cdot 0 + 16 \cdot 0.625 + 20 \cdot 0} \approx 1.88 \quad (19.25)$$

The grades of membership of the key phrases in x with respect to C_2 are $\chi_x(t_1) = 0.75, \chi_x(t_2) = 0.44, \chi_x(t_5) = 0.5, \chi_x(t_6) = 1$ and the distance of x from the cluster is:

$$d(x, C_2) = \frac{20 \cdot 0.25 + 18.3 \cdot 0.56 + 18.3 \cdot 1 + 14 \cdot 1 + 20 \cdot 0.5 + 20 \cdot 1}{20 \cdot 0.75 + 18.3 \cdot 0.44 + 18.3 \cdot 0 + 14 \cdot 0 + 20 \cdot 0.5 + 20 \cdot 0} \approx 2.35 \qquad (19.26)$$

Thus, although x could be a member of either cluster it is classified in C_1.

The clustering process, using any of the four incremental algorithms presented, depends on the order of the incoming vectors. Quite often a cluster is initiated by a wrong vector x and only after we assign several new members to the cluster, we discover that x should actually be removed. To partially overcome this problem we could adhere to the following heuristic rule: If x is classified in a cluster C with already more than one member, remove the most unfit member y from C, using a minimum search procedure. The set of unfit vectors is then reclustered until the process converges. If it does not converge, stop iterating after a prefixed number of iterations. Whenever the reclustering process is applied, the set of final clusters is less dependent on the order of the incoming vectors.

19.5 Application: Terrorist Detection System

To demonstrate the algorithm's performance with real data we chose a terrorist detection system (TDS) application . TDS is a content-based detection system developed to detect users who are interested in terror-related pages on the Web by monitoring their online activities. The system is based on real-time monitoring of Internet traffic of a well-defined group of Internet users such as students in a specific university campus or employers in a large company. The group is suspected to include one or more hidden individual terrorists or terror supporters, and the system provides a tool for detecting them. The system has the following characteristics of an ambient intelligence (AmI) technology:

- *Embedding*. The system is plugged into a computer network connected to the Internet. It has sensors (sniffers) that perform eavesdropping on all textual content downloaded by the network computers.
- *Context awareness*. The system can recognize people belonging to a certain group of Web users (e.g., college students) based on the IP addresses of their computers.
- *Personalization*. The system uses fuzzy clustering algorithms to learn the patterns of online behavior for a specific group of users rather than generic Web users.
- *Adaption*. The system can deal with changing patterns of online behavior by applying incremental clustering algorithms to the Web traffic data.

The current version of TDS refers only to the textual content of the accessed Web pages. It consists of two modules: a training module activated in batch and a real time detection module. The training module clustered a large set of Web pages downloaded from the Internet by a group of similar users (e.g., students in the same department). These pages cover many topics of interest but are not terror-related and can be regarded as normal pages. By using normalized term frequency (see Section 19.2), they are converted to vectors, which are then clustered using the FGC method. We then considered a second set of vectors obtained from Web pages that are characterized by definite features of anomaly, since they were related to a completely different topic: terror activity. Both sets of documents share many key phrases, yet there is a subset of key phrases that are definitely typical within the terror-related vectors but not among the normal ones. A good performance of the algorithm may be concluded only if, while in the detection mode, it detects most of the anomalous vectors (true positives) as such by rejecting them from all the existing clusters, while rejecting only a small number of the new normal incoming vectors (false positives) from being assigned to these clusters. The algorithm's performance is usually represented by an ROC curve.

19.5.1 The Experiment

We obtained 13,330 normal documents, each represented by a vector with maximum of 30 key phrases. The total number of distinct key phrases in all the normal vectors was 38,776. Next we downloaded 582 documents from various known terrorist Web sites. Each document of the latter set could be easily recognized and defined by experts as 'anomalous' with regard to the normal vectors. The total number of key phrases in the anomalous vectors was 5,275, out of which 2,535 key phrases already appeared in the normal set. We used 12,748 normal vectors for training the system, i.e., for obtaining the normal clusters and their centroids. The remaining 582 normal vectors were applied for validation. The set of 582 known anomalous vectors were then tested for anomaly. The fact that about 48% of the key phrases in the anomalous vectors appeared also in the normal vectors suggested that detection would not be trivial and that the algorithm must recognize that the remaining 52% of the key phrases are significantly more meaningful.

We performed the receiver-operating characteristic (ROC) analysis as follows. For each given threshold value (maximum distance that still allows membership in a cluster) we obtained the false positive (FP) vectors, i.e., the normal vectors from the subset used for validation that were not assigned to any of the normal clusters determined by the training set. This number was compared with the number of true positive (TP) vectors, i.e., the anomalous vectors that were not assigned to any of the given clusters and therefore justly recognized as anomalous.

19.5.2 The Results

We chose the 582 validation vectors randomly from the original ordered set of 13,330 vectors and performed 20 runs randomly changing the order of the training set for each run. Using $p = q = 0.5$, $r = 0.5$ and $d_0 = 5$ for clustering, we obtained about 3,000 clusters. We then calculated FP and TP for thresholds between 5 and 80 since it is impossible to know a priori which threshold value is the right one to take, i.e., which value will provide TP close to 100% and FP close to 0%. If, however, for a large range of threshold values, TP and FP are sufficiently close to 100% and 0% respectively, we may conclude that the algorithm indeed performs well. The results for this particular experiment, averaged over 20 runs, are given in Figure 19.1 and are quite encouraging. The best result, using a k-means clustering algorithm, was obtained for 100 clusters. A comparison between the two approaches, presented in

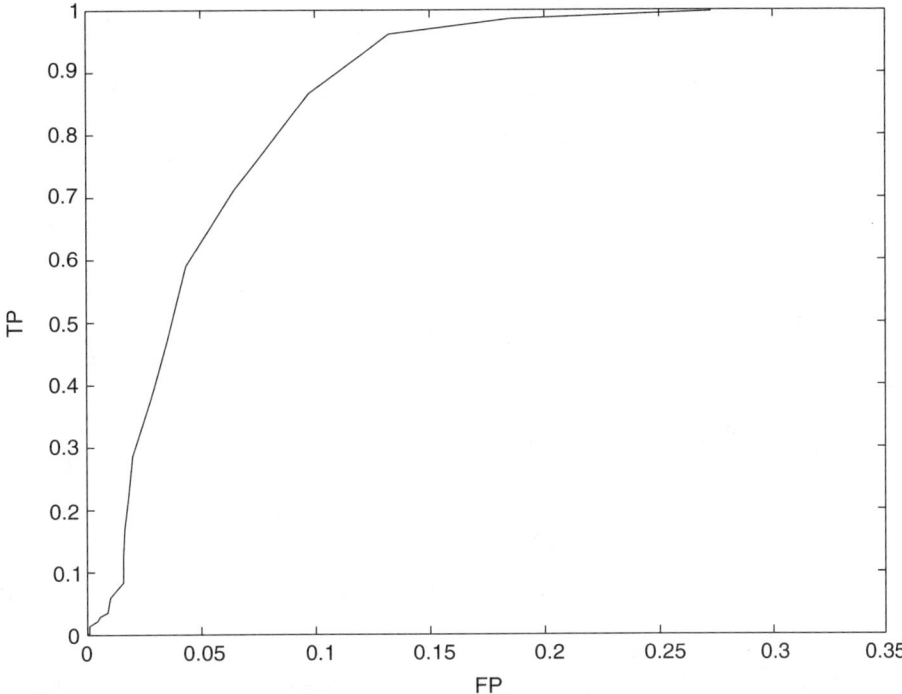

Figure 19.1 The ROC curve using the FGC method.

Figure 19.2, is clearly in favor of the FGC method. A fuzzy c-means algorithm hardly changed the results, and the FGC algorithm remained still far ahead. The reasonable trade-off between true and false positives in this preliminary experimentation is encouraging and justifies further development of the FGC algorithm for detecting anomalies. The results also showed little sensitivity to changing the order of the incoming vectors.

The large number of clusters (about 3,000) is somewhat deceiving, since only a relatively small number of them (under 50) are sizable. However, the idea of forcing a smaller number of clusters by increasing the threshold d_0 or by spreading the vectors of the real small clusters among the larger ones was not successful. The algorithm's performance dropped significantly since by applying either of these forcing methods we lost information vital for distinguishing between the features of the normal and anomalous documents. Thus, instead of artificially decreasing the number of clusters, we designed an algorithm that efficiently dealt with the problem of classification in spite of the presence of a rapidly growing number of clusters.

19.6 Conclusions

In this chapter we designed several clustering models for detecting anomalies in Web documents and consequently in the people who download them from the Internet, by means of content monitoring and analyzing. The existence of such tools could help in identifying illegal activities on the Internet and possibly preventing

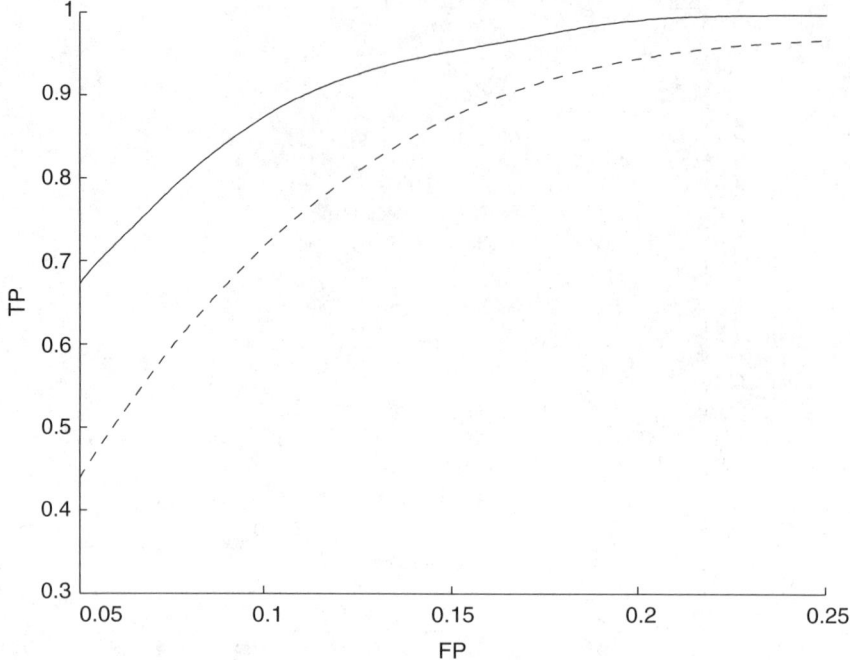

Figure 19.2 ROC curves: FGC (solid line) vs. k-means (dashed line) with 100 clusters.

future disasters. We introduced four novel algorithms for clustering documents represented by vectors of variable sizes. Due to the sparse nature of term vectors representing Web documents in the traditional vector-space model, this approach can save a significant amount of computational effort in the clustering process. Initial results, using a fuzzy-based global clustering algorithm, show that the new approach works reasonably well when applied to the problem of anomaly detection in real-world data, such as documents downloaded from the Internet. This tool may be applied successfully in every area where anomaly detection is sought, and further experimentation is expected to improve its performance.

Acknowledgments

This work was partially supported by the National Institute for Systems Test and Productivity at University of South Florida under the USA Space and Naval Warfare Systems Command Grant No. N00039-01-1-2248.

References

[1] Vasilakos, A., and W.Pedrycz, "AmI:Visions and Technologies," *Ambient Intelligence, Wireless Networking, Ubiquitous Computing*, Artech House, Norwood MA, 2006.

[2] Last, M., et al., "Content-Based Methodology for Anomaly Detection on the Web," *Advances in Web Intelligence*, Springer-Verlag, Lecture Notes in Artificial Intelligence, Vol. 2663, pp. 113–123, 2003.

[3] Jain, A. K., M. N. Murty and P. J. Flynn, "Data clustering: a Review," *ACM Computing Surveys*, Vol. 31, No. 3, 1999, pp. 264–323.

[4] Baeza-Yates, R., and B. Ribeiro-Neto, *Modern Information Retrieval*, ACM Press, 1999.

[5] Salton, G., *Automatic Text Processing: the Transformation, Analysis, and Retrieval of Information by Computer*, Addison–Wesley, Reading, 1989.

[6] Schenker, A., M. Last, and A. Kandel, "A Term-Based Algorithm for Hierarchical Clustering of Web Documents," *Proceedings of IFSA / NAFIPS 2001*, Vancouver, Canada, July 25–28, 2001. pp. 3076–3081.

[7] Dhillon, I.S., "Co-clustering Documents and Words Using Bipartite Spectral Graph Partitioning," *Proceedings of KDD01*, San-Francisco, CA, USA, 2001, pp. 269–274.

[8] Turney, P.D., "Learning Algorithms for Keyphrase Extraction," *Information Retrieval*, Vol. 2, No. 4, pp. 303–336, 2000.

[9] Klir, G. J., and Bo Yuan, *Fuzzy Sets and Fuzzy Logic*, Prentice Hall, 1995.

CHAPTER 20
Intelligent Automatic Exploration of Virtual Worlds

Dimitri Plemenos

20.1 Introduction

Virtual world exploration techniques nowadays are becoming more and more important. When, more than 10 years ago, we have proposed the very first methods to improve the knowledge of a virtual world [1, 2], many people thought that it was not an important problem. People began to understand the importance of this problem and the necessity to have fast and accurate techniques for a good exploration and understanding of various virtual worlds, only during these last years. However, there are very few articles that face this problem from the computer graphics point of view, whereas several papers have been published on the robotics artificial vision problem.

The purpose of a virtual world exploration in computer graphics is completely different from the purpose of techniques used in robotics. In computer graphics, the purpose of the program that guides a virtual camera is to allow a human being, the user, to understand a new world by using an automatically computed path, depending on the nature of the world. The main interaction is between the camera and the user, a virtual and a human agent, and not between two virtual agents or a virtual agent and his environment.

There are two kinds of virtual worlds exploration. The first one is global exploration, where the camera remains outside the world to be explored. The second kind of exploration is local exploration. In such an exploration, the camera moves inside the scene and becomes a part of the scene. Local exploration can be useful, and even necessary, in some cases, but only global exploration can give the user a general knowledge on the scene. In this chapter we are mainly concerned with global virtual world exploration, whereas interesting results have been obtained with local exploration techniques. In any case, the purpose of this chapter is visual exploration of fixed unchanging virtual worlds.

The chapter is organized in seven sections and a conclusion: In Section 20.2 the need of virtual worlds exploration is explained and justified. In Section 20.3 some early techniques for understanding simple virtual worlds, based on automatic choice of a good point of view are presented. In Section 20.4 the notion of visual complexity of a scene from a given point of view is defined and in Section 20.5 some methods to compute visual complexity are presented. In Section 20.6 some very sophisticated techniques for complex virtual worlds exploration are explained and compared. In Section 20.7 some ideas about the future evolution of virtual worlds exploration techniques are discussed. Finally, the chapter is concluded with some comments on the presented virtual world exploration techniques.

20.2 Why Explore Virtual Worlds?

Exploring virtual worlds, seen as a computer graphics problem, is quite different from the robotics problem where an autonomous virtual agent moves inside a virtual world whose it is a part of. In computer graphics, the main purpose of a virtual world exploration is to allow the user to understand this world. Here, the image of the world seen from the current point of view is very important, because it will permit the user to understand the world. In robotics, the problem is different and the use of an image is not essential because the virtual agent is not a human being.

At about the second half of the 1980's, we have identified a problem that appeared to us as very important. Very often, the rendering of a scene, obtained after a long computation time, was not possible to exploit because the choice of the angle of view was bad. In such a case, the only possibility was to choose another angle of view and to try again by running the time consuming rendering algorithm once again. We thought that it is very difficult to find a good angle of view for a 3D scene when the working interface is a 2D screen. On the other hand, we thought that the choice of an angle of view is very important for understanding a scene. In Figure 20.1, one can see two different views of the same scene. Only the second one could help to understand it.

Since the program (modeler, renderer) has a full knowledge of the geometry of the world to visualize, we thought that it could be more interesting to ask the program to find a good angle of view for this world.

Obviously, it is not possible to try all the possible angles of view and to choose the best one, because the camera's movement space is continuous. Even if we decide to sample the camera's space, we are typically in a case where there are many possibilities of solutions, and we need a method to choose a good enough solution. In

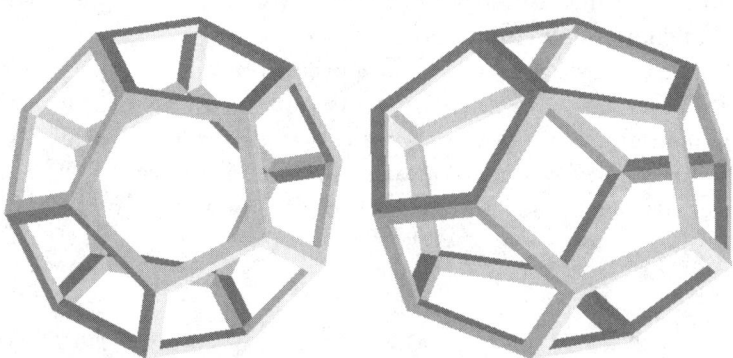

Figure 20.1 Bad and good angle of view for a scene.

such cases, the use of the artificial intelligence technique of heuristic search is recommended.

This simple computer-aided way to understand a virtual world is not always sufficient. The world to understand can be very complex and, in such a case a single point of view is often insufficient. Rather than computing several good points of view and showing the world from each of them, we think that it is much more interesting for the user to use a moving virtual camera in order to permit him to explore the world by computing an interesting path for the camera.

20.3 Simple Virtual World Understanding

The very first works in the area of understanding virtual worlds were published at the end of the 1980's and the beginning of the 1990's. There were very few works because the computer graphics community was not convinced that this area was important for computer graphics. The purpose of these works was to offer the user help to understand simple virtual worlds by computing a good point of view. In this section we will use the term of scene to point out a simple virtual world.

20.3.1 Nondegenerated View

Kamada et al. [3] consider a direction as a good point of view if it minimizes the number of degenerated images of objects when the scene is projected orthogonally. A degenerated image is an image where more than one edges belong to the same straight line. In Figure 20.2, one can see a good view of a cube together with degenerated views of it.

The method used avoids the directions parallel to planes defined by pairs of edges of the scene.

If L is the set of all the edges of the scene and T the set of unit normal vectors to planes defined by pairs of edges, let's call $\vec{P}$ the unit vector of the direction of view to be computed.

In order to minimize the number of degenerated images, the angles of the vector $\vec{P}$ with the faces of the scene must be as great as possible. This means that the angles of the vector $\vec{P}$ with the elements of T must be as small as possible. The used evaluation function is the following:

$$f(\vec{P}) = Min_{\vec{t} \in T}\left(\left|\vec{P} \cdot \vec{t}\right|\right) \tag{20.1}$$

Figure 20.2 Good and degenerated views.

Since the purpose is to minimize the angle between $\vec{P}$ and $\vec{t}$, we must maximize the function $f(\vec{P})$. To do this, a vector $\vec{P}$ which minimizes the greater angle between itself and the elements of T must be found.

In order to compute this vector, we have proposed two methods. The first one is simple and its main advantage is to be very fast. The second method is more precise, because it is based on the computation of all the normal vectors to all the faces of a scene and all the angles between these vectors. This method is time consuming. Only the first method will be explained here.

Since the purpose of this method is to decrease the computation cost of the function $f(\vec{P})$ for all the unit normal vectors, a set E of uniformly distributed unit vectors is chosen on the unitary sphere defined at the center of the scene.

The function is computed for each element of the set E and the vector with the maximum value is chosen.

This method is interesting enough due to its rapidity. Its precision depends on the number of elements of the set E.

The technique proposed by Kamada [3] is very interesting for a wire-frame display. However, it is not very useful for a more realistic display. Indeed, this technique does not take into account visibilities of the elements of the considered scene, and a big element of the scene may hide all the others in the final display.

20.3.2 Direct Approximate Viewpoint Calculation

Colin [4] has proposed a method initially developed for scenes modeled by octrees. The purpose of the method was to compute a good point of view for an octree.

The method uses the principle of direct approximate computation to compute a good direction of view. This principle can be described as follows:

1. Choose the three best directions of view among the 6 directions corresponding to the 3 coordinate axes passing through the center of the scene.
2. Compute a good direction in the pyramid defined by the 3 chosen directions, taking into account the importance of each one of the chosen directions.

How to estimate the importance of a direction of view? What is a good criterion to estimate the quality of a view? The idea is to maximize the number of visible details. A direction of view is estimated better than another one if this direction of view allows to see more details than the other.

In the method of Colin [4], each element of the scene owns a weight with reference to a point of view. This weight is defined as the ratio between the visible part of the element and the whole element. It allows to attribute an evaluation mark to each direction of view and to use this mark to choose the three best directions.

When the three best directions of view $d1$, $d2$, and $d3$ have been chosen, if $m1$, $m2$, and $m3$ are the corresponding evaluation marks, a good direction of view can be computed by linear interpolation between the three chosen directions of view, taking into account the respective evaluation marks. So, if $m1 > m2 > m3$, the final good direction is computed (Figure 20.3) by rotating the initial direction $d1$ to the direction $d2$ by an angle $\varphi 1 = 90 * m1/(m1 + m2)$, then to the direction $d2$ by an angle $\varphi 2 = 90 * m1/(m1 + m3)$.

20.3 Simple Virtual World Understanding

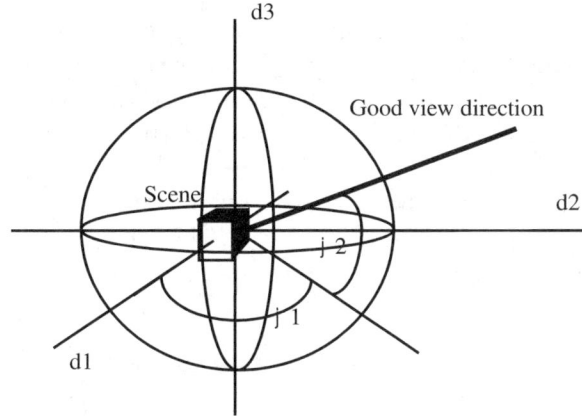

Figure 20.3 Direct approximate computation of a good direction of view.

20.3.3 Iterative Viewpoint Calculation

The good point of view computing method proposed by Plemenos [1, 2] was developed and implemented in 1987, but it was first published only in 1991.

The good view criterion used by this method is the visual complexity of the scene from a point of view. The notion of visual complexity and methods to compute it will be defined in Section 20.5.

The process used to determine a good point of view works as follows. The points of view are supposed to be on the surface of a virtual sphere whose the scene is the center. The surface of the sphere of points of view is divided in 8 spherical triangles (Figure 20.4).

The best spherical triangle is determined by positioning the camera at each intersection point of the three main axes with the sphere and computing the visual complexity of the scene from this point of view. The three intersection points with the best evaluation are selected. These three points on the sphere determine a spherical triangle, selected as the best one.

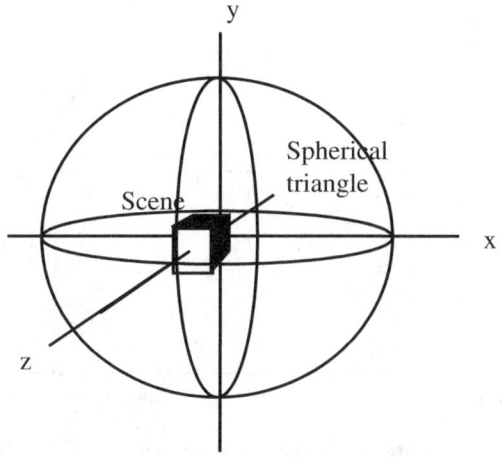

Figure 20.4 Sphere divided in 8 spherical triangles.

The next problem to resolve is selection of the best point of view on the best spherical triangle. The following heuristic search technique is used to resolve this problem. If the vertex A (Figure 20.5) is the vertex with the best evaluation of the spherical triangle ABC, two new vertices E and F are chosen at the middles of the edges AB and AC respectively and the new spherical triangle ADE becomes the current spherical triangle. This process is recursively repeated until the visual complexity of the scene from obtained points of view does not increase. The vertex of the final spherical triangle with the best evaluation is chosen as the best point of view.

20.3.4 Direct Exhaustive Viewpoint Calculation

Sbert et al. [5] proposed another method to compute a good point of view. The sphere containing the points of view is uniformly sampled and the visual complexity of the scene for each viewpoint is computed. The viewpoint corresponding to the greater visual complexity is selected as the best point of view. The main contribution of this method is the use of information theory to compute the visual complexity of the scene from a point of view.

20.4 What is Visual Complexity of a Scene?

It is generally admitted that there are scenes that are considered more complex than others. The notion of complexity of a scene is an intuitive one and, very often, given two different scenes, people are able to say which scene is more complex than the other. Another problem is that it is not always clear what kind of complexity people are speaking about. Is it computational complexity, taking into account the computational cost of rendering; geometric complexity, taking into account the complexity of each element of the scene; or quantitative complexity, depending on the number of elements of the scene?

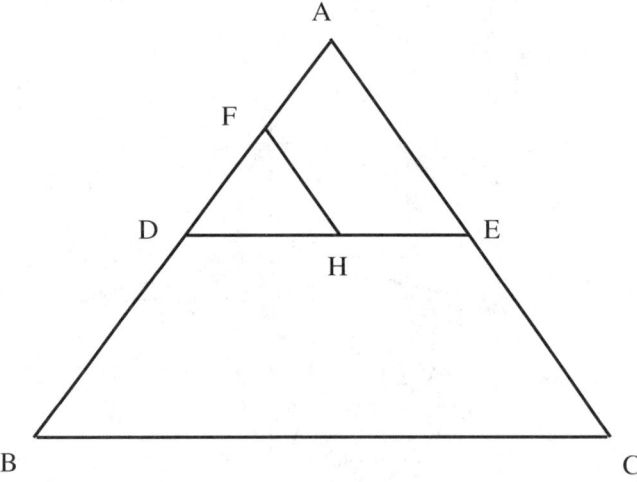

Figure 20.5 Heuristic search of the best point of view by subdivision of a spherical triangle.

We can informally define the intrinsic complexity of a scene as a quantity that does not depend on the point of view; depends on the number of details of the scene and the nature of details (convex or concave surfaces).

Some steps toward a formal definition of scene complexity have been presented [6, 7]. Unlike intrinsic complexity, the visual complexity of a scene depends on the point of view. An intrinsically complex scene, seen from a particular point of view, is not necessarily visually complex. A first measure of the notion of visual complexity of a scene from a point of view could be the number of visible details or, more precisely, the number of surfaces of the scene visible from this point of view. However, this definition of visual complexity is not very satisfactory, because the size of visible details is also important. Informally, we will define the visual complexity of a scene from a given point of view as a quantity which depends on:

- The number of surfaces visible from the point of view;
- The area of visible part of each surface of the scene from the point of view;
- The orientation of each (partially) visible surface according to the point of view;
- The distance of each (partially) visible surface from the point of view.

An intuitive idea of visual complexity of a scene is given in Figure 20.6, where the visual complexity of scene Figure 20.6(a) is less than the visual complexity of scene Figure 20.6(b) and the visual complexity of scene Figure 20.6(b) is less than the visual complexity of scene Figure 20.6(c), even if each scene contains other nonvisible surfaces.

Given a point of view, a scene may be divided in several more or less visually complex regions from this point. The visual complexity of a (part or region of) scene from a point of view, as defined above, is mainly geometry based since the elements taken into account are geometric elements. It would be possible to take into account other aspects of the scene, such as lighting, because lighting may modify the perception of a scene by a human user. However, only geometrical visual complexity will be discussed in this chapter, since the study on influence of lighting in scene perception is not yet enough advanced.

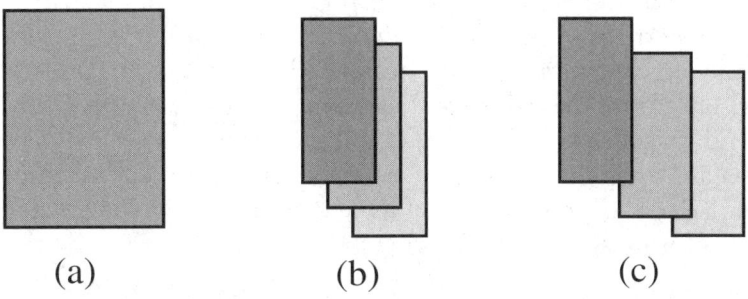

Figure 20.6 Visual complexity of scene (c) is greater than visual complexity of scene (b); visual complexity of scene (b) is greater than visual complexity of scene (a).

20.5 How to Compute Visual Complexity

Following our definition of visual complexity in this section, its calculation depends on the number of visible surfaces, the area of the visible part of each surface, and the distance and orientation of each (partially) visible surface, according to the point of view. A linear combination of these two quantities would give an accurate enough measure of visual complexity. The most important problem is the way to compute the number of visible surfaces and the visible projected area of each surface. The method used may depend on some constraints of the application using this information. Some applications require real-time calculation, whereas for others the calculation time is not an important constraint. For some applications, it is very important to have accurate visual complexity estimation and for others a fast approximate estimation is sufficient.

It is easy to see that the visible part, orientation, and distance of (partially) visible surfaces from the point of view can be accurately approximated by the projection of the visible parts of the scene on the unitary sphere centered on the point of view. This approximation will be used in this section to estimate the visual complexity of a scene from a point of view.

20.5.1 Accurate Visual Complexity Estimation

The most accurate estimation of the visual complexity of a scene can be obtained by using a hidden surface removal algorithm, working in the user space and explicitly computing the visible part of each surface of the scene. Unfortunately, it is rarely possible in practice to use such an algorithm because of the computational complexity of this kind of algorithm. For this reason, less accurate but also less complex methods have to be used.

A method proposed in [8, 9] permits use of hardware accelerated techniques in order to decrease the time complexity of estimation. This method uses image analysis to reduce the computation cost. Based on the use of the OpenGL graphical library and its integrated z-buffer, the technique used is the following.

If a distinct color is given to each surface of the scene, displaying the scene using OpenGL allows to obtain a histogram (Figure 20.7) which gives information on the number of displayed colors and the ratio of the image space occupied by each color.

Since each surface has a distinct color, the number of displayed colors is the number of visible surfaces of the scene from the current position of the camera. The fraction of the image space occupied by a color is the area of the projection of the visual part of the corresponding surface. The sum of these ratios is the projected area of the visible part of the scene. With this technique, the two visual complexity criteria are computed directly by means of an integrated fast display method.

The visual complexity of a scene from a given viewpoint can now be computed by a formula like the following one:

$$C(V) = \frac{\sum_{i=1}^{n}\left[\frac{P_i(V)}{P_i(V)+1}\right]}{n} + \frac{\sum_{i=1}^{n} P_i(V)}{r} \qquad (20.2)$$

20.5 How to Compute Visual Complexity

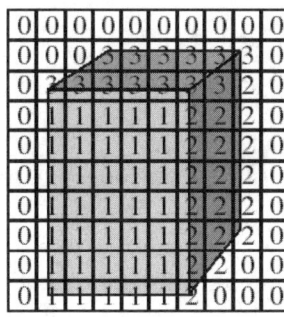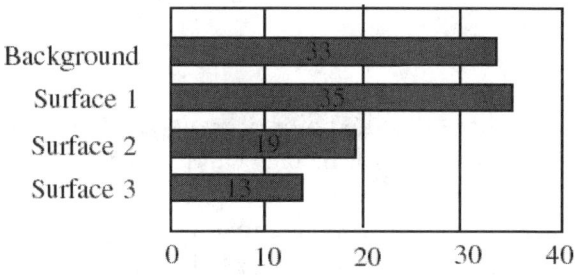

Figure 20.7 Fast computation of number of visible surfaces and area of projected visual part of the scene by image analysis.

where:

$C(V)$ is the visual complexity of the scene from the view point V,

$P_i(V)$ is the number of pixels corresponding to the polygon number i in the image obtained from the view point V,

r is the total number of pixels of the image (resolution of the image),

n is the total number of polygons of the scene.

In this formula, $[a]$ denotes the smallest integer, greater than or equal to a.

Another method to compute visual complexity has been proposed in [5, 10], based on information theory. A new measure is used to evaluate the amount of information captured from a given point of view. This measure is called viewpoint entropy. To define it, the authors use the relative area of the projected faces over the sphere of directions centered in the point of view (Figure 20.8).

In this method, the visual complexity of a scene from a given point of view is approached by computing the viewpoint entropy. The viewpoint entropy is given by the formula:

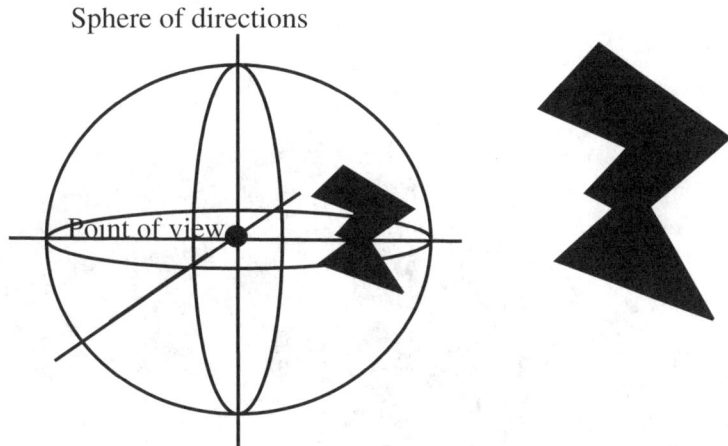

Figure 20.8 The projected areas of polygons are used as probability distribution of the entropy function.

$$H(S,P) = -\sum_{i=0}^{N_f} \frac{A_i}{A_t} \log \frac{A_i}{A_t} \qquad (20.3)$$

where P is the point of view, N_f is the number of faces of the scene S, A_i is the projected area of the face i and A_t is the total area covered over a sphere centred on the point of view.

The maximum entropy is obtained when a viewpoint can see all the faces with the same relative projected area A_i/A_t. The best viewpoint is defined as the one that has the maximum entropy.

To compute the viewpoint entropy, we use the technique explained above, based on the use of graphics hardware using OpenGL.

The selection of the best view of a scene is computed by measuring the viewpoint entropy of a set of points placed over a sphere that bounds the scene. The point of maximum viewpoint entropy is chosen as the best one. Figure 20.9 presents an example of results obtained with this method (image by Pere Pau Vazsquez).

Both methods have to compute the number of visible surfaces and the visible projected areas by using the technique described above.

20.5.2 Fast Approximate Estimation of Visual Complexity

In some cases, accurate visual complexity estimation is not requested, either because of need of real-time estimation of the visual complexity or because a less accurate estimation is enough for the application using the visual complexity [11].

In such a case, it is possible to roughly estimate the visual complexity of a scene from a given point of view, as follows.

A more or less great number of rays are randomly shot from the point of view to the scene and intersections with the surfaces of the scene are computed. Only intersections closest to the point of view surfaces are retained (Figure 20.10).

Now, we can approximate the quantities used in visual complexity calculation. We first need to define the notion of visible intersection. A visible intersection for a ray is the closest to the point of view intersection of the ray with the surfaces of the scene.

Figure 20.9 Point of view based on viewpoint entropy.

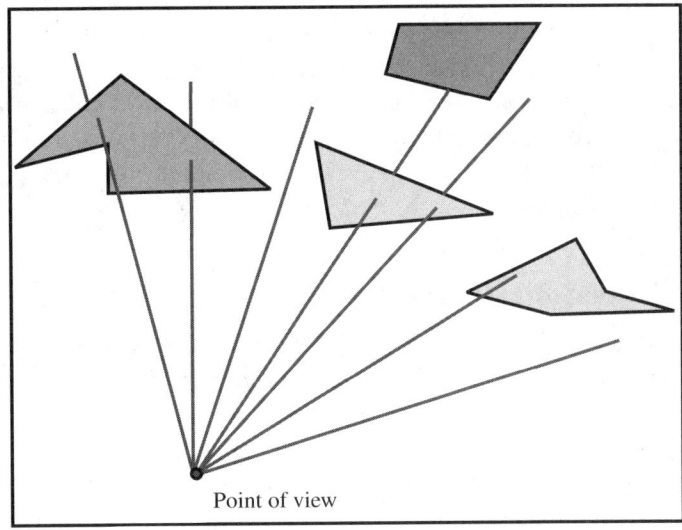

Figure 20.10 Approximate estimation of visual complexity.

- Number of visible surfaces = number of surfaces containing at least one visible intersection with a ray shot from the point of view.
- Visible projected area of a surface = number of visible intersections on the surface.
- Total visible projected area = number of visible intersections on the surfaces of the scene.
- Total projected area = total number of rays shot.

The main interest of this method is that the user can choose the degree of accuracy, which depends on the number of rays shot.

It is possible to use the same method to compute visual complexity of scenes described in a more intuitive abstraction level, the object level. An object is made of surfaces and it can be more important to see an object (that is, at least one of its surfaces) than to see many surfaces which, perhaps, belong to the same object [12].

In order to include this case in our visual complexity estimation, we have to give two more definitions:

- Number of visible objects = number of objects containing at least one visible surface.
- Visible projected area of an object = sum of visible intersections on the surfaces belonging to the object.

20.6 Virtual World Exploration

When we have to understand a complex virtual world, the knowledge of a single point of view is not enough to understand it. Computing more than one points of view is generally not a satisfactory solution in most cases, because the transition from a point of view to another one can disconcert the user, especially when the new point of view is far from the current one. Of course, the knowledge of several points of view can be used in other areas of computer graphics, such as image-based modeling and rendering [13, 14], but it is not suitable for virtual world understanding. The best solution, in the case of complex virtual worlds, is to offer an automatic exploration of the virtual world by a camera that chooses its path according to the specificities of the world to understand.

20.6.1 Incremental Outside Exploration

An important problem in automatic virtual world exploration is to make the camera able to visit the world to explore by using good points of view and, at the same time, by choosing a path that avoids brusque changes of direction.

In [8, 15] an initial idea of D. Plemenos and its implementations are described. The main principle of the proposed virtual world exploration technique is that the camera's movement must apply the following heuristic rules:

- It is important that the camera moves on positions that are good points of view.
- The camera must avoid fast returns to the starting point or to previously visited points.
- The camera's path must be as smooth as possible in order to allow the user to well understand the explored world. A movement with brusque changes of direction is confusing for the user and must be avoided.

In order to apply these heuristic rules, the next position of the camera is computed in the following way:

- The best point of view is chosen as the starting position for exploration.
- Given the current position and the current direction of the camera (the vector from the previous to the current position), only directions ensuring smooth movement are considered in computing the next position of the camera (Figure 20.11).
- In order to avoid fast returns of the camera to the starting position, the importance of the distance of the camera from the starting position must be inversely proportional to path of the camera from the starting to the current position (Figure 20.12).

Thus, the following evaluation function is used to evaluate the next position of the camera on the surface of the sphere:

$$w_c = \frac{n_c}{2}\left(1 + \frac{d_c}{p_c}\right) \qquad (20.4)$$

In this formula:

- w_c is the weight of the current camera position,
- n_c is the global evaluation of the camera's current position as a point of view,

20.6 Virtual World Exploration

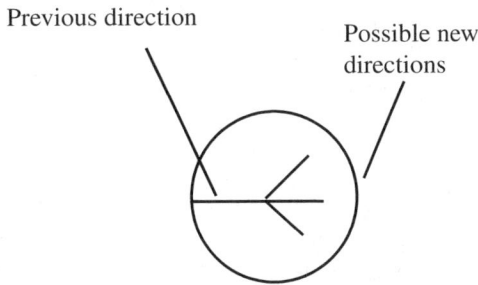

Figure 20.11 Only 3 directions are considered for a smooth movement of the camera.

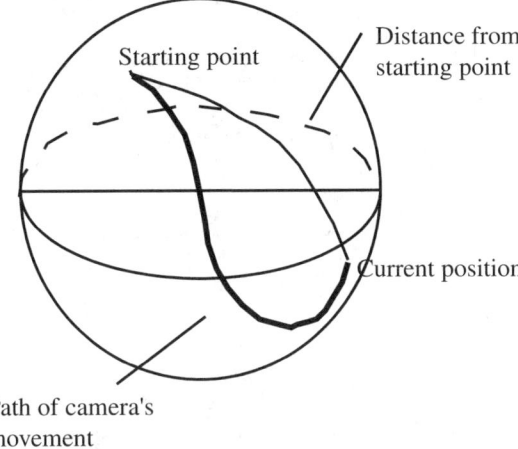

Figure 20.12 Distance of the current position of the camera from the starting point.

- p_c is the path traced by the camera from the starting point to the current position,
- *d is the distance of the current position from the starting point.*

Several variants of this technique have been proposed and applied. In Figure 20.13 one can see an example of exploration of a simple virtual world representing an office.

20.6.2 Viewpoint Entropy-Based Exploration

In [14, 16], the authors propose two methods of virtual world exploration: an outside and an indoor exploration method. The two methods are based on the notion of viewpoint entropy.

The outside exploration method is inspired from the incremental outside exploration described in the previous section. The camera is initially placed at a random position on the surface of a bounding sphere of the world to explore. An initial direction is randomly chosen for the camera. Then, the next position is computed in an incremental manner. As in [8, 15], only three possible next positions are tested, in order to ensure smooth movement of the camera (Figure 20.14).

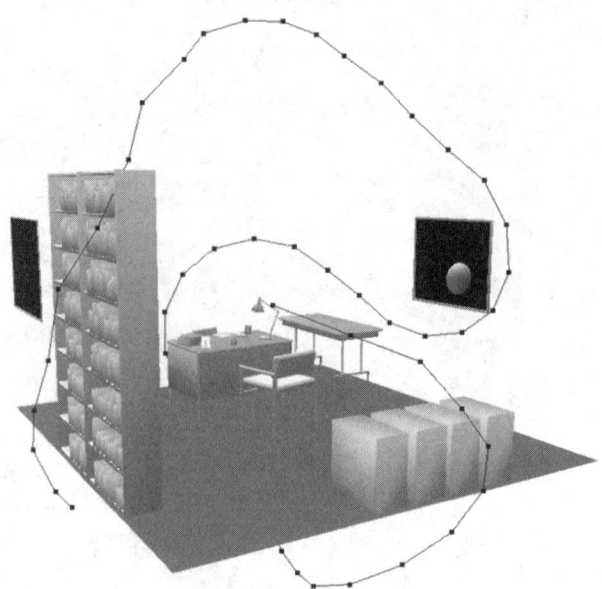

Figure 20.13 Exploration of a virtual office by incremental outside exploration.

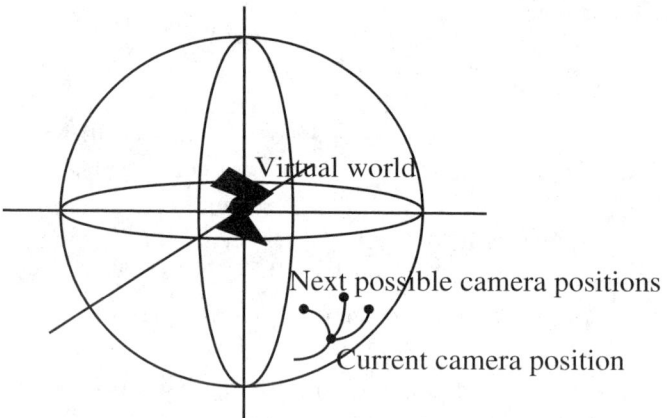

Figure 20.14 Three possible next positions evaluated.

An evaluation function computes the best next position of the camera, taking into account only the viewpoint entropy of the not yet visited polygons of the scene. The exploration is finished when a given ratio of polygons has been visited. The retained criterion of position's quality seems good enough and the presented results seem interesting. However, no measure is available to compare this method to the one discussed previously. Figure 20.15 shows viewpoint entropy-based exploration of a glass (image by Pere Pau Vazsquez).

The indoor exploration method is also an incremental exploration technique. The camera is initially placed inside the virtual world to explore, and it is supposed to move in this closed environment using each time one of the three following possi-

Figure 20.15 Viewpoint entropy-based exploration.

ble movements: move forward, turn left and turn right. Moreover, the camera is restricted to be at a constant height from the floor.

During the virtual world exploration, the path of the camera is guided by the viewpoint entropy of the polygons not yet visited. The next movement of the camera, selected from all the possible movements, is the one with maximum viewpoint entropy coming from the not yet visited polygons. In order to know, at each moment the polygons not yet visited, an array is used to encode the previously visited polygons.

The exploring algorithm is stopped when an important part of the polygons of the virtual world has been visited. This threshold value is estimated by the authors to be about 80% or 90% of the polygons of the explored world.

The main drawback of the method is that collisions are taken into account very roughly by the evaluation function of the camera movement.

20.6.3 Other Methods

There are really very few articles on the virtual world exploration area that face the problem with the computer graphics point of view. In [17] image-based techniques are used to control the camera motions in a possibly changing virtual world. The problem faced in the article is adaptation of the camera to the changes of the world. In [18] the authors propose a method allowing us to obtain a set of camera motions and virtual world motions satisfying user-defined constraints specified in an easy declarative manner. Both approaches are interesting but their purpose is quite different from what we consider as virtual world exploration techniques, because they are concerned with moving objects. The problems they try to resolve are not really computer graphics problems but rather robotic ones.

20.7 Future Issues

Virtual worlds exploration techniques are very useful techniques for the virtual world designer, as well as for the user wishing to understand a complex scene found on the Internet or to visit a virtual site such as reconstructed ancient cities, museums, and so on. We think that indoor virtual world exploration is unable to give the user a global view of the explored world. It must be used only when some parts of the explored world are not accessible with an outside exploration technique.

Two kinds of virtual world exploration can be envisaged in the future:

1. Real-time online exploration, where the virtual world is visited for the first time and the path of the camera is determined in an incremental manner. In this case, it is important to have fast exploring techniques in order to allow the user to understand in real time the explored world.
2. Offline exploration, where the user does not visit the virtual world together with the camera. The virtual world is found and analyzed by the program guiding the camera's movement, in order to determine interesting points to visit and path(s) linking these points. The user will visit the virtual world later, following the previously determined path. In such a case, it is less important to use fast techniques to determine the camera's path.

20.7.1 Online Exploration of Virtual Worlds

In online exploration it could be interesting to elaborate plans, instead of using purely incremental exploration. For example, for some scenes it could be known that the camera must reach some predefined points. In such a case, an intermediate goal of the camera could be to reach one of these points, while applying the other criteria to determine its path from the current point of view.

In Figure 20.16, the camera has to reach a predefined position. In order to reach this position, a next position P1 of the camera and an intermediate position P2 to reach are computed as close as possible to each other. Encouraging results have been obtained in this manner in the MSI laboratory of the University of Limoges.

20.7.2 Offline Exploration of Virtual Worlds

In offline exploration, the process could be decomposed in two steps: computation of a minimal number of points of view for the world to be explored and computation of an optimal path for the camera, taking into account the computed points of view.

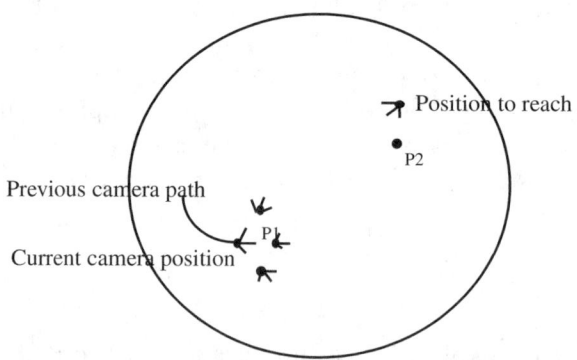

Figure 20.16 Plan-based exploration of virtual worlds.

1. Computation of a minimal number of points of view. The number of computed points of view should be enough to well understand the virtual world to explore. There are two ways to compute these points. The first one computes a big number of points of view on a sphere surrounding the virtual world and then suppresses any point of view that does not add visual information about the world to explore. The second way to compute a minimal number of points of view directly computes new points of view on the surface of the surrounding sphere, which adds visual information about the world to explore. The process is finished when no additional visual information is added. This can be done by using heuristic comparison techniques and recursive sphere subdivision.
2. Computation of an optimal path for the camera. A way to compute an optimal path for the camera is to choose as a starting position one of the points to visit and then try to reach successively the remaining points in an order determined by two criteria: distance from the current position and quality of view.

20.8 Conclusions

In this chapter the most interesting virtual world exploration techniques have been presented. We have chosen to present only techniques whose purpose is exploration for computer graphics use—that is, techniques allowing a human user to understand the virtual world visited by a camera.

The presented techniques can be divided in two groups. The first group of techniques is based on the computation of a good point of view and allows us to understand relatively simple scenes. The purpose of the second group of techniques is to understand or to visit complex virtual worlds, where a single point of view is not enough to understand them. These techniques are generally based on the movement of a camera around or inside the world to explore.

Some of the virtual world exploration techniques presented in this chapter take into account the total number of visible surfaces when computing the next camera position, whereas some other techniques take into account only the not yet seen surfaces. Both methods have advantages and drawbacks.

The main advantages of the choice to take into account the total number of visible surfaces are the following:

- It allows the user to continue exploration of the virtual world as long as the user needs it.
- Combined with the length of the camera patch, it gives interesting results.
- Processing is fast and allows real-time online exploration.

The main drawback of these methods is that we cannot be definitely sure that new parts of the virtual world will be found by the camera.

The main advantage of the choice to take into account only the not yet visited surfaces is that the computing of the next camera position allows us to discover new parts of the virtual world.

The main drawbacks of this choice are the following:

- We have to decide what to do when there are no more new polygons to discover or when no more polygons are reached. How to continue exploration in this kind of situation? It is important to remember that exploration is made for a human user who, perhaps, has not understood the virtual world, even if all its surfaces were seen.
- It is more time consuming and more difficult to implement.

Some future issues have been proposed in order to improve the exploration process which can be online or offline. Some of these proposals have been implemented and first results are available. The first results seem promising, especially with plan-based exploration, but they have yet to be well tested. From a theoretical point of view, plan-based exploration of virtual worlds has some advantages and some drawbacks, compared with previous exploration techniques, which have to be confirmed by the tests. The main advantage of this kind of exploration is that exploration is more intelligent and we can expect that visual results will be better than with other techniques. On the other hand, the method has two main drawbacks:

- Since it is important to know some interesting points of view in order to elaborate plans, the method is probably better for offline exploration, where a small number of interesting points of view can be defined offline.
- We have to decide how to continue the exploration when the last goal-viewpoint is reached.

An important problem with virtual world exploration methods is that, currently, there does not exist any measure to compare various exploration techniques. Comparison is only visual. We think that it would be very interesting to define a kind of distance, permitting us to compare all the proposed techniques by computing the distance of the computed camera paths from an ideal path. Another idea, allowing comparisons between various exploration techniques, would be an automatic or semiautomatic generation of test virtual worlds with well-chosen properties, in order to permit easy visual comparisons. For example, a sphere with some holes, containing a more or less complex virtual subworld, would be an interesting virtual world to test exploration techniques.

Of course, a lot of techniques presented in this chapter can easily be adapted to give solutions to the automatic light source placement problem. Indeed, some articles [2, 19] have proposed partial solutions to some aspects of this problem. However, the light source placement problem has first to be well formulated before being able to say what part of the problem is addressed by a proposed solution.

The virtual world exploration techniques presented here are based on the notion of visual complexity of a virtual world, or a part of a virtual world, from a point of view. This notion only takes into account the geometry of the visible parts of the virtual world. It would be interesting to take also into account the lighting—that is, the position and intensity of one or more light sources in the virtual world. There are two methods to view this problem:

1. The light sources positions and intensities are already fixed, and the problem is to define a visual complexity taking into account both geometry and lighting and then to compute good points of view.
2. The visual complexity of a virtual world from a viewpoint is computed taking into account only its geometry, and the problem is to find good positions for the light sources in order to optimize the view of the virtual world from the viewpoint.

Currently, none of the existing works in this area gives satisfactory solutions to this problem.

References

[1] Kamada, T., and S. Kawai, "A Simple Method for Computing General Position in Displaying Three-dimensional Objects," *Computer Vision, Graphics and Image Processing*, Vol. 41 (1988), pp. 43–56.

[2] Colin, C., "A System for Exploring the Universe of Polyhedral Shapes," *Eurographics'88*, Nice (France), September 1988.

[3] Plemenos, D., "A Contribution to the Study and Development of Scene Modelling, Generation and Visualisation Techniques," *The MultiFormes project*, Professorial dissertation, Nantes (France), November 1991.

[4] Plemenos, D., and M. Benayada, Intelligent Display in Scene Modeling. New Techniques to Automatically Compute Good Views, GraphiCon'96, Saint Petersburg, July 1996.

[5] Jardillier, F., and E. Languenou, "Screen-Space Constraints for Camera Movements: the Virtual Cameraman," *Computer Graphics Forum*, Vol. 17, No. 3, 1998, pp. 175–186.

[6] Barral, P., G. Dorme, D. Plemenos, "Visual Understanding of a Scene by Automatic Movement of a Camera," *GraphiCon'99*, Moscow (Russia), August 26–September 3, 1999, pp.59–65.

[7] Marchand E., and N. Courty, "Image-Bases Virtual Camera Motion Strategies," *Graphics Interface Conference, GI'2000*, Montreal (Canada), May 2000, pp. 69–76.

[8] Barral, P., G. Dorme, and D. Plemenos, "Scene Understanding Techniques Using a Virtual Camera," *Eurographics'2000*, Interlagen (Switzerland), August 20–25, 2000.

[9] Dorme, G., "Study and Implementation of 3D Scene Understanding Techniques," PhD thesis, University of Limoges (France), June 2001.

[10] Sbert, M., et al., "Applications of the Information Theory to Computer Graphics," *International Conference 3IA'2002*, Limoges (France), May 14–15, 2002, pp. 13–24.

[11] Vazquez, P.-P., M. Feixas, and W. Heidrich, "Image-Based Modeling Using Viewpoint Entropy," *Advances in Modelling, Animation and Rendering (proc. Computer Graphics International)*, Springer 2002, pp. 267–279.

[12] Gumhold, S., "Maximum Entropy Light Source Placement," *IEEE Visualization*, October 27–November 1, 2002, pp. 275–282.

[13] Vazquez, P.-P., and M. Sbert, "Automatic Indoor Scene Exploration," *Proceedings of the International Conference 3IA'2003*, Limoges, France, May 14–15, 2003, pp. 13–24.

[14] Vazquez P.-P., "On the Selection of Good Views and its Application to Computer Graphics," PhD Thesis, Barcelona (Spain), May 26, 2003.

[15] Feixas, M., et al., "An Information Theory Framework for the Analysis of Scene Complexity," *Eurographics'99, Computer Graphics Forum*, pp. 95–106.

[16] Feixas, M., "An Information Theory Framework for the Study of the Complexity of Visibility and Radiosity in a Scene," PhD thesis, Technical University of Catalonia.

[17] Plemenos D., "Exploring Virtual Worlds: Current Techniques and Future Issues," *International Conference GraphiCon'2003*, Moscow, Russia, September 5–10, 2003, pp. 30–36.

[18] Vazquez P. P.,et al., "Viewpoint Selection Using Viewpoint Entropy," *Vision, Modeling, and Visualization 2001*, Stuttgart, Germany, 2001, pp. 273–280.

[19] Plemenos D., M. Sbert M., and M. Feixas, "On Viewpoint Complexity of 3D Scenes," *International Conference GraphiCon'2004*, Moscow, Russia, September 6–10, 2004, pp. 24–31.

[20] Plemenos D., et al., "Intelligent Visibility-Based 3D Scene Processing Techniques for Computer Games," *International Conference GraphiCon'2005*, Novosibirsk, Russia, June 20–24, 2005, pp. 17–24.

Index

3DES encryption algorithm, 341
3G/WLAN
    AAA and inter-domain roaming, 335–36
    authentication architecture, 332–36
    authentication servers and proxy, 334–35
    integration, 332
    Mobile IP and, 332–33

A

Abstract Representation Of presence supporting Mutual Awareness (AROMA), 85
Accelerometers, 196
Accuracy, pose, 227–33
    dependence on image location, 229–33
    dependence on viewing angle, 227–29
    position, of DICA camera, 229
    position, of JAI camera, 230
    testing, 227
Acquisition agents, 26, 27
Actions, 67
Adaptivity, 26–32
    based on FML, 31
    defined, 14
    message sending, 27
    P2P systems, 147
Ad hoc on-demand fuzzy routing (AOFR), 312–17
    congestion loss, 321
    defuzzification, 316–17
    end-to-end delay, 320–21
    expiration sequence, 321–22
    fuzzification, 314–15
    fuzzy cost calculation, 313–17
    implementation, 313
    inference engine and knowledge base, 315–16
    normalized routing load, 322–23
    objectives, 312
    packet delivery fraction, 321, 325
    packets-per-joule, 323–24
    performance evaluation, 320–24
    route discovery phase, 313
    route reply phase, 313
    route stability, 323
    routing overhead, 325
    simulation parameters, 317–20
Ad-me, 294–96
    architecture, 295
    context, 295
    decision diagram, 296
    defined, 294
    emotion agent, 295, 296
Affective computing systems, 293–94
Affective DJ, 294
Agent-augmented meetings, 368
Agents, 14–17
    acquisition, 26
    context, 28
    creator, 24
    defined, 15
    embedded, 15, 16
    emotion, 295, 296
    FML distribution, 23–25
    intelligent, 139
    interaction, 23–24
    MyCampus, 38, 39
    registry, 24–25
    stationary, 24
Aixonix, 94–95
Ambient displays, 85
Ambient for social learning (ASL), 350–51
Ambient intelligence
    adaption, 14
    CI technologies versus, 10–11
    context awareness, 3, 13
    defined, 1, 13
    device classes, 4
    devices, 3–6
    embedding, 13
    emotional interfaces, 263–84
    enabling technology scalability, 2
    environments, 1

Ambient intelligence (continued)
  environments illustration, 2
  functions, 3–6
  fuzzy adaptive embedding agents and, 13–33
  grids, 128, 139–40
  human-machine interface functional architecture, 245–51
  learning in, 350–51
  meetings, 359–74
  metadata, 140
  MyCampus experience, 35–58
  natural interaction, 248–50
  ontologies, 140
  personalization, 13–14
  processing and communication infrastructure, 3
  processing perspective, 6–8
  software perspective, 8–9
  surroundings interaction, 3–4
  task architectural model, 239–45
  user friendliness, 167
  visions and technologies, 1–11
  wireless streaming, 3
Ambient interfaces
  consistency, 99
  constraints, 98
  defined, 84
  design recommendations, 96–99
  for distributed workgroups, 83–100
  effectiveness, 97
  EventorBot, 91–93
  EyeBot, 91
  fan, 89
  feedback, 98
  functionality, 98
  learnability, 97–98
  mapping between controls/effects, 98–99
  memorability, 97–98
  participatory environment, 99
  PRAVTA client, 93–94
  RoboDeNiro, 90
  safety, 97
  target domain, 99
  user experience, 99
  utility, 97
  water tank, 88–89
Ambient sensors, 296–300
Application-specific MPSoCs (ASMPSoCs), 7
Artificial cognitive system (ACS), 263, 268
Artificial intelligence (AI)
  agent-based approaches, 15
  techniques, 14
Associativity-based routing (ABR), 304
Attributes, FML, 19–20
Augmented reality
  defined, 289
  example, 194
  notation, 197
  pose awareness, 193–235
  problem description, 195–96
  setup, 194
  system setup, 196–97
  usability for, 233–35
Authentication
  architecture for 3G/WLAN, 332–36
  design issues, 330–32
  impact, 330
  as procedure, 330
  servers, 334–35
  system architecture, 330–32
  time measurements for security policies, 341
  UMTS, 331
  in wireless security protocols, 336–39
  See also Security
Authentication, authorization, and accounting (AAA)
  extensions, 331, 333–34
  inter-domain roaming and, 335–36
  in mobile IP networks, 334
  proxy, 334
  servers, 336
Autonomous nervous system (ANS), 272

B

Barcodes, 73–74
Base station controlled dynamic clustering (BCDCP), 180–81
Basic Support for Cooperative Work (BSCW), 87
BEACH software infrastructure, 257
Behavioral coding system, 276
Behavioral patterns, 277
Behavioral units, 276
Biosignals, 271
Blood volume pulse (BVP) sensors, 295
BodyMedia SenseWear Pro Armband, 295
Boolean gates, 281, 282
BRITE topology generator, 156

C

Calibration, 224–25
  camera, 224
  camera frame to body frame, 225

pattern pose, 224–25
CALO, 373
Camera calibration, 225
Camera model, 220–21
Camera positioning, 212–20
    Canny edge detection, 213–15
    contour detection, 215–16
    corner detection, 216–17
    fitting lines, 217–19
    marker ID determination, 219–20
    marker layout, 213
    problem specification, 212–13
    rejecting unwanted contours, 216
Canny edge detector, 213–15
    defined, 214
    hysteresis thresholding, 215
Carrier sense multiple access with collision avoidance (CSMA/CA), 318
Case-based reasoning (CBR), 36, 48
    applicability, 55
    module implementation, 56
Cell phones, 113
Cellular IP (CIP), 187
CHIL, 373
Chord, 158–61
    applications, 160–61
    defined, 158
    node join/leave operation, 159
    routing, 158–59
    routing example, 160
    routing table, 159
    See also Structured P2P systems
Chord-based DNS, 160–61
Clustering
    choice, 404
    cosine-based algorithms, 404–7
    crisp cosine (CCC), 405
    defined, 402
    of document collection, 402
    FGC, 407–11
    fuzzy-based cosine (FCC), 405
    incremental algorithms, 404
    local fuzzy-based cosine (LFCC), 405–6
    steps, 402
    TDS application, 411–13
CoBrA, 259
Community applications, 51–52
Computer intelligence (CI), 352–53
    AmI features versus, 10–11
    constructs, layered style, 357
    features, 352–53
Computing
    affective, 293–94
    grid, 129–33
    IC trends, 5–6
    mobile, 288–89, 290–92
    peer-to-peer, 139–40
    pervasive, 3
    ubiquitous, 62, 127–40, 287–89
    wearable, 103–22, 289
Condor, 132
Congestion loss, 321
Context awareness, 3
    defined, 13
    human-machine interface architecture, 247–48
    message filtering services, 50
    physical browsing and, 70–72
    recommender services, 49–50
    reminder applications, 51
Context management, 379–99
Context ontology language (CoOL), 259
Context(s), 289–90
    Ad-me, 295
    defined, 289–90
    fuzzification, 385
    fuzzifying ontology-based, 379
    fuzzy inference, 384–96
    in mobile computing, 290–92
    ontologies, extending, 380–81
    semantic modeling, 399
    spatial, 290–91
Context sensing, 117–18
Context-sensitive crime alert application, 51
Continuous bit rate (CBR), 319
Contours
    breaking, 217–18
    detection, 215–16
    unwanted, rejecting, 216
Controlled flooding, 153
CoolTown project, 70–71
Cooperative file system (CFS), 160
Core layer (e-Wallet), 44
Corner detection, 216–17
Cosine-based algorithms, 404–7
    CCC, 405
    FCC, 405
    LFCC, 405–7
Cost adaptive mechanism (CAM), 317
    implementation, 318
    load, 317
    routing metrics, 318
    transmit power, 317, 318
    wireless activity, 317

Creator agents, 24
Crisp cosine clustering (CCC), 405, 406
CyberCode, 64–65

D

Dangling String, 260
Data logging/analysis, 119–20
Decentralization, 32
Decentralized Kalman filter, 208–10
Defuzzification, 316–17
DEFuzzyfier, 279
Device profiles, 291
Devices
    classes, 4
    energy availability, 4
    everyday, 4–5
    for fixed network infrastructure, 6–7
    portable, 4
    processing perspective, 6–8
    programmable, 6
    reconfigurable, 6
    for sensor network, 7–8
    for wireless base network, 7
DICA camera, 229, 230
    pan-tilt unit measurement, 231
    position accuracy, 232
    usability, 233–34
DigitalDesk, 63
Directed diffusion, 176–78
Discrete Kalman filter, 202
Display technologies, 115–16
    bistable LCD, 115
    OLED, 116
    organic LED, 115–16
Distributed algorithmic mechanism design (DAMD), 150
Distributed coordination function (DCF), 318
Distributed hash tables (DHTs), 157
Distributed object location and routing layer (DOLR), 157
Dynamically composable human-machine interfaces, 251–53

E

Easy Meeting, 366
Edge position estimators, 225
Embedded agents, 15, 16
Embedded sensor networks, 8
Embedding
    defined, 13
    fuzzy adaptive agents, 13–33
Emotional interfaces, 263–84
    analysis, 275
    arousal detection, 272
    background, 264–68
    behavioral coding system, 276
    biosignals, 271
    current/future work, 278–83
    data, 269–71, 275
    data acquisition, 271–72
    extraction of meaningful features, 279
    face expression analysis, 279
    features to symbols, 272
    four dimensional axes, 269
    frame by frame sequences, 277
    fuzzy rules identification, 279–80
    general architecture, 271–72
    general framework, 264–65
    hybrid subsymbolic-symbolic architecture, 278
    logical architecture, 273
    materials and methods, 269
    participants, 274
    procedure, 274–78
    relevant indicator detection, 272
    signals to features, 272
    simulated experiment, 273–78
    stimuli construction, 274
    symbols to signals, 272
Emotion-enhanced learning, 353
Emotion(s)
    accurate recognition, 293
    key role, 265
    measurement approaches, 293
    no, without cognition, 267–68
    physical milieu, 265–66
    procedural actions, 265
    as process, 273
    psychological milieu, 266–67
End-to-end delay, 320–21
Energy harvesting techniques, 116
Energy management, 120
Energy model, 320
Enterprise Privacy Authorization Language (EPAL), 38
Environment
    ambient intelligence, 1, 2
    defined, 292
    FML, 19–23
    grid computing, 130–32
    user, 292
Environmental Audio Reminder (EAR) system, 85
EPICTOID, 250

Event-notification service, 396–99
   fuzzy inference, 396
   membership functions, 397
   rule surface, 397
   semantic inference, 396–99
EventorBot, 91–93
   chat request, 94
   defined, 92
   illustrated, 92, 93
   overview, 92
E-Wallet
   core layer, 44
   defined, 40
   dynamic knowledge, 41
   high-level flows/processes, 45
   idea introduction, 37
   JESS implementation, 44
   owner knowledge categories, 40–41
   privacy-enforcing rules, 45, 46
   privacy layer, 44
   privacy policies, 52
   privacy preferences, 41
   query processing, 42–44
   service invocation rules, 41
   service layer, 44
   static knowledge, 40
   three-layer implementation, 43–44
Expiration sequence, 321–22
Extended Kalman filter (EKF), 202, 208
Extensible authentication protocol (EAP), 337
Extensible Markup Language (XML)
   defined, 17–18
   environment, FML in, 19
   role, 18
Extensible stylesheet language transformations (XSLT), 18–19, 47
Eyebot, 91

F

Face expression analysis, 279
Facial action coding system (FACS), 279
Fan (TOWNER), 89
Federated Kalman filter (FKF), 210
Fixed base networks, 6–7
Flooding-based routing, 173–76
   implosion problem, 174
   overlap problem, 174
   SPIN, 175–76
   T&D, 174–75
   See also Routing
Freenet, 151

Fusion framework, 197–200
   coordinate systems, 197–99
   quaternions, 199–200
   See also Pose
Future-generation grids, 135–37
   architecture, 136–37
   requirements, 135
   services, 135–36
   See also Grid computing; grids
Fuzzification, 314–15
Fuzzy-based cosine clustering (FCC), 405, 406–7
Fuzzy-based global clustering (FGC), 403, 407–11
   cluster center, 408
   distance measure, 407–9
   general scheme, 409–11
   importance weights average, 408
   ROC curve using, 413, 414
Fuzzy controller tree, 20
Fuzzy control network (FCN), 24–25
Fuzzy induction tree
   branches, 390
   building, 389–92
   build steps, 389
   illustrated, 394
   process excerpt, 393
   training examples, 390, 391
Fuzzy inference, 383–84
   of context, 384–96
   event-notification service, 396
   on input values, 384
Fuzzy logic, 16
   background, 307–8
   cost calculation, 313–17
   extending ontologies with, 380–81
   in MANETs, 307
   in multiobjective routing, 306–7
   operation, 308
Fuzzy Markup Language (FML), 18–19
   adaptive framework, 31–32
   agent distribution, 23–25
   attributes, 19–20
   defined, 18
   editor module, 22–23
   environment description, 19–23
   framework, 18
   program contents, 19
   tags, 19–21
   values, 19–20
   visual environment, 23
   in XML environment, 19

Fuzzy ontology, 381–84
  fuzzy inference, 383–84
  FuzzyMembershipFn class, 381–83
  FuzzyRule class, 383
  Similar class, 383
Fuzzy rules, 16, 17
  automatic generation, 26
  composition, 28
  generation, 392–96
  generation example, 28
  identification, 279–80
  parameters, 316
Fuzzy sets, 16
  concept, 308
  defined, 307
FXPal, 371, 372

G

Gaia, 257
General packet radio service (GPRS), 329
Geographic adaptive fidelity (GAF), 182
Geographic energy aware routing (GEAR), 182–83
  defined, 182–83
  nodes, 183
  recursive geographic forwarding, 184
  See also Location-based routing
GesturePen, 65
Globus Toolkit, 130–31
Gnutella, 152–54
  architecture, 153
  controlled flooding, 153
  defined, 152
  inefficiencies, 154
  nodes, 152
  super-peer status, 154
  See also Unstructured P2P systems
Gradient-based routing, 176–78
  directed diffusion, 176–78
  rumor, 178
  zonal rumor (ZRR), 178
  See also Routing
Gradient Structure Tensor, 217
Greedy perimeter stateless routing (GPSR), 183–84
Grid computing, 129–33
  Condor, 132
  environments, 130–32
  Globus Toolkit, 130–31
  Legion, 131–32
  UNICORE, 131
Grids, 127–40

  ambient intelligence, 128, 139–40
  environments, 130–32
  evolution, 128
  first-generation, 133–34
  future-generation, 135–37
  heterogeneity, 129–30
  integration, 127
  site autonomy focus, 129
  ubiquitous, 128, 137–39
  user focus, 130
  uses, 127–28
  wired, 128, 138
  wireless, 128, 138

H

HAWAII protocol, 187
Head-mounted displays (HMDs), 365
Health Insurance Portability and Accountability Act (HIPAA), 122
Hierarchical-based routing, 179–81
  BCDCP, 180–81
  LEACH, 179–80
  TTDD, 181
  See also Routing
Hierarchical Mobile IP (HMIP), 187
High-level interaction protocol primitives (HLIPs), 250–51
Homogeneous transformation matrix, 222
Human-machine interface functional architecture, 245–51
  context-awareness, 247–48
  functionalities, 246–47
  illustrated, 246
  natural communication support, 245–46
  natural interaction, 248–50
  reusability, 250–51
Human-machine interfaces
  component architecture, 252
  dynamically composable, 251–53
  scenario realization, 253–56
  synthesizing, 251–56
Hybrid subsymbolic-symbolic architecture, 278

I

IBM PMH, 109–10
IC trends, for computing, 5–6
IM2 project, 372, 373
Incremental outside exploration, 428–29
Indirect Kalman filter, 202
InfoBridge, 53, 54
  screenshots, 53

usage scenario, 54
Information and communication technologies (ICT), 345
Information Society Technology (IST), 10
Information tabs, 66
Infrared technologies, 74–75
Inner cycle, 95, 96
Intelligent agents, 139
Intelligent climitization environment (ICE), 297
Inter-domain roaming, 335–36
Internet Foyer, 260
IPAQ PDA, 294
IP mobility
   CIP, 187
   HAWAII, 187
   HMIP, 187
   limitations, 188
   management, 185–88
   MIP, 186
   MIPv6, 187–88
   protocols, 186–88
IP security (IPSec), 331–32, 339, 340
   3DES encryption algorithm, 341
   TCP throughput, 342
   UDP throughput, 342
IrDA standard, 74–75
ISODATA, 402
IST Amigo project, 237
IST Ozone, 237
Itsy, 106–7

J

JADE, 49
JAI camera, 230
   pan-tilt unit measurement, 231
   position accuracy, 232
JAI CV-S3300, 196
JESS, 44, 45

K

Kalman filters, 201–12
   decentralized, 208–10
   defined, 202
   discrete, 202
   divergence problems, 207–8
   extended (EKF), 202, 208
   federated (FKF), 210
   indirect, 202
   lag and, 206–7
   measurement noise, 206
   modular, 210–12
   observation update, 206
   orientation estimation, 201
   pluggable (PKF), 211, 223
   setup, 203–4
   state update, 204
   time update, 204–5
   variants, 202
Kazaa, 150
K-means, 402
Knowledge base (KB), 315

L

Language translation, 112
LART platform, 196
Layer two tunneling protocol (L2TP), 339
LCD displays, 115
Learning, 345–57
   in AmI, 350–51
   content characteristics, 354
   emotion-enhanced, 353
   enabling technologies, 351–53
   environment design, 354
   interactive, 353
   learner characteristics, 354
   learner feedback, 353
   mobile, 347–50, 353
   models, 355–56
   paradigm, 346–50
   realization issues, 357
   technologies, 346
   technologies standards, 354–55
   theories, 355–56
Learning content management systems (LCMSs), 346
   advantages, 346
   illustrated, 347
LED displays, 115–16
Legion, 131–32
Levenberg-Marqardt algorithm, 222
Lexicalized tree adjoining grammar (LTAG), 247
   deep parsing solution, 256
   grammar, 249
Links
   aesthetics, 73
   defined, 67
   information, 72
   visualizing, 72–73
Local fuzzy-based cosine clustering (LFCC), 405–6, 407
Location-based routing, 181–85
   defined, 181

Location-based routing (continued)
   GAF, 182
   GEAR, 182–83
   GPSR, 183–84
   LCR, 184–85
   See also Routing
Logical coordinate routing (LCR), 184–85
Low energy adaptive clustering hierarchy (LEACH), 179–80

M

Mamdani MinMaxMin, 25
MANETs, 303
   AOFR, 312–17
   end-to-end delay minimization, 308
   fuzzy logic system, 307
   multiobjective routing, 305–7
   performance evaluation, 320–24
   routing, 303, 304
   simulation parameters, 317–20
   single objective routing, 304–5
   successful packet delivery probability, 308
   total battery cost, 309–12
Markers
   candidate, 221
   detection algorithms, 213
   feature points, 220–23
   ID determination, 219–20
   inner part, 219
   layout, 213
   orientation, measurements, 228
MASSIVE, 365
Maximum battery cost routing (MBCR), 317
Measurements, 225–33
The Media Equation, 273
MediWear, 105–7
Meeting processes, 362
   defined, 368
   supporting, 368–69
Meeting resources, 362
   group memory, 362
   logistics, 362
   technology and, 366–68
Meeting roles, 362–63
   supporting, 369–70
   technology and, 369, 370
Meetings, 359–74
   agent-augmented, 368
   annotations and layers of analysis, 372–73
   applications and tasks, 373
   defined, 361–62
   problems with, 363
   projects on, 371–74
   recordings and sensors, 372
   technology, 364–71
   virtuality continuum, 364–66
   VMR, 370–71
MetaPad, 108
Mica2 Motes, 297–98
Microelectromechanical sensors (MEMS), 4
MicroOptical MD-6, 104
Micro-Watt nodes, 4
Middleware components, 117–22
   context sensing, 117–18
   data logging/analysis, 119–20
   device symbiosis, 121
   energy management, 120
   privacy/security, 121–22
   sensor interfaces, 118–19
   suspend, resume, mitigation capabilities, 120–21
   See also Wearable computers
MIMOSA project, 71
MIThril Body Network, 106
Mobile ad hoc networks. *See* MANETs
Mobile computing, 288–89
   context in, 290–92
   device profile, 291
   environment, 292
   spatial context, 290–91
   user profile, 291
Mobile IP (MIP), 186, 331
   in 3G/WLAN integration, 332–33
   with AAA extensions, 333–34
   challenge/response extensions (MICRE), 333
   v6 (MIPv6), 187–88
Mobile learning, 347–50
   defined, 347
   guidelines, 348–50
   issues, 348
   levels of objectives, 348
   review, 349
   See also Learning
Mobility model, 319
Modular Kalman filter, 210–12
Moore's law, 264
Multidimensional emotional appraisal, 276
Multimodal interchange language (MMIL), 250
   language description, 250
   specification, 251
Multiobjective routing, 305–7
   complexity, 305–6

cost function, 308–12
fuzzy logic applicability, 306–7
See also MANETs
Multiple-input multiple-output (MIMO), 305, 306
Multi-processor system-on-chip (MPSoC) platforms, 7
Murata Gyrostar piezoelectric vibrating gyros, 196
MyCampus, 35–58
    conclusions, 57–58
    development experience, 53–54
    empirical evaluation, 54–57
    e-Wallet, 40–46
    infrastructure, instantiating, 49–54
    overview, 35–36
    prior work, 37–38
    resources, 58
    system architecture, 38–40
    user preference capture, 46–48
    See also E-Wallet
MyCampus applications
    collaboration, 51
    community, 51–52
    context-aware message filtering, 50
    context-aware recommender, 49–50
    context-aware reminder, 51
    context-sensitive crime alert, 51

N

Naming and discovery (ND) service, 244, 245
Napster, 151–52
    architecture, 152
    defined, 151
    design, 151
    failure, 152
    functioning, 151–52
    See also Unstructured P2P systems
Natural interaction, 248–50
    importance, 248–49
    realization, 249
Near Field Communication (NFC), 65
NEEM project, 373
NetMan, 105
.NET Passport, 37, 39
Network-on-a-chip (NoC) architectures, 7
Networks
    fixed base, 6–7
    polymorphic active, 9
    sensor, 7–8
    wireless base, 7
Neuro-fuzzy system, 281

The New Everyday, 1
Normalized routing load, 322–23
NotifyMe, 70

O

Objects, 66
Offline virtual world exploration, 432–33
OLED displays, 116
Online virtual world exploration, 432
Open Grid Services Architecture (OGSA), 130, 132–33
    defined, 132–33
    OSGI, 133
    standard mechanisms, 133
Orchestration, 242
ORESTEIA, 280
Outer cycle, 95
OWL, 38
    knowledge representation, 39, 42
    metamodel, 44, 48
    "ObjectProperty" construct, 48
    ontologies, 47, 381
Oxygen project, 1
Ozone, 248
Ozone Transport Information System (OTIS), 253–56
    accessing, 253
    client, 253, 254
    defined, 253
    screenshots, 255

P

Pacemakers, 111
Packet delivery fraction, 321, 325
Packets-per-joule, 323–24
PAC meditation, 283
Pan-tilt unit, 227, 228
    camera measurement, 231
    DICA camera measurement, 231
Pastry
    applications, 162
    defined, 161
    routing, 161–62
    See also Structured P2P systems
Pattern pose, 224–25
Peer-to-peer computing, 139–40
    applications/services using, 147–48
    challenges, 148–51
    paradigm, 148
    promises, 146–48
    as research field, 145

Peer-to-peer (P2P) systems, 143–63
    adaptability, 147
    characteristics, 144
    cost effectiveness, 147
    freeriders, 149–50
    hybrid, 146
    implementation, 145
    interest in, 143
    robustness, 147
    scalability, 147
    search and resource location, 150–51
    security, 148–49
    structured, 146, 150, 157–62
    taxonomy, 145–46
    unstructured, 151–57
Percolation, 156
Performance evaluation, 320–24
    congestion loss, 321
    end-to-end delay, 320–21
    expiration sequence, 321–22
    normalized routing load, 322–23
    packet delivery fraction, 321
    packets-per-joule, 323–24
    route stability, 323
    See also Ad hoc on-demand fuzzy routing (AOFR)
Personalization
    defined, 13–14
    technologies, 352–53
Pervasive computing, achieving, 3
Philips DICA Smart Camera, 196
Physical browsing, 61–80
    actions, 67
    context-awareness and, 70–72
    defined, 66
    demonstration applications, 75–79
    electromagnetic methods, 74
    hyperlink visualization, 72–73
    implementing, 73–75
    information tags, 66
    infrared technologies, 74–75
    links, 67
    mobile terminal, 77
    objects, 66
    physical selection, 67–70
    related work, 62–66
    research, 63–66
    research challenges, 79–80
    tangible user interfaces, 63
    technology comparison, 75
    as user interaction paradigm, 62
    visual codes, 73–74

Physical hyperlinks, visualizing, 72–73
Physical selection
    defined, 67
    methods, 67–70
    NotifyMe, 70
    PointMe, 68, 76–77
    ScanMe, 69–70, 76–77
    TouchMe, 68–69, 76–79
Plan-based exploration, 432
Pluggable Kalman filter (PKF), 211, 223
PointMe, 68, 76–77
Point-to-point tunneling protocol (PPTP), 339
Polymorphic active networks, 9
Portable devices, 4
Portable media players, 112
Portable storage devices, 116
Pose
    accuracy dependence on image location, 229–33
    accuracy dependence on viewing angle, 227–29
    awareness, 193–235
    calibration, 224–25
    camera model, 220–21
    determining from markers feature points, 220–23
    estimation, 221–23
    estimation, coordinate systems, 220
    first experiment, 223–24
    fusion framework, 197–200
    measurements, 225–33
    pattern, 224–25
    tracking, 196
Positioning technologies, 116
PResence AVailability and Task Awareness (PRAVTA) client, 93–94
Privacy layer (e-Wallet), 44
ProComp2 device, 295

Q

Q Belt-Integrated Computer (QBIC), 106, 107
Quaternions, 199–200
Query processing (e-Wallet), 42–44

R

Radio frequency identification. *See* RFID
RADIUS protocol, 334, 337, 338
Real-time online exploration, 432
Registry agents, 24–25
Remembrance agent, 112
Reusability, 250–51

RFID
    emulation demonstration, 75
    readers, 65, 75, 78
    technologies, 64
RFID tags, 61, 64
    antennas, 74
    interest, 74
    long-range, 74
RoboDeNiro, 90
Rotation matrix, 222
Route stability, 323
Routing
    associativity-based (ABR), 304
    flooding-based, 173–76
    gradient-based, 176–78
    hierarchical-based, 179–81
    location-based, 181–85
    MANETs, 303, 304
    maximum battery cost (MBCR), 317
    multiobjective, 305–7, 308–12
    normalized load, 322–23
    rumor, 178
    single objective, 304–5
    in WSNs, 173–85
    zonal rumor (ZRR), 178
Routing protocols
    comparative analysis, 167–89
    design factors, 172–73
ROWL, 58
Rumor routing, 178

S

Satellite-based augmentation systems (SBASs), 291
Scalability
    AmI software infrastructure, 9
    peer-to-peer (P2P) systems, 147
    in unstructured P2P systems, 154
    wireless sensor networks (WSN), 172–73
ScanMe, 69–70, 76–77
Screenshots, 49–50
SCRIBE, 162
Security
    authentication impact, 330
    middleware components, 121–22
    P2P, 148–49
    policies, 340–41
    wearable computers, 114–17, 121–22
Security Assertion Markup Language (SAML), 38
Security parameter index (SPI), 333
Self-organizing systems, 140

Semantic inference, 396–99
SenSay, 293–94
Sensor networks, 7–8, 140
    embedded, 8
    wireless (WSNs), 173–88, 296–97
Sensor nodes
    components, 168
    failure, 172
    WSN, 297, 298, 299
Sensors
    ambient, 296–300
    BVP, 295
    in data gathering/processing, 169
    defined, 140
    interfaces, 118–19
Service layer (e-Wallet), 44
SETI, 144
Signal processing algorithms, 9
Signal strength analysis (SSA), 304
Signal to noise ratio (SNR), 311
Simulation parameters, 317–20
    energy model, 320
    mobility model, 319
    traffic model, 319
    See also MANETs
Single objective routing, 304–5
Smart Dust, 4
SOAP, 238, 250
SOCAM, 259
SoulPad, 108–9
Spatial context, 290–91
SPIN protocol, 175–76
Splitstream, 162
Standard 802.1x authentication protocol, 337–38
Stationary fuzzy agents, 24
Structured P2P systems, 146, 150, 157–62
    background, 157–58
    Chord, 158–616
    layered view, 158
    Pastry, 161–62
    See also Peer-to-peer (P2P) systems
Subpixel edge detector, 225–27
Subscriber identity module (SIM), 330–31
Sugeno-type fuzzy system, 317
Super-peer network, 154

T

Tags
    defined, 19
    fuzzy shape, 21
    information, 66

Tags (continued)
    visual, 73
Tangible user interfaces, 63, 260
Task architectural model, 239–45
    illustrated, 243
    service/resource components, 241–42
    tasks, 242–44
    task synthesis service, 244–45
Tasks, 242–44
    end use coordination component, 242
    generic components, 244
    UI components, 243, 244
TCM2-50 liquid inclinometer, 196
Terrorist detection system (TDS), 401, 411–13
    characteristics, 411
    defined, 411
    experiment, 412
    results, 412–13
Theatre of Work Enabling Relationships.
        See TOWER
Throw and drowned (T&D) flooding, 174–75
TouchMe, 68–69, 76–79
    defined, 68
    demonstration, 77–78
    testing, 77
    See also Physical selection
TOWER, 86–87
    ambient indicators, 87–94
    ambient interfaces, 86–94
    application-independence, 86
    context models, 87
    environment, 84
    EventorBot, 91–93
    extensibility, 87
    Eyebot, 91
    fan, 89
    indicators, 87
    persistent storage, 87
    PRAVTA client, 93–94
    RoboDeNiro, 90
    sensors, 87
    strengths, 86–87
    user involvement, 94–96
    water tank, 88–89
Traffic model, 319
Transparent fuzzy control, 17–19
Twiddler keyboard, 104, 120
Two-tier data dissemination (TTDD), 181

U

Ubiquitous computing, 62, 127–40
Ubiquitous grids, 128, 137–39
    approaches, 138–39
    defined, 128
    resource trust and, 137
    See also Grids
UNIform Interface to COmputer REsources (UNICORE), 131
Universal mobile telecommunication systems (UMTS), 329
    authentication, 331
    RNS, 331
    SIM (USIM), 336
Universal remote console, 259–60
Universal resource locators (URLs), 61
Unstructured P2P systems, 151–57
    Gnutella, 152–54
    Napster, 151–52
    scalability and, 154
    topology, 154–57
    See also Peer-to-peer (P2P) systems
User interfaces, tangible, 63
User profiles, 291

V

Values, FML, 19–20
Viewpoint entropy-based exploration, 429–31
    defined, 430
    illustrated, 431
    See also Virtual world exploration
Virtuality continuum, 364–66
    computer-produced stimuli, 364
    meetings in, 365–66
Virtual meeting room (VMR), 370–71
Virtual private networks (VPNs), 339
Virtual world exploration
    algorithm, 431
    benefits, 418–19
    camera path, 431
    direct approximate viewpoint calculation, 420–21
    direct exhaustive viewpoint calculation, 422
    global, 417
    incremental outside, 428–29
    indoor, 430–31
    intelligent automatic, 417–35
    iterative viewpoint calculation, 421–22
    lighting, 434–35
    local, 417
    nondegenerated view, 419–20
    not yet visited surfaces, 433–34
    number of visible surfaces, 433
    offline, 432–33
    online, 32

plan-based, 432
problem, 434
purpose, 417
types, 417
viewpoint entropy-based, 429–31
visual complexity, 422–27
VisionPad, 106
Visual codes, 73–74
Visual complexity, 422–27
    computation, 424–27
    estimation, 424–26
    fast approximate estimation, 426–27
    illustrated, 423
    See also Virtual world exploration

W

WatchPad, 109
Water tank (TOWER), 88–89
Watt nodes, 4
Wearable computers, 103–22, 289
    aircraft maintenance applications, 112
    applications, 110–12
    CMU, 105
    context sensing, 117–18
    cost, 114
    in customer relationship management, 112
    data logging/analysis, 119–20
    defense/security, 111
    defined, 103
    device symbiosis, 121
    display technologies, 115–16
    energy harvesting techniques, 116
    energy management, 120
    flexibility, 114
    history, 104–10
    impact limiting factors, 112–15
    interaction complexity, 114
    language translation, 112
    limited quantities, 113
    medical applications, 111–12
    middleware components, 117–22
    modular, 114
    pacemakers, 111
    portable media players, 112
    portable storage devices, 116
    positioning technologies, 116
    positive loop feedback, 115–17
    power management problem, 113–14
    privacy, 121–22
    remembrance agent, 112
    security enhancements, 116–17
    security issues, 114–15, 121–22
    sensor interfaces, 118–19
    size, 114
    size limitations, 104
    suspend, resume, mitigation capabilities, 120–21
    user interfaces, 116
WearARM, 106
Web services resource framework (WSRF), 133
WebStickers, 64
Weighted Average Algorithm, 298–99
WiBro, 112
Wicked problems, 367
WiFi Protected Access (WPA), 338–39
Wired equivalent privacy (WEP) protocol, 337, 341
    control message exchange and, 341
    defined, 337
    TCP throughput, 342
    UDP throughput, 342
Wired grids, 128, 138
Wireless advertising, 292–96
    Ad-me, 294–96
    affective computing systems, 293–94
    defined, 292
    effectiveness measurement, 292
    emotions, 293
Wireless base networks, 7
Wireless body area network (BAN), 294
Wireless grids, 128, 138
Wireless local area networks (WLANs), 329, 330
    macromobility, 331
    roaming, 332
    See also 3G/WLAN
Wireless sensor networks (WSN)
    characteristics, 169–70
    communication and, 169
    communication architecture, 170–72
    deployment, 168
    fault tolerance, 172
    hardware issues, 173
    health applications, 168
    IP mobility management, 185–88
    MAC, 171
    nodes, 297, 298, 299
    power consumption, 172
    protocol stack, 117
    routing in, 173–85
    routing protocol design, 172–73
    scalability, 172–73
    sensors, 296
    topology example, 297

Wireless sensor networks (WSN) (continued)
   use of, 168
World models, 370–71
WSAMI, 238, 239
   architecture illustration, 240
   core broker (CB), 244
   infrastructure, 239
   layers, 240
   middleware-level services, 240
   naming and discovery (ND) service, 244, 245
   user interface layer, 246

WSAtkins, 95
WSDL, 238

X

XML Access Control Markup Language (XACML), 38
Xybernaut, 110

Z

Zonal rumor routing (ZRR), 178

Recent Titles in the Artech House Mobile Communications Series

John Walker, Series Editor

3G CDMA2000 Wireless System Engineering, Samuel C. Yang

3G Multimedia Network Services, Accounting, and User Profiles, Freddy Ghys, Marcel Mampaey, Michel Smouts, and Arto Vaaraniemi

802.11 WLANs and IP Networking: Security, QoS, and Mobility, Anand R. Prasad, Neeli R. Prasad

Advances in 3G Enhanced Technologies for Wireless Communications, Jiangzhou Wang and Tung-Sang Ng, editors

Advances in Mobile Information Systems, John Walker, editor

Advances in Mobile Radio Access Networks, Y. Jay Guo

Ambient Intelligence, Wireless Networking, and Ubiquitous Computing, Athanasios Vasilakos and Witold Pedrycz, editors

Applied Satellite Navigation Using GPS, GALILEO, and Augmentation Systems, Ramjee Prasad and Marina Ruggieri

CDMA for Wireless Personal Communications, Ramjee Prasad

CDMA Mobile Radio Design, John B. Groe and Lawrence E. Larson

CDMA RF System Engineering, Samuel C. Yang

CDMA Systems Capacity Engineering, Kiseon Kim and Insoo Koo

CDMA Systems Engineering Handbook, Jhong S. Lee and Leonard E. Miller

Cell Planning for Wireless Communications, Manuel F. Cátedra and Jesús Pérez-Arriaga

Cellular Communications: Worldwide Market Development, Garry A. Garrard

Cellular Mobile Systems Engineering, Saleh Faruque

The Complete Wireless Communications Professional: A Guide for Engineers and Managers, William Webb

EDGE for Mobile Internet, Emmanuel Seurre, Patrick Savelli, and Pierre-Jean Pietri

Emerging Public Safety Wireless Communication Systems, Robert I. Desourdis, Jr., et al.

The Future of Wireless Communications, William Webb

GPRS for Mobile Internet, Emmanuel Seurre, Patrick Savelli, and Pierre-Jean Pietri

GPRS: Gateway to Third Generation Mobile Networks, Gunnar Heine and Holger Sagkob

GSM and Personal Communications Handbook, Siegmund M. Redl, Matthias K. Weber, and Malcolm W. Oliphant

GSM Networks: Protocols, Terminology, and Implementation, Gunnar Heine

GSM System Engineering, Asha Mehrotra

Handbook of Land-Mobile Radio System Coverage, Garry C. Hess

Handbook of Mobile Radio Networks, Sami Tabbane

High-Speed Wireless ATM and LANs, Benny Bing

Interference Analysis and Reduction for Wireless Systems, Peter Stavroulakis

Introduction to 3G Mobile Communications, Second Edition, Juha Korhonen

Introduction to Communication Systems Simulation, Maurice Schiff

Introduction to Digital Professional Mobile Radio, Hans-Peter A. Ketterling

Introduction to GPS: The Global Positioning System, Ahmed El-Rabbany

An Introduction to GSM, Siegmund M. Redl, Matthias K. Weber, and Malcolm W. Oliphant

Introduction to Mobile Communications Engineering, José M. Hernando and F. Pérez-Fontán

Introduction to Radio Propagation for Fixed and Mobile Communications, John Doble

Introduction to Wireless Local Loop, Second Edition: Broadband and Narrowband Systems, William Webb

IS-136 TDMA Technology, Economics, and Services, Lawrence Harte, Adrian Smith, and Charles A. Jacobs

Location Management and Routing in Mobile Wireless Networks, Amitava Mukherjee, Somprakash Bandyopadhyay, and Debashis Saha

Mobile Data Communications Systems, Peter Wong and David Britland

Mobile IP Technology for M-Business, Mark Norris

Mobile Satellite Communications, Shingo Ohmori, Hiromitsu Wakana, and Seiichiro Kawase

Mobile Telecommunications Standards: GSM, UMTS, TETRA, and ERMES, Rudi Bekkers

Mobile Telecommunications: Standards, Regulation, and Applications, Rudi Bekkers and Jan Smits

Multiantenna Digital Radio Transmission, Massimiliano "Max" Martone

Multiantenna Wireless Communicatio ns Systems, Sergio Barbarossa

Multipath Phenomena in Cellular Networks, Nathan Blaunstein and Jørgen Bach Andersen

Multiuser Detection in CDMA Mobile Terminals, Piero Castoldi

OFDMA for Broadband Wireless Access,

Personal Wireless Communication with DECT and PWT, John Phillips and Gerard Mac Namee

Practical Wireless Data Modem Design, Jonathon Y. C. Cheah

Prime Codes with Applications to CDMA Optical and Wireless Networks, Guu-Chang Yang and Wing C. Kwong

QoS in Integrated 3G Networks, Robert Lloyd-Evans

Radio Engineering for Wireless Communication and Sensor Applications, Antti V. Räisänen and Arto Lehto

Radio Propagation in Cellular Networks, Nathan Blaunstein

Radio Resource Management for Wireless Networks, Jens Zander and Seong-Lyun Kim

RDS: The Radio Data System, Dietmar Kopitz and Bev Marks

Resource Allocation in Hierarchical Cellular Systems, Lauro Ortigoza-Guerrero and A. Hamid Aghvami

RF and Baseband Techniques for Software-Defined Radio Peter B. Kenington

RF and Microwave Circuit Design for Wireless Communications, Lawrence E. Larson, editor

Sample Rate Conversion in Software Configurable Radios, Tim Hentschel

Signal Processing Applications in CDMA Communications, Hui Liu

Smart Antenna Engineering, Ahmed El Zooghby

Software Defined Radio for 3G, Paul Burns

Spread Spectrum CDMA Systems for Wireless Communications, Savo G. Glisic and Branka Vucetic

Third Generation Wireless Systems, Volume 1: Post-Shannon Signal Architectures, George M. Calhoun

Traffic Analysis and Design of Wireless IP Networks, Toni Janevski

Transmission Systems Design Handbook for Wireless Networks, Harvey Lehpamer

UMTS and Mobile Computing, Alexander Joseph Huber and Josef Franz Huber

Understanding Cellular Radio, William Webb

Understanding Digital PCS: The TDMA Standard, Cameron Kelly Coursey

Understanding GPS: Principles and Applications, Second Edtion, Elliott D. Kaplan and Christopher J. Hegarty, editors

Understanding WAP: Wireless Applications, Devices, and Services, Marcel van der Heijden and Marcus Taylor, editors

Universal Wireless Personal Communications, Ramjee Prasad

WCDMA: Towards IP Mobility and Mobile Internet, Tero Ojanperä and Ramjee Prasad, editors

Wireless Communications in Developing Countries: Cellular and Satellite Systems, Rachael E. Schwartz

Wireless Intelligent Networking, Gerry Christensen, Paul G. Florack, and Robert Duncan

Wireless LAN Standards and Applications, Asunción Santamaría and Francisco J. López-Hernández, editors

Wireless Technician's Handbook, Second Edition, Andrew Miceli

For further information on these and other Artech House titles, including previously considered out-of-print books now available through our In-Print-Forever® (IPF®) program, contact:

Artech House
685 Canton Street
Norwood, MA 02062
Phone: 781-769-9750
Fax: 781-769-6334
e-mail: artech@artechhouse.com

Artech House
46 Gillingham Street
London SW1V 1AH UK
Phone: +44 (0)20 7596-8750
Fax: +44 (0)20 7630-0166
e-mail: artech-uk@artechhouse.com

Find us on the World Wide Web at: www.artechhouse.com